MEP Databook

Sidney M. Levy

McGraw-Hill

New York San Francisco Washington, D.C. Auckland Bogotá
Caracas Lisbon London Madrid Mexico City Milan
Montreal New Delhi San Juan Singapore
Sydney Tokyo Toronto

Cataloging-in-Publication Data is on file with the Library of Congress

McGraw-Hill

A Division of The **McGraw·Hill** *Companies*

1 2 3 4 5 6 7 8 9 0   KGP KGP   0 9 8 7 6 5 4 3 2 1 0

ISBN 0-07-136020-4

The sponsoring editor for this book was Larry Hager, the editing supervisor was Steven Melvin, and the production supervisor was Sherri Souffrance. It was set in ITC Century Light through the services of the PRD Group.

Quebecor/Kingsport was printer and binder.

McGraw-Hill books are available at special quantity discounts to use as premiums and sales promotions, or for use in corporate training programs. For more information, please write to the Director of Special Sales, McGraw-Hill, 2 Penn Plaza, New York, NY 10121-2298. Or contact your local bookstore.

 This book is printed on recycled, acid-free paper containing a minimum of 50% recycled, de-inked fiber.

Contents

Introduction

The MEP DataBook provides the builder, project manager, construction superintendent, design consultants and facility managers with a one-source reference guide to heating, ventilating and air-conditioning systems and related control work, electrical, plumbing, and fire protection installations, components and material specifications. Now, many of those unanswered questions can be addressed.

Valuable information including ADA guidelines, "soft" and "hard" metrification conversions charts along with useful tables and formulas also resides within the covers of this book.

The MEP DataBook contains diagrams and schematics of the most frequently encountered HVAC systems and their controls, plumbing components and product specifications for residential and commercial systems.

Key elements of the 1999 National Electric Code (NEC) are included in the Electrical Section of the DataBook along with cable, wire, conduit specifications and electrical component installation guidelines. A full range of fire protection systems, installation schematics, piping, sprinkler head configurations and other component specifications along with selected NFPA regulations will prove valuable to both contractor and designer alike.

Much of the material in the MEP DataBook has been gleaned from manufacturer's sources and trade association furnished material; some of this information may be proprietary in nature but much of it is generic in nature.

How many times during project meetings, field visits or in conversations with specialty contractors is it convenient to have a concise source of MEP product data, specifications and installation schematics readily at hand? The MEP DataBook was written with those needs in mind.

I selected those MEP components and specifications that, in my forty years in the construction industry, appear to be those for which reference material is so often required and, of course, always needed "yesterday."

For those experienced construction professionals the MEP DataBook may serve as a "refresher course" and for those new to the industry, it offers a simpler way to become familiar with the complex and often bewildering array of HVAC systems and controls, electrical, plumbing, and fire protection installations.

I hope you find the MEP DataBook a worthwhile addition to your construction library.

Sidney M. Levy

Fire Protection

Contents

1.0.0   Introduction to Fire Protection

Sprinklers and other fire-protection systems are available in many different variations, each designed for specific fire-suppression situations. Systems using water can be customized in many ways, but all maintain the same basic components. If local water systems cannot provide sufficient volume and/or pressure, water tanks are often installed to provide adequate flow, delivering the required amounts of water by gravity, air, or pump pressure.

Systems capable of delivering oxygen-starving foams or powders are frequently used if these particular agents are more effective in suppressing fire. Halogenated agents, halon, developed to replace water as the agent to extinguish fires without damaging sensitive equipment, is used in many computer rooms or other areas that contain delicate and valuable documents, fabrics, and relics. And portable fire extinguishers of various capacities are used in localized situations to extinguish combustible material, solvent oil, and electrical fires.

1.1.0   Wet-Pipe Systems

Wet pipe systems (Figs. 1.1.1 and 1.1.2) are the most common systems used in commercial and industrial construction.

- *Advantages*   Rapid response to fire control because the sprinkler pipes are always filled with water, relatively uncomplicated design, highly reliable.

- *Disadvantages*   Cannot be used where systems are to be installed in a building that is not heated and where ambient temperatures are at (or below) freezing, unless an anti-freeze solution is added to the water in the system.

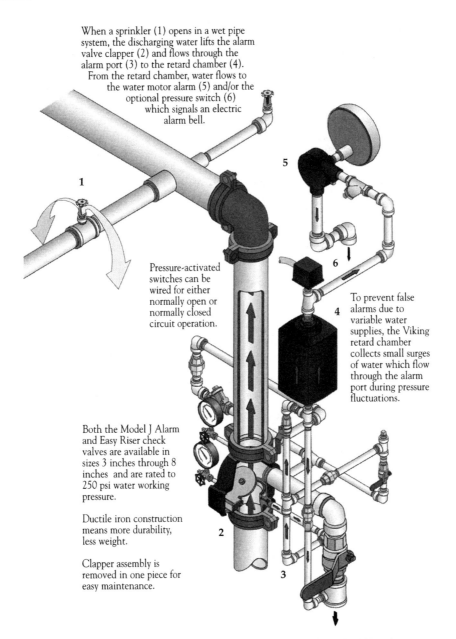

When a sprinkler (1) opens in a wet pipe system, the discharging water lifts the alarm valve clapper (2) and flows through the alarm port (3) to the retard chamber (4). From the retard chamber, water flows to the water motor alarm (5) and/or the optional pressure switch (6) which signals an electric alarm bell.

Pressure-activated switches can be wired for either normally open or normally closed circuit operation.

To prevent false alarms due to variable water supplies, the Viking retard chamber collects small surges of water which flow through the alarm port during pressure fluctuations.

Both the Model J Alarm and Easy Riser check valves are available in sizes 3 inches through 8 inches and are rated to 250 psi water working pressure.

Ductile iron construction means more durability, less weight.

Clapper assembly is removed in one piece for easy maintenance.

Viking's alarm TESTanDRAIN™ valve is available as an option. Its ball valve construction resists wear and allows testing for proper clapper movement when testing the water motor alarm.

Figure 1.1.1   Wet-pipe system sequence of operations and components. (*By permission: The Viking Corporation, Hastings, MI.*)

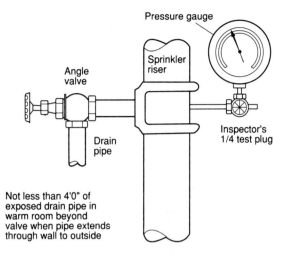

Not less than 4'0" of
exposed drain pipe in
warm room beyond
valve when pipe extends
through wall to outside

For SI Units: 1 in. = 25.4 mm; 1 ft = 0.3048 m.

Figure 1.1.2 Drain connection for a sprinkler riser. (*Reprinted with permission from NFPA 13,* Installation of Sprinkler Systems, *Copyright © 1994, National Fire Protection Association, Quincy, MA 02269. This reprinted material is not the complete and official position of the National Fire Protection Association on the referenced subject, which is represented only by the standard in its entirety.*)

1.2.0 Dry-Pipe Systems

Dry-pipe systems are (Figs. 1.2.1 through 1.2.5) used where fire protection is required to be installed in unheated spaces, where ambient temperatures will dip below the freezing mark. This system is often used in low-hazard areas.

- *Advantages*   Dry valves allow pressurized air to fill the piping until a sprinkler head requires that water enter the system; therefore, ambient freezing problems are eliminated.

- *Disadvantages*   There is a delay in response time, which this requires more water to be delivered quickly. Therefore, piping sizes are generally much larger than those in a wet system. The dry system might also require the installation of more sprinkler heads than required in a wet system. The dry system will also need an air compressor (another piece of equipment that will require maintenance) to ensure that pressure is in the system at all times.

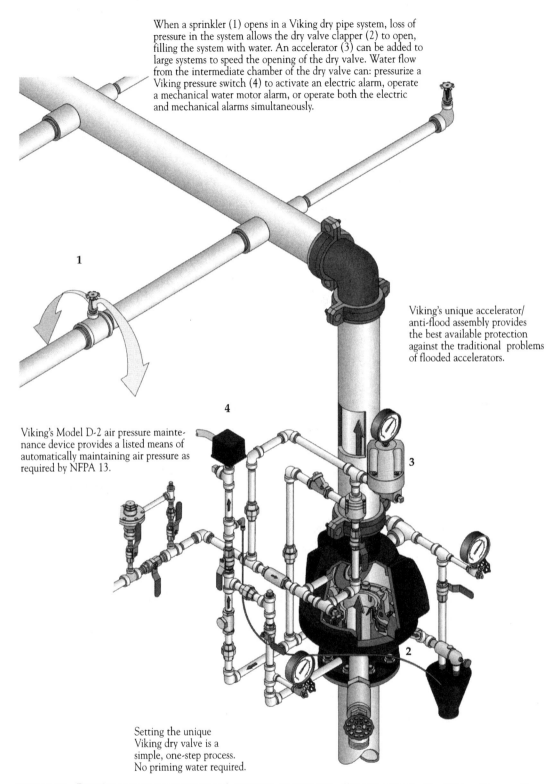

When a sprinkler (1) opens in a Viking dry pipe system, loss of pressure in the system allows the dry valve clapper (2) to open, filling the system with water. An accelerator (3) can be added to large systems to speed the opening of the dry valve. Water flow from the intermediate chamber of the dry valve can: pressurize a Viking pressure switch (4) to activate an electric alarm, operate a mechanical water motor alarm, or operate both the electric and mechanical alarms simultaneously.

Viking's unique accelerator/ anti-flood assembly provides the best available protection against the traditional problems of flooded accelerators.

Viking's Model D-2 air pressure maintenance device provides a listed means of automatically maintaining air pressure as required by NFPA 13.

Setting the unique Viking dry valve is a simple, one-step process. No priming water required.

Figure 1.2.1 Dry-pipe system components and sequence of operation. (*By permission: The Viking Corporation, Hastings, MI.*)

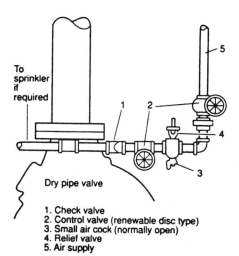

To
sprinkler
if
required

Dry pipe valve

1. Check valve
2. Control valve (renewable disc type)
3. Small air cock (normally open)
4. Relief valve
5. Air supply

Figure 1.2.2 Air supply required for dry-pipe system. (*Reprinted with permission from NFPA 13,* Installation of Sprinkler Systems, *Copyright © 1996, National Fire Protection Association, Quincy, MA 02269. This reprinted material is not the complete and official position of the National Fire Protection Association on the referenced subject, which is represented only by the standard in its entirety.*)

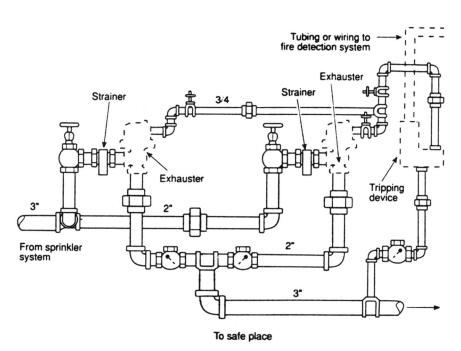

Figure 1.2.3 Arrangement of air exhaust valves for combined dry-pipe and preaction sprinkler systems. (*Reprinted with permission from NFPA 13,* Installation of Sprinkler Systems, *Copyright © 1996, National Fire Protection Association, Quincy, MA 02269. This reprinted material is not the complete and official position of the NFPA on the referenced subject, which is represented only by the standard in its entirety.*)

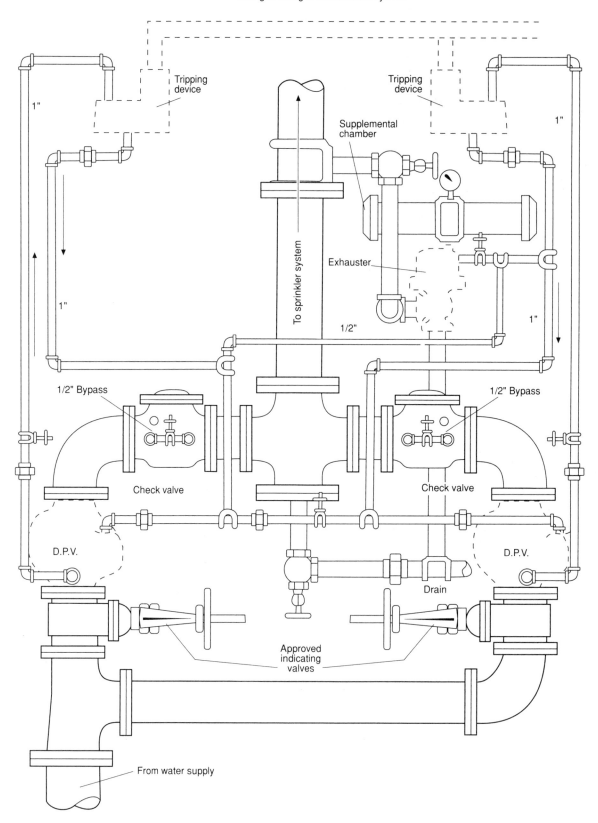

Tubing or wiring to fire detection system

Figure 1.2.4 Header for dry-pipe valves installed in parallel for combined systems; standard trimmings not shown. Arrows indicate direction of fluid flow. (*Reprinted with permission from NFPA 13,* Installation of Sprinkler Systems, *Copyright © 1996, National Fire Protection Association, Quincy, MA 02269. This reprinted material is not the complete and official position of the National Fire Protection Association on the referenced subject, which is represented only by the standard in its entirety.*)

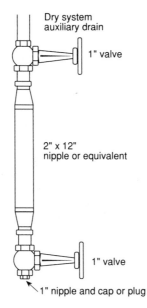

Dry system
auxiliary drain

1" valve

2" x 12"
nipple or equivalent

1" valve

1" nipple and cap or plug

Figure 1.2.5 Drain for a system installation. (*Reprinted with permission from NFPA 13,* Installation of Sprinkler Systems, *Copyright © 1996, National Fire Protection Association, Quincy, MA 02269. This reprinted material is not the complete and official position of the NFPA on the referenced subject, which is represented only by the standard in its entirety.*)

For SI Units: 1 in. = 25.4 mm; 1 ft = 0.3048 m.

1.3.0 Preaction Systems

A preaction sprinkler system (Fig. 1.3.1) is a two-stage system. The first stage is the alert and fire-notification phase. When detected, the presence of a fire activates the alarm (the first phase); the second phase is the sprinkler response.

- *Advantages*   This system combines some of the advantages of a dry system, but adds a time delay to fill the lines with water prior to the opening of any sprinkler systems.
- *Disadvantages*   There is a delay in delivering water through the sprinkler heads as the preaction valve fills the pipes with water, following the sprinkler heads to open in response to the presence of fire.

PREACTION SYSTEM

*Protection against accidental water damage in areas
like computer rooms, libraries and freezers.*

Single Interlocked Preaction System
When the detector (1) is activated by fire, a signal is sent
to the Par 3 Release Control Panel (2). The panel sends
appropriate alarm and trouble signals at the same time
that it signals the solenoid valve (3) to release. The
priming chamber (4) of the deluge valve is then vented
faster than water is supplied through the restricted orifice,
(5) allowing the deluge valve to open. The water
enters the system piping but no water is discharged until a sprinkler (6) fuses. The
pressure operated relief valve (7)
ensures continued
venting.

Viking Preaction Systems can be interfaced
with electric (as shown) or pneumatic
detection systems and can be configured for
single, double or non-interlocked preaction

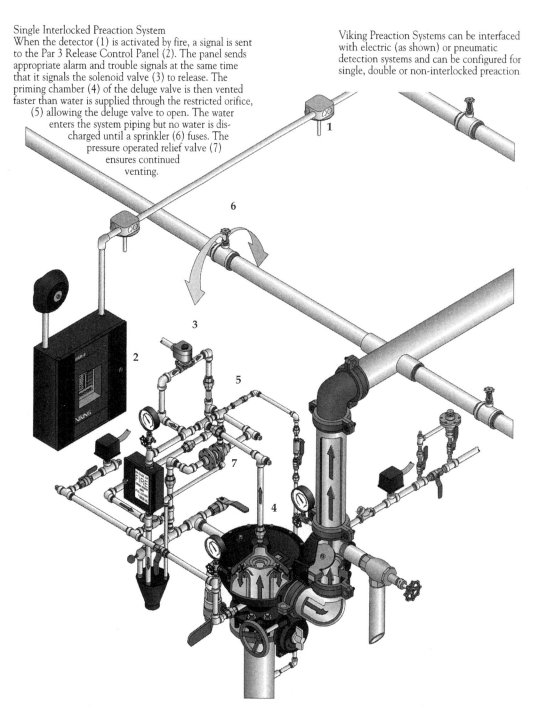

Figure 1.3.1 Preaction system—components and sequence of operation. (*By permission: The Viking Corporation, Hastings, MI.*)

1.4.0   Firecycle Systems

Firecycle systems function (Figs. 1.4.1 and 1.4.2) in much the same manner as the preaction system, except that it adds the installation of sensing devices to stop the flow of water when the fire has been extinguished.

- *Advantages*   Minimizes water damage after the fire has been extinguished.
- *Disadvantages*   Delay in delivering water to the required area(s).

FIRECYCLE SYSTEMS

Where water can be as damaging as fire.

Firecycle II and III are the only fire protection systems with the ability to turn themselves off. These systems provide fire suppression when you need it, and avoid needless water damage to valuable assets when you don't. Firecycle is ideal for enhancing the protection of priceless fine art, artifacts, documents, computer components and data. It is equally effective in manufacturing environments that require the use of toxic chemicals. A Firecycle system's automatic shut-off can reduce the risk of chemical spread.

The Firecycle system is a modification of a preaction system, which means there is no water in the piping until a fire condition exists. So the leaks and accidental water discharge common with other types of on–off systems are eliminated.

Firecycle II

Firecycle II is the only On-Off Multicycle Sprinkler System Approved by FM.

FIRECYCLE II

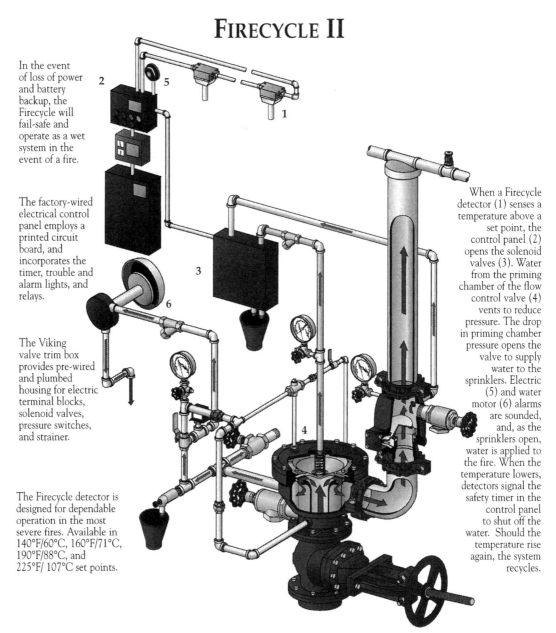

In the event of loss of power and battery backup, the Firecycle will fail-safe and operate as a wet system in the event of a fire.

The factory-wired electrical control panel employs a printed circuit board, and incorporates the timer, trouble and alarm lights, and relays.

The Viking valve trim box provides pre-wired and plumbed housing for electric terminal blocks, solenoid valves, pressure switches, and strainer.

The Firecycle detector is designed for dependable operation in the most severe fires. Available in 140°F/60°C, 160°F/71°C, 190°F/88°C, and 225°F/ 107°C set points.

When a Firecycle detector (1) senses a temperature above a set point, the control panel (2) opens the solenoid valves (3). Water from the priming chamber of the flow control valve (4) vents to reduce pressure. The drop in priming chamber pressure opens the valve to supply water to the sprinklers. Electric (5) and water motor (6) alarms are sounded, and, as the sprinklers open, water is applied to the fire. When the temperature lowers, detectors signal the safety timer in the control panel to shut off the water. Should the temperature rise again, the system recycles.

Figure 1.4.1 Firecycle system—components and sequence of operation. (*By permission: The Viking Corporation, Hastings, MI.*)

FIRECYCLE III

Firecycle III is available in preaction, wet and deluge sprinkler systems. All units are economically priced and have a UL Listing. The new solid-state design reduces system size and cost without compromising dependable operation, long life or easy installation and maintenance.

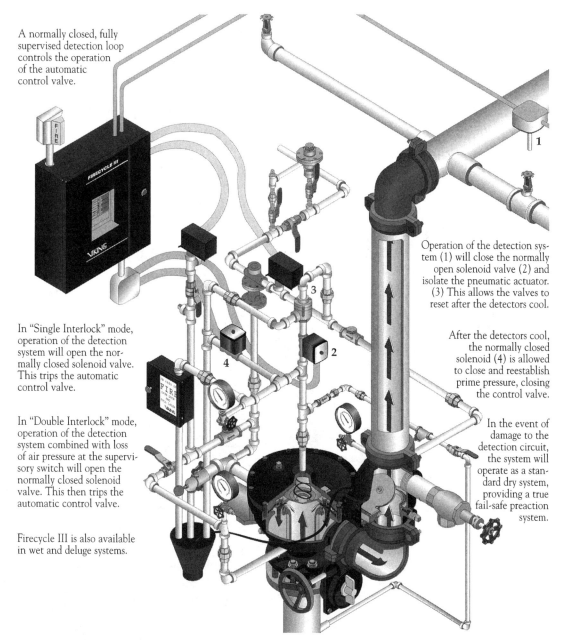

A normally closed, fully supervised detection loop controls the operation of the automatic control valve.

Operation of the detection system (1) will close the normally open solenoid valve (2) and isolate the pneumatic actuator. (3) This allows the valves to reset after the detectors cool.

In "Single Interlock" mode, operation of the detection system will open the normally closed solenoid valve. This trips the automatic control valve.

After the detectors cool, the normally closed solenoid (4) is allowed to close and reestablish prime pressure, closing the control valve.

In "Double Interlock" mode, operation of the detection system combined with loss of air pressure at the supervisory switch will open the normally closed solenoid valve. This then trips the automatic control valve.

In the event of damage to the detection circuit, the system will operate as a standard dry system, providing a true fail-safe preaction system.

Firecycle III is also available in wet and deluge systems.

Figure 1.4.2 Firecycle system available for preaction and deluge systems. (*By permission: The Viking Corporation, Hastings, MI.*)

1.5.0 Deluge Systems

The deluge system (Fig. 1.5.1) is similar to the preaction system, except that all sprinkler heads are kept open when the system is activated.

- *Advantages*   In high hazard areas, this system delivers water through all of the sprinkler heads in the system, instead of just opening selected heads in close proximity to the fire.
- *Disadvantages*   The entire area is deluged with water—even if the fire is restricted to a much smaller area.

DELUGE SYSTEM
Protection for extra-hazard risks such as aircraft hangers and petrochemical facilities.

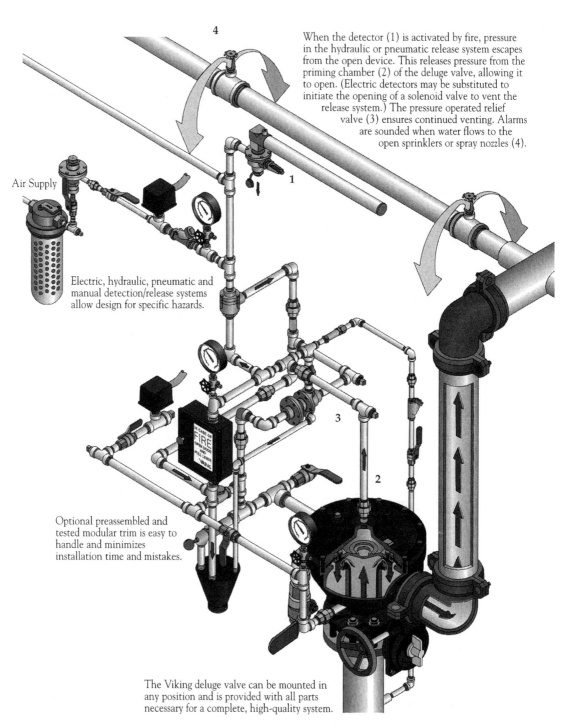

When the detector (1) is activated by fire, pressure in the hydraulic or pneumatic release system escapes from the open device. This releases pressure from the priming chamber (2) of the deluge valve, allowing it to open. (Electric detectors may be substituted to initiate the opening of a solenoid valve to vent the release system.) The pressure operated relief valve (3) ensures continued venting. Alarms are sounded when water flows to the open sprinklers or spray nozzles (4).

Air Supply

Electric, hydraulic, pneumatic and manual detection/release systems allow design for specific hazards.

Optional preassembled and tested modular trim is easy to handle and minimizes installation time and mistakes.

The Viking deluge valve can be mounted in any position and is provided with all parts necessary for a complete, high-quality system.

Figure 1.5.1 Deluge system—components and sequence of operation. (*By permission: The Viking Corporation, Hastings, MI.*)

1.6.0 Standpipes

Standpipes (Fig. 1.6.1) are installed in high-rise buildings to create an internal water supply on upper floors so that firefighters can attach their hoses to connections on the standpipe and effectively fight the fire on that floor. Standpipes are generally located in or near stair enclosures so that firefighters can have a water supply available before entering the actual floor where the fire has occurred.

Standpipe systems are usually of a wet system design and operate through an up-feed pump to ensure proper volume and pressure on the upper floors of a multistory building. However, water can be delivered through the standpipe system by the firefighter's pumper truck, connected to a city fire hydrant, delivering high-pressure, high-volume water through the standpipe via an external siamese connection.

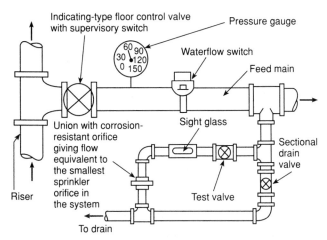

Figure 1.6.1 Sprinkler system floor control valve. (*Reprinted with permission from NFPA 13, Installation of Sprinkler Systems, Copyright © 1996, National Fire Protection Association, Quincy, MA 02269. This reprinted material is not the complete and official position of the NFPA on the referenced subject, which is represented only by the standard in its entirety.*)

1.7.0 Sprinkler Heads

Sprinkler heads (Fig. 1.7.1) are available in a variety of configurations and materials of construction, depending on their coverage and aesthetic requirements. Sprinkler heads are generally of two basic types:

- *Fusible*   Heads with soldered metal links that keep the head closed until the temperature rises to the point where the metal reaches its melting point. The solder will then yield, allowing the sprinkler head to open. Fusible metal alloy pellets can also be used. The pellet will melt at a predetermined temperature and allow the sprinkler head to open.

- *Frangible*   A breakable, transparent glass capsule containing a colored liquid that will expand to the point where the glass will break, allowing the sprinkler head to open. The liquid is color-coded so that visible inspection will confirm that the correct temperature-seeking head has been installed.

Sprinkler Discharge Characteristics Identification

Nominal Orifice Size (in.)	Orifice Type	K Factor[1]	Percent of Nominal ½ in. Discharge	Thread Type	Pintle	Nominal Orifice Size Marked On Frame
¼	Small	1.3–1.5	25	½ in. NPT	Yes	Yes
⁵⁄₁₆	Small	1.8–2.0	33.3	½ in. NPT	Yes	Yes
⅜	Small	2.6–2.9	50	½ in. NPT	Yes	Yes
⁷⁄₁₆	Small	4.0–4.4	75	½ in. NPT	Yes	Yes
½	Standard	5.3–5.8	100	½ in. NPT	No	No
¹⁷⁄₃₂	Large	7.4–8.2	140	¾ in. NPT or	No	No
				½ in. NPT	Yes	Yes
⅝	Extra Large	11.0–11.5	200	½ in. NPT or	Yes	Yes
				¾ in. NPT	Yes	Yes
¾	Very Extra Large	13.5–14.5	250	¾ in. NPT	Yes	Yes
⅝	Large-Drop	11.0–11.5	200	½ in. NPT or	Yes	Yes
				¾ in. NPT	Yes	Yes
⅝	ESFR	11.0–11.5	200	¾ in. NPT	Yes	Yes
¾	ESFR	13.5–14.5	250	¾ in. NPT	Yes	Yes

[1]K factor is the constant in the formula $Q = k\sqrt{p}$ For SI Units: $Q_m = K_m\sqrt{p_m}$
Where Q = Flow in gpm   Where Q_m = Flow in L/min
        p = Pressure in psi P_m = Pressure in bars
                                        K_m = 14 K

Figure 1.7.1 Sprinkler discharge characteristics identification.

1.8.0 Hose Stations

There is often disagreement about the advantages of installing hose cabinets, complete with reeled hoses in strategic locations throughout a building. Often, these cabinets will contain only a valved, threaded connection, but no hose or reel. Unless the hoses are inspected and maintained properly by the building's owner, they could deteriorate or even be removed from the cabinet. Many firefighters, not trusting the quality of cabinet fire hoses, bring their own hoses to the building to attach to hose connections in the cabinets.

Hose nozzles, when attached to these cabinet hoses, are available in adjustable fog, spray type, straight stream, smooth-bore, or combination solid stream and spray. Solid-stream nozzles, ranging in size from ¾ in. (1.90 cm) to 2 in. (5.08 cm), can deliver 120 to 560 gallons per minute, at pressures ranging from 10 to 40 pounds per square inch (psi).

1.9.0 Siamese Connections

A siamese connection is an external source, attached to the building, to which firefighters can attach their hoses from pumper trucks and pressurize the building's sprinkler system by drawing water from a city hydrant (Fig. 1.9.1). The location of the siamese connection, number of connections, and size and type of thread pattern varies, depending upon local fire-marshall requirements.

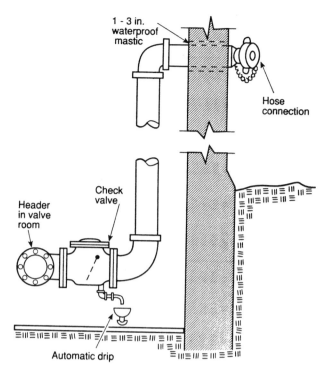

1 - 3 in.
waterproof
mastic

Hose
connection

Check
valve

Header
in valve
room

Automatic drip

For SI Units: 1 in. = 25.4 mm.

Figure 1.9.1 Illustration of a fire department connection. (*Reprinted with permission from NFPA 13,* Installation of Sprinkler Systems, *Copyright © 1996, National Fire Protection Association, Quincy, MA 02269. This reprinted material is not the complete and official position of the National Fire Protection Association on the referenced subject, which is represented only by the standard in its entirety.*)

1.10.0   Light, Ordinary, and Extra-Hazard Occupancy (Defined)

Light Hazard Occupancies include occupancies having conditions similar to:

Churches

Clubs

Eaves and overhangs, if combustible construction with no combustibles beneath

Educational

Hospitals

Institutional

Libraries, except large stack rooms

Museums

Nursing or convalescent homes

Office, including data processing

Residential

Restaurant seating area

Theaters and auditoriums excluding stages and prosceniums

Unused attics

Ordinary Hazard Occupancies (Group 1) include occupancies having conditions similar to:

Automobile parking and showrooms

Bakeries

Beverage manufacturing

Canneries

Dairy products manufacturing and processing

Electronic plants

Glass and glass products manufacturing

Laundries

Restaurant service areas

Ordinary Hazard Occupancies (Group 2) include occupancies having conditions similar to:

Cereal mills

Chemical plants—ordinary

Confectionery products

Distilleries

Dry cleaners

Feed mills

Horse stables

Leather goods manufacturing

Libraries—large stack room areas

Machine shops

Metal working

Mercantile

Paper and pulp mills

Paper process plants

Piers and wharves

Post offices

Printing and publishing

Repair garages

Stages

Textile manufacturing

Tire manufacturing

Tobacco products manufacturing

Wood machining

Wood product assembly

1.11.0 Illustration of Grid versus Looped System

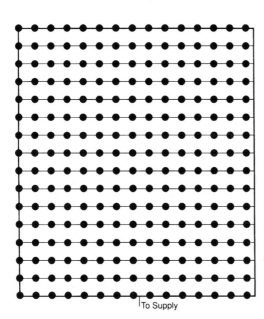

Gridded system.

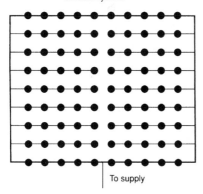

To supply

Looped system.

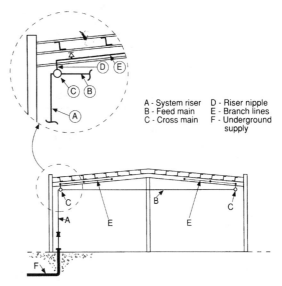

A - System riser   D - Riser nipple
B - Feed main   E - Branch lines
C - Cross main   F - Underground supply

Building elevation showing parts of sprinkler piping system.

Dry Sprinkler. Under certain ambient conditions, wet pipe systems having dry-pendent (or upright) sprinklers may freeze due to heat loss by conduction. Therefore, due consideration should be given to the amount of heat maintained in the heated space, the length of the nipple in the heated space, and other relevant factors.

(Reprinted with permission from NFPA 13, Installation of Sprinkler Systems, *Copyright © 1994, National Fire Protection Association, Quincy, MA 02269. This reprinted material is not the complete and official position of the National Fire Protection Association on the referenced subject, which is represented only by the standard in its entirety.)*

1.12.0   Placement of Sprinkler Heads in Relation to Obstructions

**Position of Sprinklers in Relation to
Obstruction Located Entirely Below the Sprinklers**

Distance of Deflector above Bottom of Obstruction	Minimum Distance to Side of Obstruction, ft (m)
Less than 6 in. (152 mm)	1½ (0.5)
6 in. (152 mm) to less than 12 in. (305 mm)	3 (0.9)
12 in. (305 mm) to less than 18 in. (457 mm)	4 (1.2)
18 in. (457 mm) to less than 24 in. (610 mm)	5 (1.5)
24 in. (610 mm) to less than 30 in. (660 mm)	6 (1.8)

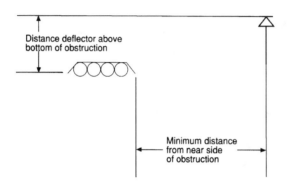

**Position of sprinklers in relation to obstructions
located entirely below the sprinklers. (To be used with Table 4-4.3.4.2.1.)**

*Exception: Where the obstruction is greater than 24 in. (610 mm)
wide, one or more lines of sprinklers shall be installed below the
obstruction.*

(c) The obstruction shall not extend more than 12 in.
(305 mm) to either side of the midpoint between sprinklers.

*Exception: When the extensions of the obstruction exceed 12 in.
(305 mm), one or more lines of sprinklers shall be installed below
the obstruction.*

(d) At least 18 in. (457 mm) clearance shall be maintained between the top of storage and the bottom of the
obstruction.

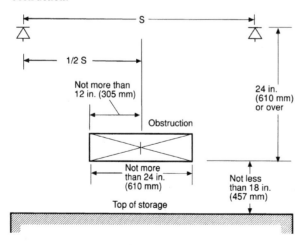

**Position of sprinklers in relation to obstructions
located 24 in. (610 mm) or more below deflectors.**

(Reprinted with permission from NFPA 13, Installation of
Sprinkler Systems, *Copyright © 1994, National Fire Protection Association, Quincy, MA 02269. This reprinted material
is not the complete and official position of the National Fire
Protection Association on the referenced subject, which is
represented only by the standard in its entirety.)*

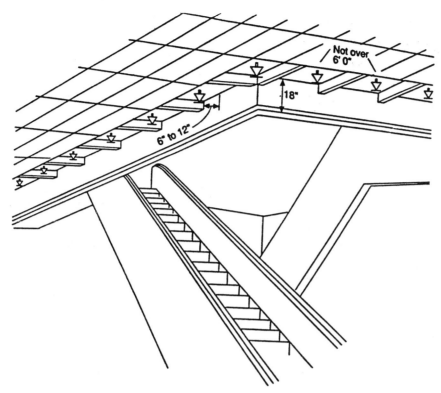

Figure 1.12.1 Configuration of sprinklers around an escalator. (*Reprinted with permission from NFPA 13,* Installation of Sprinkler Systems, *Copyright © 1996, National Fire Protection Association, Quincy, MA 02269. This reprinted material is not the complete and official position of the National Fire Protection Association on the referenced subject, which is represented only by the standard in its entirety.*)

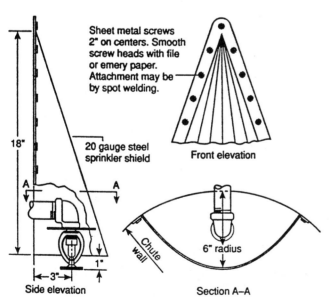

Figure 1.12.2 Protecting sprinklers in building service chutes. (*Reprinted with permission from NFPA 13,* Installation of Sprinkler Systems, *Copyright © 1996, National Fire Protection Association, Quincy, MA 02269. This reprinted material is not the complete and official position of the National Fire Protection Association on the referenced subject, which is represented only by the standard in its entirety.*)

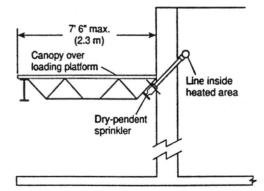

Figure 1.12.3 Dry pendent head protection at covered docks. (*Reprinted with permission from NFPA 13,* Installation of Sprinkler Systems, *Copyright © 1996, National Fire Protection Association, Quincy, MA 02269. This reprinted material is not the complete and official position of the National Fire Protection Association on the referenced subject, which is represented only by the standard in its entirety.*)

Occupancy Group Classification for Miscellaneous Storage 12 ft (3.7 m) or Less in Height*[†]

Commodity Classes I through IV

Commodity Classification	Palletized and Bin Box	Rack
I	OH-1	OH-1
II	OH-1	OH-1
III	OH-2	OH-2
IV up to 10 ft	OH-2	OH-2
IV over 10 ft to 12 ft	OH-2	EH-1

Group A Plastics

Height of Storage	Ceiling Clearance to Top of Storage	Rack-R or Palletized-P	Cartoned		Exposed	
			Solid	Expanded	Solid	Expanded
To 5 ft	No Limit	R-P	OH-2	OH-2	OH-2	OH-2
Over 5 ft to 10 ft	To 5 ft	R-P	EH-1	EH-1	EH-2	EH-2
Over 5 ft to 10 ft	Over 5 ft to 10 ft	R-P	EH-2	EH-2	EH-2	
Over 5 ft to 8 ft	Over 5 ft	P				EH-2
Over 10 ft to 12 ft	To 15 ft	P	EH-2	EH-2		
Over 10 ft to 12 ft	Over 5 ft	R	OH-2 +1 Level In Rack	OH-2 +1 Level In Rack	OH-2 +1 Level In Rack	OH-2 +1 Level In Rack
Over 10 ft to 12 ft	To 5 ft	R-P	EH-2**	EH-2**	EH-2**	EH-2

**For rack storage, OH-2 +1 Level In Rack shall also be permitted.

Figure 1.12.4 Rack storage considerations by commodity class and group A plastics. (*Reprinted with permission from NFPA 13*, Installation of Sprinkler Systems, *Copyright © 1996, National Fire Protection Association, Quincy, MA 02269. This reprinted material is not the complete and official position of the National Fire Protection Association on the referenced subject, which is represented only by the standard in its entirety.*)

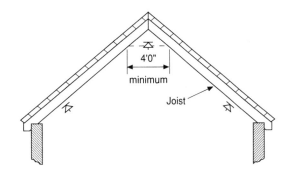

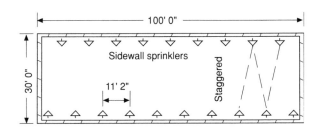

For SI Units: 1 in. = 25.4 mm; 1 ft = 0.3048 m.

For SI Units: 1 in. = 25.4 mm; 1 ft = 0.3048 m.

Figure 1.12.5 Sidewall sprinkler head guidelines. (*Reprinted with permission from NFPA 13*, Installation of Sprinkler Systems, *Copyright © 1996, National Fire Protection Association, Quincy, MA 02269. This reprinted material is not the complete and official position of the NFPA on the referenced subject, which is represented only by the standard in its entirety.*)

	Light Hazard Occupancy			Ordinary Hazard Occupancy	
	Combustible Sheathing	Combustible Construction with Noncombustible or Limited Combustible Sheathing, Wood Lath, and Plaster	Noncombustible Construction with Noncombustible or Limited Combustible Sheathing	Combustible Sheathing	Noncombustible or Limited Combustible Sheathing
Maximum distance between sprinklers on branch line	14	14	14	10	10
Maximum room width for single branch line along wall (ft)	12	12	14	10	10
Maximum area coverage (ft²)	120	168	196	80	100

For SI Units: 1 ft = 0.3048 m; 1 ft² = 0.0929 m².

Figure 1.12.6 Sidewall heads for light occupancy hazards. (*Reprinted with permission from NFPA 13*, Installation of Sprinkler Systems, *Copyright © 1996, National Fire Protection Association, Quincy, MA 02269. This reprinted material is not the complete and official position of the NFPA on the referenced subject, which is represented only by the standard in its entirety.*)

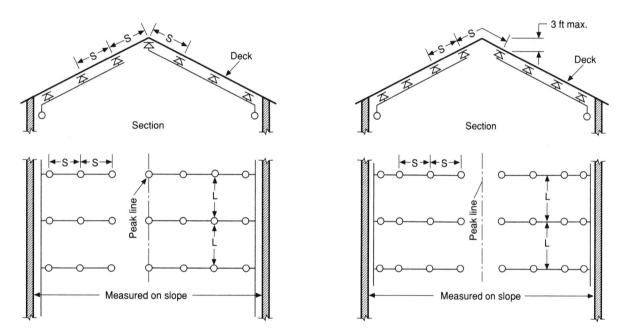

For SI Units: 1 in. = 25.4 mm; 1 ft = 0.3048 m.   For SI Units: 1 in. = 25.4 mm; 1 ft = 0.3048 m.

Figure 1.12.7 Placement of sprinkler heads in a pitched roof. (*Reprinted with permission from NFPA 13*, Installation of Sprinkler Systems, *Copyright © 1996, National Fire Protection Association, Quincy, MA 02269. This reprinted material is not the complete and official position of the NFPA on the referenced subject, which is represented only by the standard in its entirety.*)

1.13.0 Sprinkler Head Placement Requirements

Position of Deflector when Located above Bottom of Obstruction

Distance from Sprinkler to Side of Obstruction	Maximum Allowable Distance of Deflector Above Bottom of Obstruction	Maximum Allowable Distance of Deflector Above Bottom of Obstruction
	Standard Sprinklers	Extended Coverage Sprinklers
Less than 1 ft	0 in.	0 in.
1 ft to less than 1 ft 6 in.	1 in.	0 in.
1 ft 6 in. to less than 2 ft	1 in.	1 in.
2 ft to less than 2 ft 6 in.	2 in.	1 in.
2 ft 6 in. to less than 3 ft	3 in.	1 in.
3 ft to less than 3 ft 6 in.	4 in.	3 in.
3 ft 6 in. to less than 4 ft	6 in.	3 in.
4 ft to less than 4 ft 6 in.	7 in.	5 in.
4 ft 6 in. to less than 5 ft	9 in.	7 in.
5 ft to less than 5 ft 6 in.	11 in.	7 in.
5 ft 6 in. to less than 6 ft	14 in.	7 in.
6 ft to less than 6 ft 6 in.	N/A	9 in.
6 ft 6 in. to less than 7 ft	N/A	11 in.
7 ft and greater	N/A	14 in.

For SI Units: 1 in. = 25.4 mm; 1 ft = 0.3048 m

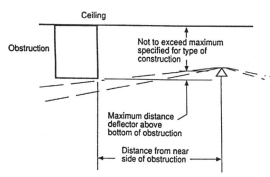

Position of deflector, upright, or pendent sprinkler when located above bottom of obstructions.

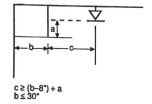

c ≥ (b–8") + a
b ≤ 30"

Horizontal obstructions against walls.

Under obstructed construction, the sprinkler deflector shall be located 1 to 6 in. (25.4 to 152 mm) below the structural members and a maximum distance of 22 in. (559 mm) below the ceiling/roof deck.

Exception No. 1:   Sprinklers shall be permitted to be installed with the deflector at or above the bottom of the structural member to a maximum of 22 in. (559 mm) below the ceiling/roof deck where the sprinkler is installed in conformance.

Horizontal and Minimum Vertical Distances for Sprinklers

Horizontal Distance	Minimum Vertical Distance below Deflector
6 in. or less	3 in.
More than 6 in. to 9 in.	4 in.
More than 9 in. to 12 in.	6 in.
More than 12 in. to 15 in.	8 in.
More than 15 in. to 18 in.	9½ in.
More than 18 in. to 24 in.	12½ in.
More than 24 in. to 30 in.	15½ in.
More than 30 in.	18 in.

For SI Units: 1 in. = 25.4 mm.

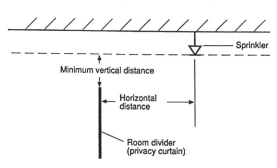

Sprinklers installed near privacy curtains, free-standing partitions, or room dividers.

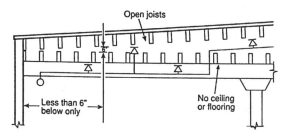

For SI Units: 1 in. = 25.4 mm.
Arrangement of sprinklers under two sets of open joists — no sheathing on lower joists.

Exception No. 2:   Where sprinklers are installed in each bay of obstructed construction, deflectors shall be a minimum of 1 in. (25.4 mm) and a maximum of 12 in. (152 mm) below the ceiling.

Exception No. 3:   Sprinklers shall only be permitted below composite wood joists where joist channels are firestopped to the full depth of the joists with material equivalent to the web construction so that individual channel areas do not exceed 300 sq ft (27.9 m²).

Exception No. 4:   Deflectors of sprinklers under concrete tee construction with stems spaced less than 7½ ft (2.3 m) but more than 3 ft (0.9 m) on centers shall, regardless of the depth of the tee, be permitted to be located at or above the plane 1 in. (25.4 mm) below the bottom of the stems of the tees and shall comply with.*

1.14.0 Sprinkler Head Requirements for Various Hazards

Hazard	Type of System	Minimum Operating Pressure,[1] psi (bar)			Hose Stream Demand gal/min (dm³/min)	Water Supply Duration, Hr
		25 (1.7)	50 (3.4)	75 (5.2)		
		Number Design Sprinklers				
Palletized[2] Storage Class I, II, and III commodities up to 25 ft (7.6 m) with maximum 10 ft (3.0 m) clearance to ceiling	Wet	15	Note 4	Note 4	500 (1900)	2
	Dry	25	Note 4	Note 4		
Class IV commodities up to 20 ft (6.1 m) with maximum 10 ft (3.0 m) clearance to ceiling	Wet	20	15	Note 4	500 (1900)	2
	Dry	Does not apply	Does not apply	Does not apply		
Unexpanded plastics up to 20 ft (6.1 m) with maximum 10 ft (3.0 m) clearance to ceiling	Wet	25	15	Note 4	500 (1900)	2
	Dry	Does not apply	Does not apply	Does not apply		
Expanded plastics commodities up to 18 ft (5.5 m) with maximum 8 ft (2.4 m) clearance to ceiling	Wet	Does not apply	15	Note 4	500 (1900)	2
	Dry	Does not apply	Does not apply	Does not apply		
Idle wood pallets up to 20 ft (6.1 m) with maximum 10 ft (3.0 m) clearance to ceiling	Wet	15	Note 4	Note 4	500 (1900)	1½
	Dry	25	Note 4	Note 4		
Solid Piled[2] Storage Class I, II, and III commodities up to 20 ft (6.1 m) with maximum 10 ft (3.0 m) clearance to ceiling	Wet	15	Note 4	Note 4	500 (1900)	1½
	Dry	25	Note 4	Note 4		
Class IV commodities and unexpanded plastics up to 20 ft (6.1 m) with maximum 10 ft (3.0 m) clearance to ceiling	Wet	Does not apply	15	Note 4	500 (1900)	1½
	Dry	Does not apply	Does not apply	Does not apply		
Double-Row Rack Storage[3] with Minimum 5.5 ft (1.7 m) Aisle Width and Multiple-Row Rack Storage with Minimum 8.0 ft (2.5 m) Aisle Width Class I and II commodities up to 25 ft (7.6 m) with maximum 5 ft (1.5 m) clearance to ceiling	Wet	20	Note 4	Note 4	500 (1900)	1½
	Dry	30	Note 4	Note 4		
Class I and II commodities up to 30 ft (9.2 m) with maximum 5 ft (1.5 m) clearance to ceiling	Wet	20 plus one level of in-rack sprinklers[5]	Note 4	Note 4	500 (1900)	1½
	Dry	30 plus one level of in-rack sprinklers[5]	Note 4	Note 4		
Class I, II, and III commodities up to 20 ft (6.1 m) with maximum 10 ft (3.0 m) clearance to ceiling	Wet	15	Note 4	Note 4	500 (1900)	1½
	Dry	25	Note 4	Note 4		
Class I, II, and III commodities up to 25 ft (7.6 m) with maximum 10 ft (3.0 m) clearance to ceiling	Wet	15 plus one level of in-rack sprinklers[5]	Note 4	Note 4	500 (1900)	1½
	Dry	25 plus one level of in-rack sprinklers[5]	Note 4	Note 4		
Class IV commodities up to 20 ft (6.1 m) with maximum 10 ft (3.0 m) clearance to ceiling	Wet	Does not apply	20	15	500 (1900)	2
	Dry	Does not apply	Does not apply	Does not apply		
Class IV commodities up to 25 ft (7.6 m) with maximum 10 ft clearance to ceiling	Wet	Does not apply	20 plus one level of in-rack sprinklers[5]	15 plus one level of in-rack sprinklers[5]	500 (1900)	2
	Dry	Does not apply	Does not apply	Does not apply		

(Reprinted with permission from NFPA 13, Installation of Sprinkler Systems, Copyright © 1994, National Fire Protection Association, Quincy, MA 02269. This reprinted material is not the complete and official position of the National Fire Protection Association on the referenced subject, which is represented only by the standard in its entirety.)

ESLO UPRIGHT, PEND. & REC. PEND.

SPECIFICATIONS	
MODEL # -	ESLO
ORIFICE SIZE -	.70
NOMINAL K FACTOR -	14.5
TEMPERATURE -	155°, 200°, 250°
THREAD SIZE -	3/4"
FINISH -	BRASS, CHROME, WHITE, BLACK

WITH THIS SPRINKLERS .70" ORIFICE SIZE, THE LARGEST AVAILABLE FOR BOTH LIGHT AND ORDINARY HAZARD APPLICATIONS, PRESSURE REQUIREMENTS ARE EXTREMELY LOW. THIS VERY EXTRA LARGE ORIFICE SPRINKLER HAS THE ABILITY TO COVER LARGE AREAS, UP TO 16'x16', WITH PRESSURES LESS THAN A 1/2" STANDARD SPRINKLER SPACED AT IT'S MAXIMUM.

ESLO-20 GB

SPECIFICATIONS	
MODEL # -	ESLO-20 GB
ORIFICE SIZE -	.70
NOMINAL K FACTOR -	14.5
TEMPERATURE -	155°, 200°, 250°
THREAD SIZE -	3/4"
FINISH -	BRASS, WHITE, BLACK

WITH THIS SPRINKLERS .70" ORIFICE SIZE, THE LARGEST AVAILABLE FOR ORDINARY HAZARD APPLICATIONS, PRESSURE REQUIREMENTS ARE EXTREMELY LOW.

ELO SW-20

SPECIFICATIONS	
MODEL # -	ELO S/W-20
ORIFICE SIZE -	.64
NOMINAL K FACTOR -	11.5
TEMPERATURE -	155°, 200°
THREAD SIZE -	3/4"
FINISH -	BRASS, WHITE, BLACK

THE ELO SW-20 HORIZONTAL SIDEWALL SPRINKLER IS DESIGNED AND LISTED FOR EXTENDED COVERAGE SPACING IN ORDINARY HAZARD APPLICATIONS. IT IS THE FIRST SIDEWALL SPRINKLER TO BE LISTED FOR EXTENDED COVERAGE IN ORDINARY HAZARD. IT UTILIZES AN EXTRA LARGE ORIFICE TO ALLOW LOW WATER PRESSURE REQUIREMENTS WHILE PROVIDING THE FLOW REQUIRED FOR EXTENDED COVERAGE. THE ELO SW-20 IS LISTED FOR UP TO A 16'-0" WIDE AND A 20'-0" THROW COVERAGE AREA.

ELO SW-24

SPECIFICATIONS	
MODEL # -	ELO S/W-24
ORIFICE SIZE -	.64
NOMINAL K FACTOR -	11.5
TEMPERATURE -	155°, 200°
THREAD SIZE -	3/4"
FINISH -	BRASS, WHITE, BLACK

THE ELO SW-24 HORIZONTAL SIDEWALL SPRINKLER IS DESIGNED AND LISTED FOR EXTENDED COVERAGE SPACING IN ORDINARY HAZARD APPLICATIONS. IT IS THE FIRST SIDEWALL SPRINKLER TO BE LISTED FOR EXTENDED COVERAGE IN ORDINARY HAZARD. IT UTILIZES AN EXTRA LARGE ORIFICE TO ALLOW LOW WATER PRESSURE REQUIREMENTS WHILE PROVIDING THE FLOW REQUIRED FOR EXTENDED COVERAGE. THE ELO SW-24 IS LISTED FOR UP TO A 16'-0" WIDE AND A 24'-0" THROW COVERAGE AREA.

ELO-16 GB FR UPRIGHT PENDENT & RECESSED PENDENT

SPECIFICATIONS	
MODEL # -	ELO-16 GB FR
ORIFICE SIZE -	.64
NOMINAL K FACTOR -	11.4
TEMPERATURE -	155°, 200°, 250°
THREAD SIZE -	3/4"
FINISH -	BRASS, CHROME, WHITE, BLACK

THE SPECIFIC ADVANTAGES FOR THIS SPRINKLER ARE MULTIPLE. THEY ARE AVAILABLE FOR COVERAGES UP TO 16'x16'.

Figure 1.14.1 Sprinkler head configurations; upright pendents and extended coverage sidewall heads. (*By permission: Central Sprinkler Company, Lansdale, PA.*)

Figure 1.14.2 Sprinkler head configurations—concealed, extended coverage uprights, pendent, and sidewall. (*By permission: Central Sprinkler Company, Lansdale, PA.*)

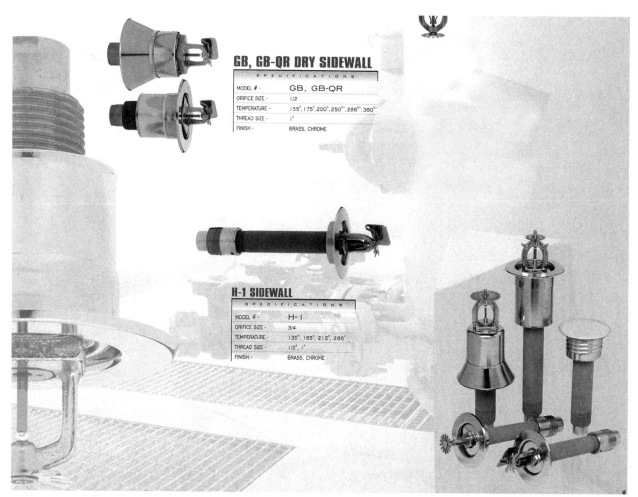

Figure 1.14.3 Sprinkler head configurations: dry pendent. (*By permission: Central Sprinkler Company, Lansdale, PA.*)

Figure 1.14.4 Sprinkler head configurations: dry pendent heads and recessed dry pendent heads. (*By permission: Central Sprinkler Company, Lansdale, PA.*)

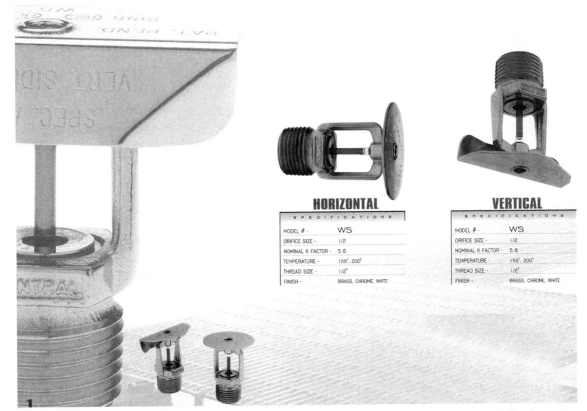

Figure 1.14.5 Sprinkler head configurations—a unique head that is capable of protecting glazing in a wall and allow it to maintain its rating up to 2 hours. (*By permission: Central Sprinkler Company, Lansdale, PA.*)

ESFR-1 WITH GUARD

WS-3 SHIELD
WITH G-3 GUARD

G-3 GUARD
FOR UPRIGHT

G-3 Guard

WSG-2 Guard &
Shield Assembly

G-2 Guard

MULTI-LEVEL GB-QR
UPRIGHT

WSG-3 GUARD & SHIELD
ASSEMBLY

Figure 1.14.6   Sprinkler head configurations: guards to protect heads in areas subject to damage. (*By permission: Central Sprinkler Company, Lansdale, PA.*)

1.15.0 Temperature Ratings of Sprinklers, Based on the Distance from the Heat Source

Temperature Ratings of Sprinklers Based on Distance from Heat Sources

Type of Heat Condition	Ordinary Degree Rating	Intermediate Degree Rating	High Degree Rating
1. Heating Ducts (a) Above	More than 2 ft 6 in.	2 ft 6 in. or less	—
(b) Side and Below	More than 1 ft 0 in.	1 ft 0 in. or less	—
(c) Diffuser Downward Discharge Horizontal Discharge	Any distance except as shown under Intermediate	*Downward:* Cylinder with 1 ft 0 in. radius from edge, extending 1 ft 0 in. below and 2 ft 6 in. above *Horizontal:* Semi-cylinder with 2 ft 6 in. radius in direction of flow, extending 1 ft 0 in. below and 2 ft 6 in. above	—
2. Unit Heater (a) Horizontal Discharge	—	*Discharge Side:* 7 ft 0 in. to 20 ft 0 in. radius pie-shaped cylinder [see Figure 4-3.1.3.2] extending 7 ft 0 in. above and 2 ft 0 in. below heater; also 7 ft 0 in. radius cylinder more than 7 ft 0 in. above unit heater	7 ft 0 in. radius cylinder extending 7 ft 0 in. above and 2 ft 0 in. below unit heater
(b) Vertical Downward Discharge [Note: For sprinklers below unit heater, see Figure 4-3.1.3.2.]	—	7 ft 0 in. radius cylinder extending upward from an elevation 7 ft 0 in. above unit heater	7 ft 0 in. radius cylinder extending from the top of the unit heater to an elevation 7 ft 0 in. above unit heater
3. Steam Mains (Uncovered) (a) Above	More than 2 ft 6 in.	2 ft 6 in. or less	—
(b) Side and Below	More than 1 ft 0 in.	1 ft 0 in. or less	—
(c) Blowoff Valve	More than 7 ft 0 in.	—	7 ft 0 in. or less

For SI Units: 1 in. = 25.4 mm; 1 ft = 0.3048 m.

Ratings of Sprinklers in Specified Locations

Location	Ordinary Degree Rating	Intermediate Degree Rating	High Degree Rating
Skylights	—	Glass or plastic	—
Attics	Ventilated	Unventilated	—
Peaked Roof: Metal or thin boards; concealed or not concealed; insulated or uninsulated	Ventilated	Unventilated	—
Flat Roof: Metal, not concealed; insulated or uninsulated	Ventilated or unventilated	Note: For uninsulated roof, climate and occupancy may necessitate intermediate sprinklers. Check on job.	—
Flat Roof: Metal; concealed; insulated or uninsulated	Ventilated	Unventilated	—
Show Windows	Ventilated	Unventilated	—

Note: A check of job condition by means of thermometers may be necessary.

Temperature Ratings, Classifications, and Color Codings

Max. Ceiling Temp. °F	°C	Temperature Rating °F	°C	Temperature Classification	Color Code	Glass Bulb Colors
100	38	135 to 170	57 to 77	Ordinary	Uncolored or Black	Orange or Red
150	66	175 to 225	79 to 107	Intermediate	White	Yellow or Green
225	107	250 to 300	121 to 149	High	Blue	Blue
300	149	325 to 375	163 to 191	Extra High	Red	Purple
375	191	400 to 475	204 to 246	Very Extra High	Green	Black
475	246	500 to 575	260 to 302	Ultra High	Orange	Black
625	329	650	343	Ultra High	Orange	Black

(Reprinted with permission from NFPA 13, Installation of Sprinkler Systems, *Copyright © 1994, National Fire Protection Association, Quincy, MA 02269. This reprinted material is not the complete and official position of the National Fire Protection Association on the referenced subject, which is represented only by the standard in its entirety.)*

1.16.0 Sprinkler Maintenance Schedules

Parts	Activity	Frequency
Flushing Piping	Test	5 years
Fire Department Connections	Inspection	Monthly
Control Valves	Inspection	Weekly—Sealed
	Inspection	Monthly—Locked
	Inspection	Monthly—Tamper Switch
	Maintenance	Yearly
Main Drain	Flow Test	Quarterly
Open Sprinklers	Test	Annual
Pressure Gauge	Calibration Test	
Sprinklers	Test	50 years
Sprinklers—High Temp	Test	5 years
Sprinklers—Residential	Test	20 years
Waterflow Alarms	Test	Quarterly
Preaction/Deluge Detection System	Test	Semiannually
Preaction/Deluge Systems	Test	Annually
Antifreeze Solution	Test	Annually
Cold Weather Valves	Open and Close Valves	Fall, Close; Spring, Open
Dry/Preaction/Deluge Systems Air Pressure and Water Pressure	Inspection	Weekly
Enclosure	Inspection	Daily—Cold Weather
Priming Water Level	Inspection	Quarterly
Low—Point Drains	Test	Fall
Dry Pipe Valves	Trip Test	Annual—Spring
Dry Pipe Valves	Full Flow Trip	3 years—Spring
Quick-Opening Devices	Test	Semiannually

1.17.0 Hangers for Sprinkler Pipes

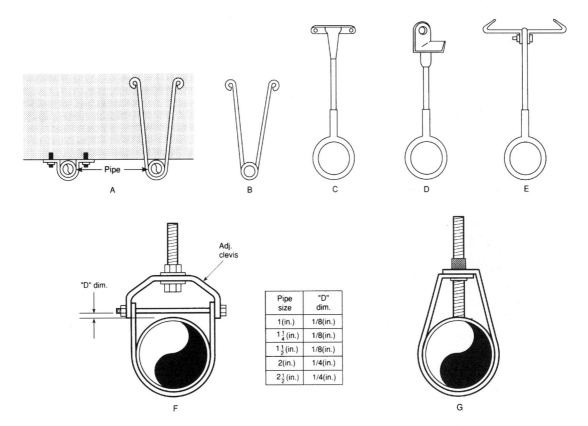

Pipe size	"D" dim.
1 (in.)	1/8 (in.)
1¼ (in.)	1/8 (in.)
1½ (in.)	1/8 (in.)
2 (in.)	1/4 (in.)
2½ (in.)	1/4 (in.)

A U-type hangers for branch lines
B Wraparound U-hook
C Adjustable clip for branch lines
D Side beam adjustable hanger
E Adjustable coach screw clip for branch lines
F Clevis hanger
G Adjustable swivel loop hanger

Examples of acceptable hangers for end of line (or armover) pendent sprinklers.

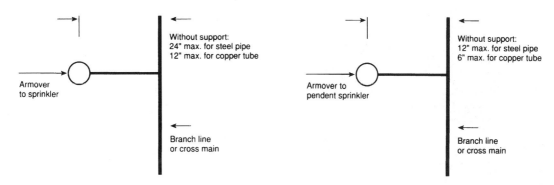

For SI Units: 1 in. = 25.4 mm; 1 ft = 0.3048 m.
Maximum length for unsupported armover.

For SI Units: 1 in. = 25.4 mm; 1 ft = 0.3048 m.
NOTE: The pendent sprinkler may be installed either directly in the fitting at the end of the armover or in a fitting at the bottom of a drop nipple.
Maximum length of unsupported armover where the maximum pressure exceeds 100 psi (6.9 bars) and a branch line above a ceiling supplies pendent sprinklers below the ceiling.

4-6.2 Pipe Support.

4-6.2.1 General.

4-6.2.1.1 Sprinkler piping shall be supported independently of the ceiling sheathing.

Exception: Toggle hangers shall be permitted only for the support of pipe 1½ in. (38 mm) or smaller in size under ceilings of hollow tile or metal lath and plaster.

4-6.2.1.2 Where sprinkler piping is installed in storage racks as defined in NFPA 231C, *Standard for Rack Storage of Materials*, piping shall be supported from the storage rack structure or building in accordance with all applicable provisions of 4-6.2 and 4-6.4.3.

4-6.2.2 Maximum Distance between Hangers.

4-6.2.2.1* The maximum distance between hangers shall not exceed that in Table 4-6.2.2.1.

Exception No. 1: The maximum distance between hangers for steel pipe and copper tube shall be modified as specified in 4-6.2.1 and 4-6.2.2.

Exception No. 2: The maximum distance between hangers for CPVC pipe and polybutylene pipe shall be modified as specified in the individual product listings.

Exception No. 3: Holes through concrete beams shall be acceptable for the support of steel pipe as a substitute for hangers.

4-6.2.3 Location of Hangers on Branch Lines. This subsection applies to the support of steel pipe or copper tube as specified in 2-3.1 and subject to the provisions of 4-6.2.2.

4-6.2.3.1 There shall be not less than one hanger for each section of pipe.

Exception No. 1*: Where sprinklers are spaced less than 6 ft (1.8 m) apart, hangers spaced up to a maximum of 12 ft (3.7 m) shall be permitted.

Exception No. 2: Starter lengths less than 6 ft (1.8 m) shall not require a hanger, unless on the end line of a sidefeed system or where an intermediate cross main hanger has been omitted.

4-6.2.3.2 The distance between a hanger and the centerline of an upright sprinkler shall not be less than 3 in. (76 mm).

4-6.2.3.3* The unsupported length between the end sprinkler and the last hanger on the line shall not be greater than 36 in. (914 mm) for 1-in. (2.5-cm) pipe or 48 in. (1219 mm) for 1¼-in. (3.2-cm) pipe, and 60 in. (15.2 cm) for 1½-in. (3.8-cm) or larger pipe. Where any of these limits is exceeded, the pipe shall be extended beyond the end sprinkler and shall be supported by an additional hanger.

Exception No. 1*: When the maximum pressure at the sprinkler exceeds 100 psi (6.9 bars), and a branch line above a ceiling supplies sprinklers in a pendent position below the ceiling, the hanger assembly supporting the pipe supplying an end sprinkler in a pendent position shall be of a type that prevents upward movement of the pipe.

Exception No. 2*: When the maximum pressure at the sprinkler exceeds 100 psi (6.9 bars), the unsupported length between the end sprinkler in a pendent position or drop nipple and the last hanger on the branch line shall not be greater than 12 in. (305 mm) for steel pipe or 6 in. (152 mm) for copper pipe. When this limit is exceeded, the pipe shall be extended beyond the end sprinkler and supported by an additional hanger. The hanger closest to the sprinkler shall be of a type that prevents upward movement of the piping.

Figure 1.17.1 Pipe supports and hanger spacing. (*Reprinted with permission from NFPA 13, Installation of Sprinkler Systems, Copyright © 1996, National Fire Protection Association, Quincy, MA 02269. This reprinted material is not the complete and official position of the National Fire Protection Association on the referenced subject, which is represented only by the standard in its entirety.*)

Nominal Pipe Size (in.)	¾	1	1¼	1½	2	2½	3	3½	4	5	6	8
Steel Pipe Except Threaded Light-wall	N/A	12-0	12-0	15-0	15-0	15-0	15-0	15-0	15-0	15-0	15-0	15-0
Threaded Light-wall Steel Pipe	N/A	12-0	12-0	12-0	12-0	12-0	12-0	N/A	N/A	N/A	N/A	N/A
Copper Tube	8-0	8-0	10-0	10-0	12-0	12-0	12-0	15-0	15-0	15-0	15-0	15-0
CPVC	5-6	6-0	6-6	7-0	8-0	9-0	10-0	N/A	N/A	N/A	N/A	N/A
Polybutylene (IPS)	N/A	3-9	4-7	5-0	5-11	N/A	N/A	N/A	N/A	N/A	N/A	N/A
Polybutylene (CTS)	2-11	3-4	3-11	4-5	5-5	N/A	N/A	N/A	N/A	N/A	N/A	N/A

For SI Units: 1 in. = 25.4 mm; 1 ft = 0.3048 m.
NOTE: (IPS) Iron Pipe Size.
    (CTS) Copper Tube Size.

4-6.2.3.4* The length of an unsupported armover to a sprinkler shall not exceed 24 in. (610 mm) for steel pipe or 12 in. (305 mm) for copper tube.

Exception: *Where the maximum pressure at the sprinkler exceeds 100 psi (6.9 bars) and a branch line above a ceiling supplies sprinklers in a pendent position below the ceiling, the length of an unsupported armover to a sprinkler and drop nipple shall not exceed 12 in. (305 mm) for steel pipe and 6 in. (152 mm) for copper tube.*

Where the limits of the unsupported armover lengths of 4-6.2.3.4 or this Exception are exceeded, the hanger closest to the sprinkler shall be of a type that prevents upward movement of the piping.

4-6.2.3.5 Wall-mounted sidewall sprinklers shall be restrained to prevent movement.

4-6.2.4 Location of Hangers on Cross Mains. This subsection applies to the support of steel pipe only as specified in 4-6.2.3, subject to the provisions of 4-6.2.2.

4-6.2.4.1 On cross mains there shall be at least one hanger between each two branch lines.

Exception No. 1: In bays having two branch lines, the intermediate hanger shall be permitted to be omitted provided that a hanger attached to a purlin is installed on each branch line located as near to the cross main as the location of the purlin permits. Remaining branch line hangers shall be installed in accordance with 4-6.2.3.

Exception No. 2: In bays having three branch lines, either side or center feed, one (only) intermediate hanger shall be permitted to be omitted provided that a hanger attached to a purlin is installed on each branch line located as near to the cross main as the location of the purlin permits. Remaining branch line hangers shall be installed in accordance with 4-6.2.3.

Exception No. 3: In bays having four or more branch lines, either side or center feed, two intermediate hangers shall be permitted to be omitted provided the maximum distance between hangers does not exceed the distances specified in 4-6.2.2.1 and a hanger attached to a purlin on each branch line is located as near to the cross main as the purlin permits.

4-6.2.4.2 Intermediate hangers shall not be omitted for copper tube.

4-6.2.4.3 At the end of the cross main, intermediate trapeze hangers shall be installed unless the cross main is extended to the next framing member with a hanger

Figure 1.17.1 *(Continued)*

installed at this point, in which event an intermediate hanger shall be permitted to be omitted in accordance with 4-6.2.4.1, Exceptions No. 1, No. 2, and No. 3.

4-6.2.5 Support of Risers.

4-6.2.5.1 Risers shall be supported by pipe clamps or by hangers located on the horizontal connections close to the riser.

4-6.2.5.2 Clamps supporting pipe by means of setscrews shall not be used.

4-6.2.5.3 In multistory buildings, riser supports shall be provided at the lowest level, at each alternate level above, above and below offsets, and at the top of the riser. Supports above the lowest level shall also restrain the pipe to prevent movement by an upward thrust where flexible fittings are used. Where risers are supported from the ground, the ground support constitutes the first level of riser support. Where risers are offset or do not rise from the ground, the first ceiling level above the offset constitutes the first level of riser support.

4-6.2.5.4 Risers in vertical shafts, or in buildings with ceilings over 25 ft (7.6 m) high, shall have at least one support for each riser pipe section.

4-6.3.6.2 Where drain pipes are buried underground, approved corrosion-resistant pipe shall be used.

4-6.3.6.3 Drain pipes shall not terminate in blind spaces under the building.

4-6.3.6.4 Where exposed to the atmosphere, drain pipes shall be fitted with a turned down elbow.

4-6.3.6.5 Drain pipes shall be arranged to avoid exposing any part of the sprinkler system to freezing conditions.

4-6.4 Protection of Piping.

4-6.4.1 Protection of Piping against Freezing.

4-6.4.1.1 Where portions of systems are subject to freezing and temperatures cannot reliably be maintained at or above 40°F (4°C), sprinklers shall be installed as a dry pipe or preaction system.

Exception: Small unheated areas are permitted to be protected by antifreeze systems or by other systems specifically listed for this purpose. (See 3-5.2.)

4-6.4.1.2 Where water-filled supply pipes, risers, system risers, or feed mains pass through open areas, cold rooms, passageways, or other areas exposed to freezing, the pipe shall be protected against freezing by insulating coverings, frostproof casings, or other reliable means capable of maintaining a minimum temperature of 40°F (4°C).

4-6.4.2 Protection of Piping against Corrosion.

4-6.4.2.1* Where corrosive conditions are known to exist due to moisture or fumes from corrosive chemicals or both, special types of fittings, pipes, and hangers that resist corrosion shall be used or a protective coating shall be applied to all unprotected exposed surfaces of the sprinkler system. *(See 2-2.4.)*

4-6.4.2.2 Where water supplies are known to have unusual corrosive properties and threaded or cut-groove steel pipe is to be used, wall thickness shall be in accordance with Schedule 30 [in sizes 8 in. (200 mm) or larger] or Schedule 40 [in sizes less than 8 in. (200 mm)].

4-6.4.2.3 Steel pipe, where exposed to weather, shall be externally galvanized or otherwise protected against corrosion.

4-6.4.2.4 Where steel pipe is used underground, the pipe shall be protected against corrosion.

4-6.4.3 Protection of Piping against Damage Where Subject to Earthquakes.

4-6.4.3.1* General. Sprinkler systems shall be protected to prevent pipe breakage where subject to earthquakes in accordance with the requirements of 4-6.4.3.

Exception: Alternative methods of providing earthquake protection of sprinkler systems based on a dynamic seismic analysis certified by a registered professional engineer such that system performance will be at least equal to that of the building structure under expected seismic forces.

4-6.4.3.2* Couplings. Listed flexible pipe couplings joining grooved end pipe shall be provided as flexure joints to allow individual sections of piping 2½ in. (64 mm) or larger to move differentially with the individual sections of the building to which it is attached. Couplings shall be arranged to coincide with structural separations within a building. They shall be installed:

(a) Within 24 in. (610 mm) of the top and bottom of all risers.

Exception No. 1: In risers less than 3 ft (0.9 m) in length, flexible couplings are permitted to be omitted.

Exception No. 2: In risers 3 to 7 ft (0.9 to 2.1 m) in length, one flexible coupling is adequate.

(b) Within 12 in. (305 mm) above and below the floor in multistory buildings such that the flexible coupling below the floor is below the main supplying that floor.

(c) On one side of concrete or masonry walls within 3 ft (0.9 m) of the wall surface.

Exception: Flexible pipe couplings are not required where clearance around the pipe is provided in accordance with 4-6.4.3.4.

(d)* At or near building expansion joints.

(e) Within 24 in. (610 mm) of the top of drops to hose lines, rack sprinklers, and mezzanines, regardless of pipe size.

(f) Within 24 in. (610 mm) of the top of drops exceeding 15 ft (4.6 m) in length to portions of systems supplying more than one sprinkler, regardless of pipe size.

(g) Above and below any intermediate points of support for a riser or other vertical pipe.

4-6.4.3.3* Seismic Separation Assembly. Seismic separation assemblies with flexible fittings shall be installed where sprinkler piping, regardless of size, crosses building seismic separation joints above ground level.

4-6.4.3.4* Clearance. Clearance shall be provided around all piping extending through walls, floors, platforms, and foundations, including drains, fire department connections, and other auxiliary piping.

4-6.4.3.4.1 Minimum clearance on all sides shall be not less than 1 in. (25.4 mm) for pipes 1 in. (25.4 mm) through 3½ in. (90 mm), and 2 in. (51 mm) for pipe sizes 4 in. (100 mm) and larger.

Exception No. 1: Where clearance is provided by a pipe sleeve, a nominal diameter 2 in. (51 mm) larger than the nominal diameter of the pipe is acceptable for pipe sizes 1 in. (25.4 mm) through 3½ in. (89 mm), and the clearance provided by a pipe sleeve of nominal diameter 4 in. (102 mm) larger than the nominal diameter of the pipe is acceptable for pipe sizes 4 in. (102 mm) and larger.

Exception No. 2: No clearance is necessary for piping passing through gypsum board or equally frangible construction that is not required to have a fire-resistance rating.

Exception No. 3: No clearance is necessary if flexible couplings are located within 1 ft (0.31 m) of each side of a wall, platform, or foundation.

4-6.4.3.4.2 Where required, the clearance shall be filled with a flexible material such as mastic.

4-6.4.3.5 Sway Bracing.

4-6.4.3.5.1 The system piping shall be supported to resist both lateral and longitudinal horizontal loads.

4-6.4.3.5.2* The assigned loads for both lateral and longitudinal sway bracing shall be determined using Table 4-6.4.3.5.2, based on a horizontal force of $F_p = 0.5 \, W_p$, where F_p is the horizontal force factor and W_p is the weight of the water-filled piping.

Figure 1.17.2 Pipe protection: freeze-ups, corrosion, and seismic damage. (*Reprinted with permission from NFPA 13*, Installation of Sprinkler Systems, *Copyright © 1996, National Fire Protection Association, Quincy, MA 02269. This reprinted material is not the complete and official position of the NFPA on the referenced subject, which is represented only by the standard in its entirety.*)

1.18.1   Piping Weights When Filled with Water

**Piping Weights for Determining
Horizontal Load**

Schedule 40 Pipe (in.)	Weight of Water-Filled Pipe (lb per ft)	½ Weight of Water-Filled Pipe (lb per ft)
1	2.05	1.03
1¼	2.93	1.47
1½	3.61	1.81
2	5.13	2.57
2½	7.89	3.95
3	10.82	5.41
3½	13.48	6.74
4	16.40	8.20
5	23.47	11.74
6	31.69	15.85
8*	47.70	23.85

Schedule 10 Pipe (in.)		
1	1.81	0.91
1¼	2.52	1.26
1½	3.04	1.52
2	4.22	2.11
2½	5.89	2.95
3	7.94	3.97
3½	9.78	4.89
4	11.78	5.89
5	17.30	8.65
6	23.03	11.52
8	40.08	20.04

* Schedule 30
For SI Units: 1 in. = 25.4 mm; 1 lb = 0.45kg; 1 ft = 0.30 m.

(Reprinted with permission from NFPA 13, Installation of Sprinkler Systems, *Copyright © 1994, National Fire Protection Association, Quincy, MA 02269. This reprinted material is not the complete and official position of the National Fire Protection Association on the referenced subject, which is represented only by the standard in its entirety.)*

Capacity of One Foot of Pipe (Based on actual internal pipe diameter)

Nominal Diameter	Gal Sch 40	Sch 10	Nominal Diameter	Gal Sch 40	Sch 10
¾ in.	0.028	—	3 in.	0.383	0.433
1 in.	0.045	0.049	3½ in.	0.513	0.576
1¼ in.	0.078	0.085	4 in.	0.660	0.740
1½ in.	0.106	0.115	5 in.	1.040	1.144
2 in.	0.174	0.190	6 in.	1.501	1.649[1]
2½ in.	0.248	0.283	8 in.	2.66[3]	2.776[2]

For SI Units: 1 in. = 25.4 mm; 1 ft = 0.3048 m; 1 gal = 3.785 L.
[1] 0.134 Wall Pipe.
[2] 0.188 Wall Pipe.
[3] Schedule 30.

Figure 1.18.2   Water capacity of 1 ft of pipe for various sizes.

Steel Pipe Dimensions

Nominal Pipe Size in.	Outside Diameter in.	Outside Diameter (mm)	Schedule 10[1] Inside Diameter in.	Schedule 10[1] Inside Diameter (mm)	Schedule 10[1] Wall Thickness in.	Schedule 10[1] Wall Thickness (mm)	Schedule 30 Inside Diameter in.	Schedule 30 Inside Diameter (mm)	Schedule 30 Wall Thickness in.	Schedule 30 Wall Thickness (mm)	Schedule 40 Inside Diameter in.	Schedule 40 Inside Diameter (mm)	Schedule 40 Wall Thickness in.	Schedule 40 Wall Thickness (mm)
1	1.315	(33.4)	1.097	(27.9)	0.109	(2.8)	—	—	—	—	1.049	(26.6)	0.133	(3.4)
1¼	1.660	(42.2)	1.442	(36.6)	0.109	(2.8)	—	—	—	—	1.380	(35.1)	0.140	(3.6)
1½	1.900	(48.3)	1.682	(42.7)	0.109	(2.8)	—	—	—	—	1.610	(40.9)	0.145	(3.7)
2	2.375	(60.3)	2.157	(54.8)	0.109	(2.8)	—	—	—	—	2.067	(52.5)	0.154	(3.9)
2½	2.875	(73.0)	2.635	(66.9)	0.120	(3.0)	—	—	—	—	2.469	(62.7)	0.203	(5.2)
3	3.500	(88.9)	3.260	(82.8)	0.120	(3.0)	—	—	—	—	3.068	(77.9)	0.216	(5.5)
3½	4.000	(101.6)	3.760	(95.5)	0.120	(3.0)	—	—	—	—	3.548	(90.1)	0.226	(5.7)
4	4.500	(114.3)	4.260	(108.2)	0.120	(3.0)	—	—	—	—	4.026	(102.3)	0.237	(6.0)
5	5.563	(141.3)	5.295	(134.5)	0.134	(3.4)	—	—	—	—	5.047	(128.2)	0.258	(6.6)
6	6.625	(168.3)	6.357	(161.5)	0.134[2]	(3.4)	—	—	—	—	6.065	(154.1)	0.280	(7.1)
8	8.625	(219.1)	8.249	(209.5)	0.188[2]	(4.8)	8.071	(205.0)	0.277	(7.0)	—	—	—	—
10	10.75	(273.1)	10.37	(263.4)	0.188[2]	(4.8)	10.14	(257.6)	0.307	(7.8)	—	—	—	—

NOTE 1: Schedule 10 defined to 5 in. (127 mm) nominal pipe size by ASTM A135.
NOTE 2: Wall thickness specified in 2-3.2.

Copper Tube Dimensions

Nominal Tube Size in.	Outside Diameter in.	Outside Diameter (mm)	Type K Inside Diameter in.	Type K Inside Diameter (mm)	Type K Wall Thickness in.	Type K Wall Thickness (mm)	Type L Inside Diameter in.	Type L Inside Diameter (mm)	Type L Wall Thickness in.	Type L Wall Thickness (mm)	Type M Inside Diameter in.	Type M Inside Diameter (mm)	Type M Wall Thickness in.	Type M Wall Thickness (mm)
¾	0.875	(22.2)	0.745	(18.9)	0.065	(1.7)	0.785	(19.9)	0.045	(1.1)	0.811	(20.6)	0.032	(0.8)
1	1.125	(28.6)	0.995	(25.3)	0.065	(1.7)	1.025	(26.0)	0.050	(1.3)	1.055	(26.8)	0.035	(0.9)
1¼	1.375	(34.9)	1.245	(31.6)	0.065	(1.7)	1.265	(32.1)	0.055	(1.4)	1.291	(32.8)	0.042	(1.1)
1½	1.625	(41.3)	1.481	(37.6)	0.072	(1.8)	1.505	(38.2)	0.060	(1.5)	1.527	(38.8)	0.049	(1.2)
2	2.125	(54.0)	1.959	(49.8)	0.083	(2.1)	1.985	(50.4)	0.070	(1.8)	2.009	(51.0)	0.058	(1.5)
2½	2.625	(66.7)	2.435	(61.8)	0.095	(2.4)	2.465	(62.6)	0.080	(2.0)	2.495	(63.4)	0.065	(1.7)
3	3.125	(79.4)	2.907	(73.8)	0.109	(2.8)	2.945	(74.8)	0.090	(2.3)	2.981	(75.7)	0.072	(1.8)
3½	3.625	(92.1)	3.385	(86.0)	0.120	(3.0)	3.425	(87.0)	0.100	(2.5)	3.459	(87.9)	0.083	(2.1)
4	4.125	(104.8)	3.857	(98.0)	0.134	(3.4)	3.905	(99.2)	0.110	(2.8)	3.935	(99.9)	0.095	(2.4)
5	5.125	(130.2)	4.805	(122.0)	0.160	(4.1)	4.875	(123.8)	0.125	(3.2)	4.907	(124.6)	0.109	(2.8)
6	6.125	(155.6)	5.741	(145.8)	0.192	(4.9)	5.845	(148.5)	0.140	(3.6)	5.881	(149.4)	0.122	(3.1)
8	8.125	(206.4)	7.583	(192.6)	0.271	(6.9)	7.725	(196.2)	0.200	(5.1)	7.785	(197.7)	0.170	(4.3)
10	10.13	(257.3)	9.449	(240.0)	0.338	(8.6)	9.625	(244.5)	0.250	(6.4)	9.701	(246.4)	0.212	(5.4)

Figure 1.18.0   Steel and copper pipe ID and OD. (*Reprinted with permission from NFPA 13,* Installation of Sprinkler Systems, *Copyright © 1996, National Fire Protection Association, Quincy, MA 02269. This reprinted material is not the complete and official position of the NFPA on the referenced subject, which is represented only by the standard in its entirety.*)

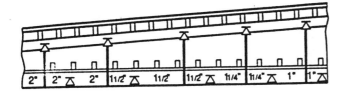

For SI Units: 1 in. = 25.4 mm.

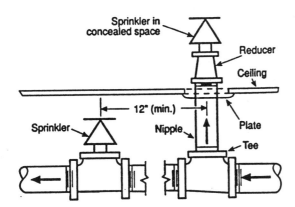

For SI Units: 1 in. = 25.4 mm.

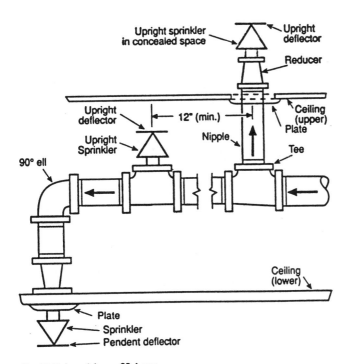

For SI Units: 1 in. = 25.4 mm.

Figure 1.19.0 Arrangement of branch lines above and below ceilings. (*Reprinted by permission of NFPA 13.*)

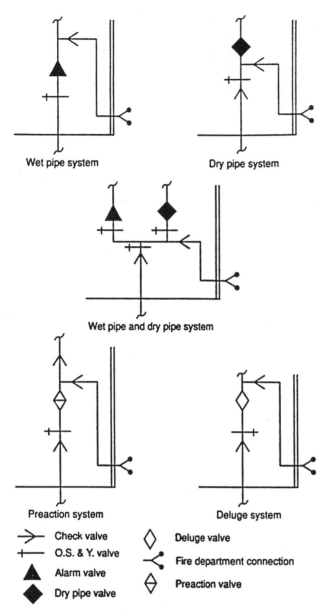

Figure 1.20.0   Acceptable valve placement in various types of installations. (*Reprinted with permission from NFPA 13*, Installation of Sprinkler Systems, *Copyright © 1996, National Fire Protection Association, Quincy, MA 02269. This reprinted material is not the complete and official position of the National Fire Protection Association on the referenced subject, which is represented only by the standard in its entirety.*)

1.21.0 Seismic Zones and Piping-Modification Requirements

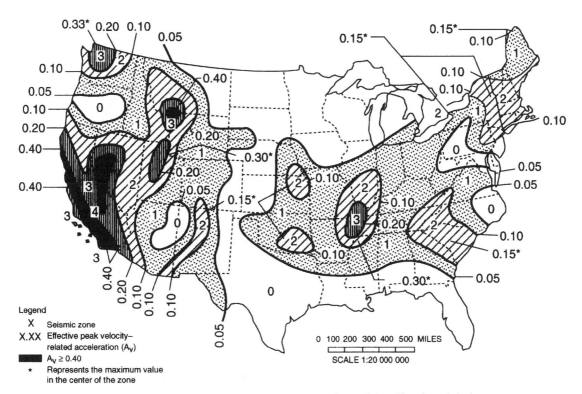

Legend

X   Seismic zone

X.XX   Effective peak velocity–
related acceleration (A_V)

█ $A_V \geq 0.40$

*   Represents the maximum value
in the center of the zone

Map of seismic zones and effective peak velocity-related acceleration (A_v) for contiguous 48 states. Linear interpolation between contours is acceptable.

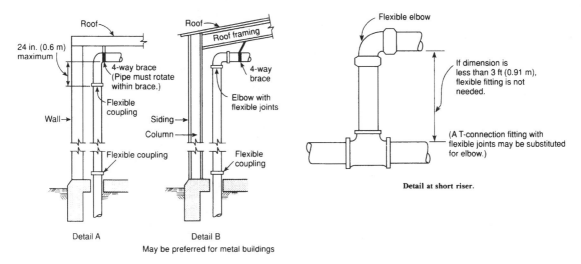

May be preferred for metal buildings

Note to Detail A: The four-way brace should be attached above the upper flexible coupling required for the riser and preferably to the roof structure if suitable. The brace should not be attached directly to a plywood or metal deck.

Riser details.

(*Reprinted with permission from NFPA 13,* Installation of Sprinkler Systems, *Copyright © 1994, National Fire Protection Association, Quincy, MA 02269. This reprinted material is not the complete and official position of the National Fire Protection Association on the referenced subject, which is represented only by the standard in its entirety.*)

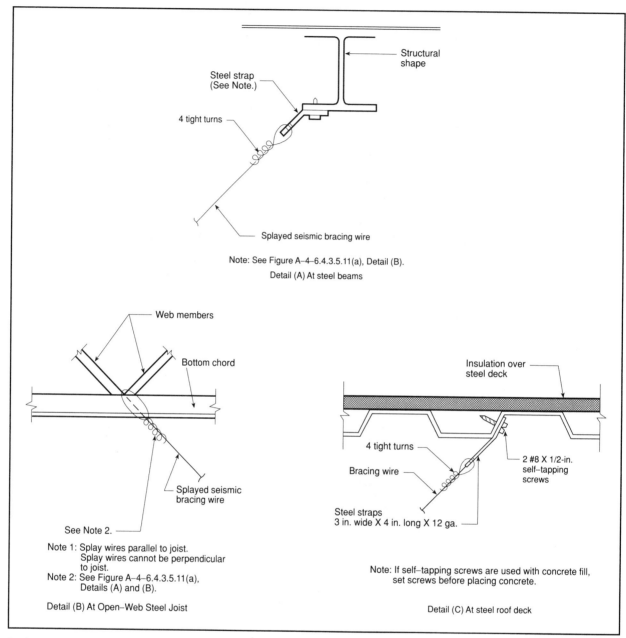

Note: See Figure A–4–6.4.3.5.11(a), Detail (B).

Detail (A) At steel beams

Note 1: Splay wires parallel to joist.
Splay wires cannot be perpendicular
to joist.
Note 2: See Figure A–4–6.4.3.5.11(a),
Details (A) and (B).

Detail (B) At Open–Web Steel Joist

Note: If self–tapping screws are used with concrete fill,
set screws before placing concrete.

Detail (C) At steel roof deck

For SI Units: 1 in. = 25.4 mm.

Figure 1.21.1   Bracing wire details for connections to steel framing. (*Reprinted with permission from NFPA 13*, Installation of Sprinkler Systems, *Copyright © 1996, National Fire Protection Association, Quincy, MA 02269. This reprinted material is not the complete and official position of the NFPA on the referenced subject, which is represented only by the standard in its entirety.*)

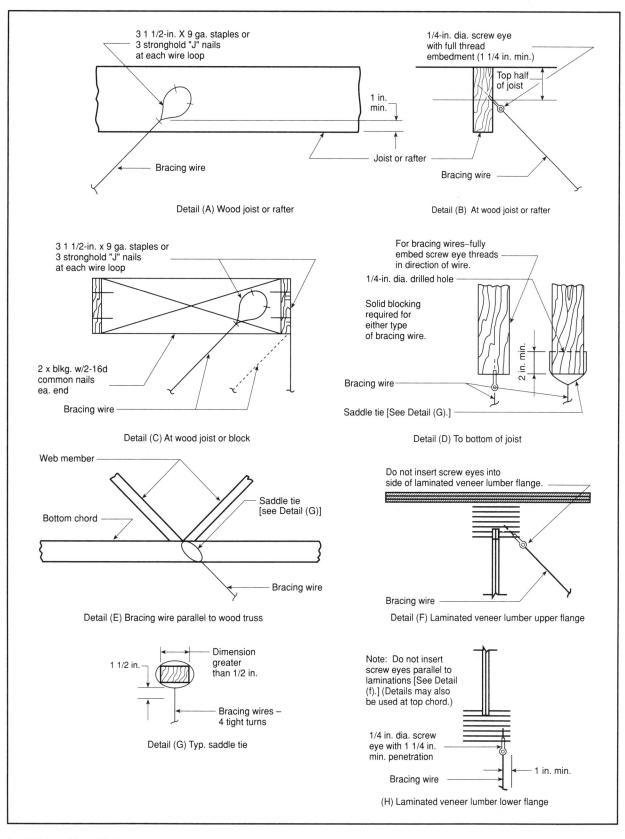

For SI Units: 1 in. = 25.4 mm.

Figure 1.21.2 Bracing wire details for wood framing. (*Reprinted with permission from NFPA 13,* Installation of Sprinkler Systems, *Copyright © 1996, National Fire Protection Association, Quincy, MA 02269. This reprinted material is not the complete and official position of the NFPA on the referenced subject, which is represented only by the standard in its entirety.*)

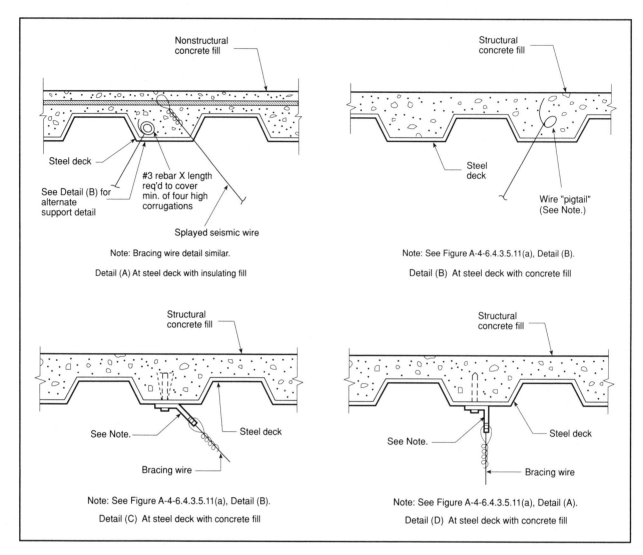

Note: Bracing wire detail similar.

Detail (A) At steel deck with insulating fill

Note: See Figure A-4-6.4.3.5.11(a), Detail (B).

Detail (B) At steel deck with concrete fill

Note: See Figure A-4-6.4.3.5.11(a), Detail (B).

Detail (C) At steel deck with concrete fill

Note: See Figure A-4-6.4.3.5.11(a), Detail (A).

Detail (D) At steel deck with concrete fill

For SI Units: 1 in. = 25.4 mm.
Note: If self-tapping screws are used with concrete fill, set screws before placing concrete.

1.21.3 Bracing wire details for concrete fill on metal deck. (*Reprinted with permission from NFPA 13,* Installation of Sprinkler Systems, *Copyright © 1996, National Fire Protection Association, Quincy, MA 02269. This reprinted material is not the complete and official position of the NFPA on the referenced subject, which is represented only by the standard in its entirety.*)

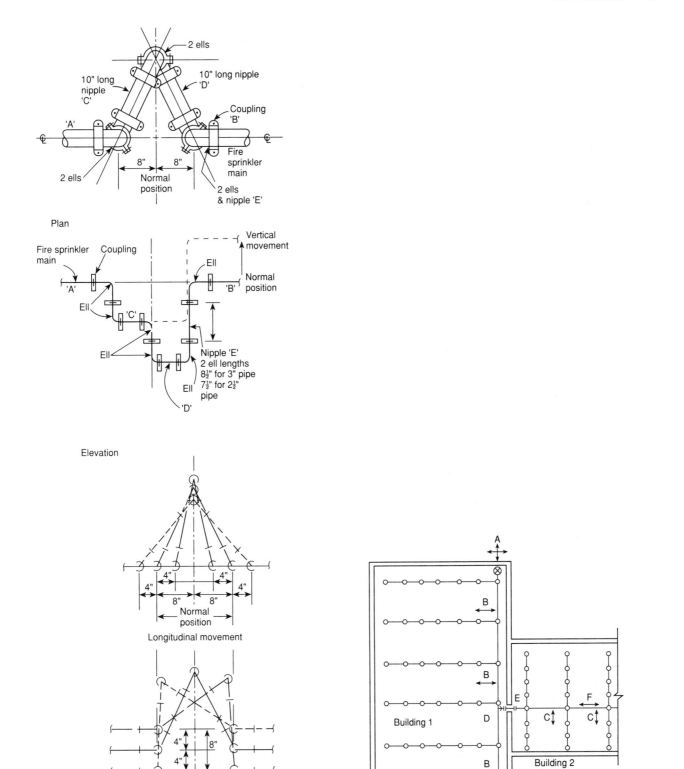

For SI Units: 1 in. = 25.4 mm; 1 ft = 0.305 m.

Figure 1.21.4 Seismic bracing details. (*Reprinted with permission from NFPA 13, Installation of Sprinkler Systems, Copyright © 1996, National Fire Protection Association, Quincy, MA 02269. This reprinted material is not the complete and official position of the NFPA on the referenced subject, which is represented only by the standard in its entirety.*)

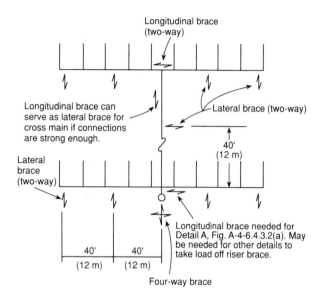

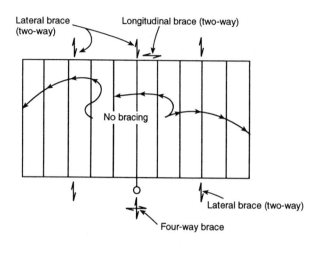

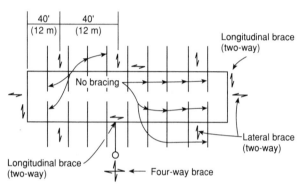

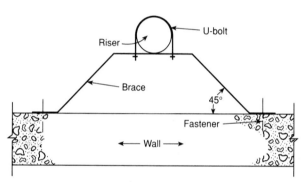

Figure 1.21.5   Sway bracing details. (*Reprinted with permission from NFPA 13*, Installation of Sprinkler Systems, *Copyright © 1996, National Fire Protection Association, Quincy, MA 02269. This reprinted material is not the complete and official position of the NFPA on the referenced subject, which is represented only by the standard in its entirety.*)

1.22.0   Unacceptable Pipe Weld Joints (Illustrated)

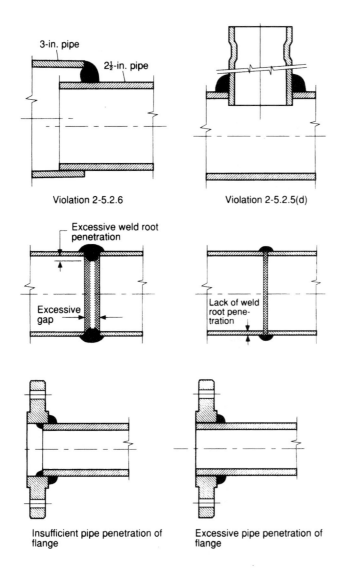

Unacceptable weld joints.

(Reprinted with permission from NFPA 13, Installation of Sprinkler Systems, Copyright © 1994, National Fire Protection Association, Quincy, MA 02269. This reprinted material is not the complete and official position of the National Fire Protection Association on the referenced subject, which is represented only by the standard in its entirety.)

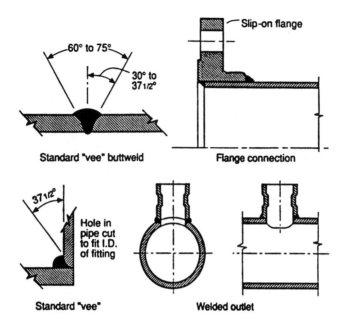

Figure 1.22.1 Acceptable weld joints. (*Reprinted with permission from NFPA 13, Installation of Sprinkler Systems, Copyright © 1996, National Fire Protection Association, Quincy, MA 02269. This reprinted material is not the complete and official position of the National Fire Protection Association on the referenced subject, which is represented only by the standard in its entirety.*)

1.23.0   Schematics of Fire Department Connections/Water Supply

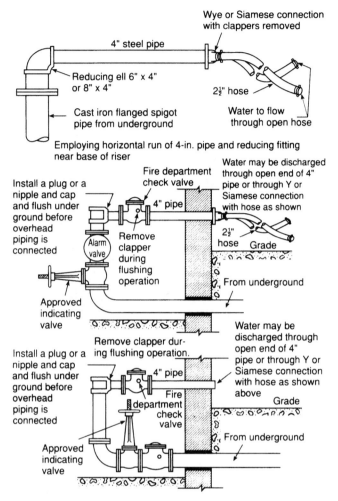

Methods of flushing water supply connections.

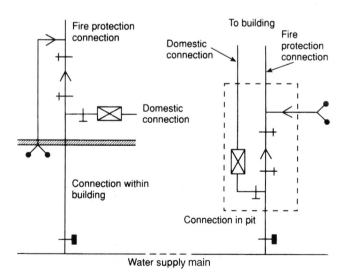

1.24.0 Schematics of Commercial Cooking Automatic Nozzle Installation

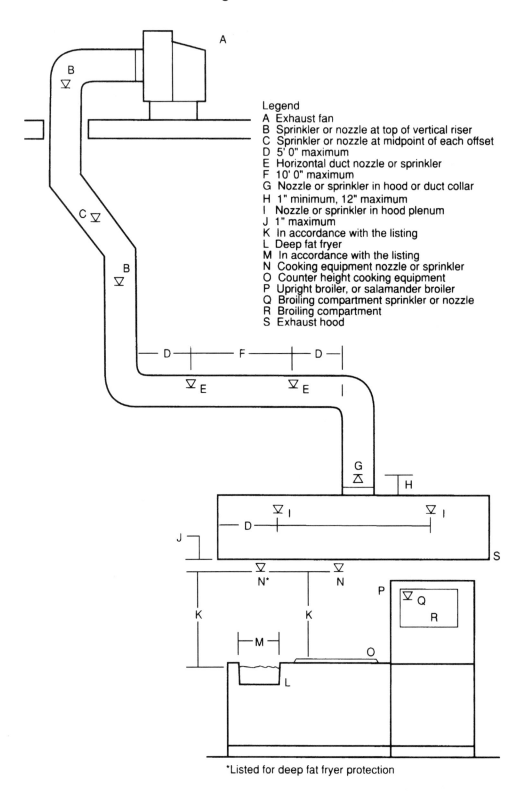

Legend
A Exhaust fan
B Sprinkler or nozzle at top of vertical riser
C Sprinkler or nozzle at midpoint of each offset
D 5' 0" maximum
E Horizontal duct nozzle or sprinkler
F 10' 0" maximum
G Nozzle or sprinkler in hood or duct collar
H 1" minimum, 12" maximum
I Nozzle or sprinkler in hood plenum
J 1" maximum
K In accordance with the listing
L Deep fat fryer
M In accordance with the listing
N Cooking equipment nozzle or sprinkler
O Counter height cooking equipment
P Upright broiler, or salamander broiler
Q Broiling compartment sprinkler or nozzle
R Broiling compartment
S Exhaust hood

*Listed for deep fat fryer protection

Typical installation showing automatic sprinklers or automatic nozzles being used for the protection of commercial cooking equipment and ventilation systems.

(Reprinted with permission from NFPA 13, Installation of Sprinkler Systems, Copyright © 1994, National Fire Protection Association, Quincy, MA 02269. This reprinted material is not the complete and official position of the NFPA on the referenced subject, which is represented only by the standard in its entirety.)

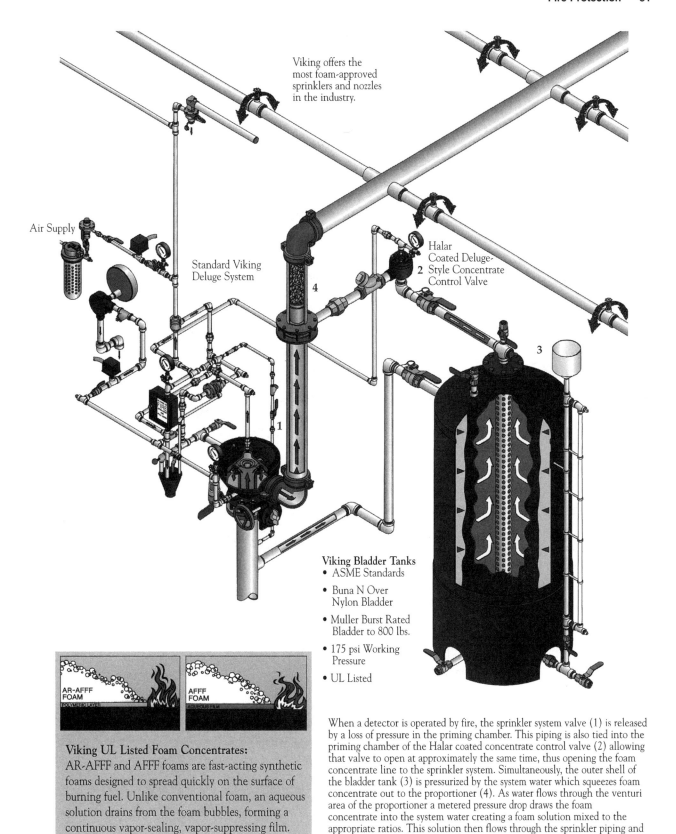

Viking offers the
most foam-approved
sprinklers and nozzles
in the industry.

Air Supply

Standard Viking
Deluge System

4

Halar
Coated Deluge-
Style Concentrate
Control Valve

2

3

1

Viking Bladder Tanks
• ASME Standards
• Buna N Over
  Nylon Bladder
• Muller Burst Rated
  Bladder to 800 lbs.
• 175 psi Working
  Pressure
• UL Listed

AR-AFFF
FOAM

AFFF
FOAM

Viking UL Listed Foam Concentrates:
AR-AFFF and AFFF foams are fast-acting synthetic
foams designed to spread quickly on the surface of
burning fuel. Unlike conventional foam, an aqueous
solution drains from the foam bubbles, forming a
continuous vapor-sealing, vapor-suppressing film.

When a detector is operated by fire, the sprinkler system valve (1) is released
by a loss of pressure in the priming chamber. This piping is also tied into the
priming chamber of the Halar coated concentrate control valve (2) allowing
that valve to open at approximately the same time, thus opening the foam
concentrate line to the sprinkler system. Simultaneously, the outer shell of
the bladder tank (3) is pressurized by the system water which squeezes foam
concentrate out to the proportioner (4). As water flows through the venturi
area of the proportioner a metered pressure drop draws the foam
concentrate into the system water creating a foam solution mixed to the
appropriate ratios. This solution then flows through the sprinkler piping and
out any open sprinklers or nozzles.

Figure 1.25.0 Foam system—components and sequence of operation. (*By permission: The Viking Corporation, Hastings, MI.*)

1.26.0   The Use of Plastic Piping in Fire Protection Installations

The National Fire Protection Association (NFPA) allows the use of certain types of plastic pipe in some sprinkler systems:

- Light hazard occupancies as defined by NFPA 13

- Residential occupancies up to four stories in height as defined by NFPA 13R

- One and two family dwellings as defined by NFPA 13D

- Air handling (plenum) spaces as defined by NFPA 90A

A comparison of properties of sprinkler materials—steel, copper, chlorinated polyvinyl chloride (CPVC), and polybutylene (PB)—was prepared by the National Institute of Standards and Testing in June 1994.

Summation of Variables Affecting Fire Sprinkler Pipe Selection

Property	Steel-Sch 40	Lightwall Steel	Copper Type M	CPVC	PB
Color	Black	Silver	Copper	Bright orange	Gray
Weight—1 in. size	2.5 (MP) 1427–1538° C, 2600–2800° F	1.8 (MP) 1427–1538° C, 2600–2800° F	0.7 (MP) 1082° C, 1980° F	0.4 (HDT) 103° C, 217° F	0.2 (MP) 122–126° C, 252–259° F
Damage susceptibility	Low	Low	Low	High with UV exposure and impact	High with UV exposure and impact
Corrosion susceptibility	High	High	Moderate	Low	Low
Occupancy classification NFPA	Not limited	Not limited	Not limited	NFPA13R, 13D, Light hazard NFPA90A	NFPA13R, 13D, Light hazard NFPA90A
Maximum ambient temperature	Not limited	Not limited	Not limited	66° C, 150°F	49°C, 120°F
Flexibility/hanger spacing—1 in.	Not flexible 3.7	Not flexible 3.7	Slightly flexible—2.4	Flexible 1.8	Very flexible 1.0
Expansion concerns	Negligible	Negligible	Negligible	Yes/offsets—changes, loops	Yes/snaking pipe
Fitting type	Threaded, grooved, flanged	Threaded, grooved flanged	Soldering, brazing grooved	Primer solvent cement	Heat fusion
Compatible antifreeze	Not limited	Not limited	Not limited	Glycerine	Ethylene, Propylene glycol, methanol, glycerine

Figure 1.26.1 Comparison of steel, copper, CPVC, and PB sprinkler piping systems.

Joining BlazeMaster® Pipe and Fittings With Red One Step Solvent Cement

Cutting. BlazeMaster pipe can be easily cut with a ratchet cutter, wheel-type plastic tubing cutter, a power saw or a fine toothed saw. To ensure the pipe is cut square, a miter box must be used when using a saw.

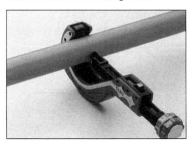

Cutting the pipe as squarely as possible provides the surface of the pipe with maximum bonding area. If any indication of damage or cracking is evident at the pipe end, cut off at least two (2) inches beyond any visible crack. Do not use ratchet cutters below 50°F.

Deburring. Burrs and filings can prevent proper contact between pipe and fitting during assembly, and must be removed from the outside and the inside of the pipe. A chamfering tool or file are suitable for this purpose. A slight bevel shall be placed at the end of the pipe to ease entry of the pipe into the socket and minimize the chances of wiping solvent cement from the fitting.

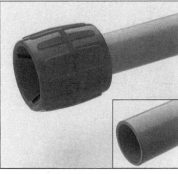

Fitting Preparation. Using a clean, dry rag, wipe loose dirt and moisture from the fitting socket and pipe end. Moisture can slow the cure time and at this stage of assembly, excessive water can reduce joint strength. Check the dry fit of the pipe and fitting. The pipe should enter the fitting socket easily 1/4 to 3/4 of the way. At this stage, the pipe should not bottom out in the socket.

Solvent Cement Application. Joining surfaces shall be penetrated and softened. Cement shall be applied (worked into pipe) with an applicator

1/2 the size of the pipe diameter. Apply a heavy, even coat of cement to the outside pipe end. Apply a medium coat to the fitting socket. Pipe sizes 1-1/4 inches and above shall always receive a second cement application on the pipe end. (Apply cement on the pipe end, in the fitting socket, and on the pipe again.) Only use solvent cements that have been specifically investigated and tested for use with BlazeMaster CPVC and approved by the pipe and fitting manufacturers.

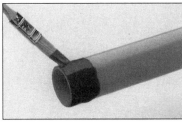

Special care shall be exercised when assembling BlazeMaster systems in extremely low temperatures (below 40°F) or extremely high temperatures (above 100°F). Extra set time shall be allowed in colder temperatures. When cementing pipe and fittings in extremely cold temperatures, make certain that the cement has not "gelled". Gelled cement must be discarded. In extremely hot temperatures, make sure both surfaces to be joined are still wet with cement when putting them together.

Assembly. After applying cement, immediately insert the pipe into the fitting socket, while rotating the pipe 1/4 turn. Properly align the fitting for the installation at this time. Pipe must bottom to the stop. Holding the assembly for 10 to 15 seconds to

ensure initial bonding. *A bead of cement should be evident around the pipe and fitting juncture. If this bead is not continuous around the socket shoulder, it may indicate that insufficient cement was applied.* If insufficient cement is applied, the fitting must be cut out and discarded.

Cement in excess of the bead can be wiped off with a rag. Care shall be exercised when installing sprinkler

heads. Sprinkler head fittings shall be allowed to cure for a minimum of 30 minutes prior to installing the sprinkler head. When installing sprinkler heads, be sure to anchor or hold the pipe drop securely to avoid rotating the pipe in previously cemented connections. Previously cemented fittings shall also be permitted to cure for a minimum of 30 minutes.

Warning: Sprinkler heads shall be installed only after all the CPVC pipe and fittings, including the sprinkler head adapters, are solvent welded to the piping and allowed to cure for a minimum of 30 minutes. Sprinkler head fittings should be visually inspected and probed with a wooden dowel in insure that the water way and threads are clear of any excess cement. Once the installation is complete and cured per Table I, II or III, (See Page 5) the system shall be hydrostatically tested. *It is an unaccepted practice to install sprinklers on the head adapter fittings and then solvent cement the assembly to the drop.* Note: Safety and Health Precautions. Prior to using CPVC solvent cements, review and follow all precautions found on the container labels, material safety data sheet, and Standard Practice for Safe Handling ASTM F402-80.

Set and Cure Times. Solvent cement set and cure times are a function of pipe size, temperature, relative humidity, and tightness of fit. Drying time is faster for drier environments, smaller pipe sizes, high temperatures and tighter fits. The assembly must be allowed to set, without any stress on the joint, for 1 to 5 minutes, depending on pipe size and temperature. Following initial set period, the assembly can be handled carefully, *avoiding significant stresses to the joint.* Refer to the following tables for minimum cure times prior to pressure testing.

Figure 1.26.2 Cutting and joining of CPVC pipe and fittings. (*By permission: BFGoodrich, Cleveland, OH.*)

Do's

- Read the manufacturer's installation instructions.
- Follow recommended safe work practices.
- Make certain that thread sealants, gasket lubricants, or fire stop materials are compatible with CPVC.
- Keep pipe and fittings in original packaging until needed.
- Cover pipe and fittings with an opaque tarp if stored outdoors.
- Follow proper handling procedures.
- Use tools specifically designed for use with plastic pipe and fittings.
- Use the proper solvent cement and follow application instructions.
- Use a drop cloth to protect interior finishes.
- Cut the pipe ends square.
- Deburr and bevel the pipe end with a chamfering tool.
- Rotate the pipe 1/4 turn when bottoming pipe in fitting socket.
- Avoid puddling of cement in fittings and pipe.
- Make certain no solvent cement is on sprinkler head and adapter threads.
- Make certain that solvent cement does not run and plug the sprinkler head orifice.
- Follow the manufacturer's recommended cure times prior to pressure testing.
- Fill lines slowly and bleed the air from the system prior to pressure testing.
- Use Teflon tape on sprinkler head threads and all other threaded connections.
- Support sprinkler head properly to prevent lift up of the head through the ceiling when activated.
- Keep threaded rod within 1/16" of the pipe or use a surge arrestor.
- Do install BlazeMaster® CPVC in wet systems only.
- Use only glycerin and water solutions for freeze protection.
- Allow for movement due to expansion and contraction.
- Renew your BlazeMaster® installation training every two years.

Don'ts

- Do not use edible oils such as Crisco as a gasket lubricant.
- Do not use petroleum or solvent based sealants, lubricants, or fire stop materials.
- Do not use any glycol based solutions as an anti-freeze.
- Do not mix glycerine and water solutions in contaminated containers.
- Do not use solvent cement that exceed it s shelf life or has become discolored or jellied.
- Do not allow solvent cement to plug the sprinkler head orifice.
- Do not connect rigid metal couplers to CPVC grooved adapters.
- Do not thread, groove, or drill CPVC pipe.
- Do not use solvent cement near sources of heat, open flame, or when smoking.
- Do not pressure test with air.
- Do not pressure test until recommended cure times are met.
- Do not use ratchet cutters below 50° F.
- Do not use CPVC pipe that has been stored outdoors, unprotected and is faded in color.
- Do not allow threaded rod to come in contact with the pipe.
- Do not install BlazeMaster® in cold weather without allowing for expansion.
- Do not install BlazeMaster® CPVC in dry systems.

Figure 1.26.3 Do's and don'ts of CPVC installations. Although this list specifically references Blazemaster CPVC piping (as do the following installation instructions), generic piping instructions may be similar; however, the specific manufacturer's installation procedures will prevail. (*Reprinted with permission from BFGoodrich, Cleveland, OH.*)

1.26.3   Installation Instructions for Aboveground CPVC Sprinkler Piping

INSTALLATION INSTRUCTIONS
FOR
BLAZEMASTER® CPVC FIRE SPRINKLER PIPE AND FITTINGS
FOR
SYSTEM RISERS

1.   Product Name:
BlazeMaster® CPVC

2.   Manufacturers:

Central Sprinkler Co.	Spears Manufacturing	IPEX	Harvel Plastics, Inc.
451 N. Cannon Ave.	15853 Olden St.	6810 Invader Crescent	Kuebler Road
Lansdale, PA 19446	Sylmar, CA 91342	Mississauga, ON L5T 2B6	P.O. Box 757
Phone: 215/362-0700	Phone: 818/364-1611	Phone: 905/670-7676	Easton, PA 18044
Fax: 215/362-5385	Fax: 818/362/1596	Fax: 905/670-5295	Phone: 610/252-7355
BlazeMaster® CPVC	BlazeMaster® CPVC	BlazeMaster® CPVC	Fax: 610/253-4436
			BlazeMaster® CPVC

3.   Product Description:

BlazeMaster® fire sprinkler pipe and fittings are produced from specialty plastic compounds known as post-chlorinated polyvinyl chloride (CPVC) manufactured by the BFGoodrich Company. The color is orange. The compound shall meet cell class 23447-B as defined by ASTM D1784, and shall be listed by the National Sanitation Foundation for use with potable water. Both base resin and compound shall be made in the U.S.A.

Pipe: Pipe shall meet or exceed the requirements of ASTM F442 in standard dimension ratio (SDR) 13.5. **Pipe for system riser service is available in size range 3/4"-3" at a maximum working pressure of 175 psi for all sizes.**

Fittings: Fittings shall meet or exceed the requirements of ASTM F437 (Sch. 80 Threaded) ASTM F438 (Sch. 50 Socket) and ASTM F459 (Sch. 80 Socket).

3.1   Solvent Cement:

All socket type joints shall be made up employing solvent cements that meet or exceed the requirements of ASTM F493. The standard practice for safe handling of solvent cements shall be in accordance with ASTM F402. Solvent cement shall be listed by the National Sanitation Foundation for use with potable water, and approved by the BlazeMaster® CPVC manufacturers.

3.2   Basic Use:

Both pipe and fittings may be used as system risers in NFPA 13R and 13D installations. Protection shall be provided for CPVC pipe and fittings. The minimum protection shall consist of either:

one layer of 3/8" gypsum wallboard
1/2" plywood soffits

NOTE: System Risers employing Listed CPVC pipe and fittings shall not be installed "exposed".

3.3   Installation Requirements:

1) NFPA 13, Sections 4-5.2.5, Support of Risers
2) The Manufacturer's installation instructions.

4. System Design:

A BlazeMaster® fire sprinkler system shall be hydraulically calculated in accordance with NFPA 13 Section 6-4, Hydraulic Calculation Procedures, using a Hazen-Williams C-Factor of 150.

5. Installation Procedures:

The installation procedures detailed here apply to Listed BlazeMaster® fire sprinkler pipe and fittings that have solvent cement joints or other approved joints in size range 3/4"-3".

5.1   Inspection

Before installation, BlazeMaster® pipe and fittings should be thoroughly inspected for cuts, scratches, gouged or split ends which me have occurred during shipment and handling.

5.2   Risers shall be supported by pipe clamps or by hangers located on the

horizontal connection close to the riser. Only Listed hangers and clamps shall be used.

5.3   Vertical lines must be supported at intervals, described in Para. 5.6 and 5.7, to avoid placing excessive load on a fitting at the lower end. This can be done by using riser clamps or double bolt pipe clamps Listed for this service. The clamps must not exert compressive stresses on the pipe. **If possible, the clamps should be located just below a coupling so that the shoulder of the coupling rests against the clamp. If necessary, a coupling can be modified and adhered to the pipe as a bearing support such that the shoulder of the fitting rests on the clamp. See Figure I. Follow the manufacturer's recommended cure time.**

5.4   Do not use riser clamps which squeeze the pipe and depend on compression of the pipe to support the weight. See recommended practice on Figure II.

5.5   Hangers and straps shall not compress, distort, cut or abrade the piping and shall allow for free movement of the pipe to allow for thermal expansion and contraction.

5.6   Maintain vertical piping in straight alignment with supports at each floor level or at 10 foot intervals whichever is less.

5.7   CPVC risers in vertical shafts or in buildings with ceilings over 25 feet, shall be aligned straightly and supported at each floor level or at 10 foot intervals whichever is less.

6. Maintenance:

Shall be in accordance with the Standard for Inspection, Testing and maintenance of Water Based Extinguishing Systems as defined by NFPA 25.

7. Warranty:

Consult the manufacturers for specific warranty information.

(By permission: BFGoodrich, Cleveland, Ohio)

TABLE I: 225 psi (maximum) Test Pressure
Ambient Temperature During Cure Period

Pipe Size	60°F to 120°F	40°F to 59°F	0°F to 39°F
³/₄"	1 hr.	4 hr.	48 hr.
1"	1¹/₂ hr.	4 hr.	48 hr.
1¹/₄" & 1¹/₂"	3 hr.	32 hr.	10 days
2"	8 hr.	48 hr.	Note 1
2¹/₂" & 3"	24 hr.	96 hr.	Note 1

TABLE II: 200 psi (maximum) Test Pressure
Ambient Temperature During Cure Period

Pipe Size	60°F to 120°F	40°F to 59°F	0°F to 39°F
³/₄"	45 min.	1¹/₂ hr.	24 hr.
1"	45 min.	1¹/₂ hr.	24 hr.
1¹/₄" & 1¹/₂"	1¹/₂ hr.	16 hr.	120 hr.
2"	6 hr.	36 hr.	Note 1
2¹/₂" & 3"	8 hr.	72 hr.	Note 1

TABLE III: 100 psi (maximum) Test Pressure
Ambient Temperature During Cure Period

Pipe Size	60°F to 120°F	40°F to 59°F	0°F to 39°F
³/₄" CTS	15 min.	15 min.	30 min.
³/₄"	15 min.	15 min.	30 min.
1"	15 min.	30 min.	30 min.
1¹/₄"	15 min.	30 min.	2 hr.

Note 1 For these sizes, the solvent cement can be applied at temperatures below 40°F, however, the sprinkler system temperature must be raised to a temperature of 40°F or above and allowed to cure per the above recommendations prior to pressure testing.

Once an installation is completed and cured, the system (except for one and two family dwellings and manufactured homes), should be pressure tested at 200 psi (Table II) for 2 hours (or at 50 psi in excess of the maximum pressure (Table I) when the maximum pressure to be maintained in the system is in excess of 150 psi) in accordance with the requirements established by NFPA Standard 13, Section 8-2.2.1. Sprinkler systems in one- and two-family dwellings and mobile homes may be tested at line pressure (Table III) in accordance with the requirements established by NFPA 13D, Section 1-5.4. When pressure testing, the sprinkler system shall be slowly filled with water and the air bled from the highest and farthest sprinkler heads before pressure testing is applied. Air must be removed from piping systems (plastic or metal) to prevent it from being locked in the system when pressure is applied. Entrapped air can generate excessive surge pressures that are potentially damaging, regardless of the piping materials used. *Air or compressed gas should never be used for pressure testing.* If a leak is found, the fitting must be cut out and discarded. A new section can be installed using couplings or a union. Unions should be used in accessible areas only.

Hangers and Supports

Because BlazeMaster pipe is rigid, it requires fewer supports than flexible plastic systems. The support spacing is shown in the following table.

Nominal Pipe Size		Maximum Support Spacing	
Inches	(millimeters)	Feet	(meters)
³/₄"	(20.0)	5¹/₂'	(1.7)
1"	(25.0)	6'	(1.8)
1¹/₄"	(32.0)	6¹/₂'	(2.0)
1¹/₂"	(40.0)	7'	(2.1)
2"	(50.0)	8'	(2.4)
2¹/₂"	(65.0)	9'	(2.7)
3"	(80.0)	10'	(3.0)

Most hangers designed for metal pipe are suitable for BlazeMaster pipe. Do not use undersized hangers. Hangers with sufficient load bearing surface shall be selected based on pipe size, i.e., 1¹/₂" hangers for 1¹/₂" pipe. The hanger shall not have rough or sharp edges which come in contact with the pipe. The pipe hangers must comply with the requirements in NFPA 13, 13R, and 13D. For Quick Response upright sprinkler heads, rigid hangers secured to the ceiling shall be used.

For installation of exposed BlazeMaster CPVC Fire Sprinkler Pipe listed support devices for thermoplastic sprinkler piping or other listed support devices shall be used which mount the piping directly to the ceiling or sidewall.

When a sprinkler head activates, a significant reactive force can be exerted on the pipe. With a pendant head, this reactive force can cause the pipe to lift vertically if it is not properly secured, especially if the sprinkler drop is from small diameter pipe. The closest hanger shall brace the pipe against vertical lift-up. See Tables A & B.

Figure 1.26.4 Pressure testing, hanger, and supports for CPVC piping. (*By permission: BFGoodrich, Cleveland, OH.*)

TABLE A: Maximum Support Spacing Distance End Line Sprinkler Head Drop Elbow

Nominal Pipe Size (IN)	Less than 100 psi	More than 100 psi
3/4"	9"	6"
1"	12"	9"
1 1/4"	16"	12"
1 1/2" - 3"	24"	12"

TABLE B: Maximum Support Spacing Distance Inline Sprinkler Head Drop Tee

Nominal Pipe Size (IN)	Less than 100 psi	More than 100 psi
3/4"	4'	3'
1"	5'	4'
1 1/4"	6'	5'
1 1/2" - 3"	7'	7'

The closest hanger shall brace the pipe against vertical lift-up. Any of a number of techniques can be used to brace the pipe. Four acceptable approaches would be to use a standard band hanger positioning the threaded support rod to 1/16 inch above the pipe, a split-ring hanger, a wrap-around U hanger, or a special escutcheon which prevents upward movement of the sprinkler through the ceiling.

Thermal Expansion. BlazeMaster CPVC, like all piping materials, expands and contracts with changes in temperature. The coefficient of linear expansion is:
0.0000340 inch/inch -°F.
A 25°F change in temperature will cause an expansion of 1/2 inch for a 50 foot straight length. For most operating and installation conditions, expansion and contraction can be accommodated at changes in direction. However, in certain instances, expansion loops or offsets may be required when installing long, straight runs of pipe.

Estimating One-Step Solvent Cement Requirements. The following guidelines are provided to allow estimation of one-step solvent cement quantities needed.

Fitting Size Inches	Solvent Cement Number of Joints Per Quart
3/4"	270
1"	180
1 1/4"	130
1 1/2"	100
2"	70
2 1/2"	50
3"	40

Antifreeze Solutions. Glycerine antifreeze solutions are acceptable for use with BlazeMaster plastic piping. Propylene glycol solutions shall not be used as it will cause damage to the BlazeMaster pipe and

Figure 1.26.4 *(Continued)*

fittings. Consult the local authority having jurisdiction before using any antifreeze solutions in fire sprinkler applications. Refer to Section A-3-5.2 of NFPA 13.

General Design Criteria. Before penetrating fire rated walls and partitions, consult building codes and authorities having jurisdiction in your area. Several UL Classified through-penetration firestop systems are available for use with CPVC pipe. Consult the pipe manufacturers or BFGoodrich for further information. *Warning: Some firestop sealants or wrap strips contain solvents or plasticizers that may be damaging to CPVC. Always consult the manufacturer of the firestop material for compatibility with BlazeMaster CPVC pipe and fittings.* BlazeMaster piping systems must be laid out so that the piping is not closely exposed to heat producing sources, such as light fixtures, ballasts and steam lines. Pipe must not be positioned directly over open ventilation grills. Finally, during periods of remodeling and renovation, appropriate steps must be taken to protect the piping from fire exposure if the ceiling is temporarily removed.

Grooved Coupling Adapters. CPVC Grooved Coupling Adapters were designed using the Victaulic style 75 flexible coupling. Other UL Listed couplings of similar design may be used. *Caution: Use of rigid style couplings may damage the grooved coupling adapter.* Lubricate the gasket with a vegetable soap base gasket lubricant — IPS Weld On Gasket/Joint Lubricant #787 or Seacord Corp. Ease-On Pipe Joint Lubricant. *Caution: Certain lubricants may contain a petroleum base or other chemicals which will cause damage to the gasket and adapter. Always consult with the manufacturer of the lubricant for compatibility with BlazeMaster CPVC pipe and fittings. DO NOT USE vegetable oils, such as Crisco, as a lubricant.*

Since BlazeMaster CPVC is more ductile than metallic sprinkler pipe, it has a greater capacity to withstand earthquake damage. In areas subject to earthquakes, BlazeMaster fire sprinkler systems shall be designed and braced in accordance with local codes or NFPA 13, Section 4-6.4.3.

1.26.5 Underground Installation of CPVC Piping

INSTALLATION INSTRUCTIONS
FOR
BLAZEMASTER® CPVC FIRE SPRINKLER PIPE AND FITTINGS
FOR
UNDERGROUND WATER PRESSURE SERVICE

1.   **Product Name:**
      BlazeMaster® CPVC

2.   **Manufacturers:**

Central Sprinkler Co.	Spears Manufacturing	IPEX	Harvel Plastics, Inc.
451 N. Cannon Ave.	15853 Olden St.	6810 Invader Crescent	Kuebler Rd., PO Box 757
Lansdale, PA 19446	Sylmar, CA 91342	Mississauga, ON L5T 2B6	Easton, PA 18044
Phone: 215/362-0700	Phone: 818/364-1611	Phone: 905/670-7676	Phone: 610/252-7355
Fax: 215/362-5385	Fax: 818/362/1596	Fax: 905/670-5295	Fax: 610/253-4436
BlazeMaster® CPVC	BlazeMaster® CPVC	BlazeMaster® CPVC	BlazeMaster® CPVC

3.   **Product Description:**

BlazeMaster® fire sprinkler pipe and fittings are produced from specialty plastic compounds known as post-chlorinated polyvinyl chloride manufactured by the BFGoodrich Company. The color is orange. The compound shall meet Cell Class 23447-B as defined by ASTM D/1784, and shall be listed by the National Sanitation Foundation (NSF) for use with potable water. Both base resin and compound shall be made in the U.S.A.

Pipe: Pipe shall meet or exceed the requirements of ASTM F442 in standard dimension ratio (SDR) 13.5. **Pipe for underground service is available in size range 3/4"-3" at a maximum working pressure of 175 psi for all sizes.**

Fittings: Fittings shall meet or exceed the requirements of ASTM F437 (Sch. 80 Threaded) ASTM F438 (Sch. 40 Socket) and ASTM F439 (Sch. 80 Socket).

Both pipe and fittings shall be tested in accordance with UL 1713, Pressure Pipe and Couplings for Underground Fire Service, and Listed by Underwriters Laboratories to supply wet automatic fire sprinkler systems and bear the logo of the Listing Agency. See UL Fire Protection Equipment Directory, category VIWT.

3.1   Solvent Cement:

All socket type joints shall be made up in accordance with the manufacturer's installation instructions using solvent cements that meet or exceed the requirement of ASTM F493. The standard practice for safe handling of solvent cements shall be in accordance with ASTM F402. Solvent cement shall be listed by the National Sanitation Foundation for use with potable water, and approved by the BlazeMaster® pipe and fittings manufacturer.

3.2   Basic Use:

BlazeMaster® fire sprinkler pipe and fittings are UL Listed for use in underground water service when installation in accordance with:

- ASTM D2774, Standard Recommended Practice for Underground Installation of Thermoplastic Pressure Piping,

- ASTM F645, Standard Guide for Selection, Design and Installation of Thermoplastic Water Pressure Piping Systems, and

- the manufacturer's instructions.

4.   System Design:

A BlazeMaster® fire sprinkler system shall be hydraulically calculated using a Hazen-Williams C-Factor of 150, and designed and installed in accordance with the Standard for Installation of Sprinkler Systems, NFPA 13.

5.   Installation Procedures:

The general installation procedures detailed here apply to BlazeMaster® CPVC pressure pipe that has solvent cement joints in size range 3/4"-3".

5.1   Inspection: Before installation, BlazeMaster® CPVC pipe and fittings should be thoroughly inspected for cuts, scratches, gouges or split ends which may have occurred to the products during shipping and handling.

5.2 **Trenching:** The trench should be of adequate width to allow convenient installation, while at the same time being as narrow as possible. Minimum trench widths may be utilized be joining pipe outside of the trench and lowering it into the trench after adequate joint strength has been achieved. (NOTE: Refer to the manufacturer's instructions for recommended set and cure time for solvent cement joints). Trench widths will have to be wider where pipe is joined in the trench or where thermal expansion and contraction is a factor. See section 5.3 titled "Snaking of Pipe".

Pipe Size	Trench Width	Light Traffic Ground Cover Minimum	Heavy Traffic Ground Cover Minimum
3" and Under	8"	12"-18"	30"-36"

- Water filled pipe should be buried at least 12" below the maximum expected frost line

- It is recommended that the thermoplastic piping be run within a metal or concrete casing when it is installed beneath surfaces that are subject to heavy-weight or constant traffic such as roadways and railroad tracks.

The trench bottom should be continuous, relatively smooth and free of rocks. Where ledge rock, hardpan or boulders are encountered, it is necessary to pad the trench bottom using a minimum of four (4) inches of tamped earth or sand beneath the pipe as a cushion and for protection of the pipe from damage.

Sufficient cover must be maintained to keep external stress levels below acceptable design stress. Reliability and safety on service is of major importance in determining minimum cover. Local, state and national codes may also govern.

5.3 **Snaking of Pipe:**

After BlazeMaster® CPVC pipe has been solvent welded, it is advisable to snake the pipe according to the below recommendations beside the trench during its required drying time. **BE ESPECIALLY CAREFUL NOT TO APPLY ANY STRESS THAT WILL DISTURB THE UNDRIED JOINT.** This snaking is necessary in order to allow for any anticipated thermal contraction that will take place in the newly joined pipeline.

Snaking is particularly necessary on the lengths that have been solvent welded during the late afternoon of a hot summer's day, because their drying time will extend through the cool of the night when thermal contraction of the pipe could stress the joints to the point of pull out. This snaking is also especially necessary with pipe that is laid in its trench (necessitating wider trenches than recommended) and is back-filled with cool earth before the joints are thoroughly dry.

Loop Offset in Inches for Contraction:

	Maximum Temperature Variation, °F, Between Time of Solvent Welding and Final Use									
Loop Length	10°	20°	30°	40°	50°	60°	70°	80°	90°	100°
20 Feet	3"	4"	5"	5"	6"	6"	7"	7"	8"	8"
50 Feet	7"	9"	11"	13"	14"	16"	17"	18"	19"	20"
100 Feet	13"	18"	22"	26"	29"	32"	35"	37"	40"	42"

5.4 Backfilling:

Note: If possible, underground pipe should be thoroughly inspected and tested for leaks prior to backfilling.

Ideally, backfilling should only be done early in the morning during hot weather when the line is fully contracted and there is no chance of insufficient dried joints being subject to contraction stresses.

The pipe should be uniformly and continuously supported over its entire length on firm, stable material. Blocking should not be used to change pipe grade or to intermittently support pipe across excavated sections.

Pipe is installed in a wide range of sub-soils. These soils should not only be stable but applied in such a manner so as to physically shield the pipe from damage. Attention should be given to local pipe laying experience which may indicate particular pipe bedding problems.

Backfill materials free of rocks with a particle size of 1/2" or less should be used to surround the pipe with 6" to 8" of cover. It should be placed in layers. Each soil layer should be sufficiently compacted to uniformly develop lateral passive soil forces during the backfill operation. It may be advisable to have the pipe under pressure, 15 to 25 psi during the backfilling.

Vibratory methods are preferred when compacting sand or gravels. Best results are obtained when the soils are in a nearly saturated condition. Where water flooding is used, the initial backfill should be sufficient to insure complete coverage of the pipe. Additional material should not be added until the water flooded backfill is firm enough to walk on. Care should be taken to avoid floating the pipe.

Sand and gravel containing a significant proportion on fine-grained material, such as silt and clay, should be compacted by hand or, <u>preferably</u> by mechanical tamper.

The remainder of the backfill should be placed and spread in approximately uniform layers in such a manner to fill the trench completely so that there will be no unfilled spaces under or about the rocks or lumps of earth in the backfill. Large or sharp rocks, frozen clods and other debris grater than 3" in diameter should be removed. Rolling equipment or heavy tampers should only be used to consolidate the final backfill.

6.   Maintenance:

Shall be in accordance with the Standard for Inspection, Testing and maintenance of Water Based Extinguishing Systems as defined by NFPA 25.

7.   Warranty:

Consult the manufacturers for specific warranty information.

(By permission: BFGoodrich, Cleveland, Ohio)

Floor/Ceiling Concrete Slabs

**407 E814 - 2hr. (F) - 2hr. (T) Closed
2hr. (F) - 2hr. (T) Vented**

3" PVC (D2665 or D1785) or 3" CPVC (F441 or F442) SDR 17 or heavier pipe penetrating a 4-1/2" concrete floor without a steel sleeve. Annular space between the pipe and through opening to be not more than 1/2". Put FS-195 wrap strip around pipe below slab: 2 layers on 3" pipe and 1 layer on 2" pipe. Caulk applied between floor and wrap strip and on exposed edges of the wrap strip. One layer of 4 mil foil tape on pipe to prevent caulk contact. Install RC-1 collar to retain wrap strip.

3M* UL XHEZ System 64-A

432 E814 - 1hr. (F) - 3/4hr. (T) Vented

3" CPVC (F442) SDR 13.5 pipe penetrating a 4-1/2" concrete floor. Annular space to be not less than 3/4" and to be filled with K2 mortar fill. Install BCF 110 B device.

Bio Fireshield* UL XHCR - R13472(N)

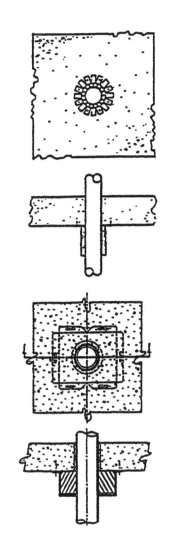

Walls -- Metal Studs

215 E814 - 2hr. (F) - 2hr. (T)

3" CPVC (F441 or F442) SDR 13.5 or heavier pipe penetrating 3-5/8" 25 ga. steel stud wall. Two layers Type "X" 5/8" gypsum wallboard fastened with 1" and 1-5/8" or longer bugle-head, Type S screws. All joints taped and filled with joint compound. Pipe installed in 8" long 30 gauge steel sleeve. Fasten FS-195 wrap strip around pipe on both sides of wall: 2 layers on 3"; 1 layer on 2". Seal with CP-25 caulk. One layer of 4 mil foil tape on pipe to prevent caulk contact.

3M* UL XHEZ System 148-C*

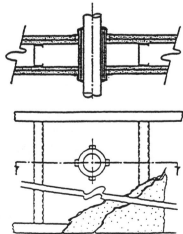

Figure 1.26.6 Sealing penetrations when using CPVC pipe. (*By permission: BFGoodrich, Cleveland, OH.*)

Walls – Wood Framing

127 E814 - 2hr. (F) - 2hr. (T)

3" CPVC (F441 or F442) SDR 13.5 or heavier pipe penetrating nominal 2" x 4" wood stud wall. Two layers Type "X" 5/8" gypsum wallboard fastened with cement coated nails. All joints taped and filled with joint compound. Pipe installed in 8" long 30 gauge steel sleeve. Fasten FS-195 wrap strip around pipe on both sides of wall: 2 layers for 3", 1 layer on 2". Seal with CP-25 caulk. One layer of 4 mil foil tape on pipe to prevent caulk contact.

3M* UL XHEZ System 148-C

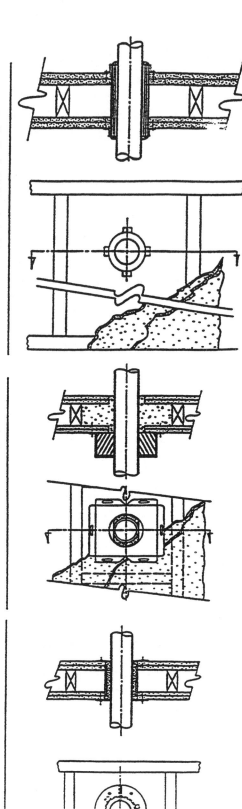

143 E814 - 1hr. (F) - 3/4hr. (T) Vented

3" CPVC (F442) SDR 13.5 pipe penetrating nominal 2" x 4" wood stud wall. Two layers Type "X" 5/8" gypsum wallboard fastened with cement coated nails. All joints taped and filled with joint compound. Pipe installed through a minimum 14" square boxed opening which is filled with K2 mortar fill. Install BCF 110B device.

Bio Fireshield* UL XHCR - R13472(N)

139 E814 - 2hr. (F) - 1-1/2hr. (T)

3" CPVC (F442) SDR 13.5 pipe penetrating nominal 2" x 4" wood stud wall. Two layers Type "X" 5/8" gypsum wallboard fastened with cement coated nails. All joints taped and filled with joint compound. Pipe installed through 5" long 30 ga. steel sleeve. Fill 3/4" annular space between pipe and sleeve with Type FSP putty. Fasten 30 ga. steel trim ring with 6 anchors each side of wall.

Nelson Electric* UL XHEZ System 189

Figure 1.26.6 *(Continued)*

Walls -- Masonry

307 E814 - 2hr. (F) - 2hr. (T) Closed
2hr. (F) - 2hr. (T) Vented

3" PVC (D2665 or D1785) Sch40 or 3" CPVC (F422) SDR17 or heavier pipe penetrating 8" wall of UL classified concrete block without a steel sleeve. Annular space between the pipe and through opening to be not more than 1/2" Put FS-195 wrap strip around pipe on both sides of wall: 2 layers on 3" pipe, and 1 layer on 2" pipe. Caulk applied between wall and wrap strip. One layer of 4 mil foil tape on pipe to prevent caulk contact. Install RC-1 collars over wrap strips on both sides of wall.

3M*   UL XHEZ System 64-A

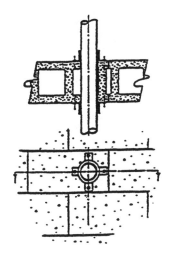

323 E814 - 2hr. (F) 1-1/2hr. (T) Closed

3" CPVC (F422) SDR 13.5 pipe penetrating an 8" concrete block wall or a 5" concrete wall through an opening that provides 3/4" annular space. Steel sleeve optional. Fill annular space around pipe with FSP putty. Fasten 30 ga. steel trim ring with 6 anchors each side of wall.

Nelson Electric* UL XHEZ System 232

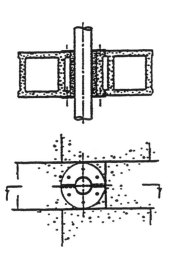

325 E814 - 1hr. (F) - 3/4hr. (T) Vented

3" CPVC (F442) SDR 13.5 pipe penetrating an 8" wall of UL classified concrete block through an opening at least 1-1/2" larger than pipe OD. Pipe centered in opening with annular space of 3/4" or more. Opening filled with K2 mortar fill. Install BCF 110B device.

Bio Fireshield* UL XHCR - R13472(N)

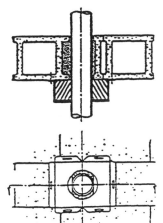

Figure 1.26.6 *(Continued)*

Friction Loss (PSI per Linear Foot) and Velocity (Feet per second)
(Hazen-Williams C Factor = 150)

Nominal Pipe Size:	¾ inch		1 inch		1¼ inches		1½ inches		2 inches		2½ inches		3 inches	
Avg. I.D. (inches)	(.884)		(1.109)		(1.400)		(1.602)		(2.003)		(2.423)		(2.951)	
GPM	Friction Loss	Velocity	Friction Loss	Velocity	Friction Loss	Velocity	Friction Loss	Velocity	Friction Loss	Velocity	Friction Loss	Velocity	Friction Loss	Velocity
1	.0008	.5	.0003	.3	.0001	.2	.0000	.1	.0000	.1	.0000	.0	.0000	.0
2	.0028	1.0	.0009	.6	.0003	.4	.0002	.3	.0001	.2	.0000	.1	.0000	.0
3	.0059	1.5	.0020	.9	.0006	.6	.0003	.4	.0001	.3	.0000	.2	.0000	.1
4	.0101	2.0	.0033	1.3	.0011	.8	.0006	.6	.0002	.4	.0001	.2	.0000	.1
5	.0152	2.6	.0051	1.6	.0016	1.0	.0008	.7	.0003	.5	.0001	.3	.0000	.2
6	.0214	3.1	.0071	1.9	.0023	1.2	.0012	.9	.0004	.6	.0002	.4	.0001	.2
7	.0284	3.6	.0094	2.3	.0030	1.4	.0016	1.1	.0005	.7	.0002	.4	.0001	.3
8	.0364	4.1	.0121	2.6	.0039	1.6	.0020	1.2	.0007	.8	.0003	.5	.0001	.3
9	.0452	4.7	.0150	2.9	.0048	1.8	.0025	1.4	.0008	.9	.0003	.6	.0001	.4
10	.0550	5.2	.0182	3.3	.0059	2.0	.0030	1.5	.0010	1.0	.0004	.6	.0002	.4
11	.0656	5.7	.0217	3.6	.0070	2.2	.0036	1.7	.0012	1.1	.0005	.7	.0002	.5
12	.0770	6.2	.0255	3.9	.0082	2.5	.0043	1.9	.0014	1.2	.0006	.8	.0002	.5
13	.0893	6.7	.0296	4.3	.0095	2.7	.0049	2.0	.0017	1.3	.0007	.9	.0003	.6
14	.1024	7.3	.0340	4.6	.0109	2.9	.0057	2.2	.0019	1.4	.0008	.9	.0003	.6
15	.1164	7.8	.0386	4.9	.0124	3.1	.0064	2.3	.0022	1.5	.0009	1.0	.0003	.7
16	.1312	8.3	.0435	5.3	.0140	3.3	.0072	2.5	.0024	1.6	.0010	1.1	.0004	.7
17	.1467	8.8	.0486	5.6	.0156	3.5	.0081	2.7	.0027	1.7	.0011	1.1	.0004	.7
18	.1631	9.4	.0541	5.9	.0174	3.7	.0090	2.8	.0030	1.8	.0012	1.2	.0005	.8
19	.1802	9.9	.0597	6.3	.0192	3.9	.0100	3.0	.0034	1.9	.0013	1.3	.0005	.8
20	.1982	10.4	.0657	6.6	.0211	4.1	.0110	3.1	.0037	2.0	.0015	1.3	.0006	.9
21	.2169	10.9	.0719	6.9	.0231	4.3	.0120	3.3	.0040	2.1	.0016	1.4	.0006	.9
22	.2364	11.5	.0783	7.3	.0252	4.5	.0131	3.5	.0044	2.2	.0017	1.5	.0007	1.0
23	.2567	12.0	.0851	7.6	.0273	4.7	.0142	3.6	.0048	2.3	.0019	1.6	.0007	1.0
24	.2777	12.5	.0920	7.9	.0296	5.0	.0153	3.8	.0052	2.4	.0020	1.6	.0008	1.1
25	.2995	13.0	.0993	8.3	.0319	5.2	.0166	3.9	.0056	2.5	.0022	1.7	.0008	1.1
26	.3220	13.5	.1067	8.6	.0343	5.4	.0178	4.1	.0060	2.6	.0024	1.8	.0009	1.2
27	.3453	14.1	.1144	8.9	.0368	5.6	.0191	4.2	.0064	2.7	.0025	1.8	.0010	1.2
28	.3693	14.6	.1224	9.3	.0394	5.8	.0204	4.4	.0069	2.8	.0027	1.9	.0010	1.3
29	.3941	15.1	.1306	9.6	.0420	6.0	.0218	4.6	.0073	2.9	.0029	2.0	.0011	1.3
30	.4196	15.6	.1391	9.9	.0447	6.2	.0232	4.7	.0078	3.0	.0031	2.0	.0012	1.4
31	.4458	16.2	.1478	10.2	.0475	6.4	.0246	4.9	.0083	3.1	.0033	2.1	.0013	1.4
32	.4728	16.7	.1567	10.6	.0504	6.6	.0261	5.0	.0088	3.2	.0035	2.2	.0013	1.5
33	.5005	17.2	.1659	10.9	.0533	6.8	.0277	5.2	.0093	3.3	.0037	2.2	.0014	1.5
34	.5289	17.7	.1753	11.2	.0564	7.0	.0292	5.4	.0098	3.4	.0039	2.3	.0015	1.5
35	.5580	18.2	.1850	11.6	.0595	7.2	.0308	5.5	.0104	3.5	.0041	2.4	.0016	1.6
36	.5879	18.8	.1949	11.9	.0626	7.5	.0325	5.7	.0109	3.6	.0043	2.5	.0017	1.6
37	.6185	19.3	.2050	12.2	.0659	7.7	.0342	5.8	.0115	3.7	.0046	2.5	.0017	1.7
38	.6498	19.8	.2154	12.6	.0692	7.9	.0359	6.0	.0121	3.8	.0048	2.6	.0018	1.7
39	.6817	20.3	.2260	12.9	.0726	8.1	.0377	6.2	.0127	3.9	.0050	2.7	.0019	1.8
40	.7144	20.9	.2368	13.2	.0761	8.3	.0395	6.3	.0133	4.0	.0053	2.7	.0020	1.8
GPM														
41			.2479	13.6	.0797	8.5	.0413	6.5	.0139	4.1	.0055	2.8	.0021	1.9
42			.2592	13.9	.0833	8.7	.0432	6.6	.0146	4.2	.0058	2.9	.0022	1.9
43			.2707	14.2	.0870	8.9	.0451	6.8	.0152	4.3	.0060	2.9	.0023	2.0
44			.2824	14.6	.0908	9.1	.0471	7.0	.0159	4.4	.0063	3.0	.0024	2.0
45			.2944	14.9	.0947	9.3	.0491	7.1	.0165	4.5	.0065	3.1	.0025	2.1
46			.3067	15.2	.0986	9.5	.0511	7.3	.0172	4.6	.0068	3.2	.0026	2.1
47			.3191	15.6	.1026	9.7	.0532	7.4	.0179	4.7	.0071	3.2	.0027	2.2
48			.3318	15.9	.1067	10.0	.0553	7.6	.0186	4.8	.0074	3.3	.0028	2.2
49			.3447	16.2	.1108	10.2	.0575	7.7	.0194	4.9	.0077	3.4	.0029	2.2
50			.3578	16.6	.1150	10.4	.0597	7.9	.0201	5.0	.0080	3.4	.0030	2.3
52			.3847	17.2	.1237	10.8	.0642	8.2	.0216	5.2	.0086	3.6	.0033	2.4
54			.4126	17.9	.1326	11.2	.0688	8.5	.0232	5.4	.0092	3.7	.0035	2.5
56			.4413	18.6	.1419	11.6	.0736	8.9	.0248	5.7	.0098	3.8	.0038	2.6
58			.4709	19.2	.1514	12.0	.0785	9.2	.0265	5.9	.0105	4.0	.0040	2.7
60			.5013	19.9	.1612	12.5	.0836	9.5	.0282	6.1	.0111	4.1	.0043	2.8
62			.5327	20.5	.1713	12.9	.0888	9.8	.0299	6.3	.0118	4.3	.0045	2.9
64			.5649	21.2	.1816	13.3	.0942	10.1	.0317	6.5	.0126	4.4	.0048	3.0
66			.5980	21.9	.1923	13.7	.0997	10.5	.0336	6.7	.0133	4.5	.0051	3.0
68			.6320	22.5	.2032	14.1	.1054	10.8	.0355	6.9	.0141	4.7	.0054	3.1
70			.6668	23.2	.2144	14.5	.1112	11.1	.0375	7.1	.0148	4.8	.0057	3.2
72			.7025	23.9	.2258	15.0	.1171	11.4	.0395	7.3	.0156	5.0	.0060	3.3
74			.7390	24.5	.2376	15.4	.1232	11.7	.0415	7.5	.0164	5.1	.0063	3.4
76			.7764	25.2	.2496	15.8	.1295	12.0	.0436	7.7	.0173	5.2	.0066	3.5
78			.8146	25.9	.2619	16.2	.1358	12.4	.0458	7.9	.0181	5.4	.0069	3.6
80			.8536	26.5	.2744	16.6	.1424	12.7	.0480	8.1	.0190	5.5	.0073	3.7

Figure 1.26.7   Friction loss in psi per lineal foot of CPVC pipe. (*By permission: BFGoodrich, Cleveland, OH.*)

GPM										
82	.2873	17.0	.1490	13.0	.0502	8.3	.0199	5.7	.0076	3.8
84	.3004	17.5	.1558	13.3	.0525	8.5	.0208	5.8	.0080	3.9
86	.3137	17.9	.1627	13.6	.0548	8.7	.0217	5.9	.0083	4.0
88	.3274	18.3	.1698	14.0	.0572	8.9	.0226	6.1	.0087	4.1
90	.3413	18.7	.1770	14.3	.0596	9.1	.0236	6.2	.0090	4.2
92	.3554	19.1	.1844	14.6	.0621	9.3	.0246	6.4	.0094	4.3
94	.3698	19.5	.1918	14.9	.0646	9.5	.0256	6.5	.0098	4.4
96	.3845	20.0	.1995	15.2	.0672	9.8	.0266	6.6	.0102	4.5
98	.3995	20.4	.2072	15.5	.0698	9.9	.0276	6.8	.0106	4.5
100	.4147	20.8	.2151	15.9	.0725	10.1	.0287	6.9	.0110	4.6
110	.4947	22.9	.2566	17.5	.0864	11.2	.0342	7.6	.0131	5.1
120	.5810	25.0	.3014	19.1	.1015	12.2	.0402	8.3	.0154	5.6
130	.6738	27.0	.3495	20.6	.1178	13.2	.0466	9.0	.0178	6.0
140	.7728	29.1	.4009	22.2	.1351	14.2	.0534	9.7	.0205	6.5
150	.8780	31.2	.4554	23.8	.1534	15.2	.0607	10.4	.0232	7.0
160			.5132	25.4	.1729	16.2	.0684	11.1	.0262	7.5
170			.5741	27.0	.1934	17.3	.0765	11.8	.0293	7.9
180			.6381	28.6	.2150	18.3	.0851	12.5	.0326	8.4
190			.7053	30.2	.2376	19.3	.0940	13.2	.0360	8.9
200			.7755	31.8	.2613	20.3	.1034	13.9	.0396	9.3
210			.8487	33.4	.2859	21.3	.1132	14.6	.0433	9.8
220			.9250	35.0	.3116	22.4	.1233	15.3	.0472	10.3
230					.3384	23.4	.1339	16.0	.0513	10.7
240					.3661	24.4	.1449	16.6	.0555	11.2
250					.3948	25.4	.1562	17.3	.0598	11.7
260					.4245	26.4	.1680	18.0	.0643	12.1
270					.4552	27.4	.1801	18.7	.0690	12.6
280					.4869	28.5	.1927	19.4	.0738	13.1
290					.5195	29.5	.2056	20.1	.0787	13.6
300					.5532	30.5	.2189	20.8	.0838	14.0
310					.5878	31.5	.2326	21.5	.0891	14.5
320					.6233	32.5	.2467	22.2	.0944	15.0
330					.6598	33.6	.2611	22.9	.1000	15.4
340					.6973	34.6	.2759	23.6	.1056	15.9
350					.7357	35.6	.2911	24.3	.1115	16.4
360					.7751	36.6	.3067	25.0	.1174	16.8
370					.8154	37.6	.3227	25.7	.1235	17.3
380					.8566	38.6	.3390	26.4	.1298	17.8
390					.8988	39.7	.3557	27.1	.1362	18.2
400					.9419	40.7	.3727	27.8	.1427	18.7
410							.3901	28.5	.1494	19.2
430							.4261	29.9	.1631	20.1
450							.4635	31.3	.1774	21.1
470							.5023	32.7	.1923	22.0
490							.5425	34.0	.2077	22.9
510							.5842	35.4	.2237	23.9
530							.6273	36.8	.2402	24.8
550							.6718	38.2	.2572	25.7
570							.7177	39.6	.2748	26.7
590							.7650	41.0	.2929	27.6
610									.3115	28.6
630									.3307	29.5
650									.3504	30.4
670									.3706	31.4
690									.3913	32.3
720									.4233	33.7
750									.4565	35.1
780									.4909	36.5
810									.5264	37.9
840									.5630	39.4
870									.6008	40.8
900									.6397	42.2
930									.6797	43.6
960									.7208	45.0
1000									.7774	46.9

**Allowance for Friction Loss in Fittings
(Equivalent Feet of Pipe)**

	¾″	1″	1¼″	1½″	2″	2½″	3″
Tee Branch	3	5	6	8	10	12	15
Tee Run	1	1	1	1	1	2	2
Elbow 90°	7	7	8	9	11	12	13
Elbow 45°	1	1	2	2	2	3	4
Coupling	1	1	1	1	1	2	2

Figure 1.26.7 (*Continued*)

1.27.0   Contractor's Material and Test Certification Forms

Contractor's Material and Test Certificate for **A**boveground Piping

PROCEDURE

Upon completion of work, inspection and tests shall be made by the contractor's representative and witnessed by an owner's representative. All defects shall be corrected and system left in service before contractor's personnel finally leave the job.

A certificate shall be filled out and signed by both representatives. Copies shall be prepared for approving authorities, owners, and contractor. It is understood the owner's representative's signature in no way prejudices any claim against contractor for faulty material, poor workmanship, or failure to comply with approving authority's requirements or local ordinances.

PROPERTY NAME		DATE

PROPERTY ADDRESS

PLANS	ACCEPTED BY APPROVING AUTHORITIES (NAMES)		
	ADDRESS		
	INSTALLATION CONFORMS TO ACCEPTED PLANS	☐ YES	☐ NO
	EQUIPMENT USED IS APPROVED	☐ YES	☐ NO
	IF NO, EXPLAIN DEVIATIONS		

INSTRUCTIONS	HAS PERSON IN CHARGE OF FIRE EQUIPMENT BEEN INSTRUCTED AS TO LOCATION OF CONTROL VALVES AND CARE AND MAINTENANCE OF THIS NEW EQUIPMENT? IF NO, EXPLAIN	☐ YES	☐ NO
	HAVE COPIES OF THE FOLLOWING BEEN LEFT ON THE PREMISES:	☐ YES	☐ NO
	1. SYSTEM COMPONENTS INSTRUCTIONS	☐ YES	☐ NO
	2. CARE AND MAINTENANCE INSTRUCTIONS	☐ YES	☐ NO
	3. NFPA 25	☐ YES	☐ NO

LOCATION OF SYSTEM	SUPPLIES BUILDINGS

SPRINKLERS	MAKE	MODEL	YEAR OF MANUFACTURE	ORIFICE SIZE	QUANTITY	TEMPERATURE RATING

PIPE AND FITTINGS	Type of Pipe _____
	Type of Fittings _____

ALARM VALVE OR FLOW INDICATOR	ALARM DEVICE			MAXIMUM TIME TO OPERATE THROUGH TEST CONNECTION	
	TYPE	MAKE	MODEL	MIN.	SEC.

DRY PIPE OPERATING TEST	DRY VALVE			Q. O. D.		
	MAKE	MODEL	SERIAL NO.	MAKE	MODEL	SERIAL NO.

DRY PIPE OPERATING TEST		TIME TO TRIP THROUGH TEST CONNECTION*		WATER PRESSURE	AIR PRESSURE	TRIP POINT AIR PRESSURE	TIME WATER REACHED TEST OUTLET*		ALARM OPERATED PROPERLY	
		MIN.	SEC.	PSI	PSI	PSI	MIN.	SEC.	YES	NO
	Without Q.O.D.									
	With Q.O.D.									
	IF NO, EXPLAIN									

*MEASURED FROM TIME INSPECTOR'S TEST CONNECTION IS OPENED.

DELUGE & PREACTION VALVES	OPERATION ☐ PNEUMATIC ☐ ELECTRIC ☐ HYDRAULIC								
	PIPING SUPERVISED ☐ YES ☐ NO DETECTING MEDIA SUPERVISED ☐ YES ☐ NO								
	DOES VALVE OPERATE FROM THE MANUAL TRIP AND/OR REMOTE CONTROL STATIONS ☐ YES ☐ NO								
	IS THERE AN ACCESSIBLE FACILITY IN EACH CIRCUIT FOR TESTING ☐ YES ☐ NO			IF NO, EXPLAIN					
	MAKE	MODEL	DOES EACH CIRCUIT OPERATE SUPERVISION LOSS ALARM		DOES EACH CIRCUIT OPERATE VALVE RELEASE		MAXIMUM TIME TO OPERATE RELEASE		
			YES	NO	YES	NO	MIN.	SEC.	

PRESSURE REDUCING VALVE TEST	LOCATION & FLOOR	MAKE & MODEL	SETTING	STATIC PRESSURE		RESIDUAL PRESSURE (FLOWING)		FLOW RATE
				INLET (PSI)	OUTLET (PSI)	INLET (PSI)	OUTLET (PSI)	FLOW (GPM)

TEST DESCRIPTION	HYDROSTATIC: Hydrostatic tests shall be made at not less than 200 psi (13.6 bars) for two hours or 50 psi (3.4 bars) above static pressure in excess of 150 psi (10.2 bars) for two hours. Differential dry-pipe valve clappers shall be left open during test to prevent damage. All aboveground piping leakage shall be stopped. PNEUMATIC: Establish 40 psi (2.7 bars) air pressure and measure drop, which shall not exceed 1-1/2 psi (0.1 bars) in 24 hours. Test pressure tanks at normal water level and air pressure and measure air pressure drop, which shall not exceed 1-1/2 psi (0.1 bars) in 24 hours.

TESTS	ALL PIPING HYDROSTATICALLY TESTED AT _____ PSI FOR _____ HRS. IF NO, STATE REASON DRY PIPING PNEUMATICALLY TESTED ☐ YES ☐ NO EQUIPMENT OPERATES PROPERLY ☐ YES ☐ NO
	DO YOU CERTIFY AS THE SPRINKLER CONTRACTOR THAT ADDITIVES AND CORROSIVE CHEMICALS, SODIUM SILICATE OR DERIVATIVES OF SODIUM SILICATE, BRINE, OR OTHER CORROSIVE CHEMICALS WERE NOT USED FOR TESTING SYSTEMS OR STOPPING LEAKS? ☐ YES ☐ NO
	DRAIN TEST READING OF GAGE LOCATED NEAR WATER SUPPLY TEST CONNECTION: _____ PSI RESIDUAL PRESSURE WITH VALVE IN TEST CONNECTION OPEN WIDE _____ PSI
	UNDERGROUND MAINS AND LEAD IN CONNECTIONS TO SYSTEM RISERS FLUSHED BEFORE CONNECTION MADE TO SPRINKLER PIPING. VERIFIED BY COPY OF THE U FORM NO. 85B ☐ YES ☐ NO OTHER EXPLAIN FLUSHED BY INSTALLER OF UNDER-GROUND SPRINKLER PIPING ☐ YES ☐ NO
	IF POWDER DRIVEN FASTENERS ARE USED IN CONCRETE, HAS REPRESENTATIVE SAMPLE TESTING BEEN SATISFACTORILY COMPLETED? ☐ YES ☐ NO IF NO, EXPLAIN

BLANK TESTING GASKETS	NUMBER USED	LOCATIONS	NUMBER REMOVED

WELDING	WELDED PIPING ☐ YES ☐ NO
	IF YES...
	DO YOU CERTIFY AS THE SPRINKLER CONTRACTOR THAT WELDING PROCEDURES COMPLY WITH THE REQUIREMENTS OF AT LEAST AWS D10.9, LEVEL AR-3? ☐ YES ☐ NO
	DO YOU CERTIFY THAT THE WELDING WAS PERFORMED BY WELDERS QUALIFIED IN COMPLIANCE WITH THE REQUIREMENTS OF AT LEAST AWS D10.9, LEVEL AR-3? ☐ YES ☐ NO
	DO YOU CERTIFY THAT WELDING WAS CARRIED OUT IN COMPLIANCE WITH A DOCUMENTED QUALITY CONTROL PROCEDURE TO INSURE THAT ALL DISCS ARE RETRIEVED, THAT OPENINGS IN PIPING ARE SMOOTH, THAT SLAG AND OTHER WELDING RESIDUE ARE REMOVED, AND THAT THE INTERNAL DIAMETERS OF PIPING ARE NOT PENETRATED? ☐ YES ☐ NO

CUTOUTS (DISCS)	DO YOU CERTIFY THAT YOU HAVE A CONTROL FEATURE TO ENSURE THAT ALL CUTOUTS (DISCS) ARE RETRIEVED? ☐ YES ☐ NO

Figure 1.27.0 (*Continued*)

HYDRAULIC DATA NAMEPLATE	NAMEPLATE PROVIDED ☐ YES ☐ NO	IF NO, EXPLAIN	
REMARKS	DATE LEFT IN SERVICE WITH ALL CONTROL VALVES OPEN:		
SIGNATURES	NAME OF SPRINKLER CONTRACTOR		
	TESTS WITNESSED BY		
	FOR PROPERTY OWNER (SIGNED)	TITLE	DATE
	FOR SPRINKLER CONTRACTOR (SIGNED)	TITLE	DATE
ADDITIONAL EXPLANATION AND NOTES			

Figure 1.27.0 (*Continued*)

This test connection should be in the upper story, and the connection should preferably be piped from the end of the most remote branch line. The discharge should be at a point where it can be readily observed. In locations where it is not practical to terminate the test connection outside the building, the test connection may terminate into a drain capable of accepting full flow under system pressure. In this event, the test connection should be made using an approved sight test connection containing a smooth bore corrosion-resistant orifice giving a flow equivalent to one sprinkler simulating the least flow from an individual sprinkler in the system. [*See Figures A-4-7.4.2(a) and A-4-7.4.2(b).*] The test valve should be located at an accessible point and preferably not over 7 ft (2.1 m) above the floor. The control valve on the test connection should be located at a point not exposed to freezing.

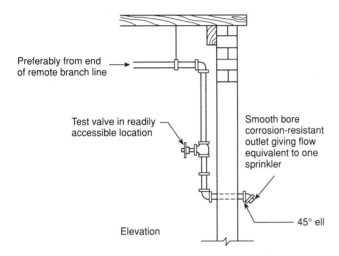

Preferably from end of remote branch line

Test valve in readily accessible location

Smooth bore corrosion-resistant outlet giving flow equivalent to one sprinkler

45° ell

Elevation

For SI Units: 1 ft = 0.3048 m.
NOTE: Not less than 4 ft (1.2 m) of exposed test pipe in warm room beyond valve where pipe extends through wall to outside.

Figure 1.27.1 Test connection requirements. (*Reprinted with permission from NFPA 13,* Installation of Sprinkler Systems, *Copyright © 1996, National Fire Protection Association, Quincy, MA 02269. This reprinted material is not the complete and official position of the NFPA, on the referenced subject, which is represented only by the standard in its entirety.*)

Identification Signs. Approved identification signs, as shown in Figure A-4-7.1.1, should be provided for outside alarm devices. The sign should be located near the device in a conspicuous position and should be worded as follows:

"SPRINKLER FIRE ALARM — WHEN BELL RINGS CALL FIRE DEPARTMENT OR POLICE."

Figure 1.27.2 Exterior sprinkler identification sign. (*Reprinted with permission from NFPA 13*, Installation of Sprinkler Systems, *Copyright © 1996, National Fire Protection Association, Quincy, MA 02269. This reprinted material is not the complete and official position of the National Fire Protection Association on the referenced subject, which is represented only by the standard in its entirety.*)

Contractor's Material and Test Certificate for Underground Piping

PROCEDURE

Upon completion of work, inspection and tests shall be made by the contractor's representative and witnessed by an owner's representative. All defects shall be corrected and system left in service before contractor's personnel finally leave the job.

A certificate shall be filled out and signed by both representatives. Copies shall be prepared for approving authorities, owners, and contractor. It is understood the owner's representative's signature in no way prejudices any claim against contractor for faulty material, poor workmanship, or failure to comply with approving authority's requirements or local ordinances.

PROPERTY NAME	DATE

PROPERTY ADDRESS

PLANS	ACCEPTED BY APPROVING AUTHORITIES (NAMES)		
	ADDRESS		
	INSTALLATION CONFORMS TO ACCEPTED PLANS	☐ YES	☐ NO
	EQUIPMENT USED IS APPROVED IF NO, STATE DEVIATIONS	☐ YES	☐ NO
INSTRUCTIONS	HAS PERSON IN CHARGE OF FIRE EQUIPMENT BEEN INSTRUCTED AS TO LOCATION OF CONTROL VALVES AND CARE AND MAINTENANCE OF THIS NEW EQUIPMENT? IF NO, EXPLAIN	☐ YES	☐ NO
	HAVE COPIES OF APPROPRIATE INSTRUCTIONS AND CARE AND MAINTENANCE CHARTS BEEN LEFT ON PREMISES? IF NO, EXPLAIN	☐ YES	☐ NO
LOCATION	SUPPLIES BUILDINGS		
UNDERGROUND PIPES AND JOINTS	PIPE TYPES AND CLASS	TYPE JOINT	
	PIPE CONFORMS TO _____ STANDARD	☐ YES	☐ NO
	FITTINGS CONFORM TO _____ STANDARD IF NO, EXPLAIN	☐ YES	☐ NO
	JOINTS NEEDING ANCHORAGE CLAMPED, STRAPPED, OR BLOCKED IN ACCORDANCE WITH _____ STANDARD IF NO, EXPLAIN	☐ YES	☐ NO

TEST DESCRIPTION	<u>FLUSHING</u>: Flow the required rate until water is clear as indicated by no collection of foreign material in burlap bags at outlets such as hydrants and blow-offs. Flush at flows not less than 390 GPM (1476 L/min) for 4-inch pipe, 880 GPM (3331 L/min) for 6-inch pipe, 1560 GPM (5905 L/min) for 8-inch pipe, 2440 GPM (9235 L/min) for 10-inch pipe, and 3520 GPM (13323 L/min) for 12-inch pipe. When supply cannot produce stipulated flow rates, obtain maximum available. <u>HYDROSTATIC</u>: Hydrostatic tests shall be made at not less than 200 psi (13.8 bars) for two hours or 50 psi (3.4 bars) above static pressure in excess of 150 psi (10.3 bars) for two hours. <u>LEAKAGE</u>: New pipe laid with rubber gasketed joints shall, if the workmanship is satisfactory, have little or no leakage at the joints. The amount of leakage at the joints shall not exceed 2 qts. per hr. (1.89 L/h) per 100 joints irrespective of pipe diameter. The leakage shall be distributed over all joints. If such leakage occurs at a few joints the installation shall be considered unsatisfactory and necessary repairs made. The amount of allowable leakage specified above may be increased by 1 fl oz per in. valve diameter per hr. (30 mL/25 mm/h) for each metal seated valve isolating the test section. If dry barrel hydrants are tested with the main valve open, so the hydrants are under pressure, an additional 5 oz per minute (150 mL/min) leakage is permitted for each hydrant.

FLUSHING TESTS	NEW UNDERGROUND PIPING FLUSHED ACCORDING TO _____ STANDARD BY (COMPANY) IF NO, EXPLAIN	☐ YES	☐ NO
	HOW FLUSHING FLOW WAS OBTAINED ☐ PUBLIC WATER ☐ TANK OR RESERVOIR ☐ FIRE PUMP	THROUGH WHAT TYPE OPENING ☐ HYDRANT BUTT. ☐ OPEN PIPE	
	LEAD-INS FLUSHED ACCORDING TO _____ STANDARD BY (COMPANY) IF NO, EXPLAIN	☐ YES	☐ NO
	HOW FLUSHING FLOW WAS OBTAINED ☐ PUBLIC WATER ☐ TANK OR RESERVOIR ☐ FIRE PUMP	THROUGH WHAT TYPE OPENING ☐ Y CONN. TO FLANGE & SPIGOT ☐ OPEN PIPE	

Figure 1.28.0 Contractor's material and test certification for underground piping. (*Reprinted with permission from NFPA 13, Installation of Sprinkler Systems, Copyright © 1996, National Fire Protection Association, Quincy, MA 02269. This reprinted material is not the complete and official position of the National Fire Protection Association on the referenced subject, which is represented only by the standard in its entirety.*)

HYDRAULIC DATA NAMEPLATE	NAMEPLATE PROVIDED ☐ YES ☐ NO	IF NO, EXPLAIN	
REMARKS	DATE LEFT IN SERVICE WITH ALL CONTROL VALVES OPEN:		
SIGNATURES	NAME OF SPRINKLER CONTRACTOR		
	TESTS WITNESSED BY		
	FOR PROPERTY OWNER (SIGNED)	TITLE	DATE
	FOR SPRINKLER CONTRACTOR (SIGNED)	TITLE	DATE
ADDITIONAL EXPLANATION AND NOTES			

Figure 1.28.0 *(Continued)*

Extinguishing Agents	Cools	Smothers	Dilutes	Emulsifies	Reacts Chemically	Solids	Liquids	Electrical	Combustible Metals	Reignition Possible	Nonconductive	Can Be Toxic	Penetrating	Corrosive	Messy	Powder/Powder	Liquid/Liquid	Liquid/Foam	Liquid/Gas	Gas/Gas	Salts/Foam	Total Flooding	Localized Spray	Hose and Standpipe	Fire Extinguisher	Contained Supply	Other
Extinguishing Action →						**Fire Class**				**Residual Effects**						**Form: Stored/Released**						**Delivery Method**					
Water: Spray	•					•									•		•					•	•	•	•		
Steam		•				•											•										•
Stream			•			•									•		•					•		•			
Wet water	•					•								•	•		•					•		•			
Thickened	•	•				•	•								•		•					•		•			
Emulsified	•			•		•								•	•		•					•		•			
Carbon dioxide	•	•					•	•		•	•	•								•		•	•		•	•	
Dry chemical																											
Sodium bicarbonate		•			•		•	•		•	•			•	•	•						•	•		•		
Ammonium phosphate		•			•	•	•	•		•	•			•	•	•						•	•		•		
Potassium bicarbonate		•			•		•	•		•	•			•	•	•						•	•		•		
Foaming agents																											
Film-forming	•	•				•	•						•	•				•					•	•		•	
Medium/high expansion	•	•				•	•					•	•		•			•					•	•			
Surfactant	•	•		•		•	•						•					•					•		•		
Halogenated agents																											
Halon 1211					•	•	•	•		•	•	•	•						•			•		•	•		•
Halon 1301					•	•	•	•		•	•	•								•		•			•	•	•
Halon 2402					•	•	•	•		•	•	•	•							•		•		•	•		•
Dry powder																											
Graphitized coke	•	•							•							•											•
NaCl w/additives		•							•							•											•
Flux; molten/crust		•							•												•						•
Trimethoxyboroxine		•							•							•					•				•		•

Figure 1.29.0 Fire extinguisher agents and when to use.

Hydraulically Designed System. A calculated sprinkler system in which pipe sizes are selected on a pressure loss basis to provide a prescribed water density, in gallons per minute per square foot [(L/min)/m^2], or a prescribed minimum discharge pressure or flow per sprinkler, distributed with a reasonable degree of uniformity over a specified area.

Limited-Combustible Material. As applied to a building construction material, a material not complying with the definition of noncombustible material that, in the form in which it is used, has a potential heat value not exceeding 3500 Btu per lb (8141 kJ/kg) and complies with one of the following paragraphs, (a) or (b). Materials subject to increase in combustibility or flame spread rating beyond the limits herein established through the effects of age, moisture, or other atmospheric condition shall be considered combustible.

(a) Materials having a structural base of noncombustible material, with a surfacing not exceeding a thickness of $\frac{1}{8}$ in. (3.2 mm) that has a flame spread rating not greater than 50.

(b) Materials, in the form and thickness used, other than as described in (a), having neither a flame spread rating greater than 25 nor evidence of continued progressive combustion and of such composition that surfaces that would be exposed by cutting through the material on any plane would have neither a flame spread rating greater than 25 nor evidence of continued progressive combustion.

Miscellaneous Storage.* Storage that does not exceed 12 ft (3.66 m) in height and is incidental to another occupancy use group as defined in 1-4.7 (see 5-2.3.1.1). Such storage shall not constitute more than 10 percent of the building area or 4,000 sq ft (372 m^2) of the sprinklered area, whichever is greater. Such storage shall not exceed 1,000 sq ft (93 m^2) in one pile or area, and each such pile or area shall be separated from other storage areas by at least 25 ft (7.62 m). Protection criteria for miscellaneous storage are within the scope of this standard.

Noncombustible Material. A material that, in the form in which it is used and under the conditions anticipated, will not ignite, burn, support combustion, or release flammable vapors when subjected to fire or heat. Materials that are reported as passing ASTM E136, *Standard Test Method for Behavior of Materials in a Vertical Tube Furnace at 750°C*, shall be considered noncombustible materials.

Pipe Schedule System. A sprinkler system in which the pipe sizing is selected from a schedule that is determined by the occupancy classification. A given number of sprinklers are allowed to be supplied from specific sizes of pipe.

Shop Welded. As used in this standard, shop in the term shop welded means either:

(a) At a sprinkler contractor's or fabricator's premise.

(b) In an area specifically designed or authorized for such work such as a detached outside location, maintenance shop, or other area (either temporary or permanent) of noncombustible or fire-resistive construction free of combustible and flammable contents and suitably segregated from adjacent areas.

Small Rooms. Rooms of Light Hazard Occupancy classification having unobstructed construction and floor areas not exceeding 800 sq ft (74.3 m^2). (*See 1-4.7.1.*)

General Definitions.

Compartment. As used in 4-3.6.3 and 6-4.4.4, a space completely enclosed by walls and a ceiling. The compartment enclosure is permitted to have openings to an adjoining space if the openings have a minimum lintel depth of 8 in. (203 mm) from the ceiling.

Drop-Out Ceiling. A suspended ceiling system with listed translucent or opaque panels that are heat sensitive and fall from their setting when exposed to heat. This ceiling system is installed below the sprinklers.

Dwelling Unit. One or more rooms arranged for the use of one or more individuals living together, as in a single housekeeping unit normally having cooking, living, sanitary, and sleeping facilities.

For purposes of this standard, dwelling unit includes hotel rooms, dormitory rooms, apartments, condominiums, sleeping rooms in nursing homes, and similar living units.

Fire Control. Limiting the size of a fire by distribution of water so as to decrease the heat release rate and pre-wet adjacent combustibles, while controlling ceiling gas temperatures to avoid structural damage.

Fire Suppression. Sharply reducing the heat release rate of a fire and preventing its regrowth by means of direct and sufficient application of water through the fire plume to the burning fuel surface

High Challenge Fire Hazard. A fire hazard typical of that produced by fires in combustible high-piled storage.

High-Piled Storage. Solid-piled, palletized, rack storage, bin box, and shelf storage in excess of 12 ft (3.7 m) in height. (*See 5-2.3.1.1.*)

Figure 1.30.0 NFPA definitions of fire protection terminology. (*Reprinted with permission from NFPA 13,* Installation of Sprinkler Systems, *Copyright © 1996, National Fire Protection Association, Quincy, MA 02269. This reprinted material is not the complete and official position of the National Fire Protection Association on the referenced subject, which is represented only by the standard in its entirety.*)

Sprinkler System.* For fire protection purposes, an integrated system of underground and overhead piping designed in accordance with fire protection engineering standards. The installation includes one or more automatic water supplies. The portion of the sprinkler system aboveground is a network of specially sized or hydraulically designed piping installed in a building, structure, or area, generally overhead, and to which sprinklers are attached in a systematic pattern. The valve controlling each system riser is located in the system riser or its supply piping. Each sprinkler system riser includes a device for actuating an alarm when the system is in operation. The system is usually activated by heat from a fire and discharges water over the fire area.

NOTE: The design and installation of water supply facilities such as gravity tanks, fire pumps, reservoirs or pressure tanks, and underground piping are covered by the following NFPA standards: NFPA 20, *Standard for the Installation of Centrifugal Fire Pumps*; NFPA 22, *Standard for Water Tanks for Private Fire Protection*; and NFPA 24, *Standard for the Installation of Private Fire Service Mains and Their Appurtenances*.

Thermal Barrier. A material that will limit the average temperature rise of the unexposed surface to not more than 250°F (121°C) after 15 minutes of fire exposure complying with the standard time-temperature curve of NFPA 251, *Standard Methods of Fire Tests of Building Construction and Materials*.

1-4.3 Sprinkler System Type Definitions.

Wet Pipe System. A sprinkler system employing automatic sprinklers attached to a piping system containing water and connected to a water supply so that water discharges immediately from sprinklers opened by heat from a fire.

Dry Pipe System. A sprinkler system employing automatic sprinklers attached to a piping system containing air or nitrogen under pressure, the release of which (as from the opening of a sprinkler) permits the water pressure to open a valve known as a dry pipe valve. The water then flows into the piping system and out the opened sprinklers.

Preaction System. A sprinkler system employing automatic sprinklers attached to a piping system containing air that may or may not be under pressure, with a supplemental detection system installed in the same areas as the sprinklers. Actuating means of the valve are described in 3-3.2.1. Actuation of the detection system opens a valve that permits water to flow into the sprinkler piping system and to be discharged from any sprinklers that are open.

Deluge System. A sprinkler system employing open sprinklers attached to a piping system connected to a water supply through a valve that is opened by the operation of a detection system installed in the same areas as the sprinklers. When this valve opens, water flows into the piping system and discharges from all sprinklers attached thereto.

Combined Dry Pipe-Preaction System. A sprinkler system employing automatic sprinklers attached to a piping system containing air under pressure with a supplemental detection system installed in the same areas as the sprinklers. Operation of the detection system actuates tripping devices that open dry pipe valves simultaneously and without loss of air pressure in the system. Operation of the detection system also opens listed air exhaust valves at the

end of the feed main, which usually precedes the opening of sprinklers. The detection system also serves as an automatic fire alarm system.

Antifreeze System. A wet pipe sprinkler system employing automatic sprinklers attached to a piping system containing an antifreeze solution and connected to a water supply. The antifreeze solution is discharged, followed by water, immediately upon operation of sprinklers opened by heat from a fire.

Circulating Closed-Loop System. A wet pipe sprinkler system having non-fire-protection connections to automatic sprinkler systems in a closed-loop piping arrangement for the purpose of utilizing sprinkler piping to conduct water for heating or cooling. Water is not removed or used from the system, but only circulated through the piping system.

Gridded System.* A sprinkler system in which parallel cross mains are connected by multiple branch lines. An operating sprinkler will receive water from both ends of its branch line while other branch lines help transfer water between cross mains.

Looped System.* A sprinkler system in which multiple cross mains are tied together so as to provide more than one path for water to flow to an operating sprinkler and branch lines are not tied together.

1-4.4* System Component Definitions.

Branch Lines. The pipes in which the sprinklers are placed, either directly or through risers.

Cross Mains. The pipes supplying the branch lines, either directly or through risers.

Feed Mains. The pipes supplying risers or cross mains.

Flexible Listed Pipe Coupling. A listed coupling or fitting that allows axial displacement, rotation, and at least 1 degree of angular movement of the pipe without inducing harm on the pipe.

Exception: For pipe diameters of 8 in. (203.2 mm) and larger, the angular movement shall be permitted to be less than 1 degree but not less than 0.5 degrees.

Risers. The vertical supply pipes in a sprinkler system.

Supervisory Devices. Devices arranged to supervise the operative condition of automatic sprinkler systems.

System Riser. The aboveground supply pipe directly connected to the water supply.

1-4.5 Sprinkler Definitions.

1-4.5.1 Sprinklers defined according to design and performance characteristics:

Spray Sprinkler. A type of sprinkler listed for its capability to provide fire control for a wide range of fire hazards.

Old-Style/Conventional Sprinkler. Sprinklers that direct from 40 to 60 percent of the total water initially in a downward direction and that are designed to be installed with the deflector either upright or pendent.

Fast-Response Sprinkler. A type of sprinkler with a high level of thermal sensitivity, enabling it to respond at an early stage of fire development. This includes ESFR, QR, QREC, QRES, and residential sprinklers.

C-1   The following documents or portions thereof are referenced within this standard for informational purposes only and thus are not considered part of the requirements of this document. The edition indicated for each reference is the current edition as of the date of the NFPA issuance of this document.

C-1.1 NFPA Publications.   National Fire Protection Association, 1 Batterymarch Park, P.O. Box 9101, Quincy, MA 02269-9101.

NFPA 14, *Standard for the Installation of Standpipe and Hose Systems*, 1993 edition.

NFPA 20, *Standard for the Installation of Centrifugal Fire Pumps*, 1993 edition.

NFPA 22, *Standard for Water Tanks for Private Fire Protection*, 1993 edition.

NFPA 24, *Standard for the Installation of Private Fire Service Mains and Their Appurtenances*, 1992 edition.

NFPA 25, *Standard for the Inspection, Testing, and Maintenance of Water-Based Fire Protection Systems*, 1992 edition.

NFPA 30, *Flammable and Combustible Liquids Code*, 1993 edition.

NFPA 30B, *Code for the Manufacture and Storage of Aerosol Products*, 1993 edition.

NFPA 40, *Standard for the Storage and Handling of Cellulose Nitrate Motion Picture Film*, 1988 edition.

NFPA 58, *Standard for the Storage and Handling of Liquefied Petroleum Gases*, 1992 edition.

NFPA 72, *National Fire Alarm Code*, 1993 edition.

NFPA 80A, *Recommended Practice for Protection of Buildings from Exterior Fire Exposures*, 1993 edition.

NFPA 81, *Standard for Fur Storage, Fumigation and Cleaning*, 1986 edition.

NFPA 96, *Standard on Ventilation Control and Fire Protection of Commercial Cooking Operations*, 1994 edition.

NFPA 220, *Standard on Types of Building Construction*, 1992 edition.

NFPA 231, *Standard for General Storage*, 1990 edition.

NFPA 231C, *Standard for Rack Storage of Materials*, 1991 edition.

NFPA 231D, *Standard for Storage of Rubber Tires*, 1989 edition.

NFPA 231F, *Standard for the Storage of Roll Paper*, 1987 edition.

NFPA 409, *Standard on Aircraft Hangars*, 1990 edition.

NFPA 703, *Standard for Fire Retardant Impregnated Wood and Fire Retardant Coatings for Building Materials*, 1992 edition.

Figure 1.31.0   Listing of NFPA publications. (*Reprinted with permission from NFPA 13*, Installation of Sprinkler Systems, *Copyright © 1996, National Fire Protection Association, Quincy, MA 02269. This reprinted material is not the complete and official position of the National Fire Protection Association on the referenced subject, which is represented only by the standard in its entirety.*)

SECTION 216 — O

OCCUPANCY is the purpose for which a building or part thereof is used or intended to be used.

OCCUPANCY CLASSIFICATION. For the purpose of this code, certain occupancies are defined as follows:

Group A Occupancies:

Group A Occupancies include the use of a building or structure, or a portion thereof, for the gathering together of 50 or more persons for purposes such as civic, social or religious functions; recreation, education or instruction; food or drink consumption; or awaiting transportation. A room or space used for assembly purposes by less than 50 persons and accessory to another occupancy shall be included as a part of that major occupancy. Assembly occupancies shall include the following:

Division 1. A building or portion of a building having an assembly room with an occupant load of 1,000 or more and a legitimate stage.

Division 2. A building or portion of a building having an assembly room with an occupant load of less than 1,000 and a legitimate stage.

Division 2.1. A building or portion of a building having an assembly room with an occupant load of 300 or more without a legitimate stage, including such buildings used for educational purposes and not classed as Group B or E Occupancies.

Division 3. A building or portion of a building having an assembly room with an occupant load of less than 300 without a legitimate stage, including such buildings used for educational purposes and not classed as Group B or E Occupancies.

Division 4. Stadiums, reviewing stands and amusement park structures not included within other Group A Occupancies.

Group B Occupancies:

Group B Occupancies shall include buildings, structures, or portions thereof, for office, professional or service-type transactions, which are not classified as Group H Occupancies. Such occupancies include occupancies for the storage of records and accounts, and eating and drinking establishments with an occupant load of less than 50. Business occupancies shall include, but not be limited to, the following:

1. Animal hospitals, kennels, pounds.
2. Automobile and other motor vehicle showrooms.
3. Banks.
4. Barber shops.
5. Beauty shops.
6. Car washes.
7. Civic administration.
8. Outpatient clinic and medical offices (where five or less patients in a tenant space are incapable of unassisted self-preservation).
9. Dry cleaning pick-up and delivery stations and self-service.
10. Educational occupancies above the 12th grade.
11. Electronic data processing.
12. Fire stations.
13. Florists and nurseries.
14. Laboratories—testing and research.
15. Laundry pick-up and delivery stations and self-service.
16. Police stations.
17. Post offices.
18. Print shops.
19. Professional services such as attorney, dentist, physician, engineer.
20. Radio and television stations.
21. Telephone exchanges.

Group E Occupancies:

Group E Occupancies shall be:

Division 1. Any building used for educational purposes through the 12th grade by 50 or more persons for more than 12 hours per week or four hours in any one day.

Division 2. Any building used for educational purposes through the 12th grade by less than 50 persons for more than 12 hours per week or four hours in any one day.

Division 3. Any building or portion thereof used for day-care purposes for more than six persons.

Group F Occupancies:

Group F Occupancies shall include the use of a building or structure, or a portion thereof, for assembling, disassembling, fabricating, finishing, manufacturing, packaging, repair or processing operations that are not classified as Group H Occupancies. Factory and industrial occupancies shall include the following:

Division 1. Moderate-hazard factory and industrial occupancies shall include factory and industrial uses which are not classified as Group F, Division 2 Occupancies, but are not limited to facilities producing the following:

1. Aircraft.
2. Appliances.
3. Athletic equipment.
4. Automobiles and other motor vehicles.
5. Bakeries.
6. Alcoholic beverages.
7. Bicycles.
8. Boats.
9. Brooms and brushes.
10. Business machines.
11. Canvas or similar fabric.
12. Cameras and photo equipment.
13. Carpets and rugs, including cleaning.
14. Clothing.
15. Construction and agricultural machinery.
16. Dry cleaning and dyeing.
17. Electronics assembly.
18. Engines, including rebuilding.
19. Photographic film.
20. Food processing.
21. Furniture.

Figure 1.32.0 Occupancy as defined by the Uniform Fire Code. (*Reproduced from the 1997 edition of the* Uniform Fire Code, *Volumes 1, 2, 3, Copyright © 1997, with the permission of the publisher, The International Fire Code Institute (IFCI). Although the International Fire Code Institute (IFI) has granted permission for reproduction of sections contained within the* Uniform Fire Code, *in conjunction with this publication, IFCI assumes no responsibility for the accuracy or completion of summaries provided therein.*)

22. Hemp products.

23. Jute products.

24. Laundries.

25. Leather products.

26. Machinery.

27. Metal.

28. Motion pictures and television filming and videotaping.

29. Musical instruments.

30. Optical goods.

31. Paper mills or products.

32. Plastic products.

33. Printing or publishing.

34. Recreational vehicles.

35. Refuse incineration.

36. Shoes.

37. Soaps and detergents.

38. Tobacco.

39. Trailers.

40. Wood, distillation.

41. Millwork (sash and door).

42. Woodworking, cabinet.

Division 2. Low-hazard factory and industrial occupancies shall include facilities producing noncombustible or nonexplosive materials which, during finishing, packing or processing, do not involve a significant fire hazard, including, but not limited to, the following:

1. Nonalcoholic beverages.

2. Brick and masonry.

3. Ceramic products.

4. Foundries.

5. Glass products.

6. Gypsum.

7. Ice.

8. Steel products—fabrication and assembly.

Group H Occupancies:

Group H Occupancies shall include buildings or structures, or portions thereof, that involve the manufacturing, processing, generation or storage of materials that constitute a high fire, explosion or health hazard. Group H Occupancies shall be:

Division 1. Occupancies with a quantity of material in the building in excess of those listed in Table 8001.15-A which present a high explosion hazard, including, but not limited to:

1. Explosives, blasting agents, Class 1.3G (Class B Special) fireworks and black powder.

> **EXCEPTIONS:** 1. Storage and the use of pyrotechnic special effect materials in motion picture, television, theatrical and group entertainment production when under permit as required by the Fire Code. The time period for storage shall not exceed 90 days.
>
> 2. Indoor storage and display of smokeless powder, black sporting powder, and primers or percussion caps exceeding the exempt amounts for Group M retail sales need not be classified as a Group H, Division 1 Occupancy where stored and displayed in accordance with the Fire Code.

2. Manufacturing of Class 1.4G (Class C Common) fireworks.

3. Unclassified detonatable organic peroxides.

4. Class 4 oxidizers.

5. Class 4 or Class 3 detonatable unstable (reactive) materials.

Division 2. Occupancies where combustible dust is manufactured, used or generated in such a manner that concentrations and conditions create a fire or explosion potential; occupancies with a quantity of material in the building in excess of those listed in Table 8001.15-A, which present a moderate explosion hazard or a hazard from accelerated burning, including, but not limited to:

1. Class I organic peroxides.

2. Class 3 nondetonatable unstable (reactive) materials.

3. Pyrophoric gases.

4. Flammable or oxidizing gases.

5. Class I, II or III-A flammable or combustible liquids which are used or stored in normally open containers or systems, or in closed containers or systems pressurized at more than 15-pounds-per-square-inch (103.4 kPa) gage.

> **EXCEPTION:** Aerosols.

6. Class 3 oxidizers.

7. Class 3 water-reactive materials.

Division 3. Occupancies where flammable solids, other than combustible dust, are manufactured, used or generated.

Division 3 Occupancies also include uses in which the quantity of material in the building in excess of those listed in Table 8001.15-A presents a high physical hazard, including, but not limited to:

1. Class II, III or IV organic peroxides.

2. Class 1 or 2 oxidizers.

3. Class I, II or III-A flammable or combustible liquids which are used or stored in normally closed containers or systems and containers or systems pressurized at 15-pounds-per-square-inch (103.4 kPa) gage or less, and aerosols.

4. Class III-B combustible liquids.

5. Pyrophoric liquids or solids.

6. Class 1 or 2 water-reactive materials.

7. Flammable solids in storage.

8. Flammable or oxidizing cryogenic fluids (other than inert).

9. Class 1 unstable (reactive) gas or Class 2 unstable (reactive) materials.

10. Storage of Class 1.4G (Class C Common) fireworks.

Division 4. Repair garages not classified as Group S, Division 3 Occupancies.

Division 5. Aircraft repair hangars not classified as Group S, Division 5 Occupancies and heliports.

Division 6. Semiconductor fabrication facilities and comparable research and development areas in which hazardous production materials (HPM) are used and the aggregate quantity of materials are in excess of those listed in Table 7902.5-A, 7903.2-B, 8001.15-A or 8001.15-B.

Division 7. Occupancies having quantities of materials in excess of those listed in Table 3-E that are health hazards, including:

1. Corrosives.

> **EXCEPTION:** Stationary lead-acid battery systems.

2. Toxic and highly toxic materials.

3. Irritants.

4. Sensitizers.

5. Other health hazards.

Figure 1.32.0 (*Continued*)

Group I Occupancies:

Group I Occupancies shall be:

Division 1.1. Nurseries for the full-time care of children under the age of six (each accommodating more than five children).

Hospitals, sanitariums, nursing homes with nonambulatory patients and similar buildings (each accommodating more than five patients).

Division 1.2. Health-care centers for ambulatory patients receiving outpatient medical care which may render the patient incapable of unassisted self-preservation (each tenant space accommodating more than five such patients).

Figure 1.32.0 *(Continued)*

Division 2. Nursing homes for ambulatory patients, homes for children six years of age or over (each accommodating more than five patients or children).

Division 3. Mental hospitals, mental sanitariums, jails, prisons, reformatories and buildings where personal liberties of inmates are similarly restrained.

> **EXCEPTION:** Group I Occupancies shall not include buildings used only for private residential purposes for a family group.

Group M Occupancies:

Group M Occupancies shall include buildings, structures, or portions thereof, used for the display and sale of merchandise.

REFERENCE TABLES FROM THE *UNIFORM BUILDING CODE*

TABLE 3-A—DESCRIPTION OF OCCUPANCIES BY GROUP AND DIVISION

GROUP AND DIVISION	SECTION	DESCRIPTION OF OCCUPANCY[1]
A-1	303.1.1	A building or portion of a building having an assembly room with an occupant load of 1,000 or more and a legitimate stage.
A-2		A building or portion of a building having an assembly room with an occupant load of less than 1,000 and a legitimate stage.
A-2.1		A building or portion of a building having an assembly room with an occupant load of 300 or more without a legitimate stage, including such buildings used for educational purposes and not classed as a Group E or Group B Occupancy.
A-3		Any building or portion of a building having an assembly room with an occupant load of less than 300 without a legitimate stage, including such buildings used for educational purposes and not classed as a Group E or Group B Occupancy.
A-4		Stadiums, reviewing stands and amusement park structures not included within other Group A Occupancies.
B	304.1	A building or structure, or a portion thereof, for office, professional or service-type transactions, including storage of records and accounts, and eating and drinking establishments with an occupant load of less than 50.
E-1	305.1	Any building used for educational purposes through the 12th grade by 50 or more persons for more than 12 hours per week or four hours in any one day.
E-2		Any building used for educational purposes through the 12th grade by less than 50 persons for more than 12 hours per week or four hours in any one day.
E-3		Any building or portion thereof used for day-care purposes for more than six persons.
F-1	306.1	Moderate-hazard factory and industrial occupancies including factory and industrial uses not classified as Group F, Division 2 Occupancies.
F-2		Low-hazard factory and industrial occupancies including facilities producing noncombustible or nonexplosive materials which during finishing, packing or processing do not involve a significant fire hazard.
H–1	307.1	Occupancies with a quantity of material in the building in excess of those listed in Table 3-D which present a high explosion hazard as listed in Section 307.1.1.
H-2		Occupancies with a quantity of material in the building in excess of those listed in Table 3-D which present a moderate explosion hazard or a hazard from accelerated burning as listed in Section 307.1.1.
H-3		Occupancies with a quantity of material in the building in excess of those listed in Table 3-D which present a high fire or physical hazard as listed in Section 307.1.1.
H-4		Repair garages not classified as Group S, Division 3 Occupancies.
H-5		Aircraft repair hangars not classified as Group S, Division 5 Occupancies and heliports.
H-6	307.1 and 307.11	Semiconductor fabrication facilities and comparable research and development areas when the facilities in which hazardous production materials are used, and the aggregate quantity of material is in excess of those listed in Table 3-D or Table 3-E.
H-7	307.1	Occupancies having quantities of materials in excess of those listed in Table 3-E that are health hazards as listed in Section 307.1.1.
I-1.1	308.1	Nurseries for the full-time care of children under the age of six (each accommodating more than five children), hospitals, sanitariums, nursing homes with nonambulatory patients and similar buildings (each accommodating more than five patients).
I-1.2		Health-care centers for ambulatory patients receiving outpatient medical care which may render the patient incapable of unassisted self-preservation (each tenant space accommodating more than five such patients).
I-2		Nursing homes for ambulatory patients, homes for children six years of age or over (each accommodating more than five persons).
I-3		Mental hospitals, mental sanitariums, jails, prisons, reformatories and buildings where personal liberties of inmates are similarly restrained.
M	309.1	A building or structure, or a portion thereof, for the display and sale of merchandise, and involving stocks of goods, wares or merchandise, incidental to such purposes and accessible to the public.
R-1	310.1	Hotels and apartment houses, congregate residences (each accommodating more than 10 persons).
R-3		Dwellings, lodging houses, congregate residences (each accommodating 10 or fewer persons).
S-1	311.1	Moderate hazard storage occupancies including buildings or portions of buildings used for storage of combustible materials not classified as Group S, Division 2 or Group H Occupancies.
S-2		Low-hazard storage occupancies including buildings or portions of buildings used for storage of noncombustible materials.
S-3		Repair garages where work is limited to exchange of parts and maintenance not requiring open flame or welding and parking garages not classified as Group S, Division 4 Occupancies.
S-4		Open parking garages.
S-5		Aircraft hangars and helistops.
U-1	312.1	Private garages, carports, sheds and agricultural buildings.
U-2		Fences over 6 feet (182.9 mm) high, tanks and towers.

[1]For detailed descriptions, see the occupancy descriptions in the noted sections.

NOTE: UFC Section 216 contains complete descriptions.

Figure 1.32.1 Occupancies defined by group and description. (*Reproduced from the 1997 edition of the* Uniform Fire Code, *Volumes 1, 2, 3, Copyright © 1997, with the permission of the publisher, The International Fire Code Institute (IFCI). Although the International Fire Code Institute (IFI) has granted permission for reproduction of sections contained within the* Uniform Fire Code, *in conjunction with this publication, IFCI assumes no responsibility for the accuracy or completion of summaries provided therein.*)

TABLE 3-B—REQUIRED SEPARATION IN BUILDINGS OF MIXED OCCUPANCY[1] (HOURS)

	A-2	A-2.1	A-3	A-4	B	E	F-1	F-2	H-2	H-3	H-4,5	H-6,7[2]	I	M	R-1	R-3	S-1	S-2	S-3
A-1	N	N	N	N	3	N	3	3	4	4	4	4	3	3	1	1	3	3	4
A-2		N	N	N	1	N	1	1	4	4	4	4	3	1	1	1	1	1	3
A-2.1			N	N	1	N	1	1	4	4	4	4	3	1	1	1	1	1	3
A-3				N	N	N	N	N	4	4	4	3	2	N	1	1	N	N	3
A-4					1	N	1	1	4	4	4	4	3	1	1	1	1	1	3
B						1	N⁵	N	2	1	1	1	2	N	1	1	N	N	1
E							1	1	4	4	4	3	1	1	1	1	1	1	3
F-1								1	2	1	1	1	3	N⁵	1	1	N	N	1
F-2									2	1	1	1	2	1	1	1	N	N	1
H-1	NOT PERMITTED IN MIXED OCCUPANCIES. SEE CHAPTER 3.																		
H-2									1	1	2	4	2	4	4	2	2	2	
H-3										1	1	4	1	3	3	1	1	1	
H-4, 5											1	4	1	3	3	1	1	1	
H-6,7[2]												4	1	4	4	1	1	1	
I													2	1	1	2	2	4	
M														1	1	1⁴	1⁴	1	
R-1															N	3	1	3	
R-3																1	1	1	
S-1																	1	1	
S-2																		1	
S-3																			
S-4	OPEN PARKING GARAGES ARE EXCLUDED EXCEPT AS PROVIDED IN SECTION 311.2.																		
S-5																			

N—No requirements for fire resistance.

[1]For detailed requirements and exceptions, see Section 302.4.

[2]For special provisions on highly toxic materials, see the Fire Code.

[3]For agricultural buildings, see also Appendix Chapter 3.

[4]See Section 309.2.2 for exception.

[5]For Group F, Division 1 woodworking establishments with more than 2,500 square feet (232.3 m²), the occupancy separation shall be one hour

Figure 1.32.2 Required separation in building of mixed occupancy expressed in hours. (*Reproduced from the 1997 edition of the* Uniform Fire Code, *Volumes 1, 2, 3, Copyright © 1997, with the permission of the publisher, The International Fire Code Institute (IFCI). Although the International Fire Code Institute (IFI) has granted permission for reproduction of sections contained within the* Uniform Fire Code, *in conjunction with this publication, IFCI assumes no responsibility for the accuracy or completion of summaries provided therein.*)

ARTICLE 10 — FIRE-PROTECTION SYSTEMS AND EQUIPMENT

SECTION 1001 — GENERAL

1001.1 Scope. Fire-protection systems and equipment shall be in accordance with Article 10. See also Appendix II-C.

Fire-protection equipment and systems shall be installed and maintained in buildings under construction in accordance with Article 87.

1001.2 Definitions. For definitions of ALARM CONTROL UNIT, ALARM-INITIATING DEVICE, ALARM SIGNAL, ALARM-SIGNALING DEVICE, ALARM SYSTEM, ALARM ZONE, ANNUNCIATOR, AUTOMATIC FIRE-EXTINGUISH-ING SYSTEM, FACILITY, FIRE DEPARTMENT INLET CON-NECTION, SMOKE DETECTOR and STANDPIPE SYSTEM, see Article 2.

1001.3 Plans. Complete plans and specifications for fire alarm systems; fire-extinguishing systems, including automatic sprin-klers and wet dry standpipes; halon systems and other special types of automatic fire-extinguishing systems; basement pipe in-lets; and other fire-protection systems and appurtenances thereto shall be submitted to the fire department for review and approval prior to system installation. Plans and specifications for fire alarm systems shall include, but not be limited to, a floor plan; location of all alarm-initiating and alarm-signaling devices; alarm control-and trouble-signaling equipment; annunciation; power connec-tion; battery calculations; conductor type and sizes; voltage drop calculations; and manufacturer, model numbers and listing information for all equipment, devices and materials.

1001.4 Installation Acceptance Testing. Fire alarm systems; fire hydrant systems; fire-extinguishing systems, standpipes, and other fire-protection systems and appurtenances thereto shall meet the approval of the fire department as to installation and loca-tion and shall be subject to such acceptance tests as required by the chief.

Condition of acceptance of halon and clean agent systems shall be satisfactory passage of a test conducted in accordance with nationally recognized standards prior to final acceptance of the system.

Fire alarm and detection systems shall be tested in accordance with UFC Standard 10-2 and nationally recognized standards.

See Section 9003, Standard n.2.5.

1001.5 Maintenance, Inspection, Testing and Systems Out of Service.

1001.5.1 Maintenance. Fire sprinkler systems, fire hydrant sys-tems, standpipe systems, fire alarm systems, portable fire extin-guishers, smoke and heat ventilators, smoke-removal systems, and other fire protective or extinguishing systems or appliances shall be maintained in an operative condition at all times, and shall be replaced or repaired where defective.

Fire-protection or fire-extinguishing systems coverage and spacing shall be maintained according to original installation standards. Such systems shall be extended, altered or augmented as necessary to maintain and continue protection whenever any building so equipped is altered, remodeled or added to. Additions, repairs, alterations and servicing shall be in accordance with rec-ognized standards.

1001.5.2 Inspection and testing. The chief is authorized to re-quire periodic inspection and testing for fire sprinkler systems, fire hydrant systems, standpipe systems, fire alarm systems, por-table fire extinguishers, smoke and heat ventilators, smoke-removal systems and other fire-protection or fire-extinguishing systems or appliances.

Automatic fire-extinguishing systems shall be inspected and tested at least annually. See Appendix III-C. Fire alarm systems shall be inspected and tested at least at frequencies specified in UFC Standard 10-2. Standpipe systems shall be inspected and tested at least every five years.

> **EXCEPTIONS:** 1. Automatic fire-extinguishing equipment asso-ciated with commercial cooking operations when in compliance with Section 1006.
>
> 2. Systems in high-rise buildings when in compliance with Sec-tion 1001.5.4.

Reports of inspections and tests shall be maintained on the premises and made available to the chief when requested.

1001.5.3 Systems out of service. The chief shall be notified when any required fire-protection system is out of service and on restoration of service.

1001.5.3.1 Problematic systems and systems out of service. In the event of a failure of a fire-protection system or an excessive number of accidental activations, the chief is authorized to require the building owner or occupant to provide firewatch personnel un-til the system is repaired.

Such individuals shall be provided with at least one approved means for notification of the fire department and their only duty shall be to perform constant patrols of the protected premises and keep watch for fires.

1001.5.4 Systems in high-rise buildings. The owner of a high-rise building shall be responsible for assuring that the fire-and life-safety systems required by the Building Code are maintained in an operable condition at all times. Unless otherwise required by the chief, quarterly tests of such systems shall be conducted by approved persons. A written record shall be maintained and shall be made available to the inspection authority. (See UBC Section 403.)

1001.5.5 Smoke-control systems. Mechanical smoke-control systems, such as those in high-rise buildings, buildings containing atria, covered mall buildings and mechanical ventilation systems utilized in smokeproof enclosures and for smoke-removal systems utilized in high-piled combustible storage occupancies, shall be maintained in an operable condition at all times. Unless otherwise required by the chief, quarterly tests of such systems shall be conducted by approved persons. A written record shall be maintained and shall be made available to the inspection author-ity.

1001.6 Tampering with Fire-protection Equipment, Barri-ers, Security Devices, Signs and Seals.

1001.6.1 Fire department property. Apparatus, equipment and appurtenances belonging to or under the supervision and control of the fire department shall not be molested, tampered with, damaged or otherwise disturbed unless authorized by the chief.

1001.6.2 Fire hydrants and fire appliances. Fire hydrants and fire appliances required by this code to be installed or maintained shall not be removed, tampered with or otherwise disturbed except for the purpose of extinguishing fire, training, recharging or making necessary repairs, or when allowed by the fire department.

Figure 1.33.0 Uniform Fire Code fire protection systems and equipment guidelines. (*Reproduced from the 1997 edition of the Uni-form Fire Code, Volumes 1, 2, 3, Copyright © 1997, with the permission of the publisher, The International Fire Code Institute (IFCI). Although the International Fire Code Institute (IFI) has granted permission for reproduction of sections contained within the Uniform Fire Code, in conjunction with this publication, IFCI assumes no responsibility for the accuracy or com-pletion of summaries provided therein.*)

When a fire appliance is removed as herein allowed, it shall be replaced or reinstalled as soon as the purpose for which it was removed has been accomplished.

1001.6.3 Barriers, security devices, signs and seals. Locks, gates, doors, barricades, chains, enclosures, signs, tags or seals which have been installed by the fire department or by its order or under its control shall not be removed, unlocked, destroyed, tampered with or otherwise molested in any manner. See also Sections 103.4.3.3 and 902.2.4.2.

> **EXCEPTION:** When authorized by the chief or performed by public officers acting within their scope of duty.

1001.6.4 Fire alarms. See Sections 1007.1.2 and 1302.

1001.7 Obstruction and Impairment of Fire Hydrants and Fire-protection Equipment.

1001.7.1 General. Posts, fences, vehicles, growth, trash, storage and other materials or things shall not be placed or kept near fire hydrants, fire department inlet connections or fire-protection system control valves in a manner that would prevent such equipment or fire hydrants from being immediately discernible. The fire department shall not be deterred or hindered from gaining immediate access to fire-protection equipment or hydrants.

1001.7.2 Clear space around hydrants. A 3-foot (914.4 mm) clear space shall be maintained around the circumference of fire hydrants except as otherwise required or approved.

1001.7.3 Fire-extinguishing equipment. Class II standpipe hose stations, Class I and Class III standpipe outlets, and portable fire extinguishers shall not be concealed, obstructed or impaired.

1001.7.4 Fire alarm equipment. Alarm-initiating devices, alarm-signaling devices and annunciators shall not be concealed, obstructed or impaired.

1001.8 Marking of Fire-protection Equipment and Fire Hydrants. Fire-protection equipment and fire hydrants shall be clearly identified in an approved manner to prevent obstruction by parking and other obstructions. See also Section 901.4.3.

1001.9 Special Hazards. For occupancies of an especially hazardous nature or where special hazards exist in addition to the normal hazard of the occupancy, or where access for fire apparatus is unduly difficult, the chief is authorized to require additional safeguards consisting of additional fire appliance units, more than one type of appliance, or special systems suitable for the protection of the hazard involved. Such devices or appliances can consist of automatic fire alarm systems, automatic sprinkler or water spray systems, standpipe and hose, fixed or portable fire extinguishers, suitable fire blankets, breathing apparatus, manual or automatic covers, carbon dioxide, foam, halogenated or dry chemical or other special fire-extinguishing systems. Where such systems are provided, they shall be designed and installed in accordance with the applicable *Uniform Fire Code Standards.* See Article 90 and Section 101.3.

1001.10 Fire Appliances. The chief is authorized to designate the type and number of fire appliances to be installed and maintained in and upon all buildings and premises in the jurisdiction other than private dwellings. This designation shall be based on the relative severity of probable fire, including the rapidity with which it could spread. Such appliances shall be of a type suitable for the probable class of fire associated with such building or premises and shall have approval of the chief.

Figure 1.33.0 *(Continued)*

SECTION 1002 — PORTABLE FIRE EXTINGUISHERS

1002.1 General. Portable fire extinguishers shall be installed in occupancies and locations as set forth in this code and as required by the chief.

Portable fire extinguishers shall be in accordance with UFC Standard 10-1.

1002.2 Prohibited Types. Vaporizing liquid extinguishers containing carbon tetrachloride or chlorobromomethane shall not be installed or used in any location for fire-protection use.

Soda-acid, foam, loaded stream, antifreeze and water fire extinguishers of the inverting types shall not be recharged or placed in service for fire-protection use.

1002.3 Sale of Defective Fire Extinguishers. Forms, types or kinds of fire extinguishers which are not approved or which are not in proper working order, or the contents of which do not meet the requirements of this code, shall not be sold or traded.

> **EXCEPTION:** The sale or trade of fire extinguishers to a person engaged in the business of selling or handling such extinguishers, and the sale or exchange of obsolete or damaged equipment for junk.

SECTION 1003 — FIRE-EXTINGUISHING SYSTEMS

1003.1 Installation Requirements.

1003.1.1 General. Fire-extinguishing systems shall be installed in accordance with the Building Code and Section 1003.

Fire hose threads used in connection with fire-extinguishing systems shall be national standard hose thread or as approved.

The location of fire department hose connections shall be approved.

In buildings used for high-piled combustible storage, fire protection shall be in accordance with Article 81.

1003.1.2 Standards. Fire-extinguishing systems shall comply with the Building Code. (See UBC Standard 9-1.)

> **EXCEPTIONS:** 1. Automatic fire-extinguishing systems not covered by the Building Code shall be approved and installed in accordance with approved standards.
>
> 2. Automatic sprinkler systems may be connected to the domestic water-supply main when approved by the building official, provided the domestic water supply is of adequate pressure, capacity and sizing for the combined domestic and sprinkler requirements. In such case, the sprinkler system connection shall be made between the public water main or meter and the building shutoff valve, and there shall not be intervening valves or connections. The fire department connection may be omitted when approved.
>
> 3. Automatic sprinkler systems in Group R Occupancies four stories or less may be in accordance with the Building Code requirements for residential sprinkler systems. (See UBC Standard 9-3.)

1003.1.3 Modifications. When residential sprinkler systems as set forth in the Building Code (see UBC Standard 9-3) are provided, exceptions to, or reductions in, Building Code requirements based on the installation of an automatic fire-extinguishing system are not allowed.

1003.2 Required Installations.

1003.2.1 General. An automatic fire-extinguishing system shall be installed in the occupancies and locations as set forth in Section 1003.2.

For provisions on special hazards and hazardous materials, see Section 1001.9 and Articles 79, 80 and 81.

1003.2.2 All occupancies except Group R, Division 3 and Group U Occupancies. Except for Group R, Division 3 and

Group U Occupancies, an automatic sprinkler system shall be installed:

1. In every story or basement of all buildings when the floor area exceeds 1,500 square feet (139.4 m²) and there is not provided at least 20 square feet (1.86 m²) of opening entirely above the adjoining ground level in each 50 lineal feet (15 240 mm) or fraction thereof of exterior wall in the story or basement on at least one side of the building. Openings shall have a minimum dimension of not less than 30 inches (762 mm). Such openings shall be accessible to the fire department from the exterior and shall not be obstructed in a manner that firefighting or rescue cannot be accomplished from the exterior.

When openings in a story are provided on only one side and the opposite wall of such story is more than 75 feet (22 860 mm) from such openings, the story shall be provided with an approved automatic sprinkler system, or openings as specified above shall be provided on at least two sides of an exterior wall of the story.

If any portion of a basement is located more than 75 feet (22 860 mm) from openings required in Section 1003.2.2, the basement shall be provided with an approved automatic sprinkler system.

2. At the top of rubbish and linen chutes and in their terminal rooms. Chutes extending through three or more floors shall have additional sprinkler heads installed within such chutes at alternate floors. Sprinkler heads shall be accessible for servicing.

3. In rooms where nitrate film is stored or handled. See also Article 33.

4. In protected combustible fiber storage vaults as defined in Article 2. See also Article 28.

5. Throughout all buildings with a floor level with an occupant load of 30 or more that is located 55 feet (16 764 mm) or more above the lowest level of fire department vehicle access.

> **EXCEPTION:** 1. Airport control towers.
>
> 2. Open parking structures.
>
> 3. Group F, Division 2 Occupancies.

1003.2.3 Group A Occupancies.

1003.2.3.1 Drinking establishments. An automatic sprinkler system shall be installed in rooms used by the occupants for the consumption of alcoholic beverages and unseparated accessory uses where the total area of such unseparated rooms and assembly uses exceeds 5,000 square feet (465 m²). For uses to be considered as separated, the separation shall not be less than as required for a one-hour occupancy separation. The area of other uses shall be included unless separated by at least a one-hour occupancy separation.

1003.2.3.2 Basements. An automatic sprinkler system shall be installed in basements classified as a Group A Occupancy when the basement is larger than 1,500 square feet (139 m²) in floor area.

1003.2.3.3 Exhibition and display rooms. An automatic sprinkler system shall be installed in Group A Occupancies which have more than 12,000 square feet (1114.8 m²) of floor area which can be used for exhibition or display purposes.

1003.2.3.4 Stairs. An automatic sprinkler system shall be installed in enclosed usable space below or over a stairway in Group A, Divisions 2, 2.1, 3 and 4 Occupancies.

1003.2.3.5 Multitheater complexes. An automatic sprinkler system shall be installed in every building containing a multitheater complex.

1003.2.3.6 Amusement buildings. An automatic sprinkler system shall be installed in all amusement buildings. The main waterflow switch shall be electrically supervised. The sprinkler main cutoff valve shall be supervised. When the amusement building is temporary, the sprinkler water-supply system may be of an approved temporary type.

> **EXCEPTION:** An automatic sprinkler system need not be provided when the floor area of a temporary amusement building is less than 1,000 square feet (92.9 m²) and the exit travel distance from any point is less than 50 feet (15 240 mm).

1003.2.3.7 Stages. All stages shall be sprinklered. Such sprinklers shall be provided throughout the stage and in dressing rooms, workshops, storerooms and other accessory spaces contiguous to such stages.

> **EXCEPTIONS:** 1. Sprinklers are not required for stages 1,000 square feet (92.9 m²) or less in area and 50 feet (15 240 mm) or less in height where curtains, scenery or other combustible hangings are not retractable vertically. Combustible hangings shall be limited to a single main curtain, borders, legs and a single backdrop.
>
> 2. Under stage areas less than 4 feet (1219 mm) in clear height used exclusively for chair or table storage and lined on the inside with ⁵/₈-inch (16 mm) Type X gypsum wallboard or an approved equal.

1003.2.3.8 Smoke-protected assembly seating. All areas enclosed with walls and ceilings in buildings or structures containing smoke-protected assembly seating shall be protected with an approved automatic sprinkler system.

> **EXCEPTION:** Press boxes and storage facilities less than 1,000 square feet (92.9 m²) in area and in conjunction with outdoor seating facilities where all means of egress in the seating area are essentially open to the outside.

1003.2.4 Group E Occupancies.

1003.2.4.1 General. An automatic fire sprinkler system shall be installed throughout all buildings containing a Group E, Division 1 Occupancy.

> **EXCEPTIONS:** 1. When each room used for instruction has at least one exterior exit door at ground level and when rooms used for assembly purposes have at least one half of the required exits directly to the exterior ground level, a sprinkler system need not be provided.
>
> 2. When area separation walls, or occupancy separations having a fire-resistive rating of not less than two hours subdivide the building into separate compartments such that each compartment contains an aggregate floor area not greater than 20,000 square feet (1858 m²), an automatic sprinkler system need not be provided.

1003.2.4.2 Basements. An automatic sprinkler system shall be installed in basements classified as Group E, Division 1 Occupancies.

1003.2.4.3 Stairs. An automatic fire sprinkler system shall be installed in enclosed usable space below or over a stairway in Group E, Division 1 Occupancies.

1003.2.5 Group F Occupancies.

1003.2.5.1 Woodworking occupancies. An automatic fire sprinkler system shall be installed in Group F woodworking occupancies over 2,500 square feet (232.3 m²) in area that use equipment, machinery or appliances which generate finely divided combustible waste or which use finely divided combustible materials.

1003.2.6 Group H Occupancies.

1003.2.6.1 General. An automatic fire-extinguishing system shall be installed in Group H, Divisions 1, 2, 3 and 7 Occupancies.

1003.2.6.2 Group H, Division 4 Occupancies. An automatic fire-extinguishing system shall be installed in Group H, Division 4

Figure 1.33.0 *(Continued)*

Occupancies having a floor area of more than 3,000 square feet (279 m²).

1003.2.6.3 Group H, Division 6 Occupancies. An automatic fire-extinguishing system shall be installed throughout buildings containing Group H, Division 6 Occupancies. The design of the sprinkler system shall not be less than that required under the Building Code (see UBC Standard 9-1) for the occupancy hazard classifications as follows:

LOCATION	OCCUPANCY HAZARD CLASSIFICATION
Fabrication areas	Ordinary Hazard Group 2
Service corridors	Ordinary Hazard Group 2
Storage rooms without dispensing	Ordinary Hazard Group 2
Storage rooms with dispensing	Extra Hazard Group 2
Corridors	Ordinary Hazard Group 2[1]

[1]When the design area of the sprinkler system consists of a corridor protected by one row of sprinklers, the maximum number of sprinklers that needs to be calculated is 13.

1003.2.7 Group I Occupancies. An automatic sprinkler system shall be installed in Group I Occupancies. In Group I, Division 1.1 and Group I, Division 2 Occupancies, approved quick-response or residential sprinklers shall be installed throughout patient sleeping areas.

> **EXCEPTION:** In jails, prisons and reformatories, the piping system may be dry, provided a manually operated valve is installed at a continuously monitored location. Opening of the valve will cause the piping system to be charged. Sprinkler heads in such systems shall be equipped with fusible elements or the system shall be designed as required for deluge systems in the Building Code (see UBC Standard 9-1).

1003.2.8 Group M Occupancies. An automatic sprinkler system shall be installed in rooms classed as Group M Occupancies where the floor area exceeds 12,000 square feet (1114.8 m²) on any floor or 24,000 square feet (2229.6 m²) on all floors or in Group M Occupancies more than three stories in height. The area of mezzanines shall be included in determining the areas where sprinklers are required.

1003.2.9 Group R, Division 1 Occupancies. An automatic sprinkler system shall be installed throughout every apartment house three or more stories in height or containing 16 or more dwelling units, every congregate residence three or more stories in height or having an occupant load of 20 or more, and every hotel three or more stories in height or containing 20 or more guest rooms. Residential or quick-response standard sprinklers shall be used in the dwelling units and guest room portions of the building.

1003.3 Sprinkler System Monitoring and Alarms.

1003.3.1 Where required. All valves controlling the water supply for automatic sprinkler systems and water-flow switches on all sprinkler systems shall be electrically monitored where the number of sprinklers are:

1. Twenty or more in Group I, Divisions 1.1 and 1.2 Occupancies.

2. One hundred or more in all other occupancies.

Valve monitoring and water-flow alarm and trouble signals shall be distinctly different and shall be automatically transmitted to an approved central station, remote station or proprietary monitoring station as defined by UFC Standard 10-2 or, when approved by the building official with the concurrence of the chief, shall sound an audible signal at a constantly attended location.

> **EXCEPTION:** Underground key or hub valves in roadway boxes provided by the municipality or public utility need not be monitored.

1003.3.2 Alarms. An approved audible sprinkler flow alarm shall be provided on the exterior of the building in an approved location. An approved audible sprinkler flow alarm to alert the occupants shall be provided in the interior of the building in a normally occupied location. Actuation of the alarm shall be as set forth in the Building Code. (See UBC Standard 9-1.)

1003.4 Permissible Sprinkler Omissions. Subject to the approval of the building official and with the concurrence of the chief, sprinklers may be omitted in rooms or areas as follows:

1. When sprinklers are considered undesirable because of the nature of the contents or in rooms or areas which are of noncombustible construction with wholly noncombustible contents and which are not exposed by other areas. Sprinklers shall not be omitted from any room merely because it is damp, of fire-resistive construction or contains electrical equipment.

2. Sprinklers shall not be installed when the application of water or flame and water to the contents may constitute a serious life or fire hazard, as in the manufacture or storage of quantities of aluminum powder, calcium carbide, calcium phosphide, metallic sodium and potassium, quicklime, magnesium powder and sodium peroxide.

3. Safe deposit or other vaults of fire-resistive construction, when used for the storage of records, files and other documents, when stored in metal cabinets.

4. Communication equipment areas under the exclusive control of a public communication utility agency, provided:

> 4.1 The equipment areas are separated from the remainder of the building by one-hour fire-resistive occupancy separation; and
>
> 4.2 Such areas are used exclusively for such equipment; and
>
> 4.3 An approved automatic smoke-detection system is installed in such areas and is supervised by an approved central, proprietary or remote station service or a local alarm which will give an audible signal at a constantly attended location; and
>
> 4.4 Other approved fire-protection equipment such as portable fire extinguishers or Class II standpipes are installed in such areas.

5. Other approved automatic fire-extinguishing systems may be installed to protect special hazards or occupancies in lieu of automatic sprinklers.

SECTION 1004 — STANDPIPES

1004.1 Installation Requirements.

1004.1.1 General. Standpipe systems shall be installed in accordance with the Building Code and Section 1004.

Fire hose threads used in connection with fire-extinguishing systems shall be national standard hose thread or as approved.

The location of fire department hose connections shall be approved.

In buildings used for high-piled combustible storage, fire protection shall be in accordance with Article 81.

1004.1.2 Standards. Standpipe systems shall comply with the Building Code. (See UBC Standard 9-2.)

1004.2 Required Installations. Standpipe systems shall be provided as set forth in Table 1004-A.

1004.3 Location of Class I Standpipe Hose Connections. There shall be a Class I standpipe outlet connection at ev-

Figure 1.33.0 *(Continued)*

ery floor-level landing of every required stairway above or below grade and on each side of the wall adjacent to the exit opening of a horizontal exit. Outlets at stairways shall be located within the exit enclosure or, in the case of pressurized enclosures, within the vestibule or exterior balcony, giving access to the stairway.

Risers and laterals of Class I standpipe systems not located within an enclosed stairway or pressurized enclosure shall be protected by a degree of fire resistance equal to that required for vertical enclosures in the building in which they are located.

> **EXCEPTION:** In buildings equipped with an approved automatic sprinkler system, risers and laterals which are not located within an enclosed stairway or pressurized enclosure need not be enclosed within fire-resistive construction.

There shall be at least one outlet above the roof line when the roof has a slope of less than 4 units vertical in 12 units horizontal (33.3% slope).

In buildings where more than one standpipe is provided, the standpipes shall be interconnected at the bottom.

1004.4 Location of Class II Standpipe Hose Connections. Class II standpipe outlets shall be accessible and shall be located so that all portions of the building are within 30 feet (9144 mm) of a nozzle attached to 100 feet (30 480 mm) of hose.

In Group A, Divisions 1 and 2.1 Occupancies with occupant loads of more than 1,000, outlets shall be located on each side of any stage, on each side of the rear of the auditorium and on each side of the balcony.

Fire-resistant protection of risers and laterals of Class II standpipe systems is not required.

1004.5 Location of Class III Standpipe Hose Connections. Class III standpipe systems shall have outlets located as required for Class I standpipes in Section 1004.3 and shall have Class II outlets as required in Section 1004.4.

Risers and laterals of Class III standpipe systems shall be protected as required for Class I systems.

> **EXCEPTIONS:** 1. In buildings equipped with an approved automatic sprinkler system, risers and laterals which are not located within an enclosed stairway or pressurized enclosure need not be enclosed within fire-resistive construction.
>
> 2. Laterals for Class II outlets on Class III systems need not be protected.

In buildings where more than one Class III standpipe is provided, the standpipes shall be interconnected at the bottom.

SECTION 1005 — BASEMENT PIPE INLETS

Basement pipe inlets shall be installed in the first floor of every store, warehouse or factory having a basement when required by the Building Code. See Appendix III-D.

SECTION 1006 — PROTECTION OF COMMERCIAL COOKING OPERATIONS

1006.1 Ventilating Hood and Duct Systems. A ventilating hood and duct system shall be provided in accordance with the Mechanical Code for commercial-type food heat-processing equipment that produces grease-laden vapors.

1006.2 Fire-extinguishing System.

1006.2.1 Where required. Approved automatic fire-extinguishing systems shall be provided for the protection of commercial-type cooking equipment.

> **EXCEPTION:** The requirement for protection does not include steam kettles and steam tables or equipment which as used does not create grease-laden vapors.

1006.2.2 Type of system. The system used for the protection of commercial-type cooking equipment shall be either a system listed for application with such equipment or an automatic fire-extinguishing system that is specifically designed for such application.

Systems shall be installed in accordance with the Mechanical Code, their listing and the manufacturer's instruction. Other systems shall be of an approved design and shall be of one of the following types:

1. Automatic sprinkler system.

2. Dry-chemical extinguishing system.

3. Carbon dioxide extinguishing system.

4. Wet-chemical extinguishing system.

1006.2.3 Extent of protection.

1006.2.3.1 General. The automatic fire-extinguishing system used to protect ventilating hoods and ducts and cooking appliances shall be installed to include cooking surfaces, deep fat fryers, griddles, upright broilers, charbroilers, range tops and grills. Protection shall also be provided for the enclosed plenum space within the hood above filters and exhaust ducts serving the hood.

1006.2.3.2 Carbon dioxide systems. When carbon dioxide systems are used, there shall be a nozzle at the top of the ventilating duct. Additional nozzles that are symmetrically arranged to give uniform distribution shall be installed within vertical ducts exceeding 20 feet (6096 mm) and horizontal ducts exceeding 50 feet (15 240 mm). Dampers shall be installed at either the top or the bottom of the duct and shall be arranged to operate automatically upon activation of the fire-extinguishing system. When the damper is installed at the top of the duct, the top nozzle shall be immediately below the damper. Carbon dioxide automatic fire-extinguishing systems shall be sufficiently sized to protect all hazards venting through a common duct simultaneously.

1006.2.4 Automatic power, fuel and ventilation shutoff.

1006.2.4.1 General. Automatic fire-extinguishing systems shall be interconnected to the fuel or current supply for cooking equipment. The interconnection shall be arranged to automatically shut off all cooking equipment and electrical receptacles which are located under the hood when the system is actuated.

Shutoff valves or switches shall be of a type that require manual operation to reset.

1006.2.4.2 Carbon dioxide systems. Commercial-type cooking equipment protected by an automatic carbon dioxide extinguishing system shall be arranged to shut off the ventilation system upon activation.

1006.2.5 Special provisions for automatic sprinkler systems. Commercial-type cooking equipment protected by automatic sprinkler systems shall be supplied from a separate, readily accessible indicating-type control valve that is identified.

Sprinklers used for the protection of fryers shall be listed for that application and installed in accordance with their listing.

1006.2.6 Manual system operation. A readily accessible manual activation device installed at an approved location shall be provided for dry chemical, wet chemical and carbon dioxide systems. The activation device is allowed to be mechanically or electrically operated. If electrical power is used, the system shall be

Figure 1.33.0 *(Continued)*

connected to a standby power system and a visual means shall be provided to show that the extinguishing system is energized. Instructions for operating the fire-extinguishing system shall be posted adjacent to manual activation devices.

1006.2.7 Portable fire extinguishers. A sodium bicarbonate or potassium bicarbonate dry-chemical-type portable fire extinguisher having a minimum rating of 40-B shall be installed within 30 feet (9144 mm) of commercial food heat-processing equipment, as measured along an unobstructed path of travel, in accordance with UFC Standard 10-1.

1006.2.8 Operations and maintenance. The ventilation system in connection with hoods shall be operated at the required rate of air movement, and classified grease filters shall be in place when equipment under a kitchen grease hood is used.

If grease extractors are installed, they shall be operated when the commercial-type cooking equipment is used.

Hoods, grease-removal devices, fans, ducts and other appurtenances shall be cleaned at intervals necessary to prevent the accumulation of grease. Cleanings shall be recorded, and records shall state the extent, time and date of cleaning. Such records shall be maintained on the premises.

Extinguishing systems shall be serviced at least every six months or after activation of the system. Inspection shall be by qualified individuals, and a Certificate of Inspection shall be forwarded to the chief upon completion.

Fusible links and automatic sprinkler heads shall be replaced at least annually, and other protection devices shall be serviced or replaced in accordance with the manufacturer's instructions.

> **EXCEPTION:** Frangible bulbs need not be replaced annually.

SECTION 1007 — FIRE ALARM SYSTEMS

1007.1 General.

1007.1.1 Applicability. Installation and maintenance of fire alarm systems shall be in accordance with Section 1007.

1007.1.2 Problematic systems and systems out of service. See Section 1001.5.3.

1007.2 Required Installations.

1007.2.1 General.

1007.2.1.1 When required. An approved manual, automatic or manual and automatic fire alarm system shall be provided in accordance with Section 1007.2.

1007.2.1.2 Use of area separation walls to define separate buildings. For the purposes of Section 1007, area separation walls shall not define separate buildings.

1007.2.2 Group A Occupancies.

1007.2.2.1 General. Group A, Divisions 1, 2 and 2.1 Occupancies shall be provided with a manual fire alarm system in accordance with Section 1007.2.2.

> **EXCEPTIONS:** 1. Manual fire alarm boxes are not required when an approved automatic fire-extinguishing system is installed which will immediately activate the prerecorded announcement upon water flow.
>
> 2. Group A Occupancy portions of Group E Occupancies are allowed to have alarms as required for the Group E Occupancy.

See also Section 1007.2.12.

1007.2.2.2 System initiation in Group A Occupancies with an occupancy load of 1,000 or more. Activation of the fire alarm in

Group A Occupancies with an occupancy load of 1,000 or more shall immediately initiate an approved prerecorded message announcement using an approved voice communication system in accordance with UFC Standard 10-2 that is audible above the ambient noise level of the occupancy.

> **EXCEPTION:** When approved, the prerecorded announcement is allowed to be manually deactivated for a period of time, not to exceed three minutes, for the sole purpose of allowing a live voice announcement from an approved, constantly attended station.

1007.2.2.3 Emergency power. Voice communication systems shall be provided with an approved emergency power source.

1007.2.3 Group B Occupancies. See Section 1007.2.12.

1007.2.4 Group E Occupancies.

1007.2.4.1 General. Group E Occupancies shall be provided with fire alarm systems in accordance with Section 1007.2.4. Group E, Division 1 Occupancies and Group E, Division 3 Occupancies having an occupant load of 50 or more shall be provided with an approved manual fire alarm system. When automatic sprinkler systems or smoke detectors provided in accordance with Section 1007.2.4.2 are installed, such systems or detectors shall be connected to the building fire alarm system, and the building fire alarm system shall be both automatic and manual. See also Section 1007.2.12.

1007.2.4.2 Smoke detectors.

1007.2.4.2.1 Increased travel distance. Smoke detectors shall be installed when required by the Building Code for increases in travel distance to exits. (See UBC Section 1007.3.)

1007.2.4.2.2 Travel through adjoining rooms. Smoke detectors shall be installed when required by the Building Code to allow the only means of egress from a room to be through adjoining or intervening rooms. (See UBC Section 1007.3.)

1007.2.4.3 Exterior alarm-signaling device. An alarm-signaling device shall be mounted on the exterior of the building.

1007.2.5 Group F Occupancies. See Section 1007.2.12.

1007.2.6 Group H Occupancies.

1007.2.6.1 General. Group H Occupancies shall be provided with fire alarm systems in accordance with Section 1007.2.6. See also Section 1007.2.12.

1007.2.6.2 Organic coatings. Organic coating manufacturing uses shall be provided with a manual fire alarm system. See Article 50.

1007.2.6.3 Group H, Division 6 Occupancies. Group H, Division 6 Occupancies shall be provided with a manual fire alarm system. See Article 51.

1007.2.6.4 Rooms used for storage, dispensing, use and handling of hazardous materials. When required by Article 80, rooms or areas used for storage, dispensing, use or handling of highly toxic compressed gases, liquid and solid oxidizers, and Class I, II, III or IV organic peroxides shall be provided with an automatic smoke-detection system.

1007.2.7 Group I Occupancies.

1007.2.7.1 Divisions 1.1, 1.2 and 2 Occupancies.

1007.2.7.1.1 System requirements. Group I, Divisions 1.1, 1.2 and 2 Occupancies shall be provided with an approved manual and automatic fire alarm system in accordance with Section 1007.2.7.1. See also Section 1007.2.12. Smoke detectors shall be provided in accordance with the Building Code as follows:

Figure 1.33.0 *(Continued)*

1. At automatic-closing doors in smoke barriers and one-hour fire-resistive occupancy separations (see UBC Sections 308.2.2.1 and 308.8),

2. In waiting areas which are open to corridors (see UBC Section 1007.5).

When actuated, alarm-initiating devices shall activate an alarm signal which is audible throughout the building.

> **EXCEPTION:** Visual alarm-signaling devices are allowed to substitute for audible devices in patient use areas.

1007.2.7.1.2 Patient room smoke detectors. Smoke detectors which receive their primary power from the building wiring shall be installed in patient sleeping rooms of hospital and nursing homes. Actuation of such detectors shall cause a visual display on the corridor side of the room in which the detector is located and shall cause an audible and visual alarm at the respective nurses' station. When single-station detectors and related devices are combined with a nursing call system, the nursing call system shall be listed for the intended combined use.

> **EXCEPTION:** In rooms equipped with automatic door closers having integral smoke detectors on the room side, the integral detector may substitute for the room smoke detector, provided it performs the required alerting functions.

1007.2.7.2 Division 3 Occupancies.

1007.2.7.2.1 General. Group I, Division 3 Occupancies shall be provided with a manual and automatic fire alarm system installed for alerting staff in accordance with Section 1007.2.7.2. See also Section 1007.2.12.

1007.2.7.2.2 System initiation. Actuation of an automatic fire-extinguishing system, a manual fire alarm box or a fire detector shall initiate an approved fire alarm signal which automatically notifies staff. Presignal systems shall not be used.

1007.2.7.2.3 Manual fire alarm boxes.

1. **General.** Manual fire alarm boxes need not be located in accordance with Section 1007.3.3.1 when they are provided at staff-attended locations having direct supervision over areas where manual fire alarm boxes have been omitted.

2. **Locking of manual fire alarm boxes.** Manual fire alarm boxes are allowed to be locked in areas occupied by detainees, provided that staff members are present within the subject area and have keys readily available to operate the manual fire alarm boxes.

1007.2.7.2.4 Smoke detection. An approved automatic smoke-detection system shall be installed throughout resident housing areas, including sleeping areas and contiguous day rooms, group activity spaces and other common spaces normally accessible to residents.

> **EXCEPTION:** Other approved smoke-detection arrangements providing equivalent protection, such as placing detectors in exhaust ducts from cells or behind protective grilles, are allowed when necessary to prevent damage or tampering.

1007.2.7.2.5 Zoning and annunciation. Alarm and trouble signals shall be annunciated at an approved constantly attended location. Such signals shall indicate the zone of origin.

Separate zones shall be provided for individual fire-protection systems, buildings, floors, cell complexes and sections of floors compartmented by smoke-stop partitions.

1007.2.7.2.6 Monitoring. The fire alarm system shall be monitored by an approved central, proprietary or remote station service or by transmission of a local alarm which will give audible and visual signals at an approved constantly attended location.

1007.2.8 Group M Occupancies. See Section 1007.2.12.

1007.2.9 Group R Occupancies.

1007.2.9.1 New Group R Occupancies.

1007.2.9.1.1 General. Group R Occupancies shall be provided with fire alarm systems in accordance with Section 1007.2.9. Group R, Division 1 Occupancies shall be provided with a manual and automatic fire alarm system in apartment houses three or more stories in height or containing 16 or more dwelling units, in hotels three or more stories in height or containing 20 or more guest rooms, and in congregate residences three or more stories in height or having an occupant load of 20 or more. See also Section 1007.2.12.

> **EXCEPTIONS:** 1. A manual fire alarm system need not be provided in buildings not over two stories in height when all individual dwelling units and contiguous attic and crawl spaces are separated from each other and public or common areas by at least one-hour fire-resistive occupancy separations and each individual dwelling unit or guest room has an exit directly to a public way, exit court or yard.
>
> 2. A separate fire alarm system need not be provided in buildings which are protected throughout by an approved supervised fire sprinkler system conforming with the Building Code and having a local alarm to notify all occupants.

1007.2.9.1.2 Manual fire alarm boxes. Manual fire alarm boxes are not required for interior corridors having smoke detectors as specified in Section 1007.2.9.1.3.

1007.2.9.1.3 Smoke detectors. Smoke detectors shall be provided in all common areas and interior corridors serving as a required means of egress for an occupant load of 10 or more.

1007.2.9.1.4 Heat detectors. Heat detectors shall be provided in common areas such as recreational rooms, laundry rooms, furnace rooms, and similar areas in accordance with UFC Standard 10-2.

1007.2.9.1.5 Visual signaling devices. Guest rooms for persons with hearing impairments shall be provided with visible and audible alarm-indicating appliances, activated by both the in-room smoke detector and the building fire alarm system.

1007.2.9.1.6 Single-station smoke detectors. Approved single-station smoke detectors shall be installed in dwelling units, congregate residences and hotel or lodging house guest rooms in accordance with the Building Code.

Single-station smoke detectors shall not be connected to a fire alarm system. See also Section 1007.2.9.1.5.

> **EXCEPTION:** Connection of such detectors for annunciation only.

1007.2.9.2 Existing Group R Occupancies.

1007.2.9.2.1 General. Existing Group R Occupancies not already provided with single-station smoke dectors shall be provided with approved single-station smoke detectors. Installation shall be in accordance with Section 1007.2.9.2.

1007.2.9.2.2 Installation. Approved single-station smoke detectors shall be installed in existing dwelling units, congregate residences, and hotel and lodging house guest rooms.

1007.2.9.2.3 Power source. In Group R Occupancies, single-station smoke detectors shall be either battery operated or may receive their primary power from the building wiring provided that such wiring is served from a commercial source. When power is provided by the building wiring, the wiring shall be permanent and without a disconnecting switch other than those required for overcurrent protection.

1007.2.9.2.4 Locations within existing Group R Occupancies. In dwelling units, detectors shall be mounted on the ceiling

Figure 1.33.0 *(Continued)*

or wall at a point centrally located in the corridor or area giving access to each separate sleeping area. Where sleeping rooms are on an upper level, the detector shall be placed at the center of the ceiling directly above the stairway. Detectors shall also be installed in the basement of dwelling units having a stairway which opens from the basement into the dwelling.

In hotel, lodging house and congregate residence sleeping rooms, detectors shall be located on the ceiling or wall of each sleeping room.

1007.2.10 Group S Occupancies. See Section 1007.2.12.

1007.2.11 Group U Occupancies. No requirements.

1007.2.12 Special uses and conditions.

1007.2.12.1 Amusement buildings.

1007.2.12.1.1 General. An approved smoke-detection system shall be provided in amusement buildings in accordance with Section 1007.2.12.1.

> **EXCEPTION:** In areas where ambient conditions will cause a smoke-detection system to alarm, an approved alternate type of automatic detector shall be installed.

1007.2.12.1.2 Alarm system. Activation of any single smoke detector, the automatic sprinkler system or other automatic fire-detection device shall immediately sound an alarm in the building at a constantly supervised location from which the manual operation of systems noted in Section 1007.2.12.1.3 can be initiated.

1007.2.12.1.3 System response. The activation of two or more smoke detectors, a single smoke detector monitored by an alarm verification zone, the automatic sprinkler system or other approved fire-detection device shall automatically:

1. Stop confusing sounds and other visual effects,

2. Activate approved directional exit marking, and

3. Cause illumination of the means of egress with light of not less than 1 footcandle (10.8 lx) at the walking surface.

1007.2.12.1.4 Public address system. The public address system is also allowed to serve as an alarm.

1007.2.12.2 High-rise buildings.

1007.2.12.2.1 General. Group B office buildings and Group R, Division 1 Occupancies, each having floors used for human occupancy located more than 75 feet (22 860 mm) above the lowest level of fire department vehicle access, shall be provided with an automatic fire alarm system and a communication system in accordance with Section 1007.2.12.2.

1007.2.12.2.2 Automatic fire alarm system. Smoke detectors shall be provided in accordance with Section 1007.2.12.2.2. Smoke detectors shall be connected to an automatic fire alarm system. The actuation of any detector required by Section 1007.2.12.2.2 shall operate the emergency voice alarm-signaling system and shall place into operation all equipment necessary to prevent the recirculation of smoke. Smoke detectors shall be located as follows:

1. In every mechanical equipment, electrical, transformer, telephone equipment, elevator machine or similar room, and in elevator lobbies. Elevator lobby detectors shall be connected to an alarm verification zone or be listed as a releasing device;

2. In the main return-air and exhaust-air plenum of each air-conditioning system. Such detectors shall be located in a serviceable area downstream of the last duct inlet;

3. At each connection to a vertical duct or riser serving two or more stories from a return-air duct or plenum of an air-conditioning system. In Group R, Division 1 Occupancies, an approved smoke detector is allowed to be used in each return-air riser carrying not more than 5,000 cubic feet per minute (2360 L/s) and serving not more than 10 air-inlet openings; and

4. For Group R, Division 1 Occupancies, in all interior corridors serving as a means of egress for an occupant load of 10 or more.

1007.2.12.2.3 Emergency voice alarm-signaling system. The operation of any automatic fire detector, sprinkler or water-flow device shall automatically sound an alert tone followed by voice instructions giving appropriate information and directions on a general or selective basis to the following terminal areas:

1. Elevators,

2. Elevator lobbies,

3. Corridors,

4. Exit stairways,

5. Rooms and tenant spaces exceeding 1,000 square feet (93 m^2) in area,

6. Dwelling units in apartment houses,

7. Hotel guest rooms or suites, and

8. Areas of refuge. (As defined in the Building Code.)

A manual override for emergency voice communication shall be provided for all paging zones.

The emergency voice alarm-signaling system shall be designed and installed in accordance with the Building Code and UFC Standard 10-2.

1007.2.12.2.4 Fire department communication system. A two-way, approved fire department communication system shall be provided for fire department use. It shall operate between the central control station and elevators, elevator lobbies, emergency and standby power rooms and at entries into enclosed stairways.

1007.2.12.3 Buildings with atriums. Actuation of an atrium smoke-control system required by the Building Code shall initiate an audible fire alarm signal in designated portions of the building.

1007.2.12.4 High-piled combustible storage uses. When required by Article 81, high-piled combustible storage uses shall be provided with an automatic fire-detection system.

1007.2.12.5 Special egress-control devices. When special egress-control devices are installed on exit doors, an automatic smoke-detection system shall be installed throughout the building. (See UBC Section 1003.3.1.10.)

1007.2.12.6 Corridors in office uses. When required by the Building Code for corridors in lieu of one-hour corridor construction, smoke detectors shall be installed within office corridors in accordance with their listing. The actuation of any detector shall activate alarms audible in all areas served by the corridor. (See UBC Section 1004.3.4.3, Exception 4.)

1007.2.12.7 Aerosol storage uses. When required by Article 88, aerosol storage rooms and general purpose warehouses containing aerosols shall be provided with an approved manual alarm system.

1007.2.12.8 Smoke-control systems. An approved automatic smoke-detection system shall be provided when required by the Building Code for automatic control of a smoke-control system. (See UBC Section 905.9.)

Figure 1.33.0 *(Continued)*

1007.2.12.9 Lumber, plywood and veneer mills. Lumber, plywood and veneer mills shall be provided with a manual fire alarm system. See Section 3004.5.2.

1007.3 General System Design and Installation Requirements.

1007.3.1 Design standards. Fire alarm systems, automatic fire detectors, emergency voice alarm communication systems and notification devices shall be designed, installed and maintained in accordance with UFC Standard 10-2 and other nationally recognized standards.

1007.3.2 Equipment. Systems and components shall be listed and approved for the purpose for which they are installed.

1007.3.3 System layout and operation.

1007.3.3.1 Manual fire alarm boxes. When a manual fire alarm system is required, manual fire alarm boxes shall be distributed throughout so that they are readily accessible, unobstructed, and are located in the normal path of exit travel from the area and as follows:

1. At every exit from every level.

2. Additional fire alarm boxes shall be located so that travel distance to the nearest box does not exceed 200 feet (60 960 mm).

When fire alarm systems are not monitored, an approved permanent sign that reads LOCAL ALARM ONLY—CALL FIRE DEPARTMENT shall be installed adjacent to each manual fire alarm box.

> **EXCEPTION:** Separate signs need not be provided when the manufacturer has permanently provided this information on the manual fire alarm box.

1007.3.3.2 Control units, annunciator panels and access keys. The alarm control unit, remote annunciator panel and access keys to locked fire alarm equipment shall be installed and maintained in an approved location.

1007.3.3.3 Alarm initiation and signal.

1007.3.3.3.1 General. When actuated, fire alarm-initiating devices shall activate an alarm signal which is audible throughout the building or in designated portions of the building when approved.

> **EXCEPTION:** Single-station detectors in dwelling units, rooms used for sleeping purposes in hotel and lodging houses, and patient sleeping rooms in hospitals and nursing homes.

1007.3.3.3.2 Audible alarm signal. The audible signal shall be the standard fire alarm evacuation signal as described in UFC Standard 10-2.

> **EXCEPTION:** This alarm signal is not required for:
>
> 1. Systems using prerecorded or live voice message announcements.
>
> 2. Patient and inmate areas of Group I Occupancies.

Figure 1.33.0 *(Continued)*

1007.3.3.3.3 Audibility. The alarm signal shall be a distinctive sound which is not used for any other purpose other than the fire alarm. Alarm-signaling devices shall produce a sound that exceeds the prevailing equivalent sound level in the room or space by 15 decibels minimum, or exceeds any maximum sound level with a duration of 30 seconds minimum by 5 decibels minimum, whichever is louder. Sound levels for alarm signals shall be 120 decibels maximum.

1007.3.3.3.4 Visual alarms. Alarm systems shall include both audible and visual alarms. Alarm devices shall be located in hotel guest rooms as required by the Building Code (see UBC Section 1105.4.6); accessible public- and common-use areas, including toilet rooms and bathing facilities; hallways; and lobbies. (See Council of American Building Officials/American National Standards Institute Standard A117.1-1992, Section 4-26.2, for additional information about visual signals.)

1007.3.3.4 Connections to other systems. A fire alarm system shall not be used for any purpose other than fire warning unless approved.

1007.3.3.5 Supervision. Means of interconnecting equipment, devices and appliances shall be supervised for the integrity of the interconnecting conductors or equivalent, as set forth in UFC Standard 10-2.

1007.3.3.6 Monitoring.

1007.3.3.6.1 General. When required by the chief, fire alarm systems shall be monitored by an approved central, proprietary or remote station service or a local alarm which gives audible and visual signals at a constantly attended location.

1007.3.3.6.2 Automatic telephone dialing devices. Automatic telephone dialing devices used to transmit an emergency alarm shall not be connected to any fire department telephone number unless approved.

1007.3.3.7 Annunciation. Fire alarm systems shall be divided into alarm zones when required by the chief. When two or more alarm zones are required, visible annunciation shall be provided in an approved location.

1007.3.4 Acceptance test and certification.

1007.3.4.1 Acceptance test. Upon completion of the installation, a satisfactory test of the entire system shall be made in the presence of the chief. All functions of the system or alteration shall be tested.

1007.3.4.2 Certification. The permittee shall provide written certification to the chief that the system has been installed in accordance with the approved plans and specifications.

1007.3.4.3 Instructions. When required by the chief, operating, testing and maintenance instructions and "as-built" drawings and equipment specifications shall be provided at an approved location.

TABLE 5-A—EXTERIOR WALL AND OPENING PROTECTION BASED ON LOCATION ON PROPERTY FOR ALL CONSTRUCTION TYPES[1,2,3]—

OCCUPANCY GROUP[4]	CONSTRUCTION TYPE	EXTERIOR WALLS		OPENINGS[5]
		Bearing	Nonbearing	
		Distances are measured to property lines (see Section 503).		
		× 304.8 for mm		
A-3	II One-hour	Two-hour N/C less than 5 feet One-hour N/C elsewhere	Same as bearing except NR, N/C 40 feet or greater	Not permitted less than 5 feet Protected less than 10 feet
	II-N	Two-hour N/C less than 5 feet One-hour N/C less than 20 feet NR, N/C elsewhere	Same as bearing	Not permitted less than 5 feet Protected less than 10 feet
	III-N	Four-hour N/C	Four-hour N/C less than 5 feet Two-hour N/C less than 20 feet One-hour N/C less than 40 feet NR, N/C elsewhere	Not permitted less than 5 feet Protected less than 20 feet
	V One-hour	Two-hour less than 5 feet One-hour elsewhere	Same as bearing	Not permitted less than 5 feet Protected less than 10 feet
	V-N	Two-hour less than 5 feet One-hour less than 20 feet NR elsewhere	Same as bearing	Not permitted less than 5 feet Protected less than 10 feet
A-4	II One-hour	One-hour N/C	Same as bearing except NR, N/C 40 feet or greater	Protected less than 10 feet
	II-N	One-hour N/C less than 10 feet NR, N/C elsewhere	Same as bearing	Protected less than 10 feet
	III-N	Four-hour N/C	Four-hour N/C less than 5 feet Two-hour N/C less than 20 feet One-hour N/C less than 40 feet NR, N/C elsewhere	Not permitted less than 5 feet Protected less than 10 feet
	V One-hour	One-hour	Same as bearing	Protected less than 10 feet
	V-N	One-hour less than 10 feet NR elsewhere	Same as bearing	Protected less than 10 feet
B F-1 M S-1, S-3	I-F.R. II-F.R. III One-hour III-N IV-H.T.	Four-hour N/C less than 5 feet Two-hour N/C elsewhere	Four-hour N/C less than 5 feet Two-hour N/C less than 20 feet One-hour N/C less than 40 feet NR, N/C elsewhere	Not permitted less than 5 feet Protected less than 20 feet
	II One-hour	One-hour N/C	Same as bearing except NR, N/C 40 feet or greater	Not permitted less than 5 feet Protected less than 10 feet
	II-N[3]	One-hour N/C less than 20 feet NR, N/C elsewhere	Same as bearing	Not permitted less than 5 feet Protected less than 10 feet
	V One-hour	One-hour	Same as bearing	Not permitted less than 5 feet Protected less than 10 feet
	V-N	One-hour less than 20 feet NR elsewhere	Same as bearing	Not permitted less than 5 feet Protected less than 10 feet
E-1 E-2[7] E-3[7]	I-F.R. II-F.R. III One-hour III-N IV-H.T.	Four-hour N/C	Four-hour N/C less than 5 feet Two-hour N/C less than 20 feet One-hour N/C less than 40 feet NR, N/C elsewhere	Not permitted less than 5 feet Protected less than 20 feet
	II One-hour	Two-hour N/C less than 5 feet One-hour N/C elsewhere	Same as bearing except NR, N/C 40 feet or greater	Not permitted less than 5 feet Protected less than 10 feet
	II-N	Two-hour N/C less than 5 feet One-hour N/C less than 10 feet NR, N/C elsewhere	Same as bearing	Not permitted less than 5 feet Protected less than 10 feet
	V One-hour	Two-hour less than 5 feet One-hour elsewhere	Same as bearing	Not permitted less than 5 feet Protected less than 10 feet
	V-N	Two-hour less than 5 feet One-hour less than 10 feet NR elsewhere	Same as bearing	Not permitted less than 5 feet Protected less than 10 feet

Figure 1.34.0 Exterior wall and opening protection based on location on property—all types of construction. (*Reproduced from the 1997 edition of the* Uniform Fire Code, *Volumes 1, 2, 3, Copyright © 1997, with the permission of the publisher, The International Fire Code Institute (IFCI). Although the International Fire Code Institute (IFI) has granted permission for reproduction of sections contained within the* Uniform Fire Code, *in conjunction with this publication, IFCI assumes no responsibility for the accuracy or completion of summaries provided therein.*)

TABLE 5-A—EXTERIOR WALL AND OPENING PROTECTION BASED ON LOCATION ON PROPERTY FOR ALL CONSTRUCTION TYPES[1,2,3]—

OCCUPANCY GROUP[4]	CONSTRUCTION TYPE	EXTERIOR WALLS Bearing	Nonbearing	OPENINGS[5]
		Distances are measured to property lines (see Section 503).		
		× 304.8 for mm		
F-2 S-2	I-F.R. II-F.R. III One-hour III-N IV-H.T.	Four-hour N/C less than 5 feet Two-hour N/C elsewhere	Four-hour N/C less than 5 feet Two-hour N/C less than 20 feet One-hour N/C less than 40 feet NR, N/C elsewhere	Not permitted less than 3 feet Protected less than 20 feet
	II One-hour	One-hour N/C	Same as bearing NR, N/C 40 feet or greater	Not permitted less than 5 feet Protected less than 10 feet
	II-N[3]	One-hour N/C less than 5 feet NR, N/C elsewhere	Same as bearing	Not permitted less than 5 feet Protected less than 10 feet
	V One-hour	One-hour	Same as bearing	Not permitted less than 5 feet Protected less than 10 feet
	V-N	One-hour less than 5 feet NR elsewhere	Same as bearing	Not permitted less than 5 feet Protected less than 10 feet
H-1[2,3]	I-F.R. II-F.R.	Four-hour N/C	NR N/C	Not restricted[3]
	II One-hour	One-hour N/C	NR N/C	Not restricted[3]
	II-N	NR N/C	Same as bearing	Not restricted[3]
	III One-hour III-N IV-H.T. V One-hour V-N	Group H, Division 1 Occupancies are not allowed in buildings of these construction types.		
H-2[2,3] H-3[2,3] H-4[3] H-6 H-7	I-F.R. II-F.R. III One-hour III-N IV-H.T.	Four-hour N/C	Four-hour N/C less than 5 feet Two-hour N/C less than 10 feet One-hour N/C less than 40 feet NR, N/C elsewhere	Not permitted less than 5 feet Protected less than 20 feet
	II One-hour	Four-hour N/C less than 5 feet Two-hour N/C less than 10 feet One-hour N/C elsewhere	Four-hour N/C less than 5 feet Two-hour N/C less than 10 feet One-hour N/C less than 20 feet NR, N/C elsewhere	Not permitted less than 5 feet Protected less than 20 feet
	II-N	Four-hour N/C less than 5 feet Two-hour N/C less than 10 feet One-hour N/C less than 20 feet NR, N/C elsewhere	Same as bearing	Not permitted less than 5 feet Protected less than 20 feet
	V One-hour	Four-hour less than 5 feet Two-hour less than 10 feet One-hour elsewhere	Same as bearing	Not permitted less than 5 feet Protected less than 20 feet
	V-N	Four-hour less than 5 feet Two-hour less than 10 feet One-hour less than 20 feet NR elsewhere	Same as bearing	Not permitted less than 5 feet Protected less than 20 feet
H-5[2]	I-F.R. II-F.R. III One-hour III-N IV-H.T.	Four-hour N/C	Four-hour N/C less than 40 feet One-hour N/C less than 60 feet NR, N/C elsewhere	Protected less than 60 feet
	II One-hour	One-hour N/C	Same as bearing, except NR, N/C 60 feet or greater	Protected less than 60 feet
	II-N	One-hour N/C less than 60 feet NR, N/C elsewhere	Same as bearing	Protected less than 60 feet
	V One-hour	One-hour	Same as bearing	Protected less than 60 feet
	V-N	One-hour less than 60 feet NR elsewhere	Same as bearing	Protected less than 60 feet
I-1.1 I-1.2 I-2 I-3	I-F.R. II-F.R.	Four-hour N/C	Four-hour N/C less than 5 feet Two-hour N/C less than 20 feet One-hour N/C less than 40 feet NR, N/C elsewhere	Not permitted less than 5 feet Protected less than 20 feet
I-1.1 I-1.2 I-3[2]	II One-hour	Two-hour N/C less than 5 feet One-hour N/C elsewhere	Same as bearing except NR, N/C 40 feet or greater	Not permitted less than 5 feet Protected less than 10 feet
	V One-hour	Two-hour less than 5 feet One-hour elsewhere	Same as bearing	Not permitted less than 5 feet Protected less than 10 feet

Figure 1.34.0 (*Continued*)

TABLE 5-A—EXTERIOR WALL AND OPENING PROTECTION BASED ON LOCATION ON PROPERTY FOR ALL CONSTRUCTION TYPES[1,2,3]—

OCCUPANCY GROUP[4]	CONSTRUCTION TYPE	EXTERIOR WALLS		OPENINGS[5]
		Bearing	Nonbearing	
		Distances are measured to property lines (see Section 503).		
		× 304.8 for mm		
U-1[3]	I-F.R. II-F.R. III One-hour III-N IV-H.T.	Four-hour N/C	Four-hour N/C less than 3 feet Two-hour N/C less than 20 feet One-hour N/C less than 40 feet NR, N/C elsewhere	Not permitted less than 3 feet Protected less than 20 feet
	II One-hour	One-hour N/C	Same as bearing except NR, N/C 40 feet or greater	Not permitted less than 3 feet
	V One-hour	One-hour	Same as bearing	Not permitted less than 3 feet
	II-N[2]	One-hour N/C less than 3 feet[3] NR, N/C elsewhere	Same as bearing	Not permitted less than 3 feet
	V-N	One-hour less than 3 feet[3] NR elsewhere	Same as bearing	Not permitted less than 3 feet
U-2	All	Not regulated		

N/C— Noncombustible.
NR — Nonrated.
H.T.— Heavy timber.
F.R. — Fire resistive.

[1]See Section 503 for types of walls affected and requirements covering percentage of openings permitted in exterior walls. For walls facing streets, yards and public ways, see also Section 601.5.

[2]For additional restrictions see Chapters 3 and 6.

[3]For special provisions and exceptions, see also Section 503.4.

[4]See Table 3-A for a description of each occupancy type.

[5]Openings requiring protection in exterior walls shall be protected by a fire assembly having at least a three-fourths-hour fire-protection rating.

[6]See Section 308.2.1, Exception 3.

[7]Group E, Divisions 2 and 3 Occupancies having an occupant load of not more than 20 may have exterior wall and opening protection as required for Group R, Division 3 Occupancies.

Figure 1.34.0 *(Continued)*

TABLE 5-B—BASIC ALLOWABLE BUILDING HEIGHTS AND BASIC ALLOWABLE FLOOR AREA FOR BUILDINGS ONE STORY IN HEIGHT[1]

		TYPE OF CONSTRUCTION								
		I	II			III		IV	V	
		F.R.	F.R.	One-hour	N	One-hour	N	H.T.	One-hour	N
		MAXIMUM HEIGHT IN FEET (mm)								
		UL	160 (48 768 mm)	65 (19 812 mm)	55 (16 764 mm)	65 (19 812 mm)	55 (16 764 mm)	65 (19 812 mm)	50 (15 240 mm)	40 (12 192 mm)
USE GROUP	HEIGHT/AREA	MAXIMUM HEIGHT (Stories) AND MAXIMUM AREA (Sq. Ft. — × 0.0929 for mm)								
A-1	H	UL	4	Not Permitted						
	A	UL	29,900							
A-2, 2.1[2]	H	UL	4	2	NP	2	NP	2	2	NP
	A	UL	29,900	13,500	NP	13,500	NP	13,500	10,500	NP
A-3, 4[2]	H	UL	12	2	1	2	1	2	2	1
	A	UL	29,900	13,500	9,100	13,500	9,100	13,500	10,500	6,000
B, F-1, M, S-1, S-3, S-5	H	UL	12	4	2	4	2	4	3	2
	A	UL	39,900	18,000	12,000	18,000	12,000	18,000	14,000	8,000
E-1, 2, 3[4]	H	UL	4	2	1	2	1	2	2	1
	A	UL	45,200	20,200	13,500	20,200	13,500	20,200	15,700	9,100
F-2, S-2	H	UL	12	4	2	4	2	4	3	2
	A	UL	59,900	27,000	18,000	27,000	18,000	27,000	21,000	12,000
H-1[5]	H	1	1	1	1	Not Permitted				
	A	15,000	12,400	5,600	3,700					
H-2[5]	H	UL	2	1	1	1	1	1	1	1
	A	15,000	12,400	5,600	3,700	5,600	3,700	5,600	4,400	2,500
H-3, 4, 5[5]	H	UL	5	2	1	2	1	2	2	1
	A	UL	24,800	11,200	7,500	11,200	7,500	11,200	8,800	5,100
H-6, 7	H	3	3	3	2	3	2	3	3	1
	A	UL	39,900	18,000	12,000	18,000	12,000	18,000	14,000	8,000
I-1.1, 1.2[6,10]	H	UL	3	1	NP	1	NP	1	1	NP
	A	UL	15,100	6,800	NP	6,800	NP	6,800	5,200	NP
I-2	H	UL	3	2	NP	2	NP	2	2	NP
	A	UL	15,100	6,800	NP	6,800	NP	6,800	5,200	NP
I-3	H	UL	2	Not Permitted[7]						
	A	UL	15,100							

		TYPE OF CONSTRUCTION								
		I	II			III		IV	V	
		F.R.	F.R.	One-hour	N	One-hour	N	H.T.	One-hour	N
		MAXIMUM HEIGHT IN FEET (mm)								
		UL	160 (48 768 mm)	65 (19 812 mm)	55 (16 764 mm)	65 (19 812 mm)	55 (16 764 mm)	65 (19 812 mm)	50 (15 240 mm)	40 (12 192 mm)
USE GROUP	HEIGHT/AREA	MAXIMUM HEIGHT (Stories) AND MAXIMUM AREA (Sq. Ft. — × 0.0929 for mm)								
R-1	H	UL	12	4	2[9]	4	2[9]	4	3	2[9]
	A	UL	29,900	13,500	9,100[9]	13,500	9,100[9]	13,500	10,500	6,000[9]
R-3	H	UL	3	3	3	3	3	3	3	3
	A	Unlimited								
S-4[3]	See Table 3-H									
U[8]	H	See Chapter 3								
	A									

A—Building area in square feet.
H—Building height in number of stories.
H.T.—Heavy timber.
NP—Not permitted.

N—No requirements for fire resistance.
F.R.—Fire resistive.
UL—Unlimited.

[1]For multistory buildings, see Section 504.2.

[2]For limitations and exceptions, see Section 303.2.

[3]For open parking garages, see Section 311.9.

[4]See Section 305.2.3.

[5]See Section 307.

[6]See Section 308.2.1 for exception to the allowable area and number of stories in hospitals, nursing homes and health care centers.

[7]See Section 308.2.2.2.

[8]For agricultural buildings, see also Appendix Chapter 3.

[9]For limitations and exceptions, see Section 310.2.

[10]For Type II F.R., the maximum height of Group I, Division 1.1 Occupancies is limited to 75 feet (22 860 mm). For Type II, One-hour construction, the maximum height of Group I, Division 1.1 Occupancies is limited to 45 feet (13 716 mm).

Figure 1.35.0   Basic allowable building heights and floor area by use group. (*Reproduced from the 1997 edition of the* Uniform Fire Code, *Volumes 1, 2, 3, Copyright © 1997, with the permission of the publisher, The International Fire Code Institute (IFCI). Although the International Fire Code Institute (IFI) has granted permission for reproduction of sections contained within the* Uniform Fire Code, *in conjunction with this publication, IFCI assumes no responsibility for the accuracy or completion of summaries provided therein.*)

TABLE 6-A—TYPES OF CONSTRUCTION—FIRE-RESISTIVE REQUIREMENTS (In Hours)
For details, see occupancy section in Chapter 3, type of construction sections in Chapter 6 and sections referenced in Table 6-A.

BUILDING ELEMENT	TYPE I Noncombustible Fire-resistive	TYPE II Noncombustible Fire-resistive	TYPE II 1-Hr.	TYPE II N	TYPE III 1-Hr.	TYPE III N	TYPE IV H.T.	TYPE V 1-Hr.	TYPE V N
1. Bearing walls— exterior	4 Sec. 602.3.1	4 Sec. 603.3.1	1	N	4 Sec. 604.3.1	4 Sec. 604.3.1	4 Sec. 605.3.1	1	N
2. Bearing walls—interior	3	2	1	N	1	N	1	1	N
3. Nonbearing walls— exterior	4 Sec. 602.3.1	4 Sec. 603.3.1	1 Sec. 603.3.1	N	4 Sec. 604.3.1	4 Sec. 604.3.1	4 Sec. 605.3.1	1	N
4. Structural frame[1]	3	2	1	N	1	N	1 or H.T.	1	N
5. Partitions— permanent	1[2]	1[2]	1[2]	N	1	N	1 or H.T.	1	N
6. Shaft enclosures[3]	2	2	1	1	1	1	1	1	1
7. Floors and floor-ceilings	2	2	1	N	1	N	H.T.	1	N
8. Roofs and roof-ceilings	2 Sec. 602.5	1 Sec. 603.5	1 Sec. 603.5	N	1	N	H.T.	1	N
9. Exterior doors and windows	Sec. 602.3.2	Sec. 603.3.2	Sec. 603.3.2	Sec. 603.3.2	Sec. 604.3.2	Sec. 604.3.2	Sec. 605.3.2	Sec. 606.3	Sec. 606.3
10. Stairway construction	Sec. 602.4	Sec. 603.4	Sec. 603.4	Sec. 603.4	Sec. 604.4	Sec. 604.4	Sec. 605.4	Sec. 606.4	Sec. 606.4

N—No general requirements for fire resistance.

H.T.—Heavy timber.

[1]Structural frame elements in an exterior wall that is located where openings are not permitted or where protection of openings is required, shall be protected against external fire exposure as required for exterior bearing walls or the structural frame, whichever is greater.

[2]Fire-retardant-treated wood (see Section 207) may be used in the assembly, provided fire-resistance requirements are maintained. See Sections 602 and 603.

[3]For special provisions, see Sections 304.6, 306.6 and 711.

Figure 1.36.0 Fire resistive requirements listed by types of construction—type I thru V. (*Reproduced from the 1997 edition of the* Uniform Fire Code, *Volumes 1, 2, 3, Copyright © 1997, with the permission of the publisher, The International Fire Code Institute (IFCI). Although the International Fire Code Institute (IFCI). Although the International Fire Code Institute (IFI) has granted permission for reproductin of sections contained within the* Uniform Fire Code, *in conjunction with this publication, IFCI assumes no responsibility for the accuracy or completion of summaries provided therein.*)

TABLE 8-A—FLAME-SPREAD CLASSIFICATION

MATERIAL QUALIFIED BY:	
Class	Flame-spread Index
I	0-25
II	26-75
III	76-200

Figure 1.37.0 Flame spread classifications. (*Reproduced from the 1997 edition of the* Uniform Fire Code, *Volumes 1, 2, 3, Copyright © 1997, with the permission of the publisher, the International Fire Code Institute (IFCI). Although the International Fire Code Institute (IFCI). Although the International Fire Code Institute (IFI) has granted permission for reproductin of sections contained within the* Uniform Fire Code, *in conjunction with this publication, IFCI assumes no responsibility for the accuracy or completion of summaries provided therein.*)

TABLE 8-B—MAXIMUM FLAME-SPREAD CLASS[1]

OCCUPANCY GROUP	ENCLOSED VERTICAL EXITWAYS	OTHER EXITWAYS[2]	ROOMS OR AREAS
A	I	II	II[3]
B	I	II	III
E	I	II	III
F	II	III	III
H	I	II	III[4]
I-1.1, I-1.2, I-2	I	I[5]	II[6]
I-3	I	I[5]	I[6]
M	I	II	III
R-1	I	II	III
R-3	III	III	III[7]
S-1, S-2	II	II	III
S-3, S-4, S-5	I	II	III
U	NO RESTRICTIONS		

[1]Foam plastics shall comply with the requirements specified in Section 2602. Carpeting on ceilings and textile wall coverings shall comply with the requirements specified in Sections 804.2 and 805, respectively.

[2]Finish classification is not applicable to interior walls and ceilings of exterior exit balconies.

[3]In Group A, Divisions 3 and 4 Occupancies, Class III may be used.

[4]Over two stories shall be of Class II.

[5]In Group I, Divisions 2 and 3 Occupancies, Class II may be used.

[6]Class III may be used in administrative spaces.

[7]Flame-spread provisions are not applicable to kitchens and bathrooms of Group R, Division 3 Occupancies.

Figure 1.38.0 Maximum flame spread class. (*Reproduced from the 1997 edition of the* Uniform Fire Code, *Volumes 1, 2, 3, Copyright © 1997, with the permission of the publisher, the International Fire Code Institute (IFCI). Although the International Fire Code Institute (IFI) has granted permission for reproduction of sections contained within the* Uniform Fire Code, *in conjunction with this publication, IFCI assumes no responsibility for the accuracy or completion of summaries provided therein.*)

TABLE 10-A—MINIMUM EGRESS REQUIREMENTS[1]

USE[2]	MINIMUM OF TWO MEANS OF EGRESS ARE REQUIRED WHERE NUMBER OF OCCUPANTS IS AT LEAST	OCCUPANT LOAD FACTOR[3] (square feet) × 0.0929 for m[2]
1. Aircraft hangars (no repair)	10	500
2. Auction rooms	30	7
3. Assembly areas, concentrated use (without fixed seats) Auditoriums Churches and chapels Dance floors Lobby accessory to assembly occupancy Lodge rooms Reviewing stands Stadiums Waiting area	50 50	7 3
4. Assembly areas, less-concentrated use Conference rooms Dining rooms Drinking establishments Exhibit rooms Gymnasiums Lounges Stages Gaming: keno, slot machine and live games area	50 50	15 11
5. Bowling alley (assume no occupant load for bowling lanes)	50	4
6. Children's homes and homes for the aged	6	80
7. Classrooms	50	20
8. Congregate residences	10	200
9. Courtrooms	50	40
10. Dormitories	10	50
11. Dwellings	10	300
12. Exercising rooms	50	50
13. Garage, parking	30	200
14. Health care facilities— Sleeping rooms Treatment rooms	8 10	120 240
15. Hotels and apartments	10	200
16. Kitchen—commercial	30	200
17. Library— Reading rooms Stack areas	50 30	50 100
18. Locker rooms	30	50
19. Malls (see Chapter 4)	—	—
20. Manufacturing areas	30	200
21. Mechanical equipment room	30	300
22. Nurseries for children (day care)	7	35
23. Offices	30	100
24. School shops and vocational rooms	50	50
25. Skating rinks	50	50 on the skating area; 15 on the deck
26. Storage and stock rooms	30	300
27. Stores—retail sales rooms Basements and ground floor Upper floors	50 50	30 60
28. Swimming pools	50	50 for the pool area; 15 on the deck
29. Warehouses[5]	30	500
30. All others	50	100

[1]Access to, and egress from, buildings for persons with disabilities shall be provided as specified in Chapter 11.

[2]For additional provisions on number of means of egress from Groups H and I Occupancies and from rooms containing fuel-fired equipment or cellulose nitrate, see Sections 1018, 1019 and 1020, respectively.

[3]This table shall not be used to determine working space requirements per person.

[4]Occupant load based on five persons for each alley, including 15 feet (4572 mm) of runway.

[5]Occupant load for warehouses containing approved high rack storage systems designed for mechanical handling may be based on the floor area exclusive of the rack area rather than the gross floor area.

Figure 1.39.0 Minimum egress requirements by use. (*Reproduced from the 1997 edition of the* Uniform Fire Code, *Volumes 1, 2, 3, Copyright © 1997, with the permission of the publisher, The International Fire Code Institute (IFCI). Although the International Fire Code Institute (IFI) has granted permission for reproduction of sections contained within the* Uniform Fire Code, *in conjunction with this publication, IFCI assumes no responsibility for the accuracy or completion of summaries provided therein.*)

CRYOGENIC FLUIDS
WEIGHT AND VOLUME EQUIVALENTS

TABLE A-VI-G-1—WEIGHT AND VOLUME EQUIVALENTS FOR COMMON CRYOGENIC FLUIDS

CRYOGEN	WEIGHT OF LIQUID OR GAS		VOLUME OF LIQUID AT NORMAL BOILING POINT		VOLUME OF GAS AT 70°F (21.1°C) AND 14.7 PSIA (101.3 kPa)	
	Pounds	Kilograms	Liters	Gallons	Cubic Feet	Cubic Meters
argon	1.000	0.454	0.326	0.086	9.67	0.274
	2.205	1.000	0.718	0.190	21.32	0.604
	3.072	1.393	1.000	0.264	29.71	0.841
	11.628	5.274	3.785	1.000	112.45	3.184
	10.340	4.690	3.366	0.889	100.00	2.832
	3.652	1.656	1.189	0.314	35.31	1.000
helium	1.000	0.454	3.631	0.959	96.72	2.739
	2.205	1.000	8.006	2.115	213.23	6.038
	0.275	0.125	1.000	0.264	26.63	0.754
	1.042	0.473	3.785	1.000	100.82	2.855
	1.034	0.469	3.754	0.992	100.00	2.832
	0.365	0.166	1.326	0.350	35.31	1.000
hydrogen	1.000	0.454	6.409	1.693	191.96	5.436
	2.205	1.000	14.130	3.733	423.20	11.984
	0.156	0.071	1.000	0.264	29.95	0.848
	0.591	0.268	3.785	1.000	113.37	3.210
	0.521	0.236	3.339	0.882	100.00	2.832
	0.184	0.083	1.179	0.311	35.31	1.000
oxygen	1.000	0.454	0.397	0.105	12.00	0.342
	2.205	1.000	0.876	0.231	26.62	0.754
	2.517	1.142	1.000	0.264	30.39	0.861
	9.527	4.321	3.785	1.000	115.05	3.250
	8.281	3.756	3.290	0.869	100.00	2.832
	2.924	1.327	1.162	0.307	35.31	1.000
nitrogen	1.000	0.454	0.561	0.148	13.80	0.391
	2.205	1.000	1.237	0.327	30.43	0.862
	1.782	0.808	1.000	0.264	24.60	0.697
	6.746	3.060	3.785	1.000	93.11	2.637
	7.245	3.286	4.065	1.074	100.00	2.832
	2.558	1.160	1.436	0.379	35.31	1.000
LNG[1]	1.000	0.454	1.052	0.278	22.968	0.650
	2.205	1.000	2.320	0.613	50.646	1.434
	0.951	0.431	1.000	0.264	21.812	0.618
	3.600	1.633	3.785	1.000	82.62	2.340
	4.356	1.976	4.580	1.210	100.00	2.832
	11.501	5.217	1.616	0.427	35.31	1.000

[1]The values listed for liquefied natural gas (LNG) are "typical" values. LNG is a mixture of hydrocarbon gases, and no two LNG steams have exactly the same composition.

To use the above table, read horizontally across the line of interest. For example, to determine the number of cubic feet of gas contained in 1.0 gallons (3.785 L) of liquid argon, find 1.000 in the column entitled "Volume of Liquid at Normal Boiling Point." Reading across the line under the column entitled "Volume of Gas at 70°F (21.1°C) - cubic feet" the value of 112.45 cubic feet (3.184 m^3) is found.

If other quantities are of interest, the numbers obtained can be multiplied or divided to obtain the quantity of interest. For example, to determine the number of cubic feet of argon gas contained in a volume of 1,000 gallons (3785 L) of liquid argon at its normal boiling point, multiply 112.45 by 1,000 to obtain 112,450 cubic feet (3184 m^3).

Figure 1.40.0 Weight and volume equivalents of common gases (cryogenic fluids). (*Reproduced from the 1997 edition of the* Uniform Fire Code, *Volumes 1, 2, 3, Copyright © 1997, with the permission of the publisher, the International Fire Code Institute (IFCI). Although the International Fire Code Institute (IFI) has granted permission for reproduction of sections contained within the* Uniform Fire Code, *in conjunction with this publication, IFCI assumes no responsibility for the accuracy or completion of summaries provided therein.*)

**TABLE 3-C—REQUIRED SEPARATION OF SPECIFIC-USE AREAS IN GROUP I,
DIVISION 1.1 HOSPITALS AND NURSING HOMES**

	DESCRIPTION	OCCUPANCY SEPARATION
1.	Employee locker rooms	None
2.	Gift/retail shops	None
3.	Handicraft shops	None
4.	Kitchens	None
5.	Laboratories which employ hazardous materials in quantities less than that which would cause classification as a Group H Occupancy	One hour
6.	Laundries greater than 100 sq. ft. (9.3 m^2)[1]	One hour
7.	Paint shops employing hazardous substances and materials in quantities less than that which would cause classification as a Group H Occupancy	One hour
8.	Physical plant maintenance shop	One hour
9.	Soiled linen room[1]	One hour
10.	Storage rooms 100 sq. ft. (9.3 m^2) or less in area storing combustible material	None
11.	Storage rooms more than 100 sq. ft. (9.3 m^2) storing combustible material	One hour
12.	Trash-collection rooms[1]	One hour

[1]For rubbish and linen chute termination rooms, see the Building Code.

Figure 1.41.0   Required separation of specific use in hospitals and nursing homes. (*Reproduced from the 1997 edition of the Uniform Fire Code, Volumes 1, 2, 3, Copyright © 1997, with the permission of the publisher, the International Fire Code Institute (IFCI). Although the International Fire Code Institute (IFI) has granted permission for reproduction of sections contained within the* Uniform Fire Code, *in conjunction with this publication, IFCI assumes no responsibility for the accuracy or completion of summaries provided therein.*)

**TABLE A-V1-F-1—TRANSITION DATES FOR EXPLOSIVES CLASSIFICTIONS
AND RELATED DOT REGULATIONS**

OCTOBER 1, 1991	DOT publishes new rules for explosives based on United Nations (UN) Recommendations. All *new* explosives must be classified under the new regulations.
OCTOBER 1, 1993	Mandatory compliance with new classification and hazard communication requirements except *placarding*.
OCTOBER 1, 1994	Mandatory use of new UN placards except existing DOT placards may continue to be used for domestic highway transport. Package manufacturers will only be permitted to make nonbulk packaging which meet United Nations performance standards.
OCTOBER 1, 1996	Mandatory use of performance-oriented packaging standards based on United Nations Recommendations for nonbulk packaging.
OCTOBER 1, 2001	Mandatory use of new United Nations placards for *all* modes of transportation.

Figure 1.42.0   Dates when rules relating to explosives classifications became (become) effective. (*Reproduced from the 1997 edition of the* Uniform Fire Code, *Volumes 1, 2, 3, Copyright © 1997, with the permission of the publisher, the International Fire Code Institute (IFCI). Although the International Fire Code Institute (IFI) has granted permission for reproduction of sections contained within the* Uniform Fire Code, *in conjunction with this publication, IFCI assumes no responsibility for the accuracy or completion of summaries provided therein.*)

Contents

2.0.0   Introduction to Plumbing

Leonardo da Vinci is credited with the design and installation of the first indoor plumbing system in Italy in the mid-sixteenth century. Other than the addition of sophisticated pumps on the supply side and designer fixtures on the other end, not much has changed, except for the materials of construction (gravity still plays as important role today as it did in 1550). A building plumbing system will generally consist of incoming domestic water service and distribution, above- and below-grade drainage systems for both sanitary and storm water, a venting system, pipe, fittings, valves, pumps, and fixtures.

2.1.0   Equivalent Length (Pipe, Elbows, Tees, and Valves)

Find the nominal pipe size being used in the left-most column. For each fitting, read the value under the appropriate heading and add this to the length of piping. This allows total system pressure drop to be calculated. (This is valid for any fluid.)

PIPE SIZE	EQUIVALENT LENGTH OF STRAIGHT PIPE (FEET)				
	STANDARD ELBOW	STANDARD TEE	GATE VALVE FULL OPEN	GLOBE VALVE FULL OPEN	ANGLE VALVE FULL OPEN
1-1/2	4	9	0.9	41	21
2	5	11	1.2	54	27
2-/1/2	6	13	1.4	64	32
3	8	16	1.6	80	40
3-1/2	9	18	2.0	91	45
4	11	21	2.2	110	55
5	13	26	2.8	140	70
6	16	32	3.4	155	81
8	20	42	4.5	210	110
10	25	55	5.5	270	140
12	30	65	6.5	320	160
14	35	75	8.0	370	190

This table contains the number of feet of straight pipe usually allowed for standard fittings and valves.

2.1.1   Equivalent Length of Pipe for 90-Degree Elbows (in Feet)

Velocity, ft/s	Pipe Size														
	1/2	3/4	1	1-1/4	1-1/2	2	2-1/2	3	3-1/2	4	5	6	8	10	12
1	1.2	1.7	2.2	3.0	3.5	4.5	5.4	6.7	7.7	8.6	10.5	12.2	15.4	18.7	22.2
2	1.4	1.9	2.5	3.3	3.9	5.1	6.0	7.5	8.6	9.5	11.7	13.7	17.3	20.8	24.8
3	1.5	2.0	2.7	3.6	4.2	5.4	6.4	8.0	9.2	10.2	12.5	14.6	18.4	22.3	26.5
4	1.5	2.1	2.8	3.7	4.4	5.6	6.7	8.3	9.6	10.6	13.1	15.2	19.2	23.2	27.6
5	1.6	2.2	2.9	3.9	4.5	5.9	7.0	8.7	10.0	11.1	13.6	15.8	19.8	24.2	28.8
6	1.7	2.3	3.0	4.0	4.7	6.0	7.2	8.9	10.3	11.4	14.0	16.3	20.5	24.9	29.6
7	1.7	2.3	3.0	4.1	4.8	6.2	7.4	9.1	10.5	11.7	14.3	16.7	21.0	25.5	30.3
8	1.7	2.4	3.1	4.2	4.9	6.3	7.5	9.3	10.8	11.9	14.6	17.1	21.5	26.1	31.0
9	1.8	2.4	3.2	4.3	5.0	6.4	7.7	9.5	11.0	12.2	14.9	17.4	21.9	26.6	31.6
10	1.8	2.5	3.2	4.3	5.1	6.5	7.8	9.7	11.2	12.4	15.2	17.7	22.2	27.0	32.0

2.2.0   Maximum Capacity of Gas Pipe (in Cubic Feet Per Hour)

Nominal Iron Pipe Size, in.	Internal Diameter, in.	Length of Pipe, ft													
		10	20	30	40	50	60	70	80	90	100	125	150	175	200
1/4	0.364	32	22	18	15	14	12	11	11	10	9	8	8	7	6
3/8	0.493	72	49	40	34	30	27	25	23	22	21	18	17	15	14
1/2	0.622	132	92	73	63	56	50	46	43	40	38	34	31	28	26
3/4	0.824	278	190	152	130	115	105	96	90	84	79	72	64	59	55
1	1.049	520	350	285	245	215	195	180	170	160	150	130	120	110	100
1-1/4	1.380	1050	730	590	500	440	400	370	350	320	305	275	250	225	210
1-1/2	1.610	1600	1100	890	760	670	610	560	530	490	460	410	380	350	320
2	2.067	3050	2100	1650	1450	1270	1150	1050	990	930	870	780	650	610
2-1/2	2.469	4800	3300	2700	2300	2000	1850	1700	1600	1500	1400	1250	1130	1050	980
3	3.068	8500	5900	4700	4100	3600	3250	3000	2800	2600	2500	2200	2000	1850	1700
4	4.026	17,500	12,000	9700	8300	7400	6800	6200	5800	5400	5100	4500	4100	3800	3500

Notes: 1. Capacity is in cubic feet per hour at gas pressures of 0.5 psig or less and a pressure drop of 0.5 in. of water; Specific gravity = 0.60.   2. Copyright by the American Gas Association and the National Fire Protection Association. Used by permission of the copyright holder.

(By permission of American Society of Heating, Refrigerating and Air-Conditioning Engineers, Inc. Atlanta, Georgia, from their 1993 ASHRAE Fundamentals Handbook)

2.3.0   Iron and Copper Elbow-Size Equivalents

Fitting	Iron Pipe	Copper Tubing
Elbow, 90°	1.0	1.0
Elbow, 45°	0.7	0.7
Elbow, 90° long turn	0.5	0.5
Elbow, welded, 90°	0.5	0.5
Reduced coupling	0.4	0.4
Open return bend	1.0	1.0
Angle radiator valve	2.0	3.0
Radiator or convector	3.0	4.0
Boiler or heater	3.0	4.0
Open gate valve	0.5	0.7
Open globe valve	12.0	17.0

[a]See Table 4 for equivalent length of one elbow.
Source: Giesecke (1926) and Giesecke and Badgett (1931, 1932).

(By permission of American Society of Heating, Refrigerating and Air-Conditioning Engineers, Inc. Atlanta, Georgia, from their 1993 ASHRAE Fundamentals Handbook)

2.4.0   Water Velocities (Types of Service)

Type of Service	Velocity, ft/s	Reference
General service	4 to 10	a, b, c
City water	3 to 7	a, b
	2 to 5	c
Boiler feed	6 to 15	a, c
Pump suction and drain lines	4 to 7	a, b

[a]Crane Co. 1976. Flow of fluids through valves, fittings, and pipe. Technical Paper 410.
[b]*System Design Manual.* 1960. Carrier Air Conditioning Co., Syracuse, NY.
[c]*Piping Design and Engineering.* 1951. Grinnell Company, Inc., Cranston, RI.

Maximum Water Velocity to Minimize Erosion

Normal Operation, h/yr	Water Velocity, ft/s
1500	15
2000	14
3000	13
4000	12
6000	10

Source: *System Design Manual,* Carrier Air Conditioning Co., 1960.

(By permission of American Society of Heating, Refrigerating and Air-Conditioning Engineers, Inc. Atlanta, Georgia, from their 1993 ASHRAE Fundamentals Handbook)

2.5.0 Flow Rates/Demand for Various Plumbing Fixtures

Proper Flow and Pressure Required during Flow for Different Fixtures

Fixture	Flow Pressure[a]	Flow, gpm
Ordinary basin faucet	8	3.0
Self-closing basin faucet	12	2.5
Sink faucet—3/8 in.	10	4.5
Sink faucet—1/2 in.	5	4.5
Dishwasher	15–25	—[b]
Bathtub faucet	5	6.0
Laundry tube cock—1/4 in.	5	5.0
Shower	12	3–10
Ball cock for closet	15	3.0
Flush valve for closet	10–20	15–40[c]
Flush valve for urinal	15	15.0
Garden hose, 50 ft, and sill cock	30	5.0

[a]Flow pressure is the pressure (psig) in the pipe at the entrance to the particular fixture considered.
[b]Varies; see manufacturers' data.
[c]Wide range due to variation in design and type of flush valve closets.

Demand Weights of Fixtures in Fixture Units[a]

Fixture or Group[b]	Occupancy	Type of Supply Control	Weight in Fixture Units[c]
Water closet	Public	Flush valve	10
Water closet	Public	Flush tank	5
Pedestal urinal	Public	Flush valve	10
Stall or wall urinal	Public	Flush valve	5
Stall or wall urinal	Public	Flush tank	3
Lavatory	Public	Faucet	2
Bathtub	Public	Faucet	4
Shower head	Public	Mixing valve	4
Service sink	Office, etc	Faucet	3
Kitchen sink	Hotel or restaurant	Faucet	4
Water closet	Private	Flush valve	6
Water closet	Private	Flush tank	3
Lavatory	Private	Faucet	1
Bathtub	Private	Faucet	2
Shower head	Private	Mixing valve	2
Bathroom group	Private	Flush valve for closet	8
Bathroom group	Private	Flush tank for closet	6
Separate shower	Private	Mixing valve	2
Kitchen sink	Private	Faucet	2
Laundry trays (1 to 3)	Private	Faucet	3
Combination fixture	Private	Faucet	3

Note: See Hunter (1941).
[a]For supply outlets likely to impose continuous demands, estimate continuous supply separately, and add to total demand for fixtures.
[b]For fixtures not listed, weights may be assumed by comparing the fixture to a listed one using water in similar quantities and at similar rates.
[c]The given weights are for total demand. For fixtures with both hot and cold water supplies, the weights for maximum separate demands can be assumed to be 75% of the listed demand for the supply.

(By permission of American Society of Heating, Refrigerating and Air-Conditioning Engineers, Inc. Atlanta, Georgia, from their 1993 ASHRAE Fundamentals Handbook)

2.5.1   Hot-Water Demand for Various Fixtures

Hot water demand per fixture for various types of buildings in gph of water per fixture, calculated at a final temperature of 140 F. (Reprinted with permission of ASHRAE.)										
Fixture	Apartment house	Club	Gymnasium	Hospital	Hotel	Industrial plant	Office building	Private residence	School	YMCA
Basins, private lavatory	2	2	2	2	2	2	2	2	2	2
Basins, public lavatory	4	6	8	6	8	12	6	—	15	8
Bath tubs[a]	20	20	30	20	20	—	—	20	—	30
Dishwashers[b]	15	50-150	—	50-150	50-200	20-100	—	15	20-100	20-100
Foot basins	3	3	12	3	3	12	—	3	3	12
Kitchen sinks	10	20	—	20	30	20	20	10	20	20
Laundry, stationary tubs	20	28	—	28	28	—	—	20	—	28
Pantry sinks	5	10	—	10	10	—	10	5	10	10
Showers	30	150	225	75	75	225	30	30	225	225
Service sinks	20	20	—	20	30	20	20	15	20	20
Hydrotherapeutic showers				400						
Hubbard baths				600						
Leg baths				100						
Arm baths				35						
Sitz baths				30						
Continuous-flow baths				165						
Circular wash sinks				20	20	30	20		30	
Semicircular wash sinks				10	10	15	10		15	
Demand factor	0.30	0.30	0.40	0.25	0.25	0.40	0.30	0.30	0.40	0.40
Storage capacity factor[c]	1.25	0.90	1.00	0.60	0.80	1.00	2.00	0.70	1.00	1.00

[a] Whirlpool baths require specific consideration based on capacity; they are not included in the bathtub category.
[b] Dishwasher requirements should be taken from this table or from manufacturers' data for the model to be used, if this is known.
[c] Ratio of storage tank capacity to probable maximum demand per hr. Storage capacity may be reduced where an unlimited supply of steam is available from a central street steam system or large boiler plant.

(By permission of American Society of Heating, Refrigerating and Air-Conditioning Engineers, Inc. Atlanta, Georgia, from their 1993 ASHRAE Fundamentals Handbook)

2.6.0   Head-of-Water Equivalents (in PSI)

Head Ft.	0	1	2	3	4	5	6	7	8	9
0	....	0.433	0.866	1.299	1.732	2.165	2.598	3.031	3.464	3.987
10	4.330	4.763	5.196	5.629	6.062	6.495	6.928	7.361	7.794	8.277
20	8.660	9.093	9.526	9.959	10.392	10.825	11.258	11.691	12.124	12.557
30	12.990	13.423	13.856	14.289	14.722	15.155	15.588	16.021	16.454	16.887
40	17.320	17.753	18.186	18.619	19.052	19.485	19.918	20.351	20.784	21.217
50	21.650	22.083	22.516	22.949	23.382	23.815	24.248	24.681	25.114	25.547
60	25.980	26.413	26.846	27.279	27.712	28.145	28.578	29.011	29.444	29.877
70	30.310	30.743	31.176	31.609	32.042	32.475	32.908	33.341	33.774	34.207
80	34.640	35.073	35.506	35.939	36.372	36.805	37.238	37.671	38.104	38.537
90	38.970	39.403	39.836	40.269	40.702	41.135	41.568	42.001	42.436	42.867

2.7.0   Pipe Sizes for Horizontal Rainwater Piping

Size of Pipe in Inches 1/8" Slope	Maximum Rainfall in Inches per Hour				
	2	3	4	5	6
3	1644	1096	822	657	548
4	3760	2506	1880	1504	1253
5	6680	4453	3340	2672	2227
6	10700	7133	5350	4280	3566
8	23000	15330	11500	9200	7600
10	41400	27600	20700	16580	13800
11	66600	44400	33300	26650	22200
15	109000	72800	59500	47600	39650

Size of Pipe in Inches 1/4" Slope	Maximum Rainfall in Inches per Hour				
	2	3	4	5	6
3	2320	1546	1160	928	773
4	5300	3533	2650	2120	1766
5	9440	6293	4720	3776	3146
6	15100	10066	7550	6040	5033
8	32600	21733	16300	13040	10866
10	58400	38950	29200	23350	19450
11	94000	62600	47000	37600	31350
15	168000	112000	84000	67250	56000

Size of Pipe in Inches 1/2" Slope	Maximum Rainfall in Inches per Hour				
	2	3	4	5	6
3	3288	2295	1644	1310	1096
4	7520	5010	3760	3010	2500
5	13660	8900	6680	5320	4450
6	21400	13700	10700	8580	7140
8	46000	30650	23000	18400	15320
10	82800	55200	41400	33150	27600
11	133200	88800	66600	53200	44400
15	238000	158800	119000	95300	79250

(By permission of Cast Iron Soil Pipe Institute)

2.8.0   Velocity/Flow in Cast-Iron Sewer Pipe of 2″ (5.08 cm) and 3″ (7.6 cm)

Pipe Size (In.)	SLOPE		¼ FULL		½ FULL		¾ FULL		FULL	
	(In./Ft.)	(Ft./Ft.)	Velocity (Ft./Sec.)	Flow (Gal./Min.)	Velocity (Ft./Sec.)	Flow (Gal./Min.)	Velocity (Ft./Sec.)	Flow (Gal./Min.)	Velocity (Ft./Sec.)	Flow (Gal./Min.)
2.0	0.0120	0.0010	0.36	0.83	0.46	2.16	0.52	3.67	0.46	4.35
	0.0240	0.0020	0.51	1.18	0.66	3.06	0.74	5.18	0.66	6.15
	0.0360	0.0030	0.62	1.45	0.80	3.75	0.90	6.35	0.80	7.53
	0.0480	0.0040	0.72	1.67	0.93	4.33	1.04	7.33	0.93	8.69
	0.0600	0.0050	0.80	1.87	1.04	4.84	1.16	8.20	1.04	9.72
	0.0720	0.0060	0.88	2.04	1.13	5.30	1.27	8.98	1.13	10.65
	0.0840	0.0070	0.95	2.21	1.23	5.72	1.38	9.70	1.23	11.50
	0.0960	0.0080	1.01	2.36	1.31	6.12	1.47	10.37	1.31	12.29
	0.1080	0.0090	1.07	2.50	1.39	6.49	1.56	11.00	1.39	13.04
	0.1200	0.0100	1.13	2.64	1.47	6.84	1.64	11.59	1.47	13.75
	0.2400	0.0200	1.60	3.73	2.07	9.67	2.33	16.39	2.07	19.44
	0.3600	0.0300	1.96	4.57	2.54	11.85	2.85	20.07	2.54	23.81
	0.4800	0.0400	2.26	5.28	2.93	13.68	3.29	23.18	2.93	27.49
	0.6000	0.0500	2.53	5.90	3.28	15.29	3.68	25.92	3.28	30.74
	0.7200	0.0600	2.77	6.47	3.59	16.75	4.03	28.39	3.59	33.67
	0.8400	0.0700	2.99	6.98	3.88	18.10	4.35	30.66	3.88	36.37
	0.9600	0.0800	3.20	7.47	4.14	19.35	4.65	32.78	4.14	38.88
	1.0800	0.0900	3.39	7.92	4.40	20.52	4.93	34.77	4.40	41.24
	1.2000	0.1000	3.58	8.35	4.63	21.63	5.20	36.65	4.63	43.47
3.0	0.0120	0.0010	0.47	2.55	0.61	6.56	0.69	11.05	0.61	13.12
	0.0240	0.0020	0.67	3.61	0.86	9.28	0.97	15.63	0.86	18.55
	0.0360	0.0030	0.82	4.42	1.06	11.36	1.19	19.14	1.06	22.72
	0.0480	0.0040	0.95	5.11	1.22	13.12	1.37	22.10	1.22	26.24
	0.0600	0.0050	1.06	5.71	1.37	14.67	1.53	24.71	1.37	29.33
	0.0720	0.0060	1.16	6.25	1.50	16.07	1.68	27.07	1.50	32.13
	0.0840	0.0070	1.25	6.75	1.62	17.35	1.81	29.24	1.62	34.71
	0.0960	0.0080	1.34	7.22	1.73	18.55	1.94	31.26	1.73	37.11
	0.1080	0.0090	1.42	7.66	1.83	19.68	2.06	33.16	1.83	39.36
	0.1200	0.0100	1.50	8.07	1.93	20.74	2.17	34.95	1.93	41.49
	0.2400	0.0200	2.21	11.42	2.73	29.33	3.07	49.43	2.73	58.67
	0.3600	0.0300	2.60	13.98	3.35	35.93	3.76	60.53	3.35	71.86
	0.4800	0.0400	3.00	16.14	3.87	41.49	4.34	69.90	3.87	82.97
	0.6000	0.0500	3.35	18.05	4.32	46.38	4.85	78.15	4.32	92.77
	0.7200	0.0600	3.67	19.77	4.74	50.81	5.31	85.61	4.74	101.62
	0.8400	0.0700	3.96	21.36	5.12	54.88	5.74	92.47	5.12	109.76
	0.9600	0.0800	4.24	22.83	5.47	58.67	6.13	98.85	5.47	117.34
	1.0800	0.0900	4.50	24.22	5.80	62.23	6.51	104.85	5.80	124.46
	1.2000	0.1000	4.74	25.53	6.11	65.29	6.86	110.52	6.11	131.19

(By permission of Cast Iron Soil Pipe Institute)

2.9.0 Expansion Characteristics of Metal and Plastic Pipe

Expansion: Allowances for expansion and contraction of building materials are important design considerations. Material selection can create or prevent problems. Cast iron is in tune with building reactions to temperature. Its expansion is so close to that of steel and masonry that there is no need for costly expansion joints and special offsets. That is not always the case with other DWV materials.

Thermal expansion of various materials.			
Material	Inches per inch 10^{-6} X per °F	Inches per 100' of pipe per 100°F.	Ratio-assuming cast iron equals 1.00
Cast iron	6.2	0.745	1.00
Concrete	5.5	0.66	.89
Steel (mild)	6.5	0.780	1.05
Steel (stainless)	7.8	0.940	1.26
Copper	9.2	1.11	1.49
PVC (high impact)	55.6	6.68	8.95
ABS (type 1A)	56.2	6.75	9.05
Polyethylene (type 1)	94.5	11.4	15.30
Polyethylene (type 2)	83.3	10.0	13.40

Here is the *actual* increase in length for 50 feet of pipe and 70° temperature rise.

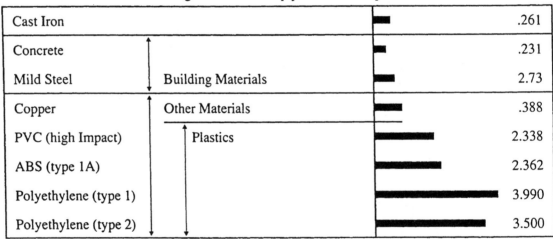

Cast Iron			.261
Concrete			.231
Mild Steel	Building Materials		2.73
Copper	Other Materials		.388
PVC (high Impact)	Plastics		2.338
ABS (type 1A)			2.362
Polyethylene (type 1)			3.990
Polyethylene (type 2)			3.500

(By permission of Cast Iron Soil Pipe Institute)

2.9.1   Expansion Characteristics of Metal and Plastic Pipe in Graph Form

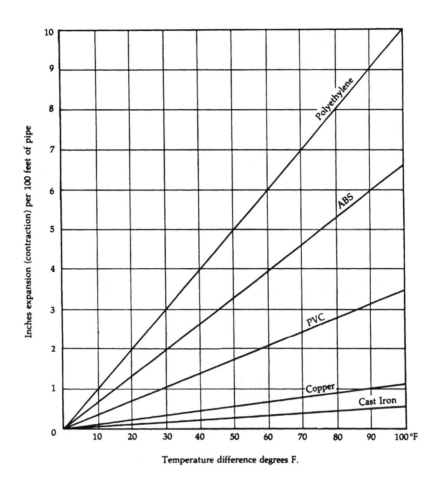

Temperature difference degrees F.

Example: Find the expansion allowance required for a 120 ft. run of ABS pipe in a concrete & masonry building and for a temperature difference of 90°F.

Answer: At a temperature difference of 90°F read from the chart, ABS expands 6″ and concrete expands ¾″.

(6 - ¾) x $\frac{120}{100}$ = 5¼ x $\frac{120}{100}$ = 6.3 inches

(By permission of Cast Iron Soil Pipe Institute, Chattanooga, TN)

2.10.0   Size of Roof Drains for Varying Amounts of Rainfall (in Square Feet)

Rain Fall in Inches	Size of Drain or Leader in Inches*					
	2	3	4	5	6	8
1	2880	8800	18400	34600	54000	116000
2	1440	4400	9200	17300	27000	58000
3	960	2930	6130	11530	17995	38660
4	720	2200	4600	8650	13500	29000
5	575	1760	3680	6920	10800	23200
6	480	1470	3070	5765	9000	19315
7	410	1260	2630	4945	7715	16570
8	360	1100	2300	4325	6750	14500
9	320	980	2045	3845	6000	12890
10	290	880	1840	3460	5400	11600
11	260	800	1675	3145	4910	10545
12	240	730	1530	2880	4500	9660

*Round, square or rectangular rainwater pipe may be used and are considered equivalent when closing a scribed circle quivalent to the leader diameter.

Source: Uniform Plumbing Code (IAPMO) 1985 Edition

(By permission of Cast Iron Soil Pipe Institute)

2.11.0   Comparative Costs of Steam-Condensate Lines

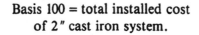

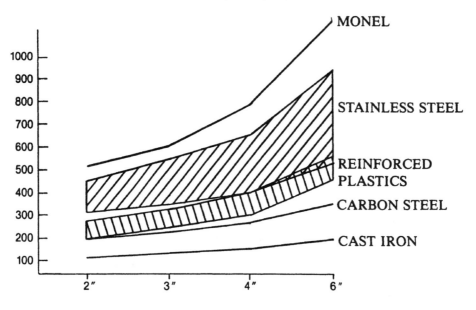

PIPE SIZE, O.D.

(By permission of Cast Iron Soil Pipe Institute)

2.12.0   Supports for Pipe Risers (Illustrated)

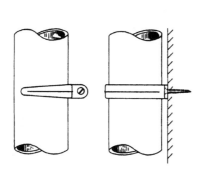

Bracket for Vertical Pipe

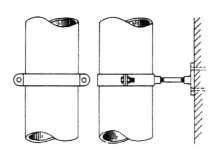

One Hole Strap for Vertical Pipe

(By permission of Cast Iron Soil Pipe Institute)

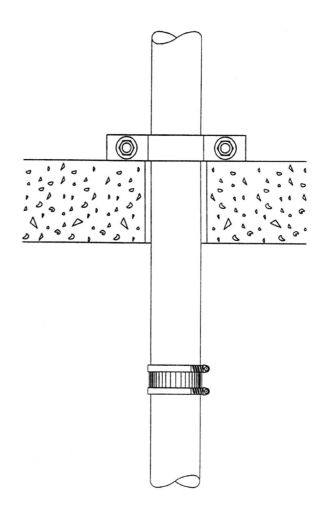

Method of clamping
the Pipe at Each Floor,
Using a Friction Clamp
or Floor Clamp

2.12.1 Supports for Horizontal Pipe Runs (Illustrated)

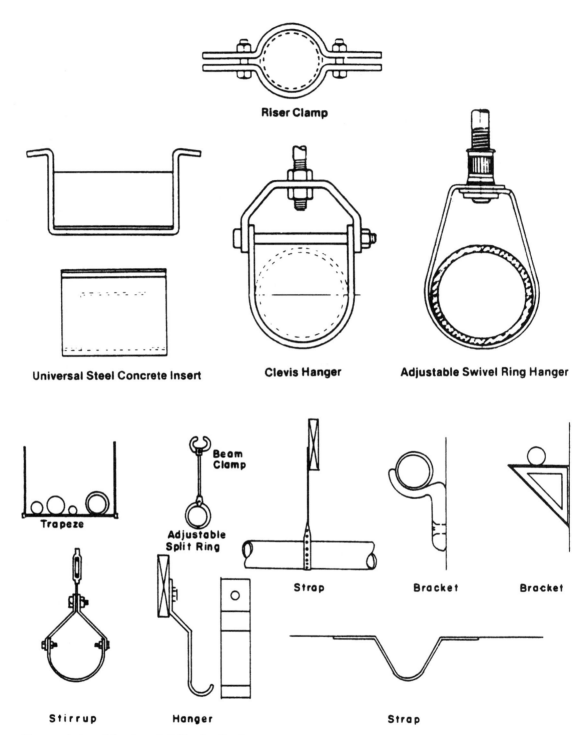

(By permission of Cast Iron Soil Pipe Institute)

2.13.0 Cast-Iron Pipe Hub-Barrel Dimensions

The following dimensions are given for use as convenient information on details of the hub barrel and spigot, and are not requirements of this specification.

Outside Dimensions of Hub, Barrel, and Spigot for Detailing, in.

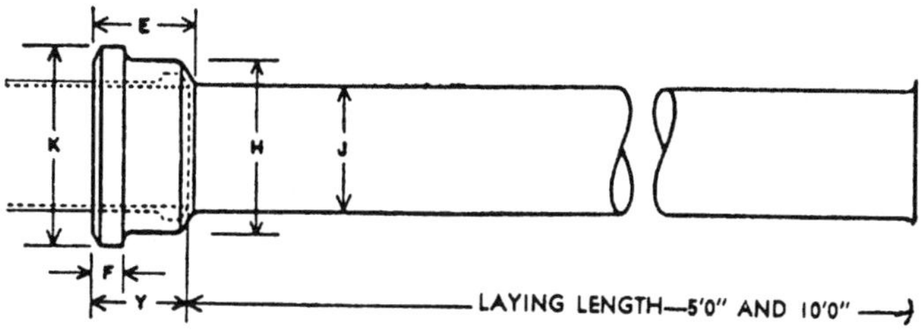

NOTE 1— in. = 25.4 mm.

Size (nominal ID)	Extra-Heavy Pipe 'XH'					
	K max	*H* max	*J*	*F*	*Y*	*E*
2	4 1/8	3 5/8	2 3/8	3/4	2 1/2	2 3/4
3	5 3/8	4 15/16	3 1/2	13/16	2 3/4	3 1/4
4	6 3/8	5 15/16	4 1/2	7/8	3	3 1/2
5	7 3/8	6 15/16	5 1/2	7/8	3	3 1/2
6	8 3/8	7 15/16	6 1/2	7/8	3	3 1/2
8	11 1/16	10 7/16	8 5/8	1 3/16	3 1/2	4 1/8
10	13 5/16	12 11/16	10 3/4	1 3/16	3 1/2	4 1/8
12	15 7/16	14 13/16	12 3/4	1 7/16	4 1/4	5
15	18 13/16	18 3/16	15 7/8	1 7/16	4 1/4	5

Size (nominal ID)	Service-Pipe 'SV'					
	K max	*H* max	*J*	*F*	*Y*	*E*
2	3 15/16	3 3/8	2 1/4	3/4	2 1/2	2 3/4
3	5	4 1/2	3 1/4	13/16	2 3/4	3 1/4
4	6	5 1/2	4 1/4	7/8	3	3 1/2
5	7	6 1/2	5 1/4	7/8	3	3 1/2
6	8	7 1/2	6 1/4	7/8	3	3 1/2
8	10 1/2	9 7/8	8 3/8	1 3/16	3 1/2	4 1/8
10	12 13/16	12 3/16	10 1/2	1 3/16	3 1/2	4 1/8
12	14 15/16	14 5/16	12 1/2	1 7/16	4 1/4	5
15	18 5/16	17 5/8	15 5/8	1 7/16	4 1/4	5

(By permission of Cast Iron Soil Pipe Institute)

2.14.0   Pipe Diameters and Trench Widths (U.S. and Metric Sizes)

Pipe Diameter (millimeters)	Trench Width (millimeters)	Pipe Diameter (millimeters)	Trench Width (millimeters)
100	470	1500	2500
150	540	1650	2800
200	600	1800	3000
250	680	1950	3200
300	800	2100	3400
375	910	2250	3600
450	1020	2400	3900
525	1100	2550	4100
600	1200	2700	4300
675	1300	2850	4500
825	1600	3000	4800
900	1700	3150	5000
1050	1900	3300	5200
1200	2100	3450	5400
1350	2300	3600	5600

NOTE: Trench widths based on 1.25 Bc + 300 where Bc is the outside diameter of the pipe in millimeters.

Pipe Diameter (inches)	Trench Width (feet)	Pipe Diameter (inches)	Trench Width (feet)
4	1.6	60	8.5
6	1.8	66	9.2
8	2.0	72	10.0
10	2.3	78	10.7
12	2.5	84	11.4
15	3.0	90	12.1
18	3.4	96	12.9
21	3.8	102	13.6
24	4.1	108	14.3
27	4.5	114	14.9
33	5.2	120	15.6
36	5.6	126	16.4
42	6.3	132	17.1
48	7.0	138	17.8
54	7.8	144	18.5

NOTE: Trench widths based on 1.25 Bc + 1 ft where Bc is the outside diameter of the pipe in inches.

2.15.0   Pipe Test Plugs (Illustrated)

Typical test plugs used for air/water tests.

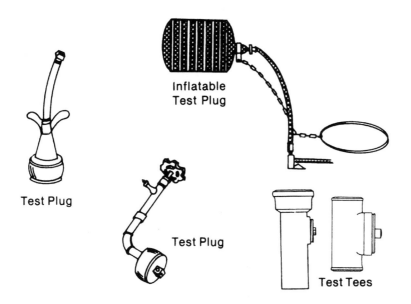

Test Plug

Inflatable Test Plug

Test Plug

Test Tees

2.16.0   Thrust Pressures When Hydrostatically Testing Soil Pipe

PIPE SIZE		1½″	2″	3″	4″	5″	6″	8″	10″
HEAD, Feet of Water	PRESSURE PSI	THRUST lb.	THRUST lb.	THRUST lb.	THRUST lb.	THRUST lb.	THRUST lb.	THRUST lb.	THRUST lb.
10	4.3	12	19	38	65	95	134	237	377
20	8.7	25	38	77	131	192	271	480	762
30	13.0	37	56	115	196	287	405	717	1139
40	17.3	49	75	152	261	382	539	954	1515
50	21.7	62	94	191	327	479	676	1197	1900
60	26.0	74	113	229	392	574	810	1434	2277
70	30.3	86	132	267	457	668	944	1671	2654
80	34.7	99	151	306	523	765	1082	1914	3039
90	39.0	111	169	344	588	860	1216	2151	3416
100	43.4	123	188	382	654	957	1353	2394	3801
110	47.7	135	208	420	719	1052	1487	2631	4178
120	52.0	147	226	458	784	1147	1621	2868	4554
AREA, OD. in.²		2.84	4.34	8.81	15.07	22.06	31.17	55.15	87.58

Thrust = Pressure x Area

(By permission of Cast Iron Soil Pipe Institute)

2.17.0   Piping Schematics (Vent and Stack Installations)

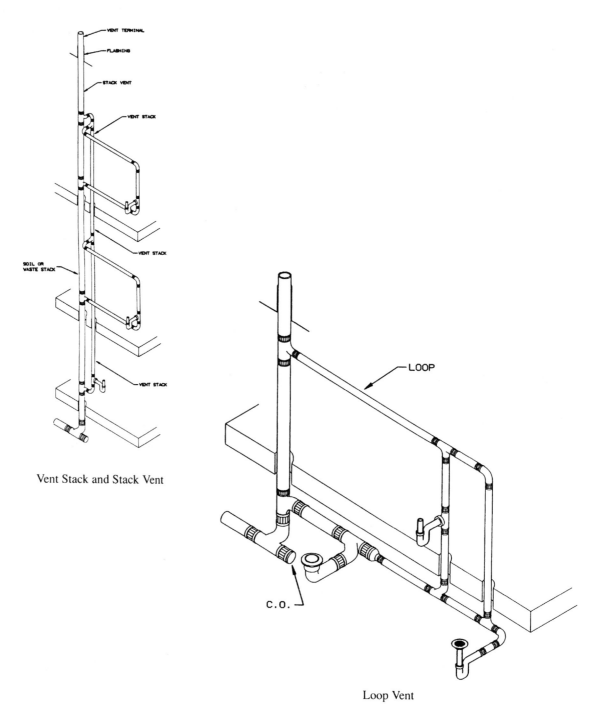

Vent Stack and Stack Vent

Loop Vent

(By permission of Cast Iron Soil Pipe Institute)

2.17.1   Piping Schematics (Continuous- and Looped-Vent System)

TYPICAL LAYOUTS

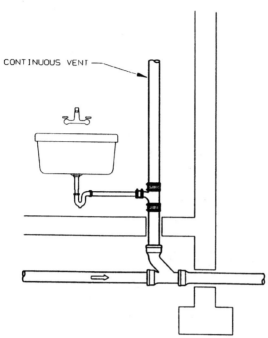

Continuous Vent

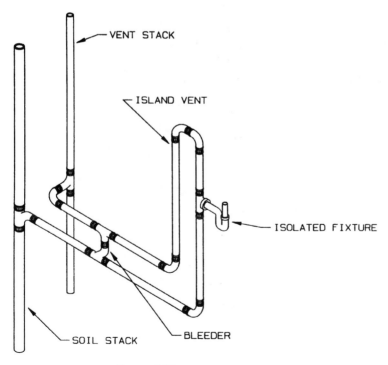

Looped Vent

(By permission of Cast Iron Soil Pipe Institute)

2.17.2   Piping Schematics (Stacked Fixture Installation)

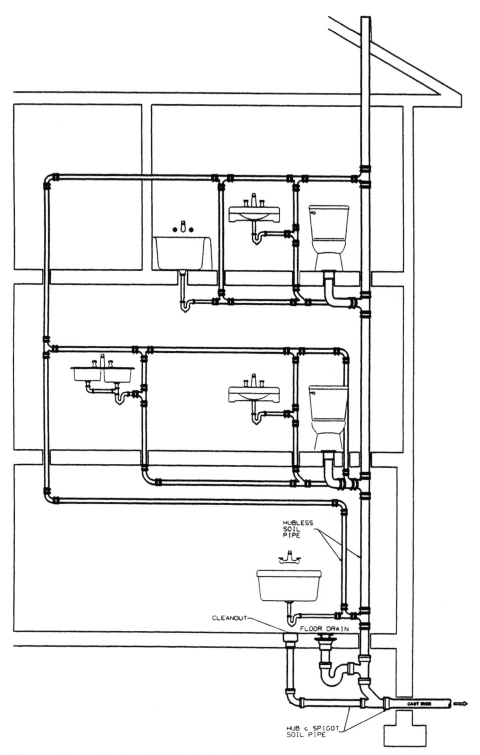

(By permission of Cast Iron Soil Pipe Institute)

2.17.3 Piping Schematics (Roof Drain and Leader, Hubless/Hub Pipe)

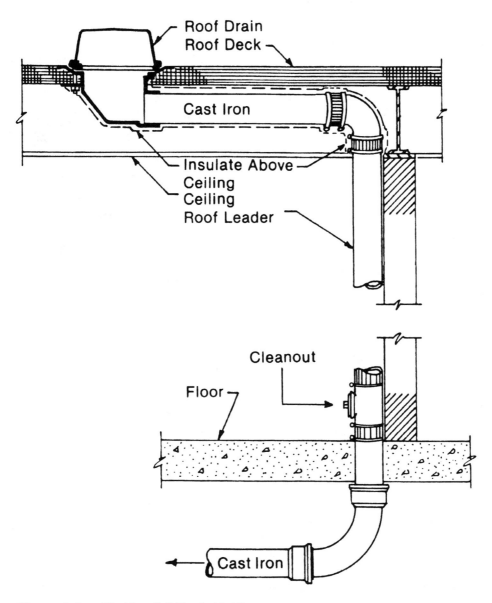

(By permission of Cast Iron Soil Pipe Institute)

2.17.4   Piping Schematics (Battery of Fixtures with a Common Vent)

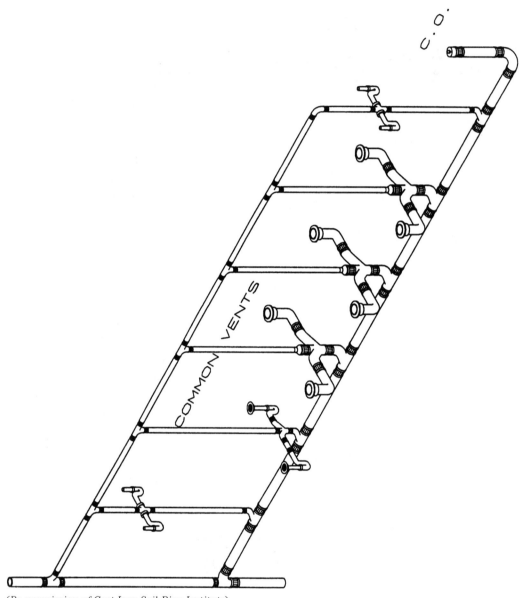

(By permission of Cast Iron Soil Pipe Institute)

2.17.5 Piping Schematics (Circuit Venting/Wet Venting)

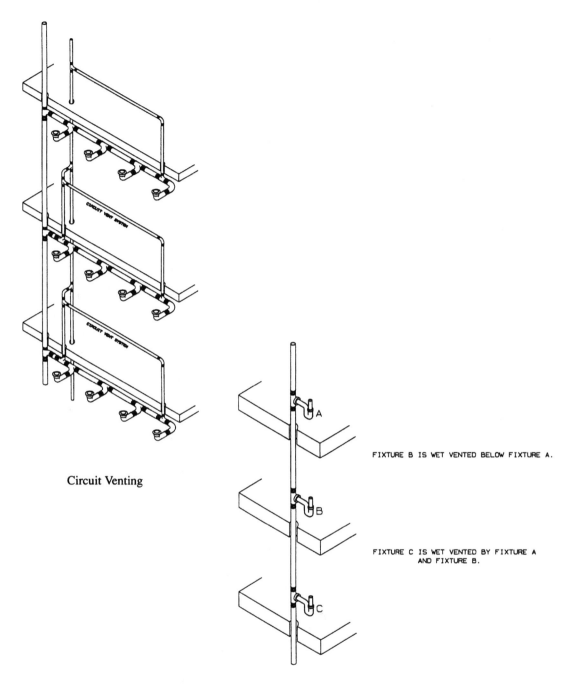

Circuit Venting

FIXTURE B IS WET VENTED BELOW FIXTURE A.

FIXTURE C IS WET VENTED BY FIXTURE A
AND FIXTURE B.

Wet Vent

(By permission of Cast Iron Soil Pipe Institute)

2.17.6 Piping Schematics (Typical Waste and Vent Installation)

Typical waste and vent pipe installation for plumbing fixtures

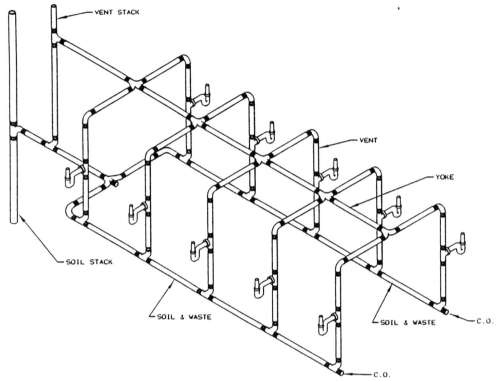

Drainage for a Battery of Fixtures with a Wide Pipe Space Available

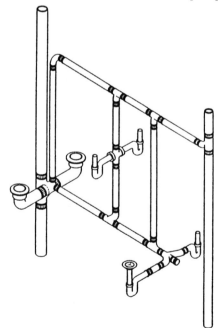

**PIPING FOR TUB, LAVATORY & WATER CLOSET
EACH FIXTURE VENTED**

Typical Piping Arrangement for a Water Closet, Lavatory and Tub.
Piping may be either Hubless or Hub and Spigot.

(By permission of Cast Iron Soil Pipe Institute)

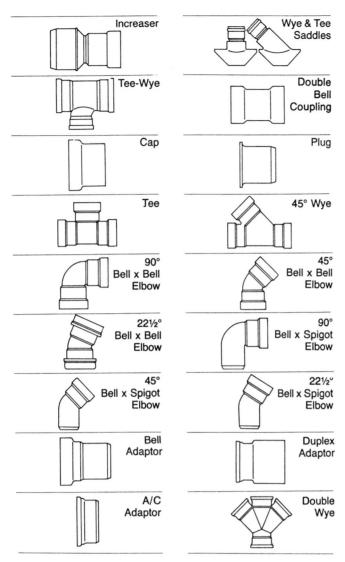

Figure 2.18.0 Various fitting configurations.

Materials and products	ANSI	ASTM	FS	IAPMO	Other standards	Footnote remarks
FERROUS PIPE AND FITTINGS:						
Cast Iron Screwed Fittings (125 & 250 lbs) (56.8 & 113.5 Kg)	B16.4-1963	A 126-66				
Cast Iron Soil Pipe and Fittings		A 74-82				Note 4
Cast Iron Soil Pipe and Fittings for Hubless Cast Iron Sanitary Systems					CISPI 301-85	Note 4
Cast Iron Threaded Drainage Fittings	B16.12-1971					Note 4
Gray Iron and Ductile Iron Pressure Pipe		A 377-66				
Hubless Cast Iron Sanitary And Rainwater Systems (Installation)				IS 6-89		
Malleable Iron Threaded Fittings (150 & 300 lb) (68.1 & 136.2 Kg)	B16.3-1977					
Neoprene Rubber Gaskets for Hub and Spigot Cast Iron Soil Pipe and Fittings					CISPI HSN-85	
Pipe, Steel, Black and Hot-Dipped, Zinc-Coated Welded and Seamless		A 53-83				
Pipe, Steel, Black and Hot-Dipped, Zinc-Coated (Galvanized) Welded and Seamless, For Ordinary Uses		A 120-82				
Pipe Threads, General Purpose (Inch)	B1.20.1-83					
Roof Drains	A112.21.2M-1983					
Shielded Couplings for Use with Hubless Cast Iron Soil Pipe and Fittings				PS 35-89		

Figure 2.19.0 ANSI, ASTM, other standard designations for ferrous and nonferrous plastic pipe and fittings. (*By permission: McGraw-Hill Inc.* Plumber's and Pipefitter's Calculations Manual, *R. Dodge Woodson.*)

Special Cast Iron Fittings.			PS 5-84
Subdrains For Built-up Shower Pans			PS16-90
Threaded Cast Iron Pipe For Drainage, Vent and Waste Services.	A40.5-1943		
Welded and Seamless Carbon Steel and Austenitic Stainless Steel Pipe Nipples		A 733-76	
NONFERROUS PIPE AND FITTINGS:			
Brass-, Copper-, and Chromium-Plated Pipe Nipples.		B 687-81	
Bronze Pipe Flanges and Flanged Fittings (Class 150 & 300)	B16.24-1979		
Cast Brass and Tubing P-Traps			PS 2-89
Cast Copper Alloy Fittings for Flared Copper Tubes	B16.26-1987		
Cast Bronze Threaded Fittings (Classes 125 & 250).	B16.15-1985		
Cast Copper Alloy Solder-Joint Drainage Fittings-DWV.	B16.23-1984		
Cast Copper Alloy Solder-Joint Pressure Fittings	B16.18-1984		Note 4

Figure 2.19.0 (*Continued*)

Welded Copper Tube		B 447-86	
Welded Copper Water Tube		B 716-86	
Wrought Copper and Copper Alloy Solder-Joint Pressure Fittings	B16.22-1986		
Wrought Copper and Wrought Copper Alloy Solder-Joint Drainage Fittings-DWV	B16.29-1986		Note 4
NON-METALLIC PIPE:			
Acrylonitrile-Butadiene-Styrene (ABS) Building Drain, Waste and Vent Pipe and Fittings (Installation)		IS 5-89	Note 4
Acrylonitrile-Butadiene-Styrene (ABS) Schedule 40 Plastic Drain Waste and Vent Pipe		D 2661-87a	Note 4
Acrylonitrile-Butadiene-Styrene (ABS) Schedule 40 Plastic Drain Waste and Vent Pipe With a Cellular Core		F 628-88	Note 4
Acrylonitrile-Butadiene-Styrene (ABS) Sewer Pipe and Fittings		D 2751-88	Note 4

Figure 2.19.0 *(Continued)*

Description			
Drain, Waste and Vent (DWV) Plastic Fittings Patterns	D 3311-86		Note 4
Extra Strength Vitrified Clay Pipe in Building Drains (Installation)		IS 18-85	
Fittings for Joining Polyethylene Pipe for Water Service and Yard Piping		PS 25-84	
Joints for IPS PVC Pipe Using Solvent Cement	D 2672-88		
Non-Metallic Building Sewers (Installation)		IS 1-90	
Plastic Insert Fittings For Polybutylene (PB) Tubing	F 845-88		Note 4
Plastic Insert Fittings for Polyethylene (PE) Plastic Pipe	D 2609-88		Note 6
Polybutylene (PB) Cold Water Building Supply and Yard Piping and Tubing (Installation)		IS 17-90	
Polybutylene Hot and Cold Water Distribution Tubing Systems Using Insert Fittings (Installation)		IS 22-90	
Polybutylene Hot and Cold Water Distribution Tubing Systems Using Compression Joints (Installation)		IS 25-90	

Figure 2.19.0 (*Continued*)

Materials and products	ANSI	ASTM	FS	IAPMO	Other standards	Footnote remarks
NONFERROUS PIPE FITTINGS:						
Copper Drainage Tube (DWV)............		B 306-86				
Copper Plumbing Tube, Pipe and Fittings (Installation)......................				IS 3-89		
Diversion Tees and Twin Waste Elbow.......				PS 9-84		
Drains for Prefabricated and Precast Showers ..				PS 4-90		
Flexible Metallic Water Connectors.........				PS 14-89		
General Requirements for Wrought Seamless Copper and Copper-Alloy Tube		B 251-87				
Seamless Brass Tube....................		B 135-86(a)				
Seamless Copper Pipe, Standard Sizes		B 42-87				
Seamless Copper Tube....................		B 75-86				
Seamless Copper Water Tube		B 88-86				
Seamless Red Brass Pipe, Standard Sizes		B 43-87				
Seamless and Welded Copper Distribution Tube (Type D)............................		B 641-86				
Threadless Copper Pipe....................		B 302-87				
Tubing Trap Wall Adapters				PS 7-84		
Welded Brass Tube......................		B 587-88				
Welded Copper-Alloy UNS No. C21000 Water Tube................................		B 642-86				
Welded Copper and Copper Alloy Water Tube (Installation)...........................				IS 21-89		

Figure 2.19.0 (*Continued*)

Materials and products	ANSI	ASTM	FS	IAPMO	Other standards	Footnote remarks
NON-METALLIC PIPE:						
Acrylonitrile-Butadiene-Styrene (ABS) Sewer Pipe and Fittings (Installation)				IS 11-87		Notes 1 & 3
Asbestos-Cement Nonpressure Sewer Pipe		C 428-74				
Asbestos Cement Pressure Pipe		C 296-73				
Asbestos-Cement Pressure Pipe For Water and other Liquids					AWWA C400-72	
Asbestos Cement Pressure Pipe For Water Service and Yard Piping (Installation)				IS 15-82		
Borosilicate Glass Pipe and Fittings for Drain, Waste and Vent (DWV) Applications		C 1053-85				
Chlorinated Poly (Vinyl Chloride) (CPVC) Plastic Pipe, Schedules 40 and 80		F 441-88				
Chlorinated Poly (Vinyl Chloride) (CPVC) Plastic Hot and Cold Water Distribution Systems		D 2846-89[e1]				
Chlorinated Poly (Vinyl Chloride) (CPVC) Solvent Cemented Hot and Cold Water Distribution Systems (Installation)				IS 20-89		
Coextruded Poly (Vinyl Chloride) Plastic Pipe with a Cellular Core		F891-86[e1]				
Concrete Drain Tile		C 412-80				Note 3
Concrete Sewer, Storm Drain and Culvert Pipe		C 14-80				

Figure 2.19.0 (*Continued*)

Materials and products	ANSI	ASTM	FS	IAPMO	Other standards	Footnote remarks
NON-METALLIC PIPE:						
Polybutylene Hot and Cold Water Distribution Pipe, Tubing and Fitting Systems Using Heat Fusion (Installation)				IS 23-90		
Polybutylene Hot and Cold Water Distribution Pipe, Tubing and Fitting systems Using Pressure-Lock Fittings (Installation)				IS 24-90		
Polybutylene (PB) Plastic Hot- & Cold-Water Distribution Systems		D 3309-88a				
Polybutylene (PB) Plastic Pipe (SIDR-PR) Based on Controlled Inside Diameter		D 2662-88				
Polybutylene (PB) Plastic Tubing		D 2666-88				
Polyethylene (PE) Cold Water Building Supply and Yard Piping (Installation)				IS 7-90		
Polyethylene (PE) For Gas Yard Piping (Installation)				IS 12-90		
Polyethylene (PE) Plastic Pip (SIDR-PR) Based on Controlled Inside Diameter		D 2239-88				
Poly (Vinyl Chloride) (PVC) Building Drain, Waste and Vent Pipe and Fittings (Installation)				IS 9-90		
Poly (Vinyl Chloride) (PVC) Cold Water Building Supply and Yard Piping (Installation)				IS 8-89		

Figure 2.19.0 (*Continued*)

Materials and products	ANSI	ASTM	FS	IAPMO	Other standards	Footnote remarks
NON-METALLIC PIPE:						
Socket-Type Chlorinated Poly (Vinyl Chloride) (CPVC) Plastic Pipe Fittings, Schedule 40......		F 438-88				
Socket-Type Chlorinated Poly (Vinyl Chloride) (CPVC) Plastic Pipe Fittings, Schedule 80		F 439-88				
Socket-Type Poly (Vinyl Chloride) (PVC) Plastic Pipe Fittings Schedule 80............		D 2467-88				Note 4
Solvent Cement For Acrylonitrile-Butadiene-Styrene (ABS) Plastic Pipe and Fittings.........		D 2235-88				
Solvent Cements For Chlorinated Poly (Vinyl Chloride) (CPVC) Plastic Pipe and Fittings......		F 493-88				
Solvent Cements For Poly (Vinyl Chloride) (PVC) Plastic Pipe and Fittings..............		D 2564-88				
Thermoplastic Accessible and Replaceable Plastic Tube and Tubular Fittings.............		F 409-88				Note 4
Thermoplastic Gas Pressure Pipe, Tubing and Fittings...............		D 2513-88b				
Type PS-46 Poly (Vinyl Chloride) (PVC) Plastic Gravity Flow Sewer Pipe and Fittings.........		F 789-85				
Type PSM Poly (Vinyl Chloride) (PVC) Sewer Pipe and Fittings.............		D 3034-88				
Threaded Poly (Vinyl Chloride) (PVC) Plastic Pipe Fittings Schedule 80.............		D 2464-88				Note 4
Vitrified Clay Pipe, Extra Strength, Standard Strength and Perforated............		C 700-78				

Figure 2.19.0 (*Continued*)

Poly (Vinyl Chloride) (PVC) Corrugated Sewer Pipe with a Smooth Interior and Fittings	F 949-86a	Note 4
Poly (Vinyl Chloride) (PVC) Natural Gas Yard Piping (Installation)	IS 10-90	
Poly (Vinyl Chloride) (PVC) Plastic Drain, Waste and Vent Pipe and Fittings	D 2665-88	
Poly (Vinyl Chloride) (PVC) Pressure-Rated Pipe (SDR Series)	D 2241-88	
Poly (Vinyl Chloride) (PVC) Plastic Pipe, Schedules 40, 80 and 120................	D 1785-88	
Poly (Vinyl Chloride) (PVC) Plastic Pipe Fittings (Schedule 40)	D 2466-88	Note 4
Primers For Use in Solvent Cement Joints of Poly (Vinyl Chloride) (PVC) Plastic Pipe and Fittings	F 656-88	
Rubber Rings for Asbestos-Cement Pipe.	D 1869-79	
Safe Handling Of Solvent Cements, Primers, and Cleaners Used For Joining Thermoplastic Pipe and Fittings.....................	F 402-88	
Smoothwall Polyethylene (PE) Pipe for Use in Drainage and Waste Disposal Absorption Fields	F810-85	

Explanation of Notes:
Note 1: Limited to domestic sewage.
Note 2: Alloy C85200 for cleanout plugs.
Note 3: Type II only.
Note 4: Some of the tubes or fittings are not acceptable for use under the Uniform Plumbing Code.
Note 5: PDI Standard G101 by reference.
Note 6: Limited to nylon material only.

Figure 2.19.0 (*Continued*)

PLUMBING FIXTURES:

Enameled Cast Iron Plumbing Fixtures	A112.19.1M-1987
Jetted Whirlpool Bathtubs	A112.19.7-87
Plastic Bathtub Units	Z124.1-1987
Plastic Lavatories	Z124.3-1986
Plastic Shower Receptors and Shower Stalls....	Z124.2-1987
Plastic Water Closet Bowls and Tanks	Z124.4-1986
Plumbing Fixtures For Land Use	
Plumbing Requirements for Diverters for Plumbing Faucets with Hose Spray, Anti-Siphon Type Resiential Application.............	ASSE/ANSI 1025-78

WWP-541-71

WWP-541-71
WWP-541-71
WWP-541-71
WWP-541-71
WWP-541-71

Figure 2.19.1 ANSI, ASTM designations for plumbing fixtures. (*By permission: McGraw-Hill Inc. Plumber's and Pipefitter's Calculations Manual, R. Dodge Woodson.*)

Flange standards. All dimensions are in inches.

125 lb. CAST IRON **ASA B16.1**

Pipe Size	1/2	3/4	1	1 1/4	1 1/2	2	2 1/2	3	3 1/2	4	5	6	8	10	12
Diameter of Flange			4 1/4	4 5/8	5	6	7	7 1/2	8 1/2	9	10	11	13 1/2	16	19
Thickness of Flange (min)[1]			7/16	1/2	9/16	5/8	11/16	3/4	13/16	15/16	15/16	1	1 1/8	1 3/16	1 1/4
Diameter of Bolt Circle			3 1/8	3 1/2	3 7/8	4 3/4	5 1/2	6	7	7 1/2	8 1/2	9 1/2	11 3/4	14 1/4	17
Number of Bolts			4	4	4	4	4	4	8	8	8	8	8	12	12
Diameter of Bolts			1/2	1/2	1/2	5/8	5/8	5/8	5/8	5/8	3/4	3/4	3/4	7/8	7/8

[1] 125 lb. flanges have plain faces.

250 lb CAST IRON **ASA B16.2**

Pipe Size	1/2	3/4	1	1 1/4	1 1/2	2	2 1/2	3	3 1/2	4	5	6	8	10	12
Diameter of Flange			4 7/8	5 1/4	6 1/8	6 1/2	7 1/2	8 1/4	9	10	11	12 1/2	15	17 1/2	20 1/2
Thickness of Flange (min)[2]			11/16	3/4	13/16	7/8	1	1 1/8	1 3/16	1 1/4	1 3/8	1 7/16	1 5/8	1 7/8	2
Diameter of Raised Face			2 11/16	3 1/16	3 9/16	4 3/16	4 15/16	6 5/16	6 5/16	6 15/16	8 5/16	9 11/16	11 15/16	14 1/16	16 7/16
Diameter of Bolt Circle			3 1/2	3 7/8	4 1/2	5	5 7/8	6 5/8	7 1/4	7 7/8	9 1/4	10 5/8	13	15 1/4	17 3/4
Number of Bolts			4	4	4	8	8	8	8	8	8	12	12	16	16
Diameter of Bolts			5/8	5/8	3/4	5/8	3/4	3/4	3/4	3/4	3/4	3/4	7/8	1	1 1/8

[2] 250 lb. flanges have a 1/16″ raised face which is included in the flange thickness dimensions.

150 lb BRONZE **ASA B16.24**

Pipe Size	1/2	3/4	1	1 1/4	1 1/2	2	2 1/2	3	3 1/2	4	5	6	8	10	12
Diameter of Flange	3 1/2	3 7/8	4 1/4	4 5/8	5	6	7	7 1/2	8 1/2	9	10	11	13 1/2	16	19
Thickness of Flange (min)[3]	5/16	11/32	3/8	13/32	7/16	1/2	9/16	5/8	11/16	11/16	3/4	13/16	15/16	1	1 1/16
Diameter of Bolt Circle	2 3/8	2 3/4	3 1/8	3 1/2	3 7/8	4 3/4	5 1/2	6	7	7 1/2	8 1/2	9 1/2	11 3/4	14 1/4	17
Number of Bolts	4	4	4	4	4	4	4	4	8	8	8	8	8	12	12
Diameter of Bolts	1/2	1/2	1/2	1/2	1/2	5/8	5/8	5/8	5/8	5/8	3/4	3/4	3/4	7/8	7/8

[3] 150 lb. bronze flanges have plain faces with two concentric gasket-retaining grooves between the port and the bolt holes.

300 lb. BRONZE **ASA B16.24**

Pipe Size	1/2	3/4	1	1 1/4	1 1/2	2	2 1/2	3	3 1/2	4	5	6	8	10	12
Diameter of Flange	3 3/4	4 5/8	4 7/8	5 1/4	6 1/8	6 1/2	7 1/2	8 1/4	9	10	11	12 1/2	15		
Thickness of Flange (min)[4]	1/2	17/32	19/32	5/8	11/16	3/4	13/16	29/32	31/32	1 1/16	1 1/8	1 3/16	1 3/8		
Diameter of Bolt Circle	2 5/8	3 1/4	3 1/2	3 7/8	4 1/2	5	5 7/8	6 5/8	7 1/4	7 7/8	9 1/4	10 5/8	13		
Number of Bolts	4	4	4	4	4	8	8	8	8	8	8	12	12		
Diameter of Bolts	1/2	5/8	5/8	5/8	3/4	5/8	3/4	3/4	3/4	3/4	3/4	3/4	7/8		

[4] 300 lb. bronze flanges have plain faces with two concentic gasket-retaining grooves between the port and the bolt holes.

150 lb. STEEL **ASA B16.5**

Pipe Size	1/2	3/4	1	1 1/4	1 1/2	2	2 1/2	3	3 1/2	4	5	6	8	10	12
			4 1/4	4 5/8	5	6	7	7 1/2	8 1/2	9	10	11	13 1/2	16	19
Thickness of Flange (min)[5]			7/16	1/2	9/16	5/8	13/16	3/4	13/16	15/16	15/16	1	1 1/8	1 3/16	1 1/2
Diameter of Raised Face			2	2 1/2	2 7/8	3 5/8	4 1/8	5	5 1/2	6 3/16	7 5/16	8 1/2	10 5/8	12 3/4	15
Diameter of Bolt Circle			3 1/8	3 1/2	3 7/8	4 3/4	5 1/2	6	7	7 1/2	8 1/2	9 1/2	11 3/4	14 1/4	17
Number of Bolts			4	4	4	4	4	4	8	8	8	8	8	12	12
Diameter of Bolts			1/2	1/2	1/2	5/8	5/8	5/8	5/8	5/8	3/4	3/4	3/4	7/8	7/8

[5] 150 lb. steel flanges have a 1/16″ raised face which is included in the flange thickness dimensions.

300 lb. STEEL **ASA B16.5**

Pipe Size	1/2	3/4	1	1 1/4	1 1/2	2	2 1/2	3	3 1/2	4	5	6	8	10	12
Diameter of Flange			4 7/8	5 1/4	6 1/8	6 1/2	7 1/2	8 1/4	9	10	11	12 1/2	15	17 1/2	20 1/2
Thickness of Flange (min)[6]			11/16	3/4	13/16	7/8	1	1 1/8	1 3/16	1 1/4	1 3/8	1 7/16	1 5/8	1 7/8	2
Diameter of Raised Face			2	2 1/2	2 7/8	3 5/8	4 1/8	5	5 1/2	6 3/16	7 5/16	8 1/2	10 5/8	12 3/4	15
Diameter of Bolt Circle			3 1/2	3 7/8	4 1/2	5	5 7/8	6 5/8	7 1/4	7 7/8	9 1/4	10 5/8	13	15 1/4	17 3/4
Number of Bolts			4	4	4	8	8	8	8	8	8	12	12	16	16
Diameter of Bolts			5/8	5/8	3/4	5/8	3/4	3/4	3/4	3/4	3/4	3/4	7/8	1	1 1/8

[6] 300 lb. steel flanges have a 1/16″ raised face which is included in the flange thickness dimensions.

Figure 2.20.0 Dimensions of cast iron, steel, and bronze flanges.

400 lb. STEEL **ASA B16.5**

Pipe Size	½	¾	1	1¼	1½	2	2½	3	3½	4	5	6	8	10	12
Diameter of Flange	3¾	4⅝	4⅞	5¼	6⅛	6½	7½	8¼	9	10	11	12½	15	17½	20½
Thickness of Flange (min)[7]	9/16	⅝	11/16	13/16	⅞	1	1⅛	1¼	1⅜	1⅜	1½	1⅝	1⅞	2⅛	2¼
Diameter of Raised Face	1⅜	1 11/16	2	2½	2⅞	3⅝	4⅛	5	5½	6 3/16	7 5/16	8½	10⅝	12¾	15
Diameter of Bolt Circle	2⅝	3¼	3½	3⅞	4½	5	5⅞	6⅝	7¼	7⅞	9¼	10⅝	13	15¼	17¾
Number of Bolts	4	4	4	4	4	8	8	8	8	8	8	12	12	16	16
Diameter of Bolts	½	⅝	⅝	⅝	¾	⅝	¾	¾	⅞	⅞	⅞	⅞	1	1⅛	1¼

[7] 400 lb. steel flanges have a ¼″ raised face which is NOT included in the flange thickness dimensions.

600 lb. STEEL **ASA B16.5**

Pipe Size	½	¾	1	1¼	1½	2	2½	3	3½	4					
Diameter of Flange	3¾	4⅝	4⅞	5¼	6⅛	6½	7½	8¼	9	10¾	13	14	16½	10	12
Thickness of Flange (min)[8]	9/16	⅝	11/16	13/16	⅞	1	1⅛	1¼	1⅜	1½	1¾	1⅞	2 3/16	2½	2⅝
Diameter of Raised Face	1⅜	1 11/16	2	2½	2⅞	3⅝	4⅛	5	5½	6 3/16	7 5/16	8½	10⅝	12¾	15
Diameter of Bolt Circle	2⅝	3¼	3½	3⅞	4½	5	5⅞	6⅞	7¼	8½	10½	11½	13¼	17	19½
Number of Bolts	4	4	4	4	4	8	8	8	8	8	8	12	12	16	20
Diameter of Bolts	½	⅝	⅝	⅝	¾	⅝	¾	¾	⅞	⅞	1	1	1⅛	1¼	1¼

[8] 600 lb. steel flanges have a ¼″ raised face which is NOT included in the flange thickness dimensions.

Figure 2.20.0 *(Continued)*

COPPER TUBE
TYPE L

NOMINAL PIPE SIZE IN INCHES	OUTSIDE DIAMETER IN INCHES	INSIDE DIAMETER IN INCHES	WEIGHT PER LINEAL FOOT IN POUNDS
1/4	.375	.315	.126
3/8	.500	.430	.198
1/2	.625	.545	.285
5/8	.750	.666	.362
3/4	.875	.785	.455
1	1.125	1.025	.655
1 1/4	1.375	1.265	.884
1 1/2	1.625	1.505	1.111
2	2.125	1.985	1.750
2 1/2	2.625	2.465	2.480
3	3.125	2.945	3.333
3 1/2	3.625	3.425	4.290
4	4.125	3.905	5.382
5	5.125	4.875	7.611
6	6.125	5.845	10.201
8	8.125	7.725	19.301
10	10.125	9.625	30.060
12	12.125	11.565	40.390

Figure 2.21.0   L-type copper pipe—ID, OD, weights per foot. (*By permission: McGraw-Hill Inc.* 1998 Plumber's Business Planner *by R. Dodge Woodson.*)

COPPER TUBE
TYPE K

NOMINAL PIPE SIZE IN INCHES	OUTSIDE DIAMETER IN INCHES	INSIDE DIAMETER IN INCHES	WEIGHT PER LINEAL FOOT IN POUNDS
1/4	.375	.305	1.45
3/8	.500	.402	.269
1/2	.625	.527	3.44
5/8	.750	.652	.418
3/4	.875	.745	.641
1	1.125	.995	.839
1 1/4	1.375	1.245	1.040
1 1/2	1.625	1.481	1.360
2	2.125	1.959	2.060
2 1/2	2.625	2.435	2.932
3	3.125	2.907	4.000
3 1/2	3.625	3.385	5.122
4	4.125	3.857	6.511
5	5.125	4.805	9.672
6	6.125	5.741	13.912
8	8.125	7.583	25.900
10	10.125	9.449	40.322
12	12.125	11.315	57.802

Figure 2.21.1 K-type copper pipe, ID, OD, weights per foot. (*By permission: McGraw-Hill Inc.* 1998 Plumber's Business Planner *by R. Dodge Woodson.*)

POLYVINYL CHLORIDE PLASTIC PIPE (PVC)

NOMINAL PIPE SIZE IN INCHES	OUTSIDE DIAMETER IN INCHES	INSIDE DIAMETER IN INCHES
1/2	.840	.750
3/4	1.050	.940
1	1.315	1.195
1 1/4	1.660	1.520
1 1/2	1.900	1.740
2	2.375	2.175
2 1/2	2.875	2.635
3	3.500	3.220
4	4.500	4.110

Figure 2.22.0 ID, OD dimensions of PVC pipe (nominal size). (*By permission: McGraw-Hill Inc.* 1998 Plumber's Business Planner *by R. Dodge Woodson.*)

DIMENSIONS OF BRASS PIPE

NOMINAL DIMENSIONS (INCHES)			
NOMINAL PIPE SIZE	OUTSIDE DIAMETER	INSIDE DIAMETER	WALL THICKNESS
1/8	.405	.281	.062
1/4	.376	.376	.082
3/8	.675	.495	.090
1/2	.840	.626	.107
3/4	1.050	.822	.144
1	1.315	1.063	.126
1 1/4	1.660	1.368	.146
1 1/2	1.900	1.600	.150
2	2.375	2.063	.156
2 1/2	2.875	2.501	.187
3	3.500	3.062	.219

Figure 2.23.0 ID, OD dimensions of brass pipe. (*By permission: McGraw-Hill Inc.* 1998 Plumber's Business Planner *by R. Dodge Woodson.*)

Nominal pipe size (NPS), in IP	ASHRAE std. wt. size, mm	AWWA pipe size, mm	NFPA pipe size, mm	ASTM copper tube size, mm	Nominal pipe size DN, mm
1/8	—	—	—	6	6
3/16	—	—	—	8	8
1/4	8	—	—	10	10
3/8	10	—	—	12	12
1/2	15	12.7 & 13	12	15	15
5/8	—	—	—	18	18
3/4	20	—	—	22	20
1	25	25	25 & 25.4	28	25
1 1/4	32	—	33	35	32
1 1/2	40	45	38 & 38.1	42	40
2	50	50 & 50.8	51	54	50
2 1/2	65	63 & 63.5	63.5 & 64	67	65
3	80	75	76 & 80	79	80
3 1/2	—	—	89	—	90
4	100	100	102	105	100
4 1/2	—	114.3			115
5	—	—	127	130	125
6	150	150	152	156	150
8	200	200	203	206	200
10	250	250	—	257	250
12	300	300	305	308	300
14	—	350	—		350
18	—	400	—		400
18	—	—	—		450
20	—	500	—		500
24	—	600	—		600
28					700
30					750
32					800
36					900
40					1000
44					1100
48					1200
52					1300
56					1400
60					1500

Figure 2.24.0 Metric equivalent of NPS, ASHRAE, AWWA, NFPA, and ASTM tube and pipe sizes. (*By permission: McGraw-Hill Inc.* Plumber's and Pipefitter's Calculations Manual, *R. Dodge Woodson.*)

Pipe material	Coefficient in/in/°F	(°C)
Metallic pipe		
Carbon steel	0.000005	(14.0)
Stainless steel	0.000115	(69)
Cast iron	0.0000056	(1.0)
Copper	0.000010	(1.8)
Aluminum	0.0000980	(1.7)
Brass (yellow)	0.000001	(1.8)
Brass (red)	0.000009	(1.4)
Plastic pipe		
ABS	0.00005	(8)
PVC	0.000060	(33)
PB	0.000150	(72)
PE	0.000080	(14.4)
CPVC	0.000035	(6.3)
Styrene	0.000060	(33)
PVDF	0.000085	(14.5)
PP	0.000065	(77)
Saran	0.000038	(6.5)
CAB	0.000080	(14.4)
FRP (average)	0.000011	(1.9)
PVDF	0.000096	(15.1)
CAB	0.000085	(14.5)
HDPE	0.00011	(68)
Glass		
Borosilicate	0.0000018	(0.33)

Figure 2.25.0 Thermal expansion of metallic and plastic pipe. (*By permission: McGraw-Hill Inc.* Plumber's and Pipefitter's Calculations Manual, *R. Dodge Woodson.*)

Nominal rod diameter, in	Root area of thread, in^2	Maximum safe load at rod temperature of 650°F, lb
1/4	0.027	240
5/16	0.046	410
3/8	0.068	610
1/2	0.126	1,130
5/8	0.202	1,810
3/4	0.302	2,710
7/8	0.419	3,770
1	0.552	4,960
1 1/8	0.693	6,230
1 1/4	0.889	8,000
1 3/8	1.053	9,470
1 1/2	1.293	11,630
1 5/8	1.515	13,630
1 3/4	1.744	15,690
1 7/8	2.048	18,430
2	2.292	20,690
2 1/4	3.021	27,200
2 1/2	3.716	33,500
2 3/4	4.619	41,600
3	5.621	50,600
3 1/4	6.720	60,500
3 1/2	7.918	71,260

Figure 2.26.0 Load rating of various sized threaded rods. (*By permission: McGraw-Hill Inc.* Plumber's and Pipefitter's Calculations Manual, *R. Dodge Woodson.*)

Pipe size, in	Rod size, in
2 and smaller	$\frac{3}{8}$
$2\frac{1}{2}$ to $3\frac{1}{2}$	$\frac{1}{2}$
4 and 5	$\frac{5}{8}$
6	$\frac{3}{4}$
8 to 12	$\frac{7}{8}$
14 and 16	1
18	$1\frac{1}{8}$
20	$1\frac{1}{4}$
24	$1\frac{1}{2}$

Figure 2.27.0 Recommended threaded rod size for pipe from less than 2″ to 24″. (*By permission: McGraw-Hill Inc.* Plumber's and Pipefitter's Calculations Manual, *R. Dodge Woodson.*)

	Diameter (inches)	Service weight (lb)	Extra heavy weight (lb)
Double hub, 5-ft lengths	2	21	26
	3	31	47
	4	42	63
	5	54	78
	6	68	100
	8	105	157
	10	150	225
Double hub, 30-ft length	2	11	14
	3	17	26
	4	23	33
Single hub, 5-ft lengths	2	20	25
	3	30	45
	4	40	60
	5	52	75
	6	65	95
	8	100	150
	10	145	215
Single hub, 10-ft lengths	2	38	43
	3	56	83
	4	75	108
	5	98	133
	6	124	160
	8	185	265
	10	270	400
No-hub pipe, 10-ft lengths	$1\frac{1}{2}$	27	
	2	38	
	3	54	
	4	74	
	5	95	
	6	118	
	8	180	

Figure 2.28.0 Weights of cast iron pipe from 2″ to 10″ diameter. (*By permission: McGraw-Hill Inc.* Plumber's and Pipefitter's Calculations Manual, *R. Dodge Woodson.*)

Oil Piping Pressure Drop (Viscosity = 40 SSU and Specific Gravity = 0.9)

FUEL OIL FLOW RATE (GPH)	NOMINAL PIPE SIZE (INCHES)							
	0.5	0.75	1	1.5	2	2.5	3	4
25	0.3	0.1	0.0	0.01	0.00	0.001	0.000	0.000
50	0.6	0.2	0.1	0.01	0.00	0.002	0.001	0.000
75	0.9	0.3	0.1	0.02	0.01	0.003	0.001	0.000
100	1.1	0.4	0.1	0.03	0.01	0.005	0.002	0.001
150	3.6	0.9	0.2	0.04	0.01	0.007	0.003	0.001
200	6.0	1.6	0.5	0.05	0.02	0.009	0.004	0.001
250	8.9	2.3	0.7	0.06	0.02	0.011	0.005	0.002
300	12.3	3.2	1.0	0.13	0.03	0.014	0.006	0.002
400	20.3	5.3	1.7	0.22	0.007	0.018	0.008	0.003
500	30.5	7.9	2.5	0.32	0.10	0.042	0.010	0.003
600	42.2	11.0	3.5	0.45	0.13	0.058	0.020	0.004
700	55.6	14.5	4.6	0.59	0.18	0.076	0.027	0.007

NOTE: Pressure Drop (psig) per 100 equivalent ft of pipe for a fuel oil viscosity of 40 SSU, specific gravity of 0.9

Oil Piping Pressure Drop (Viscosity = 100 SSU and Specific Gravity = 0.94)

FUEL OIL FLOW RATE (GPH)	NOMINAL PIPE SIZE (INCHES)							
	0.5	0.75	1	1.5	2	2.5	3	4
25	3.9	1.3	0.5	0.1	0.03	0.02	0.007	0.002
50	7.8	2.5	1.0	0.2	0.06	0.03	0.013	0.004
75	11.6	3.8	1.4	0.3	0.10	0.05	0.020	0.007
100	15.5	5.0	1.9	0.3	0.13	0.06	0.026	0.009
150	23.3	7.6	2.9	0.5	0.19	0.09	0.039	0.013
200	31.1	10.1	3.8	0.7	0.25	0.13	0.052	0.018
250	38.8	12.6	4.8	0.9	0.32	0.16	0.066	0.022
300	46.6	15.1	5.8	1.0	0.38	0.19	0.079	0.027
400	62.1	20.2	7.7	1.4	0.51	0.25	0.10	0.035
500	77.6	25.2	9.6	1.7	0.64	0.31	0.13	0.044
600	93.2	30.2	11.5	2.1	0.76	0.38	0.16	0.053
700	108.7	35.3	13.4	2.4	0.89	0.44	0.18	0.062

NOTE: Pressure Drop (psig) per 100 equivalent ft of pipe for a fuel oil viscosity of 100 SSU, specific gravity of 0.94

Oil Piping Pressure Drop (Viscosity = 500 SSU and Specific Gravity = 0.94)

FUEL OIL FLOW RATE (GPH)	NOMINAL PIPE SIZE (INCHES)							
	0.5	0.75	1	1.5	2	2.5	3	4
25	7.8	2.5	1.0	0.2	0.06	0.03	0.013	0.004
50	15.6	5.1	1.9	0.3	0.13	0.06	0.026	0.009
75	23.5	7.6	2.9	0.5	0.19	0.019	0.040	0.013
100	31.3	10.2	3.9	0.7	0.26	0.13	0.053	0.018
150	46.9	15.2	5.8	1.0	0.38	0.19	0.079	0.027
200	62.6	20.3	7.7	1.4	0.51	0.25	0.106	0.036
250	78.2	25.4	9.7	1.7	0.64	0.32	0.132	0.045
300	93.9	30.5	11.6	2.1	0.77	0.38	0.159	0.053
400	125.1	40.6	15.5	2.8	1.03	0.50	0.21	0.071
500	156.4	50.8	19.3	3.5	1.28	0.63	0.26	0.089
600	187.7	69.9	23.2	4.2	1.54	0.76	0.32	0.107
700	219.0	71.1	27.1	4.9	1.80	0.88	0.37	0.125

NOTE: Pressure Drop (psig) per 100 equivalent ft of pipe for a fuel oil viscosity of 500 SSU, specific gravity of 0.94

Figure 2.29.0 Oil pressure drop by fuel oil flow rate and pipe size.

Friction loss for water in feet per 100 feet of Schedule 40 teel pipe.

U.S. Gal./Min.	Vel. Ft./Sec.	hf Friction	U.S. Gal./Min.	Vel. Ft./Sec.	hf Friction
3/8" PIPE			1/2" PIPE		
1.4	2.25	9.03	2	2.11	5.50
1.6	2.68	11.6	2.5	2.64	8.24
1.8	3.02	14.3	3	3.17	11.5
2.0	3.36	17.3	3.5	3.70	15.3
2.5	4.20	26.0	4.0	4.22	19.7
3.0	5.04	36.6	5	5.28	29.7
3.5	5.88	49.0	6	6.34	42.0
4.0	6.72	63.2	7	7.39	56.0
5.0	8.40	96.1	8	8.45	72.1
6	10.08	136	9	9.50	90.1
7	11.8	182	10	10.56	110.6
8	13.4	236	12	12.7	156
9	15.1	297	14	14.8	211
10	16.8	364	16	16.9	270
3/4" PIPE			1" PIPE		
4.0	2.41	4.85	6	2.23	3.16
5	3.01	7.27	8	2.97	5.20
6	3.61	10.2	10	3.71	7.90
7	4.21	13.6	12	4.45	11.1
8	4.81	17.3	14	5.20	14.7
9	5.42	21.6	16	5.94	19.0
10	6.02	26.5	18	6.68	23.7
12	7.22	37.5	20	7.42	28.9
14	8.42	50.0	22	8.17	34.8
16	9.63	64.8	24	8.91	41.0
18	10.8	80.9	26	9.65	47.8
20	12.0	99.0	28	10.39	55.1
22	13.2	120	30	11.1	62.9
24	14.4	141	35	13.0	84.4
26	15.6	165	40	14.8	109
28	16.8	189	45	16.7	137
			50	18.6	168
1 1/4" PIPE			1 1/2" PIPE		
12	2.57	2.85	16	2.52	2.26
14	3.00	3.77	18	2.84	2.79
16	3.43	4.83	20	3.15	3.38
18	3.86	6.00	22	3.47	4.05
20	4.29	7.30	24	3.78	4.76

Figure 2.30.0 Frictional loss for water piped through Schedule 40 steel pipe.

Flow of water through Schedule 40 steel pipe

Discharge Gals. per Min.	1" Vel. Ft. per Sec.	1" Pressure Drop	1¼" Vel. Ft. per Sec.	1¼" Pressure Drop	1½" Vel. Ft. per Sec.	1½" Pressure Drop	2" Vel. Ft. per Sec.	2" Pressure Drop	2½" Vel. Ft. per Sec.	2½" Pressure Drop	3" Vel. Ft. per Sec.	3" Pressure Drop	3½" Vel. Ft. per Sec.	3½" Pressure Drop	4" Vel. Ft. per Sec.	4" Pressure Drop	5" Vel. Ft. per Sec.	5" Pressure Drop	6" Vel. Ft. per Sec.	6" Pressure Drop	8" Vel. Ft. per Sec.	8" Pressure Drop
1	.37	0.49																				
2	.74	1.70	0.43	0.45																		
3	1.12	3.53	0.64	0.94	0.47	0.44																
4	1.49	5.94	0.86	1.55	0.63	0.74																
5	1.86	9.02	1.07	2.36	0.79	1.12																
6	2.24	12.25	1.28	3.30	0.95	1.53	.57	0.46														
8	2.98	21.1	1.72	5.52	1.26	2.63	.76	0.75														
10	3.72	30.8	2.14	8.34	1.57	3.86	.96	1.14	.67	0.48												
15	5.60	64.6	3.21	17.6	2.36	8.13	1.43	2.33	1.00	0.99												
20	7.44	110.5	4.29	29.1	3.15	13.5	1.91	3.86	1.34	1.64	.87	0.59										
25			5.36	43.7	3.94	20.2	2.39	5.81	1.68	2.48	1.08	0.67	.81	0.42								
30			6.43	62.9	4.72	29.1	2.87	8.04	2.01	3.43	1.30	1.21	.97	0.60								
35			7.51	82.5	5.51	38.2	3.35	10.95	2.35	4.49	1.52	1.58	1.14	0.79	.88	0.42						
40					6.30	47.8	3.82	13.7	2.68	5.88	1.74	2.06	1.30	1.00	1.01	0.53						
45					7.08	60.6	4.30	17.4	3.00	7.14	1.95	2.51	1.46	1.21	1.13	0.67						
50					7.87	74.7	4.78	20.6	3.35	8.82	2.17	3.10	1.62	1.44	1.26	0.80						
60							5.74	29.6	4.02	12.2	2.60	4.29	1.95	2.07	1.51	1.10						
70							6.69	38.6	4.69	15.3	3.04	5.84	2.27	2.71	1.76	1.50	1.12	0.48				
80							7.65	50.3	5.37	21.7	3.48	7.62	2.59	3.53	2.01	1.87	1.28	0.63				
90							8.60	63.6	6.04	26.1	3.91	9.22	2.92	4.46	2.26	2.37	1.44	0.80				
100							9.56	75.1	6.71	32.3	4.34	11.4	3.24	5.27	2.52	2.81	1.60	0.95	1.11	0.39		
125									8.38	48.2	5.45	17.1	4.05	7.86	3.15	4.38	2.00	1.48	1.39	0.56		
150									10.06	60.4	6.51	23.3	4.86	11.3	3.78	6.02	2.41	2.04	1.67	0.78		
175									11.73	90.0	7.59	32.0	5.67	14.7	4.41	8.20	2.81	2.78	1.94	1.06		
200											8.68	39.7	6.48	19.2	5.04	10.2	3.21	3.46	2.22	1.32		
225											9.77	50.2	7.29	23.1	5.67	12.9	3.61	4.37	2.50	1.66	1.44	0.44
250											10.85	61.9	8.10	28.5	6.30	15.9	4.01	5.14	2.78	2.05	1.60	0.55
275											11.94	75.0	8.91	34.4	6.93	18.3	4.41	6.22	3.06	2.36	1.76	0.63
300											13.02	84.7	9.72	40.9	7.56	21.8	4.81	7.41	3.33	2.80	1.92	0.75

Pressure Drop per 1000 Feet of Schedue 40 Steel Pipe, in Pounds per Square Inch.

Figure 2.31.0   Pressure drop in flow of water through Schedule 40 steel pipe.

Velocity and Flow in Cast Iron Soil Pipe Sewers and Drains
(Based on Mannings Formula with N = .012)

Pipe Size (In.)	SLOPE (In./Ft.)	SLOPE (Ft./Ft.)	¼ FULL Velocity (Ft./Sec.)	¼ FULL Flow (Gal./Min.)	½ FULL Velocity (Ft./Sec.)	½ FULL Flow (Gal./Min.)	¾ FULL Velocity (Ft./Sec.)	¾ FULL Flow (Gal./Min.)	FULL Velocity (Ft./Sec.)	FULL Flow (Gal./Min.)
4.0	0.0120	0.0010	0.57	5.45	0.74	14.08	0.83	23.63	0.74	28.12
	0.0240	0.0020	0.81	7.70	1.05	19.91	1.17	33.42	1.05	39.77
	0.0360	0.0030	0.99	9.44	1.28	24.38	1.44	40.92	1.28	48.71
	0.0480	0.0040	1.15	10.90	1.48	28.16	1.66	47.26	1.48	56.25
	0.0600	0.0050	1.28	12.18	1.65	31.48	1.85	52.83	1.65	62.88
	0.0720	0.0060	1.40	13.34	1.81	34.48	2.03	57.88	1.81	68.89
	0.0840	0.0070	1.51	14.41	1.96	37.25	2.19	62.51	1.96	74.41
	0.0960	0.0080	1.62	15.41	2.09	39.82	2.34	66.83	2.09	79.54
	0.1080	0.0090	1.72	16.34	2.22	42.23	2.49	70.88	2.22	84.37
	0.1200	0.0100	1.81	17.23	2.34	44.52	2.62	74.72	2.34	88.93
	0.2400	0.0200	2.56	24.36	3.31	62.96	3.71	105.67	3.31	125.77
	0.3600	0.0300	3.14	29.84	4.05	77.11	4.54	129.42	4.05	154.04
	0.4800	0.0400	3.62	34.46	4.68	89.04	5.24	149.44	4.68	177.86
	0.6000	0.0500	4.05	38.52	5.23	99.55	5.86	167.08	5.23	198.86
	0.7200	0.0600	4.43	42.20	5.73	109.05	6.42	183.02	5.73	217.84
	0.8400	0.0700	4.79	45.58	6.19	117.79	6.94	197.69	6.19	235.29
	0.9600	0.0800	5.12	48.73	6.62	125.92	7.41	211.34	6.62	251.54
	1.0800	0.0900	5.43	51.68	7.02	133.56	7.86	224.15	7.02	266.80
	1.2000	0.1000	5.73	54.48	7.40	140.78	8.29	236.28	7.40	281.23
5.0	0.0120	0.0010	0.67	9.94	0.86	25.71	0.96	43.15	0.86	51.37
	0.0240	0.0020	0.94	14.06	1.22	36.35	1.36	61.02	1.22	72.65
	0.0360	0.0030	1.15	17.22	1.49	44.52	1.67	74.74	1.49	88.98
	0.0480	0.0040	1.33	19.88	1.72	51.41	1.93	86.30	1.72	102.75
	0.0600	0.0050	1.49	22.23	1.92	57.48	2.15	96.49	1.92	114.87
	0.0720	0.0060	1.63	24.35	2.11	62.97	2.36	105.70	2.11	125.84
	0.0840	0.0070	1.76	26.30	2.28	68.01	2.55	114.17	2.28	135.92
	0.0960	0.0080	1.88	28.12	2.43	72.71	2.72	122.05	2.43	145.31
	0.1080	0.0090	2.00	29.82	2.58	77.12	2.89	129.45	2.58	154.12
	0.1200	0.0100	2.10	31.44	2.72	81.29	3.05	136.45	2.72	162.46
	0.2400	0.0200	2.97	44.46	3.85	114.96	4.31	192.97	3.85	229.75
	0.3600	0.0300	3.64	54.45	4.71	140.80	5.28	236.34	4.71	281.38
	0.4800	0.0400	4.21	62.88	5.44	162.58	6.09	272.91	5.44	324.91
	0.6000	0.0500	4.70	70.30	6.08	181.77	6.81	305.12	6.08	363.26
	0.7200	0.0600	5.15	77.01	6.66	199.12	7.46	334.24	6.66	397.94
	0.8400	0.0700	5.56	83.18	7.19	215.07	8.06	361.02	7.19	429.82
	0.9600	0.0800	5.95	88.92	7.69	229.92	8.62	385.95	7.69	459.50
	1.0800	0.0900	6.31	94.31	8.16	243.92	9.14	409.36	8.16	487.37
	1.2000	0.1000	6.65	99.42	8.60	257.06	9.63	431.50	8.60	513.73

Figure 2.32.0 Velocity and flow through cast iron soil pipe—sizes 4″ through 15″. (*By permission: Cast Iron Soil Pipe Institute, Chattanooga, TN.*)

Velocity and Flow in Cast Iron Soil Pipe Sewers and Drains
(Based on Mannings Formula with N = .012)

Pipe Size (In.)	SLOPE (In./Ft.)	SLOPE (Ft./Ft.)	¼ FULL Velocity (Ft./Sec.)	¼ FULL Flow (Gal./Min.)	½ FULL Velocity (Ft./Sec.)	½ FULL Flow (Gal./Min.)	¾ FULL Velocity (Ft./Sec.)	¾ FULL Flow (Gal./Min.)	FULL Velocity (Ft./Sec.)	FULL Flow (Gal./Min.)
6.0	0.0120	0.0010	0.75	16.23	0.97	41.98	1.09	70.55	0.97	83.96
	0.0240	0.0020	1.06	22.95	1.37	59.37	1.54	99.77	1.37	118.74
	0.0360	0.0030	1.30	28.11	1.68	72.71	1.89	122.20	1.68	145.42
	0.0480	0.0040	1.50	32.46	1.94	83.96	2.18	141.10	1.94	167.92
	0.0600	0.0050	1.68	36.29	2.17	93.87	2.44	157.76	2.17	187.74
	0.0720	0.0060	1.84	39.75	2.38	102.83	2.67	172.81	2.38	205.66
	0.0840	0.0070	1.99	42.94	2.57	111.07	2.88	186.66	2.57	222.13
	0.0960	0.0080	2.13	45.90	2.75	118.74	3.08	199.55	2.75	237.47
	0.1080	0.0090	2.26	48.69	2.92	125.94	3.27	211.65	2.92	251.88
	0.1200	0.0100	2.38	51.32	3.07	132.75	3.44	223.10	3.07	265.50
	0.2400	0.0200	3.36	72.58	4.35	187.74	4.87	315.51	4.35	375.47
	0.3600	0.0300	4.12	88.89	5.32	229.93	5.97	386.42	5.32	459.86
	0.4800	0.0400	4.75	102.64	6.15	265.50	6.89	446.20	6.15	531.00
	0.6000	0.0500	5.32	114.76	6.87	296.84	7.70	498.87	6.87	593.68
	0.7200	0.0600	5.82	125.71	7.53	325.17	8.44	546.27	7.53	650.34
	0.8400	0.0700	6.29	135.78	8.13	351.22	9.11	590.27	8.13	702.45
	0.9600	0.0800	6.72	145.16	8.70	375.47	9.74	631.02	8.70	750.95
	1.0800	0.0900	7.13	153.96	9.22	398.25	10.33	669.30	9.22	796.50
	1.2000	0.1000	7.52	162.29	9.72	419.79	10.89	705.51	9.72	839.59
8.0	0.0120	0.0010	0.91	35.25	1.18	91.04	1.32	153.06	1.18	182.09
	0.0240	0.0020	1.29	49.85	1.67	128.75	1.87	216.46	1.67	257.51
	0.0360	0.0030	1.58	61.05	2.04	157.69	2.29	265.11	2.04	315.38
	0.0480	0.0040	1.83	70.50	2.36	182.09	2.64	306.12	2.36	364.17
	0.0600	0.0050	2.04	78.82	2.64	203.58	2.96	342.26	2.64	407.16
	0.0720	0.0060	2.24	86.34	2.89	223.01	3.24	374.92	2.89	446.02
	0.0840	0.0070	2.42	93.26	3.12	240.88	3.50	404.96	3.12	481.75
	0.0960	0.0080	2.58	99.70	3.34	257.51	3.74	432.92	3.34	515.02
	0.1080	0.0090	2.74	105.75	3.54	273.13	3.97	459.18	3.54	546.26
	0.1200	0.0100	2.89	111.47	3.73	287.90	4.18	484.02	3.73	575.81
	0.2400	0.0200	4.08	157.64	5.28	407.16	5.91	684.51	5.28	814.32
	0.3600	0.0300	5.00	193.06	6.46	498.66	7.24	838.35	6.46	997.33
	0.4800	0.0400	5.78	222.93	7.46	575.81	8.36	968.05	7.46	1151.62
	0.6000	0.0500	6.46	249.24	8.34	643.77	9.35	1082.31	8.34	1287.55
	0.7200	0.0600	7.07	273.03	9.14	705.22	10.24	1185.61	9.14	1410.44
	0.8400	0.0700	7.64	294.91	9.87	761.72	11.06	1280.60	9.87	1523.45
	0.9600	0.0800	8.17	315.27	10.55	814.31	11.83	1369.02	10.55	1628.63
	1.0800	0.0900	8.66	334.40	11.19	863.71	12.54	1452.07	11.19	1727.42
	1.2000	0.1000	9.13	352.48	11.80	910.43	13.22	1530.61	11.80	1820.86

Figure 2.32.0 (*Continued*)

Velocity and Flow in Cast Iron Soil Pipe Sewers and Drains
(Based on Mannings Formula with N = .012)

Pipe Size (In.)	SLOPE (In./Ft.)	(Ft./Ft.)	¼ FULL Velocity (Ft./Sec.)	Flow (Gal./Min.)	½ FULL Velocity (Ft./Sec.)	Flow (Gal./Min.)	¾ FULL Velocity (Ft./Sec.)	Flow (Gal./Min.)	FULL Velocity (Ft./Sec.)	Flow (Gal./Min.)
10.0	0.0120	0.0010	1.06	64.08	1.37	165.75	1.54	278.56	1.37	331.51
	0.0240	0.0020	1.50	90.62	1.94	234.41	2.17	393.95	1.94	468.83
	0.0360	0.0030	1.84	110.99	2.37	287.10	2.66	482.48	2.37	574.19
	0.0480	0.0040	2.12	128.16	2.74	331.51	3.07	557.12	2.74	663.02
	0.0600	0.0050	2.37	143.29	3.07	370.64	3.43	622.88	3.07	741.28
	0.0720	0.0060	2.60	156.96	3.36	406.01	3.76	682.33	3.36	812.03
	0.0840	0.0070	2.80	169.54	3.63	438.55	4.06	737.01	3.63	877.09
	0.0960	0.0080	3.00	181.24	3.88	468.82	4.34	787.89	3.88	937.65
	0.1080	0.0090	3.18	192.24	4.11	497.26	4.61	835.69	4.11	994.53
	0.1200	0.0100	3.35	202.64	4.33	524.16	4.86	880.89	4.33	1048.32
	0.2400	0.0200	4.74	286.57	6.13	741.28	6.87	1245.77	6.13	1482.55
	0.3600	0.0300	5.80	350.98	7.51	907.88	8.41	1525.75	7.51	1815.75
	0.4800	0.0400	6.70	405.27	8.67	1048.32	9.71	1761.78	8.67	2096.65
	0.6000	0.0500	7.49	453.11	9.69	1172.06	10.86	1969.73	9.69	2344.13
	0.7200	0.0600	8.21	496.36	10.62	1283.93	11.90	2157.74	10.62	2567.86
	0.8400	0.0700	8.87	536.12	11.47	1386.80	12.85	2330.62	11.47	2773.61
	0.9600	0.0800	9.48	573.14	12.26	1482.55	13.74	2491.54	12.26	2965.11
	1.0800	0.0900	10.05	607.91	13.00	1572.49	14.57	2642.67	13.00	3144.97
	1.2000	0.1000	10.60	640.79	13.71	1657.55	15.36	2785.62	13.71	3315.09
12.0	0.0120	0.0010	1.20	104.53	1.55	270.34	1.74	454.27	1.55	540.68
	0.0240	0.0020	1.69	147.83	2.19	382.32	2.45	642.43	2.19	764.63
	0.0360	0.0030	2.07	181.05	2.68	468.24	3.01	786.82	2.68	936.48
	0.0480	0.0040	2.40	209.06	3.10	540.68	3.47	908.54	3.10	1081.35
	0.0600	0.0050	2.68	233.74	3.46	604.49	3.88	1015.78	3.46	1208.99
	0.0720	0.0060	2.93	256.05	3.79	662.19	4.25	1112.73	3.79	1324.38
	0.0840	0.0070	3.17	276.56	4.10	715.25	4.59	1201.88	4.10	1430.50
	0.0960	0.0080	3.39	295.66	4.38	764.63	4.91	1284.87	4.38	1529.27
	0.1080	0.0090	3.59	313.59	4.65	811.01	5.21	1362.81	4.65	1622.03
	0.1200	0.0100	3.79	330.56	4.90	854.88	5.49	1436.53	4.90	1709.77
	0.2400	0.0200	5.36	467.48	6.93	1208.99	7.76	2031.55	6.93	2417.98
	0.3600	0.0300	6.56	572.54	8.48	1480.71	9.50	2488.14	8.48	2961.41
	0.4800	0.0400	7.58	661.11	9.80	1709.77	10.98	2873.05	9.80	3419.54
	0.6000	0.0500	8.47	739.14	10.95	1911.58	12.27	3212.17	10.95	3823.17
	0.7200	0.0600	9.28	809.69	12.00	2094.03	13.44	3518.76	12.00	4188.07
	0.8400	0.0700	10.02	874.57	12.96	2261.81	14.52	3800.69	12.96	4523.63
	0.9600	0.0800	10.71	934.95	13.86	2417.98	15.52	4063.11	13.86	4835.96
	1.0800	0.0900	11.36	991.67	14.70	2564.65	16.46	4309.57	14.70	5129.30
	1.2000	0.1000	11.98	1045.31	15.49	2703.38	17.35	4542.69	15.49	5406.76

Figure 2.32.0 (Continued)

Velocity and Flow in Cast Iron Soil Pipe Sewers and Drains
(Based on Mannings Formula with N = .012)

Pipe Size (In.)	SLOPE (In./Ft.)	(Ft./Ft.)	¼ FULL Velocity (Ft./Sec.)	Flow (Gal./Min.)	½ FULL Velocity (Ft./Sec.)	Flow (Gal./Min.)	¾ FULL Velocity (Ft./Sec.)	Flow (Gal./Min.)	FULL Velocity (Ft./Sec.)	Flow (Gal./Min.)
15.0	0.0120	0.0010	1.39	192.03	1.80	496.67	2.02	834.85	1.80	993.34
	0.0240	0.0020	1.97	271.58	2.55	702.40	2.86	1180.65	2.55	1404.79
	0.0360	0.0030	2.42	332.61	3.12	860.25	3.50	1445.99	3.12	1720.51
	0.0480	0.0040	2.79	384.07	3.61	993.34	4.04	1669.69	3.61	1986.67
	0.0600	0.0050	3.12	429.40	4.03	1110.58	4.52	1866.77	4.03	2221.17
	0.0720	0.0060	3.42	470.38	4.42	1216.58	4.95	2044.95	4.42	2433.17
	0.0840	0.0070	3.69	508.07	4.77	1314.06	5.35	2208.79	4.77	2628.12
	0.0960	0.0080	3.94	543.15	5.10	1404.79	5.72	2361.30	5.10	2809.58
	0.1080	0.0090	4.18	576.10	5.41	1490.01	6.06	2504.54	5.41	2980.01
	0.1200	0.0100	4.41	607.26	5.70	1570.60	6.39	2640.01	5.70	3141.21
	0.2400	0.0200	6.24	858.80	8.06	2221.17	9.04	3733.54	8.06	4442.34
	0.3600	0.0300	7.64	1051.81	9.88	2720.37	11.07	4572.64	9.88	5440.73
	0.4800	0.0400	8.82	1214.52	11.41	3141.21	12.78	5280.03	11.41	6282.41
	0.6000	0.0500	9.86	1357.88	12.75	3511.98	14.29	5903.25	12.75	7023.95
	0.7200	0.0600	10.80	1487.48	13.97	3847.18	15.65	6466.69	13.97	7694.35
	0.8400	0.0700	11.67	1606.66	15.09	4155.43	16.91	6984.82	15.09	8310.85
	0.9600	0.0800	12.47	1717.60	16.13	4442.33	18.07	7467.07	16.13	8884.66
	1.0800	0.0900	13.23	1821.78	17.11	4711.80	19.17	7920.03	17.11	9423.61
	1.2000	0.1000	13.94	1920.33	18.03	4966.68	20.21	8348.44	18.03	9933.35

Figure 2.32.0 (*Continued*)

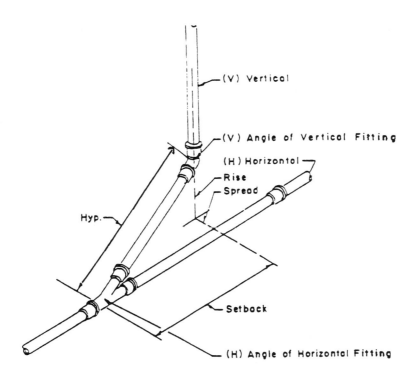

Compound Offsets

		$60°(\frac{1}{6}bd)V$ $45°(\frac{1}{5}bd)H$	$45°(\frac{1}{8}bd)V$ $72°(\frac{1}{5}bd)H$	$72°(\frac{1}{5}bd)V$ $60°(\frac{1}{6}bd)H$	$60°(\frac{1}{6}bd)V$ $60°(\frac{1}{6}bd)H$	$45°(\frac{1}{8}bd)V$ $60°(\frac{1}{6}bd)H$	$72°(\frac{1}{5}bd)V$ $45°(\frac{1}{5}bd)H$	$60°(\frac{1}{6}bd)V$ $45°(\frac{1}{8}bd)H$	
Spread	X	1.24	1.57	1.23	1.41	2.	1.57	2.	= Hyp
Rise	X	2.	1.41	3.323	2.	1.41	3.358	2.	= Hyp
Setback	X	3.25	3.25	1.96	2.	2.	1.41	1.41	= Hyp
Hyp	X	.807	.684	.81	.71	.5	.634	.5	= Spread
Rise	X	1.61	.90	2.66	1.41	.707	2.	1.	= Spread
Setback	X	2.62	2.	1.60	1.41	1.	.90	.71	= Spread
Hyp	X	.5	.70	.305	.5	.71	.307	.5	= Rise
Spread	X	.62	1.11	.3697	.71	1.41	.469	1.	= Rise
Setback	X	1.62	2.29	.584	1.	1.41	.437	.71	= Rise
Hyp	X	.307	.307	.587	.5	.5	.71	.71	= Setback
Spread	X	.38	.48	.625	.71	1.	1.114	1.41	= Setback
Rise	X	.615	.437	1.70	1.	.71	2.377	1.41	= Setback

Figure 2.33.0 Determining hypotenuse, spread, rise, and setback for compound pipe offsets. (*By permission: Cast Iron Soil Pipe Institute, Chattanooga, TN.*)

Finding the Length of Pipe Needed to Connect Two Ends, Offset, and in the Same Plane

Degree of Offset "a"	When A = 1 B is	When B = 1 A is	When A = 1 C is
72° (1/5 bd)	.3249	3.077	1.0515
60° (1/6 bd)	.5773	1.732	1.1547
45° (1/8 bd)	1.0000	1.000	1.4142
22 1/2° (1/16 bd)	2.414	.4142	2.6131

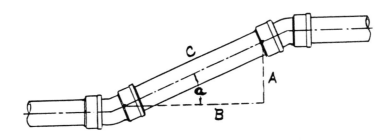

Simple Offset

EXAMPLE: When Angle a = 22.50° and Side A = 12″
  Side C = 2.6131 × 12″
  C = 31.36″

When Angle a = 45° and Side A = 12″
Side C = 1.4142 × 12″
C = 16.97″

Figure 2.34.0 Finding length of pipe needed to connect two ends offset in the same plane. (*By permission: Cast Iron Soil Pipe Institute, Chattanooga, TN.*)

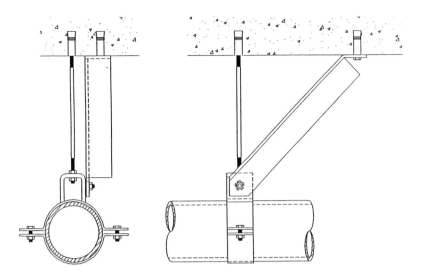

Horizontal Installation of Large Diameter Pipe.

Horizontal pipe and fittings five (5) inches and larger must be suitably braced to prevent horizontal movement. This must be done at every branch opening or change of direction by the use of braces, blocks, rodding or other suitable method, to prevent movement or joint separation. Figure 5 illustrates several methods of bracing.

Suggested Installation of Horizontal Fittings.

(a) Hangers should be provided as necessary to provide alignment and grade. Hangers should be provided at each horizontal branch connection. Hangers should be adequate to maintain alignment and prevent sagging and should be placed adjacent to the coupling. By placing the hangers properly, the proper grade will be maintained. Adequate provision should be made to prevent shear. Where pipe and fittings are suspended in excess of eighteen inches by means of non-rigid hangers they should be suitably braced against movement horizontally, often called sway bracing. Refer to Figures 3 and 4 for illustrations.

(b) Closet bends, traps, trap-arms and similar branches must be firmly secured against movement in any direction. Closet bends installed above ground should be stabilized. Where vertical closet studs are used they must be stabilized against horizontal or vertical movement. In Figures 6 and 7 see illustration for strapping a closet bend under a sub-floor and how a clevis type hanger has been used to an advantage.

(c) When a hubless blind plug is used for a required cleanout, the complete coupling and plug must be accessible for removal and replacement.

(d) The connection of closet rings, floor and shower drains and similar "slip-over" fittings and the connection of hubless pipe and fittings to soil pipe hubs may be accomplished by the use of caulked lead and oakum or compression joints.

Figure 2.35.0 Suggested installation of cast iron horizontal fittings. (*By permission: Cast Iron Soil Pipe Institute, Chattanooga, TN.*)

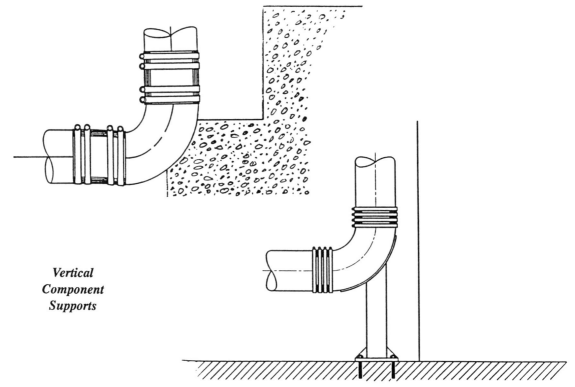

Vertical
Component
Supports

Figure 2.35.1   Vertical supports for cast iron pipe. (*By permission: Cast Iron Soil Pipe Institute, Chattanooga, TN.*)

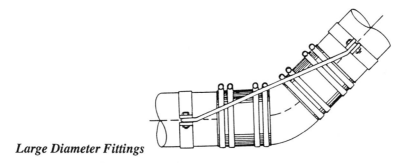

Large Diameter Fittings

Horizontal pipe and fittings five (5) inches and larger shall be suitably braced to prevent horizontal movement. This shall be done at every branch opening or change of direction by the use of braces, blocks, rodding or other suitable method, to prevent movement.

Figure 2.35.2   Installation of large diameter pipe and fittings. (*By permission: Cast Iron Soil Pipe Institute, Chattanooga, TN.*)

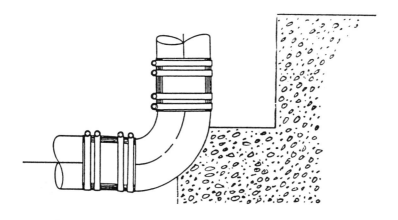

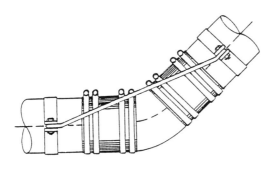

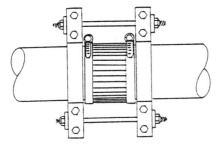

Large Diameter Pipe

Figure 2.35.2 *(Continued)*

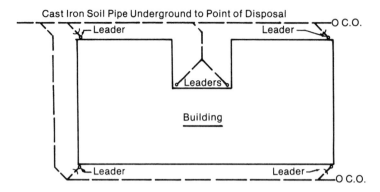

Roof Leaders and Drains Outside Building

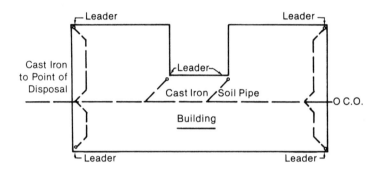

Roof Leaders and Drains Inside Building

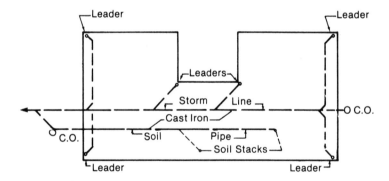

Combination Sewer (Sanitary and Storm) Where Permitted by Code

Figure 2.36.0 Schematics of cast iron roof leaders, storm, and sanitary pipe installations. (*By permission: Cast Iron Soil Pipe Institute, Chattanooga, TN.*)

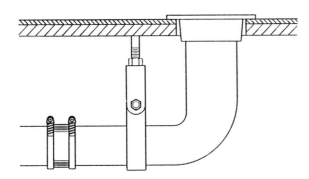

Closet bends, traps, trap-arms and similar branches must be secured against movement in any direction. Closet bends installed above ground shall be stabilized by firmly strapping and blocking. Where vertical closet stubs are used they must be stabilized against horizontal or vertical movements.

GENERAL INSTALLATION INSTRUCTIONS

A. *Vertical Piping:*
   (1) Secure vertical piping at sufficiently close intervals to keep the pipe in alignment and to support the weight of the pipe and its contents. Support stacks at their bases and at sufficient floor intervals to meet the requirements of local codes. Approved metal clamps or hangers should be used for this purpose.
   (2) If vertical piping is to stand free of any support or if no structural element is available for support and stability during construction, secure the piping in its proper position by means of adequate stakes or braces fastened to the pipe.

B. *Horizontal Piping, Suspended:*
   (1) Support horizontal piping and fittings at sufficiently close intervals to maintain alignment and prevent sagging or grade reversal. Support each length of pipe by an approved hanger located not more than 18 inches from the joint.
   (2) Support terminal ends of all horizontal runs or branches and each change of direction or alignment with an approved hanger.
   (3) Closet bends installed above ground should be firmly secured.

C. *Horizontal Piping, Underground:*
   (1) To maintain proper alignment during backfilling, stabilize the pipe in proper position by partial backfilling and cradling.
   (2) Piping laid on grade should be adequately secured to prevent misalignment when the slab is poured.
   (3) Closet bends installed under slabs should be adequately secured.

D. *Installation Inside the Building:*
   (1) Installation suggestions. According to most authorities and plumbing codes, it is sufficient to support horizontal pipe at each joint, i.e. 5′ pipe should be supported at five foot intervals, 10′ in length may be supported at ten foot intervals. Supports should be adequate to maintain alignment and prevent sagging and should be placed within eighteen inches of the joint.

Figure 2.37.0   General instructions for installation of cast iron pipe. (*By permission: Cast Iron Soil Pipe Institute, Chattanooga, TN.*)

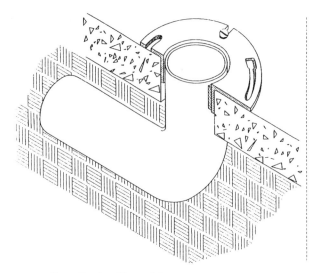

Cross Section View of Closet Bend Showing Flange Properly Secured

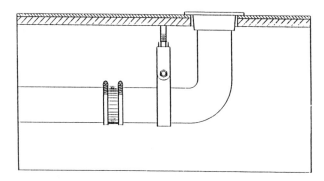

Method of Using Hanger for Closet Bend

Figure 2.37.1   Securing closet bends. (*By permission: Cast Iron Soil Pipe Institute, Chattanooga, TN.*)

Seismic Restraints

The following recommendations are some of the factors to consider when installing cast iron pipe in seismically active areas. All installations must comply with local codes and instructions of architects or engineers who are responsible for the piping design.

A) Brace all pipe 2″ and larger.

Exceptions:

Seismic braces may be omitted when the top of the pipe is suspended 12″ or less from the supporting structure member and the pipe is suspended by an individual hanger.

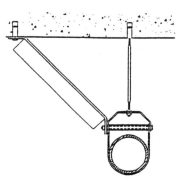

Figure 2.38.0   Seismic restraints for cast iron pipe. (*By permission: Cast Iron Soil Pipe Institute, Chattanooga, TN.*)

Thrust or Displacement Forces Encountered in
Hydrostatic Testing of No-Hub Cast Iron Soil Pipe

PIPE SIZE		1½"	2"	3"	4"	5"	6"	8"	10"
HEAD, Feet of Water	PRESSURE PSI	THRUST lb.	THRUST lb.	THRUST lb.	THRUST lb.	THRUST lb.	THRUST lb.	THRUST lb.	THRUST lb.
10	4.3	12	19	38	65	95	134	237	377
20	8.7	25	38	77	131	192	271	480	762
30	13.0	37	56	115	196	287	405	717	1139
40	17.3	49	75	152	261	382	539	954	1515
50	21.7	62	94	191	327	479	676	1197	1900
60	26.0	74	113	229	392	574	810	1434	2277
70	30.3	86	132	267	457	668	944	1671	2654
80	34.7	99	151	306	523	765	1082	1914	3039
90	39.0	111	169	344	588	860	1216	2151	3416
100	43.4	123	188	382	654	957	1353	2394	3801
110	47.7	135	208	420	719	1052	1487	2631	4178
120	52.0	147	226	458	784	1147	1621	2868	4554
AREA, OD. in.²		2.84	4.34	8.81	15.07	22.06	31.17	55.15	87.58

Thrust = Pressure x Area

Figure 2.39.0 Thrust and displacement forces encountered in hydrostatic testing of no-hub cast iron pipe. (*By permission: Cast Iron Soil Pipe Institute, Chattanooga, TN.*)

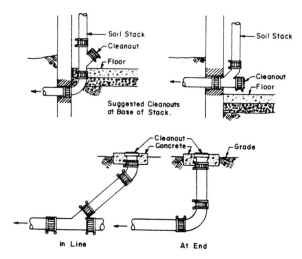

—Suggested Cleanouts Using Cast Iron Soil Pipe

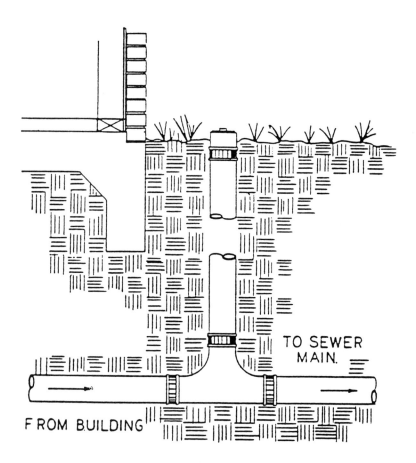

Twin Cleanout

Figure 2.40.0 Diagrams of suggested clean-out arrangements in cast iron pipe. (*By permission: Cast Iron Soil Pipe Institute, Chattanooga, TN.*)

PVC SCHEDULE 40 PIPE GRAY AND WHITE

40 PIPE SIZE	O.D.	AVE. I.D.	MIN. WALL THICK.	APPROX WT. PER 100 FT.	MAX. WORK PRESS. PLAIN ENDS	
1/4"	.540	.354	.088	7.88	780	|
3/8"	.675	.483	.091	10.53	620	|
1/2"	.840	.608	.109	15.63	600	|
3/4"	1.050	.810	.113	20.78	480	|
1"	1.315	1.033	.133	30.85	450	|
1 1/4"	1.660	1.364	.140	41.76	370	|
1 1/2"	1.900	1.592	.145	49.94	330	|
2"	2.375	2.049	.154	67.92	280	|
2 1/2"	2.875	2.445	.203	105.46	300	|
3"	3.500	3.042	.216	137.91	260	|
4"	4.500	3.998	.237	194.58	220	|
6"	6.625	6.031	.280	338.88	180	|

PVC SCHEDULE 80 PIPE GRAY AND WHITE

80 PIPE SIZE	O.D.	AVE. I.D.	MIN. WALL THICK.	APPROX WT. PER 100 FT.	WORK PRESS. PLAIN ENDS	MAX. WORK PRESS. THRD. ENDS	
1/4"	.540	.302	.119	9.4	1130	550	|
3/8"	.675	.423	.126	13.0	920	450	|
1/2"	.840	.546	.147	19.2	850	420	|
3/4"	1.050	.742	.154	25.9	690	340	|
1"	1.315	.957	.179	38.2	630	320	|
1 1/4"	1.660	1.278	.191	52.7	520	260	|
1 1/2"	1.900	1.500	.220	63.9	470	240	|
2"	2.375	1.939	.218	88.3	400	210	|
2 1/2"	2.875	2.323	.276	134.6	420	200	|
3"	3.500	2.900	.300	180.4	370	190	|
4"	4.500	3.826	.337	263.4	320	160	|
6"	6.625	5.761	.432	502.6	280	140	|

(Chlorinated Polyvinyl Chloride) conforming to ASTM D-1784 Class 23447-B, formerly designated Type IV, Grade 1, CPVC has physical properties at 73 degrees F similar to those of PVC, and its chemical resistance is similar to that of PVC. CPVC, with a design stress of 2,000 psi and maximum service temperature of 210 degrees F has, over a period of about 25 years, proven to be an excellent material for hot corrosive liquids, hot and cold water distribution and similar applications above the temperature range of PVC. CPVC is joined by solvent cementing, threading or flanging.

Figure 2.41.0 Schedule 40/80 PVC pipe sizes, weights, and working pressures—similarities between CPVC and PVC.

Installation Tips for PVC and CPVC Pipe

1. Protect plastic pipe from contact with hard and pointed objects. Impact resistance is lower than metals.

2. Avoid bending pipe. Pipe should not be bent in trenches or in above ground installations. Pipe and joints that are stressed reduce pressure rating and can cause failure.

3. Protect pipe from extreme heat and cold. Extremes of heat and cold can cause failure. Allowing liquids to freeze inside PVC or CPVC and metallic piping can cause the pipe and/or joints to crack. Freeze protection should be designed into the system. Heat beyond design limits can also cause failures.

4. Protect pipe from sunlight. PVC and CPVC pipe compounds normally do not provide extended protection from ultraviolet rays of the sun. Therefore, unless the material has been specially formulated for protection from sunlight, damage to the pipe may occur after years of exposure.

Application

1. Never use PVC and CPVC piping materials to transport compressed air or gases. Compressed air or gases can surge to high pressures and cause explosive failures that could seriously injure personnel. PVC and CPVC pipe and fittings are excellent products for transporting weather and corrosive chemicals, but there are a number of other piping products that are especially suited for transporting compressed air and gases.

2. Only use approved chemicals. Certain chemicals, especially petroleum distillates and derivatives, can cause failure. Every chemical should be verified and approved in the manufacturer's chemical resistance chart.

System Design

1. Allow for flexibility in the design of the system. Expansion and contraction is greater than that for metals. This can cause leaks and breaks if the system design is not flexible to permit movement. When installing smaller diameters of pipe below grade, the pipe should be "snaked" in the trench to allow for expansion and contraction. If solvent cement welding is used for the method of joining, snaking, pressure testing, and pipe movement should not be done until after the joints have been given sufficient time to dry.

2. Design safeguards into the system to prevent excessive surge pressures. Water hammer (surge) in a PVC or CPVC system can cause pipe, fittings and valves to burst. Liquid velocities should not exceed 5 feet per second maximum

Installation

1. Carefully follow solvent cement welding instructions. Failure to follow application procedures correctly can reduce strength and integrity of joints and fittings and can cause joint/fitting failure. Most PVC and CPVC installation failures occur because of shortcuts and/or improper joining techniques.

2. Remove rocks and other debris that can rupture pipe when burying pipe in trenches. When laying PVC and CPVC pipe below grade, care must be taken to remove all rocks, boards, empty primer and cement cans, brushes, bottles, and other debris from the trench. Backfilling and top loading should be watched very carefully.

3. Follow recommended support spacing for PVC and CPVC piping systems. The modulus of elasticity of PVC and CPVC is smaller than it is for metal pipes. Maximum working temperatures and room temperature should be considered when determining the required support spacing.

Testing

1. Never use compessed air or gas or air-over-water boosters to pressure test PVC or CPVC piping systems. Only hydraulic pressure testing is to be conducted on PVC and CPVC piping systems. Compressed air or gases can surge to high pressures and cause explosive failures that could seriously injure personnel.

2. Carefully follow all instructions for hydrostatic pressure testing. Failure to follow these instructions can result in a system failure.

3. Before water testing a system, always bleed all entrapped air from system. Entrapped air is a major cause of surge and burst failure in plastic piping systems.

Figure 2.41.1 Installation tips for PVC and CPVC pipe.

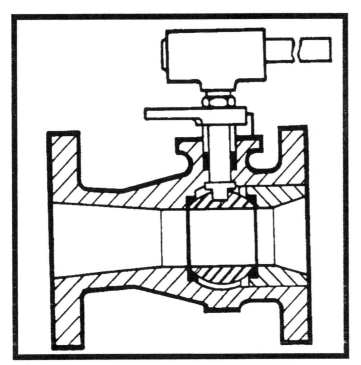

Figure 2.42.0 Valves—ball type—section through. Ball valves are low-torque, quarter-turn, rotary action valves with low resistance to flow. They are designed for straight-through configuration. Most ball valves have seats made of nylon or PTFE, which provides a bubble-tight seal but limits the maximum working temperature. These relatively soft seats limit the ability of the ball valve to handle abrasive liquids.

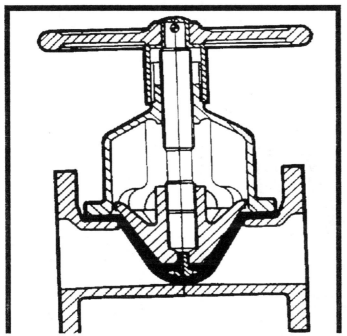

Figure 2.42.1 Valves—straight through diaphragm valve—section through. Diaphragm valves have either flow-through or weir-type configurations and are used to regulate the flow of liquids and gases. They are used in low-pressure applications and operate effectively only within a limited temperature range. When used as a straight-through flow valve, it has long diaphragm movements that decrease the diaphragm's life, and because the function of the valve depends on the flexibility of the diaphragm, limited choices of elastomers limit the applications in which it can be used. Because diaphragms in the flow-through configuration are subject to wear, they require frequent maintenance when used on a regular basis.

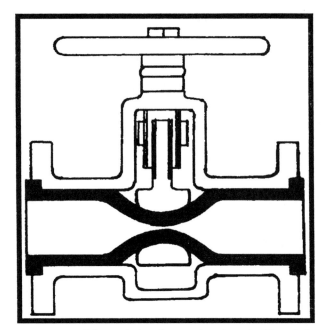

Figure 2.42.2 Valves—pinch valve—section through. The pinch valve contains a flexible tube, either exposed or enclosed in its body, and this tube is pinched to close mechanically or by application of fluid pressure on the body. They can be controlled either pneumatically or electrically. These valves are considered linear action valves and are similar in concept to diaphragm valves. They function well for on–off operations on abrasive slurries, fluids with suspended particles, powders, and corrosive liquids. Pinch valves are easily maintained, but occasional replacement of the sleeve is to be expected.

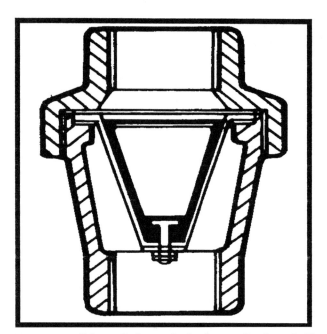

Figure 2.42.3 Valves—diaphragm check valve—section through. Although the diaphragm check valve is not commonly used in a purely valve operation, it is worthy of consideration because it offers the following advantages: free unrestricted movement; fast closure and full closure; and ability to handle viscous and abrasive fluids and slurries more reliably than other types of check valves. The design of the diaphragm check valve provides a stable operation with pressure variations, pulsing flow, and frequent flow reversals. When the pressure and temperature limitations of the flexible membrane permit its use, it provides a reliable alternative to other valve types. One disadvantage of the diaphragm check valve is the need to properly select durable material for the membrane.

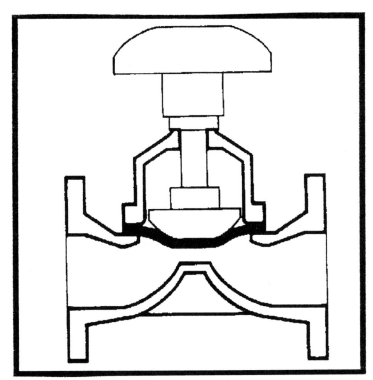

Figure 2.42.4 Valves—weir type—section through. A weir type valve provides complete shutoff with a relatively low operating force. This valve has a rather short diaphragm movement that over an extended period of time increases the life of the diaphragm element. Therefore, weir valves have a history of long life with little maintenance. The weir valve, unlike its close relative, the straight-through valve, is better at throttling flow but the control at very low flow rates is poor. This is one of the less than desirable qualities of the weir valve.

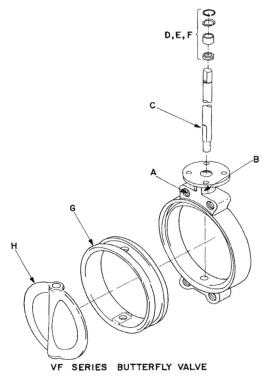

VF SERIES BUTTERFLY VALVE

Figure 2.42.5 Valves—butterfly—exploded view. (*By permission: Johnson Controls, Milwaukee, WI.*)

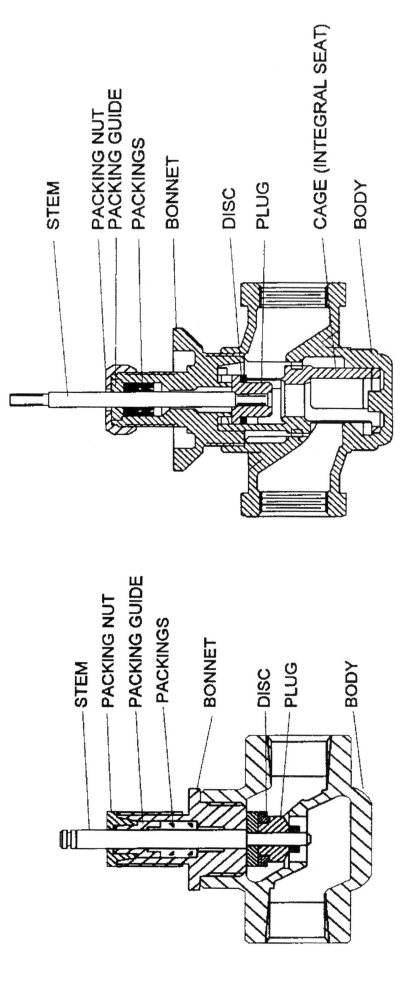

Figure 2.42.6 Globe valves—sections through. (*By permission: Johnson Controls, Milwaukee, WI.*)

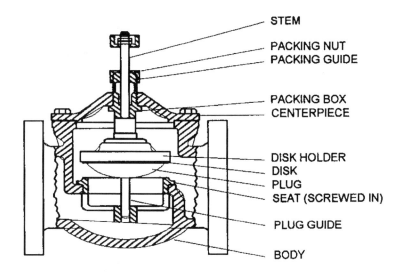

N.O. Cast Iron Flanged Style Valve Body

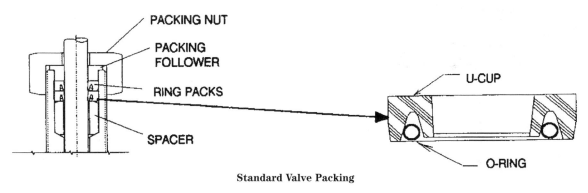

Standard Valve Packing

Figure 2.42.7   Globe valve and stem packing parts identified. (*By permission: Johnson Controls, Milwaukee, WI.*)

Introduction

As we enter the new millennium, code-making organizations are realigning. At this writing, three code-making bodies, SBCCI, BOCA, and ICBO, have consolidated into one code organization, the ICC, and produced the International Plumbing Code (IPC), the International Fuel Gas Code (IFGC), and the International Mechanical Code (IMC).

Another code-making organization, the International Association of Plumbing and Mechanical Officials (IAPMO), will continue to publish its Uniform Plumbing Code (UPC) and its Uniform Mechanical Code (UMC).

Check with your local Administrative Authority (building department) to see which code(s) they have adopted.

The IPC and the IMC codes have very different approaches to several key issues, like sizing water lines and vents. The older UPC is based on a proscriptive approach (established practice), while the IPC is rooted in a performance approach (engineering specifications).

As a rule, the UPC is more restrictive, but easier to understand, than the IPC. The UPC language has been refined over the course of 22 editions since 1945.

The IPC is a completely new approach to plumbing codes. It allows materials and methods not accepted by the UPC. However, the IPC has not had the review and refinement that years of use provide. Ultimately, it is more flexible than the UPC but somewhat more difficult to understand.

It is one of the goals of this book to make the IPC a little more user-friendly. Another goal of this book is to help provide data where it is needed, on site. We like to think of the *Code Check* series more as tools than books.

Origin of the Uniform Plumbing Code:
IAPMO began in 1926 as the Plumbing Inspectors' Association of Los Angeles. Forty-two plumbing inspectors "banded together to bring about an improvement in the application of common-sense codification and application of ordinances based on scientific knowledge."—IAPMO

Origin of the International Codes:
On December 9, 1994, the International Code Council (ICC) was established as a nonprofit organization dedicated to developing a single set of comprehensive and coordinated national codes.—ICC

Code July 1997

Key

- **[301.4.1] [912.3]** = Code references are followed by two bracketed numbers
- **[301.4.1] [....]** = The first bracket contains a reference number to the 1997 IPC or IMC codes according to the elected discipline
- **[....] [912.3]** = The second bracket is the equivalent code number or reference for areas covered by the UPC or UMC
- □ = Applies to both IPC and UPC
- ❖ = Only applies to IPC
- ➤ = Only applies to UPC
- **[F101]** = Letter refers to an appendix (except "T" for table)
- **Pt-3** = Plumbing Table 3
- **Fig. 3** = Illustration 3
- **[local]** = This citation is under discussion. Opinions vary widely. Check with your Administrative Authority (building department) for their interpretation.
- **[energy]** = Energy conservation measures modeled on California Title 24 or the Model Energy Code
- **[mfr.]** = Follow manufacturer's installation instructions
- **[N/A]** = Not addressed
- **[T 310.5]** = Table in cited code book
- **[G 1201]** = Appendix G in the IPC
- **[trade]** = Unwritten trade practice – "The way it's always been done"
- **(** ½in. larger than**)** = UPC difference from IPC
- **EXC** = Exception to rule follows

Abbreviations

To maximize space, many abbreviations are used in *Code Check*.
- **AA** = Administrative Authority
- **ABS** = acrylonitrile-butadiene-styrene
- **Al** = aluminum
- **AGA** = American Gas Association: www.aga.com
- **ANSI** = American National Standards Institute: www.ansi.org
- **Appl.** = appliance
- **ASHRAE** = American Society of Heating, Refrigeration, and Air Conditioning Engineers: www.ashrae.org
- **AWG** = American Wire Gauge
- **B** (vent) = double-walled gas appliance flue
- **BW** (vent) = oval double-walled gas appliance flue (rated for use within walls)
- **Bldg.** = building
- **BTU** = British Thermal Unit
- **C.I.** = cast iron
- **Cu** = copper
- **EXC** = exception to preceding rule
- **Ext.** = exterior
- **Fe** = iron or steel pipe
- **IFGC** = International Fuel Gas Code 1998: www.intlcode.org
- **IMC** = International Mechanical Code 1998: www.intlcode.org
- **IPC** = International Plumbing Code 1997: www.intlcode.org
- **Lav, Lavies** = bathroom lavatory
- **PB** = polybutylene pipe
- **PE** = polyethylene pipe
- **PVC** = polyvinyl chloride
- **Sch.** = schedule of pipe
- **SW** = single-wall gas flue pipe
- **UBC** = Uniform Building Code 1997 published by the International Conference of Building Officials: www.icbo.org
- **UL** = Underwriters Laboratory: www.ul.com
- **UMC** = Uniform Mechanical Code 1997, published by the International Association of Plumbing and Mechanical Officials: www.iapmo.org
- **UPC** = Uniform Plumbing Code, published by the International Association of Plumbing and Mechanical Officials: www.iapmo.org

Figure 2.43.0 Introduction to residential plumbing code orientation. (*By permission: McGraw-Hill, Taunton Press subsidiary,* Code Check Plumbing—A Field Guide to the Plumbing Codes, *Redwood Kardon and Jeff Hutcher.*)

Pt-7 Sizing Drains [UPC T 7-5]						
Pipe size		1¼in.	1½in.	2in.	3in.	4in.
Drain units*	Vert.	1	2**	16***	48****	256
	Horiz.	1	1	8***	35	216*****
Drain feet	Vert.	45	65	85	212	300
	Horiz.	unlimited				

* Excluding trap arm
** Except sinks and dishwashers
*** Except water closets
**** Only 4 water closets on vert., 3 on horiz.
***** ¼in./ft. slope

Pt-8 Fixture Unit Load and Trap Size [IPC T709.1] [UPC T7-3]				
	Fixture units	Fixture units	Min. trap size (in.)	Min. trap size (in.)
	IPC	UPC	IPC***	UPC
Bar sink	N/A	1	N/A	1 1/12
Bath/shower	2	3	1½	1½**
Bathroom group*	6	N/A	–	N/A
Bidet	2	1	1¼	1¼
Clothes washer	2	3	2	2
Kitchen sink	2	2	1½	1½**
Kitchen sink with disposal and dishwasher	2	3	1½	1½**
Kitchen sink with disposal only	2	2	1½	1½**
Kitchen sink with dishwasher only	2	3	1½	1½**
Lavatory	1	1	1¼	1¼
Laundry sink (1 or 2 compartment)	2	2	1½	1½
Shower stall	2	2	2	2
Water closet	4	3	3	3

* IPC uses a bathroom group designation which includes a water closet, lavatory, bidet and bathtub/shower. The bathroom group becomes important because it allows for horizontal wet venting discussed later.
** UPC requires a 2in. drain beyond the trap for these fixtures.
*** IPC – Minimum trap size IS minimum drain size for all fixtures.

Pt-10 Trap Loading and Maximum Trap Arm Distance [UPC 702.0 T10-1]			
Trap size	Unit loading	Max. length of trap arm to vent	Slope per foot (in.)
1¼	1	2ft. 6in.	¼
1½	3	3ft. 6in.	¼
2	4	5ft.	¼
3	6	6ft.	¼
4 & larger	8	10ft.*	¼

* The developed length of the trap arm for a water closet, measured from the top of the closet ring to the inner edge of vent, shall not exceed 6ft.

Figure 2.43.1 Sizing drains and traps for residential plumbing fixtures. (*By permission: McGraw-Hill, Taunton Press subsidiary,* Code Check Plumbing—A Field Guide to the Plumbing Codes, *Redwood Kardon and Jeff Hutcher.*)

VENTS

❖ Main vent req'd on any drainage system receiving discharge from toilet, typically the continuation of the waste stack through the roof. An undiminished 1½ in. vent through the roof for a toilet satisfies this requirement. . . . .**Fig. 24** [903.1] [NA]

❖ Main vent through roof is the only vent req'd to terminate above roof. . . . .**Fig. 24** [903.1] [NA]

➢ All vents to terminate above roof. . . . . [NA] [905.4]

➢ Total cross sectional area of vents through roof to be equal to the req'd building drain size. . . . .**Pt-14** [NA] [904.1]

Fig. 24 Main Vent—IPC

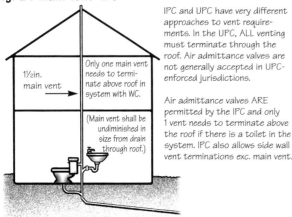

IPC and UPC have very different approaches to vent requirements. In the UPC, ALL venting must terminate through the roof. Air admittance valves are not generally accepted in UPC-enforced jurisdictions.

Air admittance valves ARE permitted by the IPC and only 1 vent needs to terminate above the roof if there is a toilet in the system. IPC also allows side wall vent terminations exc. main vent.

1½ in. main vent

Only one main vent needs to terminate above roof in system with WC.

(Main vent shall be undiminished in size from drain through roof.)

Fig. 28 No Flat Venting

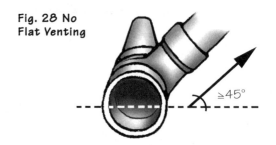

≥45°

☐ No flat venting (take off above horiz. centerline). . . . . . **Fig. 28** [905.4] [905.2]

❖ Horiz. venting and the use of side inlet ¼ bends are allowed when used as a wet vent. (Vent kept clear by washing from the upstream fixture.) . . . . [905.6] [NA]

Pt-15 Size and Length of Stack Vents [IPC T916]						
Size of waste stack (in.)	Total fixture units vented	Maximum length of vent/pipe size				
		1¼ in.	1½ in.	2 in.	3 in.	4 in.
1¼	2	30ft.				
1½	8	50ft.	150ft.			
1½	10	30ft.	100ft.			
2	12	30ft.	75ft.	200ft.		
2	20	26ft.	50ft.	150ft.		
3	10		42ft.	150ft.	1,040ft.	
3	21		32ft.	110ft.	810ft.	
3	53		27ft.	94ft.	680ft.	
3	102		25ft.	86ft.	620ft.	
4	43			35ft.	250ft.	980ft.

* Vents shall not be less than half the size of drain to which they are connected. Min. vent size for drain with toilet – 1½ in.

Pt-14 Aggregate Vent Area (must = req'd building drain size) [UPC 904.1]				
Vent size (in.)	Area sq. in.	# vents	Net vent area	
1¼	1.23			Example:
1½	1.77			Req'd 4in. bldg.drain = 12.57sq.in.
2	3.14			Three 2in. vents = 9.42sq.in.
3	7.07			One 1¼in. vent = 1.23sq.in.
4	12.57			1.23 + 9.42 = 10.65sq.in.
				Thus, more venting would be req'd.

Figure 2.43.2 Sizing and installation of vent lines in residential projects. (*By permission: McGraw-Hill, Taunton Press subsidiary,* Code Check Plumbing—A Field Guide to the Plumbing Codes, *Redwood Kardon and Jeff Hutcher.*)

Wet Vents IPC Fig. 32

❖ Any combination of fixtures within 2 bathroom groups on the same level may be connected to a wet vent. Only fixtures within the bathroom group shall connect to the wet vent. Other fixtures to connect downstream of the bathroom group. . . . . Pt-18 [909.1] [NA]

❖ The dry vent connection to the wet vent shall be an indiv. or common vent connected to the lav, bidet, bath, or shower. . . . . [909.2] [N/A]

Pt-18 Wet Vent Sizing [IPC T909]	
IPC wet vent pipe size (in.)	Drainage fixture unit loading
1½	1
2	4
3	12

Fig. 32 Wet Venting IPC

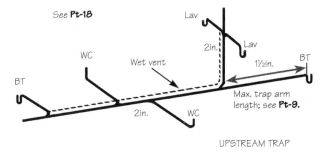

Combination Drain &Vent (CD&V) IPC Fig. 33

(not to be confused with UPC version seldom used in residential structures)

❖ Limited to sinks, lavs, standpipes, and floor drains . . . . [912.1] [N/A]

❖ Only the sink, standpipe or lav fixture connection is allowed to have vert. piping (max. 8ft.) connecting to the horiz. CD&V . . .[912.2] [NA]

❖ Max. slope of CD&V piping is ½in./ft . . . . [912.2.1] [NA]

❖ CD&V shall connect to a vented horiz. branch or shall have a vent connected to the drain and vent that rises vertically ≥6in. above the fixture flood rim before making a horiz. offset. . . . . [912.2.2] [NA]

❖ The fixture branch (trap arm) shall attach to the combination drain and vent within the distances in Pt-9 [trap arm length]. The drain and vent shall be considered the vent. . . . . [912.2.4] [NA]

Waste Stack Venting IPC Fig. 34

❖ The waste stack shall be vertical and no offsets allowed until 6in. above the highest fixture served. No toilets allowed on waste stack. . .[910.2&.3] [N/A]

❖ The waste & vent stack shall be the same size entire length. . . .[910.4] [N/A]

Fig. 34 IPC Waste Stack Vent

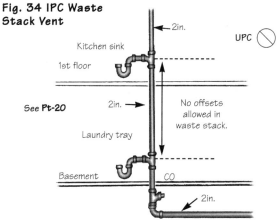

Island Sink (Western Plumbing)

IPC ISLAND Fig. 35

❖ Island venting limited to sinks and lavies. . . . .[913.1] [NA]

❖ Island vents need only rise vert. above the drain outlet before offsetting horiz. and down. . . . .[913.2] [NA]

❖ The lowest part of the island vent shall connect full size to a vert. drain or top half of horiz. drain. . . . .[913.3] [NA]

❖ Cleanout required in island fixture drain. . . . .[913.3] [NA]

UPC ISLAND Fig. 35a

➤ Island sink vented with drainage pattern fittings only. . . . .[NA] [909]

➤ Drain that serves island shall serve no other fixtures upstream from return vent. . . . .[NA] [909]

➤ Cleanout req'd in vert. section of foot vent. . . . . .Fig. 35a [NA] [909]

Fig. 35 IPC Island Venting   Fig. 35a UPC Island Venting

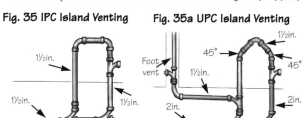

Figure 2.43.3 Wet venting procedures and waste stack venting in residential projects. (*By permission: McGraw-Hill, Taunton Press subsidiary,* Code Check Plumbing—A Field Guide to the Plumbing Codes, *Redwood Kardon and Jeff Hutcher.*)

Gas Piping

Pt-30 Piping System (Excluding Connectors) See Below for Exceptions and Conditions		
Materials	IPC T-g103.2	UPC-T 1210.0
Aluminum-alloy pipe and tubing	OK	N/A
Brass pipe	OK	OK
Copper pipe	OK	N/A
Copper tube (type k and l)	OK	N/A
Copper seamless tube (acr)	OK	N/A
Corrugated stainless-steel tubing	OK	N/A
Plastic pipe and tubing	OK	OK
Steel pipe	OK	OK
Steel tubing	OK	N/A
For capacities not listed below, see mfr. listings		

General

☐ Shall be new or previously used for gas piping only ......[G103.8][1210.2]
☐ Bends in plastic pipe to mfr. instructions ....[G104.14.2][1202.2]
❖ Not be run through air duct, chimney, vent, or plenum .....[G104.1][N/A]
➤ Bends in PE pipe radius (20 times dia. of pipe). ....[N/A][1211.17]

Material of Pipe and Tubing

(Excluding Connectors)
☐ Plastic underground outside only. ....EXC [G103.4][1210.1]
❖ EXC – May terminate aboveground, outside, when in accordance with mfr.
  instructions. ....[G103.4] [N/A]
❖ Cu & brass not allowed where gas contains >0.3 grain of hydrogen
  sulfide per 100cu.ft. (check/supplier). ....[G103.5][N/A]
❖ Al not allowed outside or underground. ....[G103.6][N/A]
❖ Corrugated stainless steel allowed in interior locations in accordance
  with mfr. instructions. ....[IFGC 403.2.6.4] [N/A]
➤ Yellow brass shall contain not more than 75% Cu. ....[N/A][1210.1]

Joints in Pipe or Tubing

Always check mfrs. listing and with your AA when joining unfamiliar or
dissimilar materials.
☐ Fe to be threaded or welded. ....[G103.18][12.11.1,.2]
❖ Cu and brass shall be threaded, brazed, mechanical,
  or welded...................................[G103.14,.13][N/A]
❖ Al pipe or tubing shall be flared or mechanical. ....[G103.12][N/A]
❖ Cu tubing shall be brazed, mechanical, or flared. ....[G103.15][N/A]
❖ PE or PB shall be heat-fused. ....[G103.17.2][N/A]
❖ Unions and right/left nipples ok in exposed locations only. [IFGC 404.3][N/A]
➤ No unions EXC. ....[N/A][1211.10]
➤ EXC – right/left nipple and couplings ....[N/A][1211.10]
➤ EXC – at exposed appliance connection ....[N/A][1211.10]
➤ EXC – at exposed exterior locations on the discharge side
  of a bldg. shutoff ....[N/A][1211.10]

Figure 2.43.4 Gas piping guidelines in residential projects. (*By permission: McGraw-Hill, Taunton Press subsidiary,* Code Check Plumbing—A Field Guide to the Plumbing Codes, *Redwood Kardon and Jeff Hutcher.*)

☐ Max. 3ft. ....PT-31 EXC. [G109.1.2][1212.0 .1]
☐ EXC – may be 6ft. for a range or dryer ....PT-31 [G109.1.2][1212.0.1]
☐ Shall be immediately upstream of the shutoff ....[G109.2][1212.0.3]
☐ Sized to demand of appliance ....PT-31 & 32 [G109.1.2][1212.0.4]
☐ Al in dry locations only & no contact with masonry,
  insulation, or plaster ....[IFGC403.2.6.3][1212.0.5]
➤ Gas hose <15ft. ok outdoor portable appl. only. ....[N/A][1212.0.7]

Pt-31 Maximum Capacities of Connectors in Thousands BTU/hr. [IPC TG109.1.2(2)] [UPC T1210]									
Semirigid connector outside dia. (in.)									
	Flexible connector nominal inside dia. (in.)								
		All gas appliances					Ranges and domestic clothes dryers **only**		
		1ft.	**1½ft.**	**2ft.**	**2½ft.**	**3ft.**	**4ft.**	**5ft.**	**6ft.**
⅜in.	¼in.	28	23	20	19	17			
½in.	⅜in.	66	54	47	44	41	72	63	57
⅝in.	½in.	134	110	95	88	82			
—	¾in.	285	233	202	188	174			
—	1	567	467	405	378	353			
(On pressure drop of 0.4in. water column) natural gas demand of 1,100 BTU per cu. ft.									

Pt-32 Common Appliance Demand Table [IPC TG102.3] (typical residential appliances) [UPC T12.1]	
Appliance	BTU/hr.
Clothes dryer	35,000
Oven (built in)	25,000
Range (free standing)	65,000
Cooktop only	40,000
Log lighter	25,000
Water heater (30 gallons)	30,000
Water heater (40 – 50 gallons)	50,000 – 55,000
Barbecue	50,000

Pt-33 Sizing Gas Example
1. Meter to most remote outlet 75ft.
2. Size outlet A and section 1,2,3 using 80ft. column.
3. Use 70ft. column to size outlet B.
4. Use 50ft. column to size outlet C, D, E, & sec. 4.
5. Use 30ft. column to size outlet F.
Based on 1100 BTU/cu.ft. – CFH = cu. ft. per hr. – Sched. 40 Fe. See Pt-34.

Figure 2.43.5 Common appliance gas demand tables and connector tables. (*By permission: McGraw-Hill, Taunton Press subsidiary,* Code Check Plumbing—A Field Guide to the Plumbing Codes, *Redwood Kardon and Jeff Hutcher.*)

☐ DWV piping shall be tested with 10ft. head of water for
15 minutes or 5 PSI air for 15 minutes . . . .[312.2,.3] [712.2,.3]
☐ Water piping shall be tested with working pressure or 50 PSI
for 15 minutes . . . .[312.5] [609.4]
❖ Fuel gas piping shall be tested with 3 PSI air for
10 minutes . . . .[IFGC 406.4.2,.3] [N/A]
➤ Fuel gas piping shall be tested with 10 PSI air or 6in. of
mercury for 15 minutes . . . .[NA] [1204.3.2]

Notches in Joists and Rafters Fig. 83
☐ Not in ends exceeding one-quarter depth
of member . . . .[F101.1] [UBC2320.12.4]
☐ OK in top or bottom if <one-sixth depth
of member . . . .[F101.1] [UBC2320.12.4]
☐ Not in middle third of span . . . .[F101.1] [UBC2320.12.4]

Holes in Joists and Rafters Fig. 83
☐ ≤one-third depth of member . . . .[F101.1] [UBC2320.12.4]
☐ No holes <2in. to top or bottom of member . . . . .[F101.1] [UBC2320.12.4]

Fig. 82 Boring in Wood I-Joists

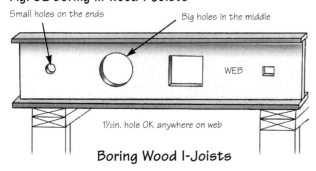

Small holes on the ends

Big holes in the middle

WEB

1½in. hole OK anywhere on web

Boring Wood I-Joists

The bearing ends of manufactured wood I-joists are weakened more by
boring than are the center sections. Notches are rarely acceptable. It is
essential to follow the joist mfr. instructions

Notching Studs
☐ Ext. and bearing walls ≤25% of stud . . . . . .Fig. 84 [F101.2] [UBC2326.11.9]
☐ Non-bearing partitions ≤40% of stud . . . .[F101.2] [UBC2326.11.9]

Holes in Studs
☐ <40% of bearing stud width allowed EXC. Fig. 84 [F101.3] [UBC2326.11.9]
☐ EXC – holes (60% of stud OK if stud is doubled – max. 2
successive studs . . . .[F101.3] [UBC2326.11.9]
☐ <60% of non-bearing stud width allowed . . . .[F101.3] [UBC2326.11.9]
☐ ≤⅝in. from edge of stud . . . .[F101.3] [UBC2326.11.9]
☐ Not in same stud section as cuts or notches . . . . . .[F101.3] [UBC2326.11.9]

Fig. 83 Joist Notching

Holes min. 2in. from top or bottom
Max. one-third depth

Notch max. one-sixth depth

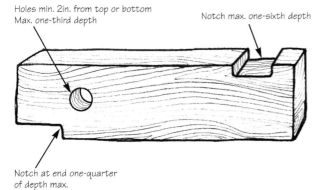

Notch at end one-quarter
of depth max.

Fig. 84 Notching & Boring Studs

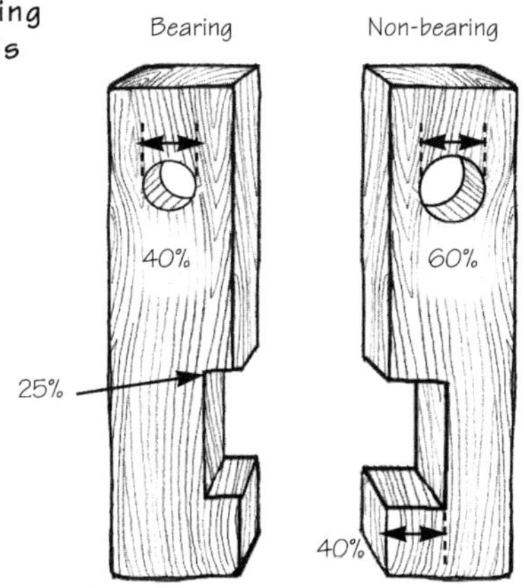

Bearing

Non-bearing

40%

60%

25%

40%

Pt-38 Notching & Boring Studs [UBC2326.11.9]				
		Notch	Bore	2x*
2x4	Bearing	⅞in.	1⁷⁄₁₆in.	2in.
	Non-bearing	1⁷⁄₁₆in.	2in.	N/A
3x4	Bearing	⅞in.	⅞in.	2in.
	Non-bearing	1⁷⁄₁₆in.	2in.	N/A
2x6	Bearing	1⅜in.	2¼in.	3⅜in.
	Non-bearing	2¼in.	3⅜in.	N/A
* 2 doubled consecutive studs				

Pt-39 Notching & Boring Joists [UBC2320.12.4]			
	Notching		Boring
	End	Outer third	2in. to edge
2x6	1⅜in.	⅞in.	1½in.
2x8	1⅞in.	1¼in.	2⅜in.
2x10	2⅜in.	1½in.	3in.
2x12	2⅞in.	1⅞in.	3¾in.

Figure 2.43.6 Joist and stud notching procedures for pipe installations. (*By permission: McGraw-Hill, Taunton Press subsidiary,* Code Check Plumbing—A Field Guide to the Plumbing Codes, *Redwood Kardon and Jeff Hutcher.*)

Temperature & Pressure Relief

- Water can scald on contact at temperatures as low as 110°F.
- Water, when it changes state to a gas (steam), expands 1,700 times.
- At sea level atmospheric pressure, water boils at 212°F.
- Temperature relief valves are set below 210°F.
- Pressure relief valves are set to 150 PSI.
- Under pressure, the boiling temperature of water goes up.
- At 150 PSI, water boils at 358°F.

Without protection, a domestic hot-water heater whose thermostat has failed would see a continuous rise in temperature and pressure (from the expanding water). This temperature and pressure would continue to rise until the pressure exceeds the pressure capacity of the tank (300 PSI). In the same instant the tank bursts, the superheated water would instantly boil and expand with explosive force. The danger of superheated water like this flashing over into steam inside a steel container is obvious. It is a bomb. **Fig. 65.**

To prevent such catastrophic failures, water heaters are required to be protected for both over temperature and over pressure. Usually this is done with a temperature and pressure (T&P) relief valve.

Fig. 65 Why T&P?

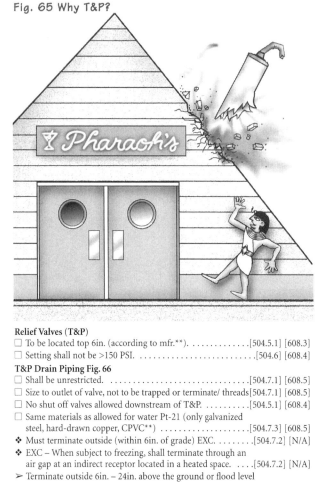

Relief Valves (T&P)
- ☐ To be located top 6in. (according to mfr.**). . . . .[504.5.1] [608.3]
- ☐ Setting shall not be >150 PSI. . . . .[504.6] [608.4]

T&P Drain Piping Fig. 66
- ☐ Shall be unrestricted. . . . .[504.7.1] [608.5]
- ☐ Size to outlet of valve, not to be trapped or terminate/ threads[504.7.1] [608.5]
- ☐ No shut off valves allowed downstream of T&P. . . . .[504.5.1] [608.4]
- ☐ Same materials as allowed for water Pt-21 (only galvanized steel, hard-drawn copper, CPVC**) . . . .[504.7.3] [608.5]
- ❖ Must terminate outside (within 6in. of grade) EXC. . . . .[504.7.2] [N/A]
- ❖ EXC – When subject to freezing, shall terminate through an air gap at an indirect receptor located in a heated space. . . . .[504.7.2] [N/A]
- ➤ Terminate outside 6in. – 24in. above the ground or flood level of the area receiving the discharge & point down . . . .[N/A] [608.5]

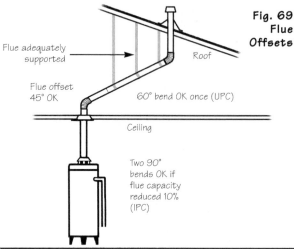

Fig. 69 Flue Offsets

Pt-37 B&BW Vent (with listed cap) Terminations	
Pitch	Height above roof
6/12	1ft.
7/12 (30°)	1ft. 3in.
8/12	1ft. 6in.
9/12	2ft.
10/12	2ft. 6in.
11/12	3ft. 3in.
12/12 (45°)	4ft.
18/12	7ft. 6in.
(20/12 (60°)	8ft.

Termination of "B" Vent Fig. 71
- ☐ 8ft. min. from vert. surface EXC . . . .[802.6*] [517.3]
- ☐ EXC – Direct vented appliances . . . .[805*] [517.5]
- ☐ 2ft. above highest point of roof penetration but not less than 2ft. higher than any part of building within 10ft. EXC. . . . . .[802.6*] [517.3]
- ☐ EXC – Direct vented appliances . . . .[805*] [517.5]
- ☐ Min. 3ft. above forced air inlet within 10ft. . . . .[805.3.4*] [517.6]
- ➤ Min. 4ft. from property line except a public way . . . .[N/A] [517.6]

Fig. 70 Vent Connectors

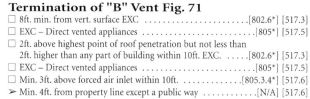

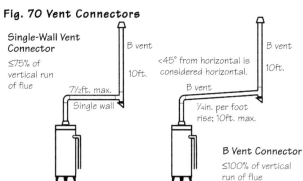

Figure 2.43.7 Water heater combustion air, vent, and flue information. (*By permission: McGraw-Hill, Taunton Press subsidiary,* Code Check Plumbing—A Field Guide to the Plumbing Codes, *Redwood Kardon and Jeff Hutcher.*)

➢ Min. door size opening 24in. wide x 30in. high ....[N/A] [511.1]

➢ Attic & underfloor – light (to be switched at opening or door) and outlet at or near heater. ....[N/A] [511.5]

➢ Req'd working space at heater to be solid flooring ....[N/A] [511.5]

Required Pans & Drain

☐ Water heaters in attic or other areas that can be damaged due to leakage shall be installed in a watertight pan. ......**Fig. 56** [504.8] [510.7]

☐ Pan shall be galvanized steel or other corrosion-resistant metal (may be plastic**) with a min. 24-gauge thickness. ....[504.8] [510.7]

☐ Drain size 1in. min. (¾in.**). ....[504.8.1] [510.7]

☐ Drain shall be run to (approved location**) indirect waste or to outside terminating 6in.–24in. (no specs**) above grade. ....[504.8.2] [510.7]

❖ Pan shall be min. 1½ in. in depth. ....[504.8.1] [N/A]

Fig. 56 Water Heater in Attic

Roof line

Unions required by UPC

20ft. Max. from Water Heater to Attic Opening

Type B vent in attic min. 5ft.

T & P relief valve drain to exterior of bldg. or other approved location

Gas supply

24in. walkway all around

30in.

30in. min. opening

30in. 24in. min. Catwalk

Pan for collecting water overflow

Drain to approved location

Heat Exchangers

☐ Using toxic fluids shall be double walled and vented to atmosphere (visible leak detection) ....[608.16.3] [603.4.4]

❖ With non-toxic fluids may be single wall ....[608.16.3] [N/A]

➢ May be single wall if all the following req's are met: ....[N/A] [L3.0]

· Transfer fluid is potable water or FDA approved

· Transfer fluid pressure < potable water side of exchanger

· Labeled to permit only FDA approved additives on heat transfer side

COMBUSTION AIR

Openings And Ducts

General

☐ Min. cross sectional area ≥3in. ....[709.1*] [507.3.1]

☐ Shall serve a single enclosure ....[709.1*] [507.3.4]

☐ Ducts shall not slope downward toward the source of combustion air ....[709.1*] [703.1.2**]

☐ Openings shall be ≥¼in. screening except to attic. ....[709.1*] [507.4]

☐ Separation shall be maintained between upper and lower combustion air ducts from enclosure to combustion air source. ....[709.1*] [507.3.5]

➢ Openings to terminate ≥3in. in front of fire box. This space to be maintained from floor to ceiling. ....[N/A] [507.3.2]

Fig. 60   **Fig. 61**   **Fig. 62**

Two horiz. ducts 1sq.in. per 2,000 BTU/hr. each

Two openings in ext. wall 1 sq.in. per 5,000 BTU/hr. each

Two openings to ventilated attic 1sq.in. per 4,000 BTU/hr. each

Crawl Space Source

☐ Lower combustion air only ....**Fig. 63** [*701.4.1] [T5-1]

☐ Ventilation in crawl to be 2x req'd combustion air ....[*701.4.1] [T5-1]

☐ Opening shall be 1sq.in. per 4000 BTU ....**Fig. 63** [*701.4.1] [T5-1]

☐ Crawl space to meet min. bldg. code requirements. (18in. clearance from joists to earth) ....[* 704.1.1] [T5-1]

Fig. 63   **Fig. 64**

Attic & crawl space min. 1sq.in. per 4,000 BTU/hr. each

Two vert. ducts min. 1sq.in. per 4,000 BTU/hr. each

Attic Space Source

☐ Attic height min. 30in. ....[701.4.2*] [703.0.2.1**]

☐ Opening in attic to be sleeved 6in. above ceiling joists / 26-gauge galvanized metal or other approved material. ....**Fig. 63** [701.4.2*] [703.1.2.3**]

☐ Openings into attic shall not be screened. ....[709.1*] [507.3.7]

☐ Attic ventilation openings to outside min. 1sq.ft. per 300sq.ft. area of attic ....[704.4.2*] [T5-1]

☐ Ducts min. 1sq.in. per 4,000 BTU. ....**Fig. 63** [704.4.2*] [T5-1]

☐ Other methods of supplying comb. air to meet engineering principles and approval of AA ....[701.2*] [507.5]

Figure 2.43.8 Water heater pans, drains, and combustion air openings and ducts. (*By permission: McGraw-Hill, Taunton Press subsidiary*, Code Check Plumbing—A Field Guide to the Plumbing Codes, *Redwood Kardon and Jeff Hutcher.*)

WATER

Pt-21 Materials for Water Service Piping

	IPC T 605.4	UPC sec. 604.1
ABS plastic pipe	OK	N/A
Asbestos-cement pipe	OK	OK
Brass pipe	OK	OK
Copper type K,L,M	OK	OK
CPVC plastic	OK	OK
Ductile iron	OK	OK
Galvanized steel pipe	OK	OK
PB plastic pipe	OK	N/A
PE plastic pipe	OK	OK
PE plastic tubing	OK	OK
PEX plastic tubing	OK	N/A
PEX-AL-PEX pipe	OK	N/A
PE-AL-PE pipe	OK	N/A
PVC plastic pipe	OK	OK

• IPC requires that plastic water service pipe terminate within 5ft. inside the point of entry into a building. This does not apply if the piping is approved within the building (see chart below).
• Approval may be required when replacing metallic water service pipe with plastic piping. An alternate grounding method would be required.

Pt-22 Materials for Water Distribution within a Building or Structure

	IPC T 605.5	UPC sec 604.1-604.2
Brass	OK	OK
CPVC	OK	OK
Copper type K,L,M *	OK	OK*
PEX	OK	N/A
PEX-AL-PEX	OK	N/A
Galvanized steel**	OK	OK
PB plastic pipe and tubing	OK	N/A

* UPC allows type M copper in all locations except underground within the bldg.
** UPC Ferrous piping within the building under or in a slab to be machine-wrapped pipe. Joints to be wrapped to same level of protection. (40 mil) [UPC 609.3.1]

Pt-23 Minimum Required Air Gaps [IPC T608.15.1] [UPC T6-3]

Fixtures	When not affected by side walls* [in.]		When affected by side walls [in.]	
	IPC	UPC	IPC	UPC
Effective opening ½in. or smaller [lav.]	1	1	1½	1½
Effective opening not exceeding ¾in. [sink laundry tray]	1½	1½	2½	2¼
Effective opening not exceeding 1in. [bath filler]	2	2	3	3
Effective opening exceeding 1in.	2 times diameter of opening	2 times diameter of opening	3 times diameter of opening	3 times diameter of opening

* Where the distance from inner edge of spout to wall is greater than 3 times the diameter of the effective opening, or greater than four times the opening for two intersecting walls. See Fig. 50.

Fig. 50 Sink and Faucet with Air Gap between Spout and Flood Rim

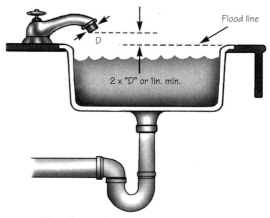

Vacuum Breakers Required On

☐ Hose connections except where built into equipment such as washing machine . . . .[608.15.4.2] [603.4.7]
☐ Faucets with pullout spouts . . . .[608.15.4.1] [603.4.8]
☐ Irrigation system 6in. above highest head . . . . .**Fig. 51** [608.16.5] [603.4.6.1]
☐ If sprinkler head upstream or chemicals additives introduced, a reduced pressure backflow device req'd. . . . .**Fig. 52** [608.16.5] [603.4.6.4]
☐ Hand showerheads <6in. of tub shall be protected by vacuum breaker. . . . .[608.15.4] [603.4.7]

Figure 2.43.9 Residential construction water distribution materials, air gaps, and vacuum breakers. (*By permission: McGraw-Hill, Taunton Press subsidiary,* Code Check Plumbing—A Field Guide to the Plumbing Codes, *Redwood Kardon and Jeff Hutcher.*)

TRAPS

Fig. 36 Trap Seal

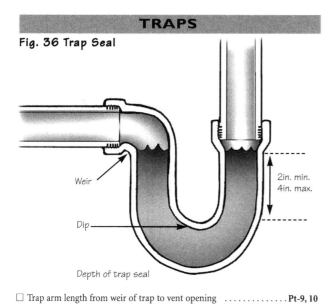

Weir

2in. min.
4in. max.

Dip

Depth of trap seal

Fig. 40 Common Trap Arm IPC

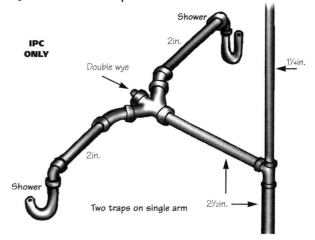

Shower

IPC ONLY

2in.

Double wye

1¼in.

2in.

Shower

Two traps on single arm

2½in.

Fig. 41 Fixture tailpiece max. 24in.

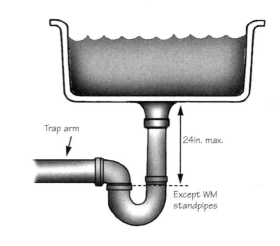

Trap arm

24in. max.

Except WM standpipes

☐ Trap arm length from weir of trap to vent opening Pt-9, 10
☐ All fixtures req. separate water seal trap. ....[1002.1] [1001.1]
☐ Trap seal shall be 2in. – 4in. deep. ....Fig. 36 [1002.4] [1005.0]
☐ Fixture tailpiece max. 24in. .... Fig. 41 EXC [1002.1] [1001.4]
☐ EXC washing machine standpipes ....Fig. 73 [1002.1] [1001.4]
☐ Traps shall be set level and protected from freezing where
   necessary. ....[1002.7] [1005.0]
☐ Slip joints – only one allowed on outlet side of trap Fig. 44 [1002.2] [1003.3]
☐ "S" traps, bell traps, crown vented traps, and drum traps
   are prohibited. ....Fig. 43 [1002.3] [1004.0]
☐ Fixture shall not be double trapped. ....[1002.1] [1004.0]
☐ Sized to drain rapidly and not less than size in table . . Pt-8 [1002.5] [1003.3]
☐ Min. distance from trap weir to vent inlet 2x trap
   arm dia. ....Fig. 26 [906.2] [1002.2]
☐ All fittings must be drainage type. ....Fig. 38, 39 [1004.1] [701.2.1]
❖ One trap may serve a combination fixture if there is <6in. diff. in depth of
   compartments and the waste outlets ≤30in. apart). . . Fig. 37 [1002.1.2] [N/A]
➤ Trap shall not be >1 pipe size of tailpiece. ....[N/A] [1003.3]
➤ Only one trap permitted on trap arm. ....[N/A] [1001.1]
➤ 1 trap may serve 3 sinks, laundry tubs, or lavies of the same depth,
   in the same room, max. 30in. apart. ....Fig. 37 [N/A] [1001.2]
➤ No tubular trap shall be installed w/out listed
   trap adapter. ....Fig. 42 [N/A] [1003.2]
➤ Not more than one trap permitted on trap arm ....[N/A] [1001.1]

Figure 2.43.10 Schematics of trap installations for residential projects. (*By permission: McGraw-Hill, Taunton Press subsidiary,* Code Check Plumbing—A Field Guide to the Plumbing Codes, *Redwood Kardon and Jeff Hutcher.*)

➤ Trap shall be same size as arm. . . . .[N/A] [1003.3]
➤ Trap arms <3 in. req. cleanout if direction change >90°
   (3in. req. cleanout if direction change >135°. . . . .[N/A] [1002.3]
➤ Tubular brass traps shall be a min. 17 gauge . . . .[N/A] [1003.1]

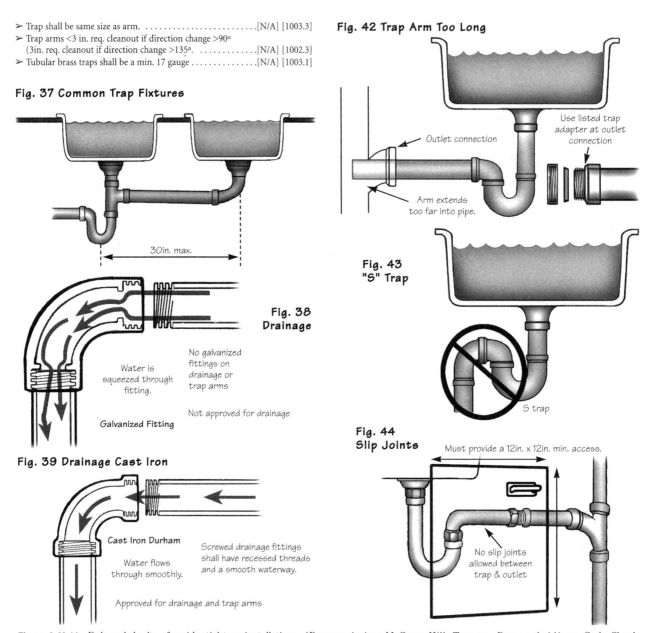

Fig. 37 Common Trap Fixtures

30in. max.

Fig. 38 Drainage

Water is squeezed through fitting.

Galvanized Fitting

No galvanized fittings on drainage or trap arms

Not approved for drainage

Fig. 39 Drainage Cast Iron

Cast Iron Durham

Water flows through smoothly.

Screwed drainage fittings shall have recessed threads and a smooth waterway.

Approved for drainage and trap arms

Fig. 42 Trap Arm Too Long

Outlet connection

Use listed trap adapter at outlet connection

Arm extends too far into pipe.

Fig. 43 "S" Trap

S trap

Fig. 44 Slip Joints

Must provide a 12in. x 12in. min. access.

No slip joints allowed between trap & outlet

Figure 2.43.11 Do's and don'ts of residential trap installations. (*By permission: McGraw-Hill, Taunton Press subsidiary,* Code Check Plumbing—A Field Guide to the Plumbing Codes, *Redwood Kardon and Jeff Hutcher.*)

Showers Fig. 74

Shower Pan

☐ Min. area 900sq.in. (1,024 sq.in.**) min. diameter 30in.
measured from finished wall to center of threshold. ....[417.4] [412.7]

☐ Min. area excludes shower heads, valves, and soap dishes ....[417.4] [412.7]

❖ Square shower units with a min. 31½in. length on each
exterior base side of pan may be used if made to fit in
a 32in. rough opening. ....[417.4.1] [N/A]

➢ Min. shower area to be maintained to a point 70in. above drain ..[N/A] [412.7]

➢ Finished threshold height min. 2in. – max. 9in.
(from top of drain) ....[N/A] [412.6]

➢ Threshold able to accommodate 22in. door ....[N/A] [412.6]

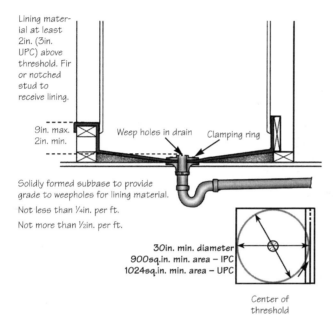

Lining material at least 2in. (3in. UPC) above threshold. Fir or notched stud to receive lining.

9in. max. 2in. min.

Weep holes in drain   Clamping ring

Solidly formed subbase to provide grade to weepholes for lining material.

Not less than ¼in. per ft.

Not more than ½in. per ft.

30in. min. diameter
900sq.in. min. area – IPC
1024sq.in. min. area – UPC

Center of threshold

Pan Liner Fig. 74

☐ Must conform to approved standards acceptable to the AA ...[417.1] [412.8]

☐ Shower liners shall turn up on sides a min. 2in. (3in.**) above
finished threshold height ....[417.5.2] [412.8]

☐ PVC&PE sheet lining min. 040in. thick and shall be installed
according to mfr. instructions (approved standards**) ..[417.5.3&.4] [412.8]

❖ Shower valves need not be pressure balanced or thermostatic
type in residential bldg. ....[424.4] [N/A]

➢ Slope min. ¼in. per ft. to weep holes in drain. ....[N/A] [412.8]

➢ No penetrations <1in. of finished threshold. ....[N/A] [412.8]

➢ Roll over top of rough threshold & fastened to outside edge. ..[N/A] [412.8]

➢ No penetrations allowed on top of rough threshold. ....[N/A] [412.8]

➢ Shower valves must be pressure balanced or thermostatic type ..[N/A] [420.0]

Figure 2.43.12 Showers—related code sections and minimum shower dimensions. (*By permission: McGraw-Hill, Taunton Press subsidiary,* Code Check Plumbing—A Field Guide to the Plumbing Codes, *Redwood Kardon and Jeff Hutcher.*)

NOTE: While not specifically stated that floor flanges be flush with finished floor, the wording in both codes intimates that this is clearly intended.

Kitchens
Dishwasher
❖ Water supply req. air gap or backflow device ....[409.2] [N/A]
❖ May discharge separately to a trap, trapped fixture, branch tailpiece on kitchen sink, or directly to a garbage disposal .....[409.3] [N/A]

NOTE: The piping from the air gap to disposal or branch tailpiece is considered drainage piping and shall be sloped to drain.

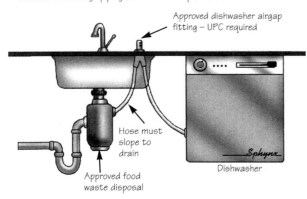

➤ Dishwasher to connect to drainage system through air gap fitting installed on top of sink or drainboard, whichever is higher. .....**Fig. 79** [N/A] [807.4]
➤ Not to connect to drain on discharge side of disposal. .**Fig. 79** [N/A] [405.4]
➤ Continuous wastes receiving the discharge from disposal or dishwasher shall use directional fittings (i.e. wyes, combos, or tees with baffles). ....**Fig. 80** [N/A] [405.4]

Bathrooms
☐ Floor flanges req'd for floor outlets. ....**Fig. 78** [405.4.1&.2] [408.4&.5]
☐ Floor flange to structure w/corrosion resistant fasteners ...[405.4.1&.2] [408.5]
☐ Toilets and bidets clearance (15in. center-side wall or obstruction). ....**Fig. 76** [405.3.1] [408.6]
☐ 18in. min. (24in.**) in front of any toilet or lav ....**Fig. 77** [405.3.1] [408.6]
➤ Reducing and offset floor flanges are prohibited. ....[N/A] [408.5]

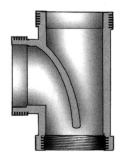

Center outlet T

End outlet T

Figure 2.43.13 Concealed toilet flanges and dishwasher connects for residential fixtures/appliances. (*By permission: McGraw-Hill, Taunton Press subsidiary,* Code Check Plumbing—A Field Guide to the Plumbing Codes, *Redwood Kardon and Jeff Hutcher.*)

| Pt-11 Minimum Capacity of Sewage Ejector or Pump [IPC T712.4.1] ||
Discharge Pipe Diameter	Capacity of Ejector
2in.	21GPM
3in.	46GPM

Fig. 22 Cleanout Requirements IPC

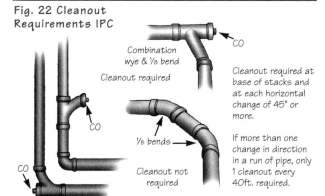

Combination wye & ⅛ bend

Cleanout required

CO

⅛ bends

Cleanout not required

CO

CO

Cleanout required at base of stacks and at each horizontal change of 45° or more.

If more than one change in direction in a run of pipe, only 1 cleanout every 40ft. required.

| Pt-13 Where Cleanouts Are Required ||
IPC 708	UPC 707
• Horiz. branches not more than 100ft. apart. • Horiz. change of direction exceeding 45 degrees (where more than one change of direction in a run of pipe, only 1 cleanout every 40ft. required). **Fig. 22** • At the base of each stack. **Fig. 22** • Cleanout required within 10ft. of the connection between building sewer and building drain. • Cleanouts in concealed piping or piping under floor with less than 24in. vert. clearance shall be extended above floor or to outside of building.	• Upper terminal of horiz. branches not to exceed 100ft. w/o cleanout. • Cleanouts not required on horiz. branch less than 5ft. (unless branch serves sink). Fig. 23 • NO cleanout required on branch 72 degrees or less from the vert. (⅕ bend). • NO cleanout required above first floor. • Cleanout required for horizontal change of direction exceeding 135 degrees. Fig. 23 • Unless end of line, cleanouts shall be installed above flow line. • Cleanouts in underfloor piping shall have 18in. vert. and 30in. horiz. clearance from the means of access. • Cleanouts in underfloor piping shall not be located more than 20ft. from the access to the underfloor area.

For cleanout clearances and sizes, see Pt-12.

Figure 2.43.14 Cleanout requirements and illustrations. (*By permission: McGraw-Hill, Taunton Press subsidiary,* Code Check Plumbing—A Field Guide to the Plumbing Codes, *Redwood Kardon and Jeff Hutcher.*)

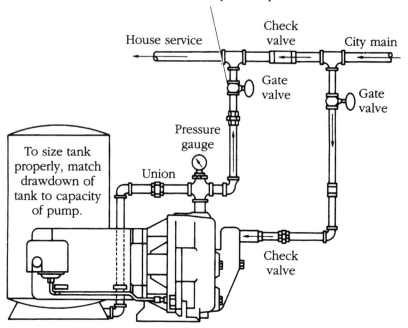

Use flow control or manual valve on discharge to throttle pump, must be sized, or set, to load motor below max. nameplate amps.

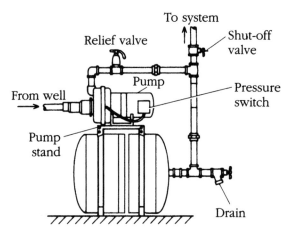

Figure 2.44.0 Domestic well-water jet pump installation schematics. (*By permission: McGraw-Hill, Inc.* Plumber's and Pipefitter's Calculations Manual, *R. Dodge Woodson.*)

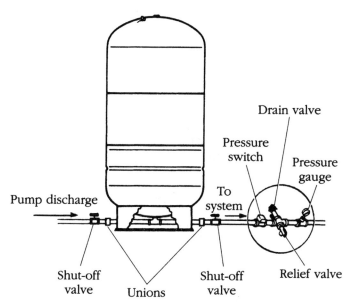

Figure 2.45.0   Standard pressure tank installed for well-water storage. (*By permission: McGraw-Hill, Inc.* Plumber's and Pipefitter's Calculations Manual, *R. Dodge Woodson.*)

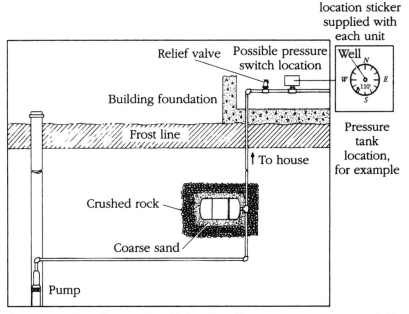

Figure 2.46.0   An underground installation of a well-water storage pressure vessel. (*By permission: McGraw-Hill Inc.* Plumber's and Pipefitter's Calculations Manual, *R. Dodge Woodson.*)

1. What well conditions might possibly limit the capacity of the pump?	The rate of flow from the source of supply, the diameter of a cased deep well, and the pumping level of the water in a cased deep well.
2. How does the diameter of a cased deep well and pumping level of the water affect the capacity?	They limit the size pumping equipment which can be used.
3. If there are no limiting factors, how is capacity determined?	By the maximum number of outlets or faucets likely to be in use at the same time.
4. What is suction?	A partial vacuum, created in the suction chamber of the pump, obtained by removing pressure due to atmosphere, thereby allowing greater pressure outside to force something (air, gas, water) into the container.
5. What is atmospheric pressure?	The atmosphere surrounding the earth presses against the earth and all objects on it, producing what we call atmospheric pressure.
6. How much is the pressure due to atmosphere?	This pressure varies with elevation or altitude. It is greatest at sea level (14.7 pounds per square inch) and gradually decreases as elevation above sea level is increased. The rate is approximately 1 foot per 100 feet of elevation.
7. What is maximum theoretical suction lift?	Since suction lift is actually that height to which atmospheric pressure will force water into a vacuum, theoretically we can use the maximum amount of this pressure 14.7 pounds per square inch at sea level which will raise water 33.9 feet. From this, we obtain the conversion factor of 1 pound per square inch of pressure equals 2.31-feet head.
8. How does friction loss affect suction conditions?	The resistance of the suction pipe walls to the flow of water uses up part of the work which can be done by atmospheric pressure. Therefore, the amount of loss due to friction in the suction pipe must be added to the vertical elevation which must be overcome, and the total of the two must not exceed 25 feet at sea level. This 25 feet must be reduced 1 foot for every 1000-feet elevation above sea level, which corrects for a lessened atmospheric pressure with increased elevation.

Figure 2.47.0 Domestic water pump questions and answers. (*By permission: McGraw-Hill Inc.* Plumber's and Pipefitter's Calculations Manual, *R. Dodge Woodson.*)

9. When and why do we use a deep-well jet pump?	The resistance of the suction pipe walls to below the pump because this is the maximum practical suction lift which can be obtained with a shallow-well pump at sea level.
10. What do we mean by water systems?	A pump with all necessary accessories, fittings, etc., necessary for its completely automatic operation.
11. What is the purpose of a foot value?	It is used on the end of a suction pipe to prevent the water in the system from running back into the source of supply when the pump isn't operating.
12. Name the two basic parts of a jet assembly.	Nozzle and diffuser.
13. What is the function of the nozzle?	The nozzle converts the pressure of the driving water into velocity. The velocity thus created causes a vacuum in the jet assembly or suction chamber.
14. What is the purpose of the diffuser?	The diffuser converts the velocity from the nozzle back into pressure.
15. What do we mean by "driving water"?	That water which is supplied under pressure to drive the jet.
16. What is the source of the driving water?	The driving water is continuously recirculated in a closed system.
17. What is the purpose of the centrifugal pump?	The centrifugal pump provides the energy to circulate the driving water. It also boosts the pressure of the discharged capacity.
18. Where is the jet assembly usually located in a shallow-well jet system?	Bolted to the casing of the centrifugal pump.
19. What is the principal factor which determines if a shallow-well jet system can be used?	Total suction lift.
20. When is a deep-well jet system used?	When the total suction sift exceeds that which can be overcome by atmospheric pressure.
21. Can a foot valve be omitted from a deep-well jet system? Why or why not?	No, because there are no valves in the jet assembly, and the foot valve is necessary to hold water in the system when it is primed. Also, when the centrifugal pump isn't running, the foot valve prevents the water from running back into the well.

Figure 2.47.0 *(Continued)*

22. What is the function of a check valve in the top of a submersible pump?	To hold the pressure in the line when the pump isn't running.
23. A submersible pump is made up of two basic parts. What are they?	Pump end and motor.
24. Why did the name submersible pump come into being?	Because the whole unit, pump and motor, is designed to be operated under water.
25. Can a submersible pump be installed in a 2-inch well?	No, they require a 4-inch well or larger for most domestic use. Larger pumps with larger capacities require 6-inch wells or larger.
26. A stage in a submersible pump is made up of three parts. What are they?	Impeller, diffuser, and bowl.
27. Does a submersible pump have only one pipe connection?	Yes, the discharge pipe.
28. What are two reasons we should always consider using a submersible first?	It will pump more water at higher pressure with less horsepower. It also has easier installation.
29. The amount of pressure a pump is capable of making is controlled by what?	The diameter of the impeller.
30. What do the width of an impeller and guide vane control?	The amount of water or capacity the pump is capable of pumping.

Figure 2.47.0 (*Continued*)

Single-phase control boxes

Checking and repairing procedures
(Power on)

Caution: Power must be on for these tests. Do not touch any live parts.

A. General procedures:
   1. Establish line power.
   2. Check no load voltage (pump not running).
   3. Check load voltage (pump running).
   4. Check current (amps) in all motor leads.

B. Use of volt/amp meter:
   1. Meter such as Amprobe Model RS300 or equivalent may be used.
   2. Select scale for voltage or amps depending on tests.
   3. When using amp scales, select highest scale to allow for inrush current, then select for midrange reading.

C. Voltage measurements:
   Step 1, no load.
   1. Measure voltage at L1 and L2 of pressure switch or line contractor.
   2. Voltage Reading: Should be ±10% of motor rating.
   Step 2, load.
   1. Measure voltage at load side of pressure switch or line contractor with pump running.
   2. Voltage Reading: Should remain the same except for slight dip on starting.

D. Current (amp) measurements:
   1. Measure current on all motor leads. Use 5 conductor test cord for Q.D. control boxes.
   2. Amp Reading: Current in Red lead should momentarily be high, then drop within one second. This verifies relay or solid state relay operation.

E. Voltage symptoms:
   1. Excessive voltage drop on starting.
   2. Causes: Loose connections, bad contacts or ground faults, or inadequate power supply.

F. Current symptoms:
   1. Relay or switch failures will cause Red lead current to remain high and overload tripping.
   2. Open run capacitor(s) will cause amps to be higher than normal in the Black and Yellow motor leads and lower than normal or zero amps in the Red motor lead.
   3. Relay chatter is caused by low voltage or ground faults.
   4. A bound pump will cause locked rotor amps and overloading tripping.
   5. Low amps may be caused by pump running at shutoff, worn pump or stripped splines.
   6. Failed start capacitor or open switch/relay are indicated if the red lead current is not momentarily high at starting.

Figure 2.48.0 Troubleshooting water pump motors. (*By permission: McGraw-Hill Inc. Plumber's and Pipefitter's Calculations Manual, R. Dodge Woodson.*)

Single-phase control boxes

Checking and repairing procedures
(Power off)

Caution: Turn power off at the power supply panel and discharge capacitors
before using ohmmeter.

A. General procedures:
   1. Disconnect line power.
   2. Inspect for damaged or burned parts, loose connections, etc.
   3. Check against diagram in control box for misconnections.
   4. Check motor insulation and winding resistance.

B. Use of ohmmeter:
   1. Ohmmeter such as Simpson Model 372 or 260. Triplet Model 630 or
      666 may be used.
   2. Whenever scales are changed, clip ohmmeter lead together and "zero
      balance" meter.

C. Ground (insulation resistance) test:
   1. Ohmmeter Setting: Highest scale R × 10K, or R × 100K
   2. Terminal Connections: One ohmmeter lead to "Ground" terminal or
      Q.D. control box lid and touch other lead to the other terminals on the
      terminal board.
   3. Ohmmeter Reading: Pointer should remain at infinity (∞).

Additional tests

Solid state capacitor run
(CRC) control box

A. Run capacitor
   1. Meter setting: R × 1,000
   2. Connections: Red and Black leads
   3. Correct meter reading: Pointer should swing toward zero, then drift
      back to infinity.

B. Inductance coil
   1. Meter setting: R × 1
   2. Connections: Orange leads
   3. Correct meter reading: Less than 1 ohm.

C. Solid state switch
   Step 1 triac test
   1. Meter setting: R × 1,000
   2. Connections: R(Start) terminal and Orange lead on start switch.
   3. Correct meter reading: Should be near infinity after swing.
   Step 2 coil test
   1. Meter setting: R × 1
   2. Connections: Y(Common) and L2.
   3. Correct meter reading: Zero ohms

Figure 2.48.1 Troubleshooting water pump motor control boxes. (*By permission: McGraw-Hill Inc.* Plumber's and Pipefitter's Calculations Manual, *R. Dodge Woodson.*)

Normal ohm and megohm values between all leads and ground		
Insulation resistance varies very little with rating. Motors of all hp, voltage, and phase rating have similar values of insulation resistance.		
Condition of motor and leads	Ohm value	Megohm value
A new motor (without drop cable).	20,000,000 (or more)	20.0 (or more)
A used motor which can be reinstalled in the well.	10,000,000 (or more)	10.0 (or more)
Motor in well. Ohm readings are for drop cable plus motor.		
A new motor in the well.	2,000,000 (or more)	2.0 (or more)
A motor in the well in reasonably good condition.	500,000–2,000,000	0.5–2.0
A motor which may have been damaged by lightning or with damaged leads. Do not pull the pump for this reason.	20,000–500,000	0.02–0.5
A motor which definitely has been damaged or with damaged cable. The pump should be pulled and repairs made to the cable or motor replaced. The motor will not fail for this reason alone, but it will probably not operate for long.	10,000–20,000	0.01–0.02
A motor which has failed or with completely destroyed cable insulation. The pump must be pulled and the cable repaired or the motor replaced.	less than 10,000	0–0.01

Figure 2.48.2 Ohm and megohm values between leads and grounds on pump motors. (*By permission: McGraw-Hill Inc.* Plumber's and Pipefitter's Calculations Manual, *R. Dodge Woodson.*)

Preliminary tests—all sizes—single and three phase	
What is to be done	What it means
Measure resistance from any cable to ground (insulation resistance)	1. If the ohm value is normal, the motor windings are not grounded and the cable insulation is not damaged.
	2. If the ohm value is below normal, either the windings are grounded or the cable insulation is damaged. Check the cable at the well seal as the insulation is sometimes damaged by being pinched.
Measure winding resistance (resistance between leads)	1. If all ohm values are normal, the motor windings are neither shorted nor open, and the cable colors are correct.
	2. If any one ohm value is less than normal, the motor is shorted.
	3. If any one ohm value is greater than normal, the winding or the cable is open, or there is a poor cable joint or connection.
	4. If some ohm values are greater than normal and some less on single phase motors, the leads are mixed.

Figure 2.48.3 Preliminary tests on single-phase and three-phase pump motors. (*By permission: McGraw-Hill Inc.* Plumber's and Pipefitter's Calculations Manual, *R. Dodge Woodson.*)

Motor runs continuously		
Cause of trouble	Checking procedure	Correction action
A. Pressure switch.	Switch contacts may be "welded" in closed position. Pressure switch may be set too high.	Clean contacts replace switch, or readjust setting.
B. Low level well.	Pump may exceed well capacity. Shut off pump, wait for well to recover. Check static and drawdown level from well head.	Throttle pump output or reset pump to lower level. Do not lower if sand may clog pump.
C. Leak in system.	Check system for leaks.	Replace damaged pipes or repair leaks.
D. Worn pump, motor shaft.	Symptoms of worn pump are similar to those of drop pipe leak or low water level in well. Reduce pressure switch setting. If pump shuts off, worn parts may be at fault. Sand is usually present in tank.	Pull pump and replace worn impellers, casing or other close fitting parts.
E. Loose or broken.	No or little water will be delivered if coupling between motor and pump shaft is loose or if a jammed pump has caused the motor shaft to shear off.	Check for damaged shafts if coupling is loose and replace worn or defective units.
F. Pump screen blocked.	Restricted flow may indicate a clogged intake screen on pump. Pump may be installed in mud or sand.	Clean screen and reset at less depth. It may be necessary to clean well.
G. Check valve stuck closed.	No water will be delivered if check valve is in closed position.	Replace if defective.
H. Control box malfunction.		Repair or replace.
Motor runs but overload protector trips		
A. Incorrect voltage.	Using voltmeter, check the line terminals. Voltage must be within $\pm$ 10% of rated voltage.	Contact power company if voltage is incorrect.
B. Overheated protectors.	Direct sunlight or other heat source can make control box hot causing protectors to trip. The box must not be hot to touch.	Shade box, provide ventilation or move box away from heat source.
C. Defective control box.		Repair or replace.
D. Defective motor or cable.		Repair or replace.
E. Worn pump or motor.		Replace pump and/or motor.

Figure 2.48.4   Pump motors that run continuously or trip on overload. (*By permission: McGraw-Hill Inc.* Plumber's and Pipefitter's Calculations Manual, *R. Dodge Woodson.*)

Motor runs continuously		
Causes of trouble	Checking procedure	Corrective action
Pressure switch.	Switch contacts may be "welded" in closed position. Pressure switch may be set too high.	Clean contacts, replace switch, or readjust setting.
Low-level well.	Pump may exceed well capacity. Shut off pump, wait for well to recover. Check static and draw-down level from well head.	Throttle pump output or reset pump to lower level. Do not lower if sand may clog pump.
Leak in system.	Check system for leaks.	Replace damaged pipes or repair leaks.
Worn pump.	Symptoms of worn pump are similar to that of drop pipe leak or low water level in well. Reduce pressure switch setting. If pump shuts off, worn parts may be at fault. Sand is usually present in tank.	Pull pump and replace worn impellers, casing, or other close fitting parts.
Loose or broken motor shaft.	No or little water will be delivered if coupling between motor and pump shaft is loose or if a jammed pump has caused the motor shaft to shear off.	Check for damaged shafts if coupling is loose, and replace worn or defective units.
Pump screen blocked.	Restricted flow may indicate a clogged intake screen on pump. Pump may be installed in mud or sand.	Clean screen and reset at less depth. It may be necessary to clean well.
Check valve stuck closed.	No water will be delivered if check valve is in closed position.	Replace if defective.
Control box malfunction.		Repair or replace.
Motor runs but overload protector tips		
Incorrect voltage.	Using voltmeter, check the line terminals. Voltage must be within ± 10% of rated voltage.	Contact power company if voltage is incorrect.
Overheated protectors.	Direct sunlight or other heat source can make control box hot, causing protectors to trip. The box must not be hot to touch.	Shade box, provide ventilation, or move box away from heat source.
Defective control box.		Repair or replace.
Defective motor or cable.		Repair or replace.
Worn pump or motor.		Replace pump and/or motor.

SOURCE: A. Y. McDonald Manufacturing Co.

Figure 2.48.4 *(Continued)*

Motor does not start		
Cause of trouble	Checking procedure	Correction action
A. No power or incorrect voltage.	Using voltmeter check the line terminals Voltage must be ±10% of rated voltage.	Contact power company if voltage is incorrect.
B. Fuses blown or circuit breakers tripped.	Check fuses for recommended size and check for loose, dirty or corroded connections in fuse receptacle. Check for tripped circuit breaker.	Replace with proper fuse or reset circuit breaker.
C. Defective pressure switch.	Check voltage at contact points. Improper contact of switch points can cause voltage less than line voltage.	Replace pressure switch or clean points.
D. Control box malfunction.		Repair or replace.
E. Defective wiring.	Check for loose or corroded connections. Check motor lead terminals with voltmeter for power.	Correct faulty wiring or connections.
F. Bound pump.	Locked rotor conditions can result from misalignment between pump and motor or a sand bound pump. Amp readings 3 to 6 times higher than normal will be indicated.	If pump will not start with several trials it must be pulled and the cause corrected. New installations should always be run without turning off until water clears.
G. Defective cable or motor.		Repair or replace.
Motor starts too often		
A. Pressure switch.	Check setting on pressure switch and examine for defects.	Reset limit or replace switch.
B. Check valve, stuck open.	Damaged or defective check valve will not hold pressure.	Replace if defective.
C. Waterlogged tank, (air supply)	Check air charging system for proper operation.	Clean or replace.
D. Leak in system.	Check system for leaks.	Replace damaged pipes or repair leaks.

Figure 2.48.5 Troubleshooting a motor that does not start. (*By permission: McGraw-Hill Inc.* Plumber's and Pipefitter's Calculations Manual, *R. Dodge Woodson.*)

Motor does not start		
Cause of trouble	Checking procedure	Corrective action
No power or incorrect voltage.	Using voltmeter, check the line terminals. Voltage must be ± 10% of rated voltage.	Contact power company if voltage is incorrect.
Fuses blown or circuit fuse breakers tripped.	Check fuses for recommended size and check for loose, dirty, or corroded connections in fuse receptacle. Check for tripped circuit breaker.	Replace with proper or reset circuit breaker.
Defective pressure switch.	Check voltage at contact points. Improper contact of switch points can cause voltage less than line voltage.	Replace pressure switch or clean points.
Control box malfunction.	For detailed procedure, see***	Repair or replace.
Defective wiring.	Check for loose or corroded connections. Check motor lead terminals with voltmeter for power.	Correct faulty wiring or connections.
Bound pump.	Locked rotor conditions can result from misalignment between pump and motor or a sand bound pump. Amp readings 3 to 6 times higher than normal will be indicated.	If pump will not start with several trials, it must be pulled and the cause corrected. New installations should always be run without turning off until water clears.
Defective cable or motor.		Repair or replace.
Motor starts too often		
Pressure switch.	Check setting on pressure switch and examine for defects.	Reset limit or replace switch.
Check valve, stuck open.	Damaged or defective check valve will not hold pressure.	Replace if defective.
Waterlogged tank (air supply).	Check air-charging system for proper operation.	Clean or replace.
Leak in system.	Check system for leaks.	Replace damaged pipes or repair leaks.

SOURCE: A. Y. McDonald Manufacturing Co.

Figure 2.48.5 (*Continued*)

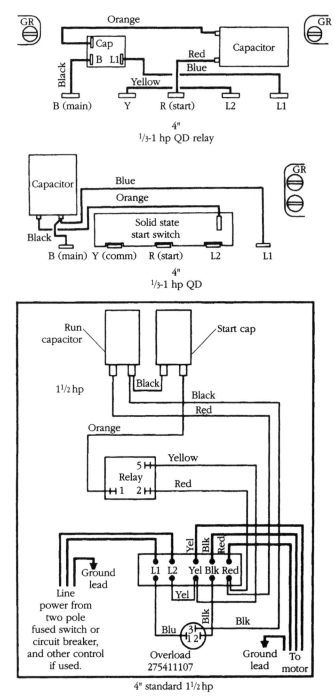

Figure 2.48.6 Wiring diagrams for pump motors. (*By permission: Mc-Graw-Hill Inc.* Plumber's and Pipefitter's Calculations Manual, *R. Dodge Woodson.*)

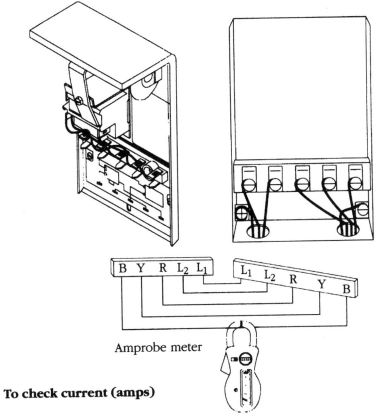

To check current (amps)

1. Turn power OFF

2. Connect test cord as shown.

3. Turn power ON.

4. Use hook-on type ammeter as shown.

Figure 2.48.7 Tips on checking amperage on motors. (*By permission: McGraw-Hill Inc.* Plumber's and Pipefitter's Calculations Manual, *R. Dodge Woodson.*)

Integral horsepower control box parts					
Motor rating hp dia.	Control box (1) model no.	Part no. (2)	Capacitors MFD	Volts	Qty.
5–6″	282 2009 202	275 468 117 S	130–154	330	2
		275 479 103 (5)	15	370	2
	282 2009 203	275 468 117 S	130–154	330	2
		155 327 101 R	30	370	1
5–6″ DLX	282 2009 303	275 468 117 S	130–154	330	2
		155 327 101 R	30	370	1
7½–6″	282 2019 210	275 468 119 S	270–324	330	1
		275 468 117 S	130–154	330	1
		155 327 109 R	45	370	1
	282 2019 202	275 468 117S	130–154	330	3
		275 479 103 R (5)	15	370	3
	282 2019 203	275 468 117 S	130–154	330	3
		155 327 101 R	30	370	1
		155 328 101 R	15	370	1
7½–6″ DLX	282 2019 310	275 468 119 S	270–324	330	1
		275 468 117 S	130–154	330	1
		155 327 109 R	45	370	1
	282 2019 303	275 468 117 S	130–154	330	3
		155 327 101 R	30	370	1
		155 328 101 R	15	370	1
10–6″	282 2029 210	275 468 119 S	270–324	330	2
		155 327 102 R	35	370	2
	282 2029 202	275 468 117 S	130–154	330	4
		275 479 103 R (5)	15	370	5
	282 2029 203	275 468 117 S	130–154	330	4
		155 327 101 R	30	370	2
		155 328 101 R	15	370	1
	282 2029 207	275 468 119 S	270–324	330	2
		155 327 101 R	30	370	2
		155 328 101 R	15	370	1

Figure 2.48.8 Single-phase motor control box parts. (*By permission: Mc-Graw-Hill Inc.* Plumber's and Pipefitter's Calculations Manual, *R. Dodge Woodson.*)

Meter connections for motor testing

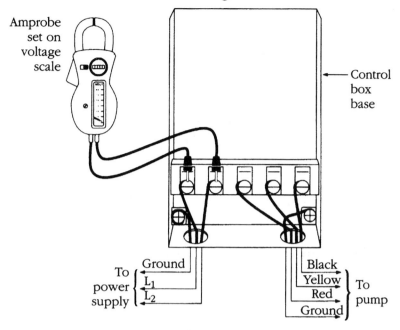

To check voltage

1. Turn power OFF

2. Remove QD cover to break all motor connections.

Caution: L1 and L2 are still connected to the power supply.

3. Turn power ON.

4. Use voltmeter as shown.

Caution: Both voltage and current tests require live circuits with power ON.

Figure 2.48.9 Checking voltage on motors used on residential pumps. (*By permission: McGraw-Hill Inc.* Plumber's and Pipefitter's Calculations Manual, *R. Dodge Woodson.*)

MAXIMUM DISTANCE FROM A FIXTURE TRAP TO ITS VENT

TRAP SIZE (INCHES)	DRAIN SIZE (INCHES)	PITCH (INCH PER FOOT)	DISTANCE ALLOWED (FEET)
1 1/4	1 1/4	1 1/4	3 1/2
1 1/4	1 1/2	1/4	5
1 1/2	1 1/2	1/4	5
1 1/2	2	1/4	8
2	2	1/4	6
3	3	1/8	10
4	4	1/8	12

Figure 2.49.0 Distance required from trap to vent. (*By permission: McGraw-Hill Inc.* 1998 Plumber's Business Planner *by R. Dodge Woodson.*)

MINIMUM PITCH REQUIRED ON HORIZONTAL DRAINS

PIPE SIZE IN INCHES	MINIMUM PITCH (INCH PER FOOT)
1 1/2	1/4
2	1/4
3	1/8
4	1/8

Figure 2.50.0 Minimum pitch preferred from trap to vent. (*By permission: McGraw-Hill Inc.* 1998 Plumber's Business Planner *by R. Dodge Woodson.*)

DRAINAGE FIXTURE UNITS

TYPE OF FIXTURE	FIXTURE UNIT VALUE
Washing machine	3
Laundry tub	2
Bathtub	2
Shower	2
Lavatory	1
Bidet	1
Toilet	4
Kitchen sink	2
Bar sink	1
Floor drain	2

Figure 2.51.0 Drainage fixture units. (*By permission: McGraw-Hill Inc.* 1998 Plumber's Business Planner *by R. Dodge Woodson.*)

RECOMMENDED TRAP SIZES

TYPE OF FIXURE	TRAP SIZE IN INCHES
Washing machine	2
Laundry tub	1 1/2
Bathtub	1 1/2
Shower	2
Lavatory	1 1/4
Bidet	1 1/4
Kitchen sink	1 1/2
Bar sink	1 1/2
Floor drain	2

Figure 2.52.0 Guide to trap sizes. (*By permission: McGraw-Hill Inc.* 1998 Plumber's Business Planner, *R. Dodge Woodson.*)

SCREW LENGTHS AND AVAILABLE GAUGE NUMBERS

LENGTH	GAUGE NUMBERS
1/4"	0 to 3
3/8"	2 to 7
1/2"	2 to 8
5/8"	3 to 10
3/4"	4 to 11
7/8"	6 to 12
1"	6 to 14
1 1/4"	7 to 16
1 1/2"	6 to 18
1 3/4"	8 to 20
2"	8 to 20
2 1/4"	9 to 20
2 1/2"	12 to 20
2 3/4"	14 to 20
3"	16 to 20
3 1/2"	18 to 20
4"	18 to 20

Figure 2.53.0 Length and gauge size of fasteners used in residential work. (*By permission: McGraw-Hill Inc.* 1998 Plumber's Business Planner, *R. Dodge Woodson.*)

Diameter of pipe[5] (in)	Maximum no. of fixture units that may be connected to:			
	Any horizontal fixture branch[1,4]	One stack of 3 stories or 3 intervals maximum	More than 3 stories in height	
			Total for stack	Total at one story or branch interval
1¼	1	2	2	1
1½	3	4	8	2
2	6	10	24	6
2½	12	20	42	9
3	20[2]	30[2]	60[3]	16[2]
4	160	240	500	90
5	360	540	1100	200
6	620	960	1900	350
8	1400	2200	3600	600
10	2500	3800	5600	1000
12	3900	6000	8400	1500
15	7000			

[1]Does not include branches of the building drain.
[2]Not over two water closets.
[3]Not over six water closets.
[4]50% less for battery vented fixture branches, no size reduction permitted for battery vented branches throughout the entire branch length.
[5]The minimum size of any branch or stack serving a water closet shall be 3″.

Figure 2.54.0   Maximum number of fixtures to be connected to stacks, one or more stories. (*By permission: McGraw-Hill Inc.* Plumber's and Pipefitter's Calculations Manual, *R. Dodge Woodson.*)

Number of wet-vented fixtures	Diameter of vent stacks (in)
1 or 2 bathtubs or showers	2
3 to 5 bathtubs or showers	2½
6 to 9 bathtubs or showers	3
10 to 16 bathtubs or showers	4

Figure 2.55.0   Vent sizes based on number of fixtures. (*By permission: McGraw-Hill Inc.* Plumber's and Pipefitter's Calculations Manual, *R. Dodge Woodson.*)

Size of fixture drain (in)	Size of trap (in)	Fall (in/ft)	Max. distance from trap
1¼	1¼	¼	3 ft 6 in
1½	1¼	¼	5 ft
1½	1½	¼	5 ft
2	1½	¼	8 ft
2	2	¼	6 ft
3	3	⅛	10 ft
4	4	⅛	12 ft

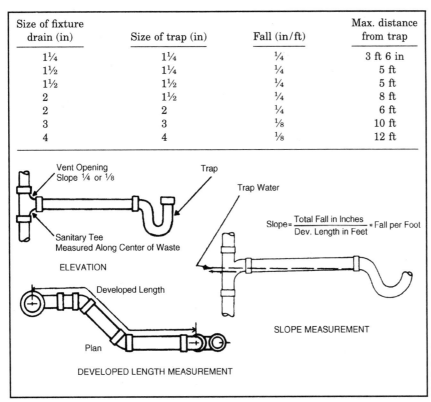

Figure 2.56.0 Vent to trap distances—tables and illustrations. (*By permission: McGraw-Hill Inc. Plumber's and Pipefitter's Calculations Manual, R. Dodge Woodson.*)

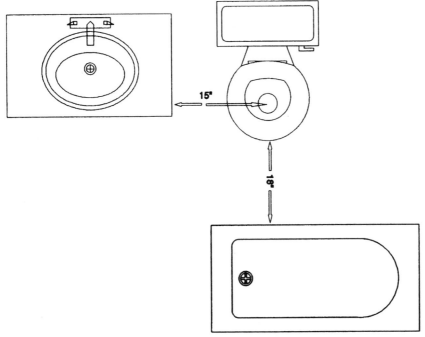

Figure 2.57.0 Distance required from toilet to tub and tub to lavatory. (*By permission: McGraw-Hill Inc. Plumber's and Pipefitter's Calculations Manual, R. Dodge Woodson.*)

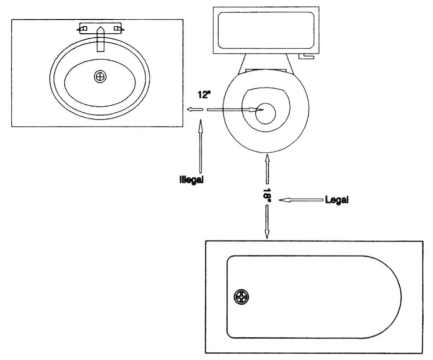

Figure 2.57.1 Illustration of noncomplying fixture spacing. (*By permission: McGraw-Hill Inc.* Plumber's and Pipefitter's Calculations Manual, *R. Dodge Woodson.*)

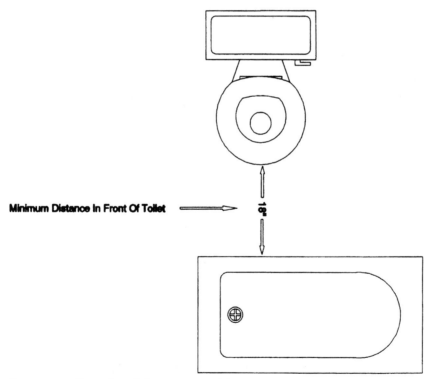

Figure 2.57.2 Illustration of distance required between toilet and tub. (*By permission: McGraw-Hill Inc.* Plumber's and Pipefitter's Calculations Manual, *R. Dodge Woodson.*)

RADIANT HEAT FACTS

RADIATION
Three feet of 1-in. pipe equal 1 square foot of radiation.
Two and one-third lineal feet of 1 1/4 in. pipe equal 1 square foot of radiation.
Hot water radiation gives off 150 B.T.U. per square foot of radiation per hour.
Steam radiation gives off 240 B.T.U. per square foot of radiation per hour.
On greenhouse heating, figure 2/3 square foot of radiation per square foot of glass.
One square foot of direct radiation condenses .25 pound of water per hour.

Figure 2.58.0 Radiant heat facts. (*By permission: McGraw-Hill Inc.* 1998 Plumber's Business Planner, *R. Dodge Woodson.*)

ATMOSPHERIC PRESSURE PER SQUARE INCH

BAROMETER (INS. OF MERCURY)	PRESSURE (LBS. PER SQ. IN.)
28.00	13.75
28.25	13.88
28.50	14.00
28.75	14.12
29.00	14.24
29.25	14.37
29.50	14.49
29.75	14.61
29.921	14.696
30.00	14.74
30.25	14.86
30.50	14.98
30.75	15.10
31.00	15.23

RULE: Barometer in inches of mercury $\times$ .49116 = lbs. per sq. in.

Figure 2.59.0 Inches of mercury and corresponding pressure in psi. (*By permission: McGraw-Hill Inc.* 1998 Plumber's Business Planner, *R. Dodge Woodson.*)

FACTS ABOUT WATER

A cubic foot of water contains 7 1/2 gallons, 1728 cubic inches, and weighs 62 1/2 pounds.
A gallon of water weighs 8 1/3 pounds and contains 231 cubic inches.
Water expands 1/23 of its volume when heated from 40° to 212°
The height of a column of water, equal to a pressure of 1 pound per square inch, is 2.31 feet.
To find the pressure in pounds per square inch of a column of water, multiply the height of the column in feet by .434.
The average pressure of the atmosphere is estimated at 14.7 pounds per square inch so that with a perfect vacuum it will sustain a column of water 34 feet high.
The friction of water in pipes varies as the square of the velocity.
To evaporate 1 cubic foot of water requires the consumption of 7 1/2 pounds of ordinary coal or about 1 pound of coal to 1 gallon of water.
A cubic inch of water evaporated at atmospheric pressure is converted into approximately 1 cubic foot of steam.

Figure 2.60.0 Facts about water—weight, volume, and evaporation. (*By permission: McGraw-Hill Inc.* 1998 Plumber's Business Planner, *R. Dodge Woodson.*)

MINIMUM SIZING FOR SEWER PUMPS

DISCHARGE-PIPE SIZE IN INCHES	GALLONS-PER-MINUTE OF PUMP
2	21
2 1/2	30
3	46

Figure 2.61.0 Sizing of pipes for sewer pumps. (*By permission: McGraw-Hill Inc.* 1998 Plumber's Business Planner, *R. Dodge Woodson.*)

SIZING FOR MINIMAL WATER SUPPLY REQUIREMENTS

FIXTURE	MINIMUM PIPE SIZE IN INCHES
Bathtub	1/2
Bidet	3/8
Dishwasher	1/2
Hose bibb	1/2
Kitchen sink	1/2
Laundry tub	1/2
Lavatory	3/8
Shower	1/2
Toilet (two-piece)	3/8
Toilet (one-piece)	1/2

Figure 2.62.0 Sizing of pipe based on type of fixture to be connected. (*By permission: McGraw-Hill Inc.* 1998 Plumber's Business Planner, *R. Dodge Woodson.*)

WATER PRESSURE IN POUNDS WITH EQUIVALENT HEADS

POUNDS PER SQUARE INCH	FEET HEAD
1	2.31
2	4.62
3	6.93
4	9.24
5	11.54
6	13.85
7	16.16
8	18.47
9	20.78
10	23.09
15	34.63
20	46.18
25	57.72
30	69.27
40	92.36
50	115.45
60	138.54
70	161.63
80	184.72
90	207.81
100	230.90

Figure 2.63.0 Converting feet of head to pounds per square inch for water. (*By permission: McGraw-Hill Inc.* 1998 Plumber's Business Planner, *R. Dodge Woodson.*)

BOILING POINT OF WATER AT VARIOUS PRESSURES

VACUUM IN INCHES OF MERCURY	BOILING POINT
29	76.62
28	99.93
27	114.22
26	124.77
25	133.22
24	140.31
23	146.45
22	151.87
21	156.75
20	161.19
19	165.24
18	169.00
17	172.51
16	175.80
15	178.91
14	181.82
13	184.61
12	187.21
11	189.75
10	192.19
9	194.50
8	196.73
7	198.87
6	200.96
5	202.25
4	204.85
3	206.70
2	208.50
1	210.25

Figure 2.64.0   Boiling point of water in relationship to inches of vacuum. (*By permission: McGraw-Hill Inc.* 1998 Plumber's Business Planner, *R. Dodge Woodson.*)

WATER-DEMAND AT INDIVIDUAL OUTLETS

TYPE OF OUTLET	DEMAND (GPM)*
Ordinary lavatory faucet	2.0
Self-closing lavatory faucet	2.5
Sink faucet, 3/8" or 1/2"	4.5
Sink faucet, 3/4"	6.0
Bath faucet, 1/2"	5.0
Shower head, 1/2"	5.0
Laundry faucet, 1/2"	5.0
Ballcock in water closet flush tank	3.0
1" flush valve (25 psi-flow pressure)	35.0
1" flush valve (15 psi flow pressure)	27.0
3/4" flush valve (15 psi flow pressure)	15.0
Drinking fountain jet	0.75
Dishwashing machine (domestic)	4.0
Laundry machine (8 or 16 lb.)	4.0
Aspirator (operating room or laboratory)	2.5
Hose bib or sill cock (1/2")	5.0

* Demands do not take into account the use of water-conservation devices.

Figure 2.65.0 Typical fixture water demands. (*By permission: McGraw-Hill Inc.* 1998 Plumber's Business Planner, *R. Dodge Woodson.*)

AVERAGE WATER USAGE OF FIXTURES IN GALLONS PER MINUTE

FIXTURE	FLOW RATE (GPM)
·Bathtub	4
Washing machine	4
Dishwasher	3
Kitchen sink	2 1/2
Laundry tub	4
Lavatory	2
Shower	3
Toilet	3

Figure 2.66.0   Average water usage for residential appliances and fixtures. (*By permission: McGraw-Hill Inc.* 1998 Plumber's Business Planner, *R. Dodge Woodson.*)

BUILDING DRAINS AND SEWERS

NUMBER OF FIXTURES ALLOWED ON A SEWER OR BUILDING DRAIN (The fixture unit amounts are based on a pipe with a quarter of an inch to the foot in pitch.)	
PIPE SIZE IN INCHES	MAXIMUM FIXTURE UNITS ALLOWED
2	21
3	42 (not more than two toilets)
4	216

Figure 2.67.0   Number of fixtures allowed per 2, 3, and 4 inch diameter pipe waste line. (*By permission: McGraw-Hill Inc.* 1998 Plumber's Business Planner, *R. Dodge Woodson.*)

WATER PIPE FIXTURE UNIT VALUES

FIXTURE	HOT	COLD	TOTAL UNIT VALUES
Bathtub	1 1/2	1 1/2	2 (when combined)
Bidet	1 1/2	1 1/2	2 (when combined)
Kitchen sink	1 1/2	1 1/2	2 (when combined)
Laundry tub	2	2	3 (when combined)
Lavatory	1 1/2	1 1/2	2 (when combined)
Shower head	3	3	4 (when combined)
Toilet	0	5	5

Figure 2.68.0   Water pipe fixture unit values. (*By permission: McGraw-Hill Inc.* 1998 Plumber's Business Planner, *R. Dodge Woodson.*)

WATER FLOW RATE

FIXTURE	FLOW RATE (GPM)*
Ordinary basin faucet	2.0
Self-closing basin faucet	2.5
Sink faucet, 3/8"	4.5
Sink faucet, 1/2"	4.5
Bathtub faucet	6.0
Laundry tub cock, 1/2"	5.0
Shower	5.0
Ballcock for water closet	3.0
Flushometer valve for water closet	15-35
Flushometer valve for urinal	15.0
Drinking fountain	.75
Sillcock or wall hydrant	5.0

* Figures do not represent the use of water-conservation devices.

Figure 2.69.0 Flow rates for various fixtures. (*By permission: McGraw-Hill Inc.* 1998 Plumber's Business Planner, *R. Dodge Woodson.*)

COMMON SEPTIC TANK CAPACITIES

SINGLE FAMILY DWELLINGS NUMBER OF BEDROOMS	MULTIPLE DWELLING UNITS OR APARTMENTS ONE BEDROOM EACH	OTHER USES; MAXIMUM FIXTURE UNITS SERVED	MINIMUM SEPTIC TANK CAPACITY IN GALLONS
1-3		20	1000
4	2	25	1200
5 or 6	3	33	1500
7 or 8	4	45	2000
	5	55	2250
	6	60	2500
	7	70	2750
	8	80	3000
	9	90	3250
	10	100	3500

Figure 2.70.0 Capacities of septic tank requirements based on number of bedrooms. (*By permission: McGraw-Hill Inc.* 1998 Plumber's Business Planner, *R. Dodge Woodson.*)

POTENTIAL SEWAGE FLOWS ACCORDING TO TYPE OF ESTABLISHMENT

TYPE OF ESTABLISHMENT	GALLONS PER DAY PER PERSON
Schools (toilet and lavatories only)	15 per day per person
Schools (with above plus cafeteria)	25 per day per person
Schools (with above plus cafeteria and showers)	35 per day per person
Day workers at schools and offices	15 per day per person
Day camps	25 per day per person
Trailer parks or tourist camps (with built-in bath)	50 per day per person
Trailer parks or tourist camps (with central bathhouse)	35 per day per person
Work or construction camps	50 per day per person
Public picnic parks (toilet wastes only)	5 per day per person
Public picnic parks (bathhouse, showers and flush toilets)	10 per day per person
Swimming pools and beaches	10 per day per person
Country clubs	25 per locker
Luxury residences and estates	150 per day per person
Rooming houses	40 per day per person
Boarding houses	50 per day per person
Hotels (with connecting baths)	50 per day per person
Hotels (with private baths 2 persons per room)	100 per day per person
Boarding schools	100 per day per person
Factories (gallons per person per shift—exclusive of industrial wastes)	25 per day per person
Nursing homes	75 per day per person
General hospitals	150 per day per person
Public institutions (other than hospitals)	100 per day per person
Restaurants (toilet and kitchen wastes per unit of serving capacity)	25 per day per person
Kitchen wastes from hotels, camps, boarding houses, etc. Serving 3 meals per day	10 per day per person
Motels	50 per bed space

Figure 2.71.0   Guidelines for sewage flow per type of establishment. (*By permission: McGraw-Hill Inc.* 1998 Plumber's Business Planner, *R. Dodge Woodson.*)

WHERE TO LOOK FOR CAUSES OF
WATER-PRESSURE PROBLEMS

	NO WATER PRESSURE	NO WATER PRESSURE AT FIXTURE	LOW WATER PRESSURE TO FIXTURE
Street water main	X		X
Curb stop	X		X
Water service	X		X
Branches		X	X
Valves	X	X	X
Stems, washers (hot & cold)		X	X
Aerator		X	X
Water meter	X	X	X

Figure 2.72.0   Troubleshooting water pressure problems. (*By permission: McGraw-Hill Inc.* 1998 Plumber's Business Planner, *R. Dodge Woodson.*)

TROUBLESHOOTING TOILETS

SYMPTOMS	PROBABLE CAUSES
Will not flush	• No water in tank • Stoppage in drainage system
Flushes poorly	• Clogged flush holes • Flapper or tank ball is not staying open long enough • Not enough water in tank • Partial drain blockage • Defective handle • Bad connection between handle and flush valve • Vent is clogged
Water droplets covering tank	• Condensation
Tank fills slowly	• Defective ballcock • Obstructed supply pipe • Low water pressure • Partially closed valve • Partially frozen pipe
Makes unusual noises when flushed	• Defective ballcock
Water runs constantly	• Bad flapper or tank ball • Bad ballcock • Float rod needs adjusting • Float is filled with water • Ballcock needs adjusting • Pitted flush valve • Undiscovered leak • Cracked overflow tube
Water seeps from base of toilet	• Bad wax ring • Cracked toilet bowl
Water dripping from tank	• Condensation • Bad tank-to-bowl gasket • Bad tank-to-bowl bolts • Cracked tank • Flush-valve nut is loose
No water comes into the tank	• Closed valve • Defective ballcock • Frozen pipe • Broken pipe

Figure 2.72.1   Troubleshooting toilet operating problems. (*By permission: McGraw-Hill Inc.* 1998 Plumber's Business Planner, *R. Dodge Woodson.*)

TROUBLESHOOTING LAVATORIES

SYMPTOMS	PROBABLE CAUSES
Faucet drips from spout	• Bad washers or cartridge • Bad faucet seats
Faucet leaks at base of spout	• Bad "O" ring
Faucet will not shut off	• Bad washers or cartridge • Bad faucet seats
Poor water pressure	• Partially closed valve • Clogged aerator • Not enough water pressure • Blockage in the faucet • Partially frozen pipe
No water	• Closed valve • Broken pipe • Frozen pipe
Drains slowly	• Hair on pop-up assembly • Partial obstruction in drain or trap • Pop-up needs to be adjusted
Will not drain	• Blocked drain or trap • Pop-up is defective
Gurgles as it drains	• Partial drainage blockage • Partial blockage in the vent
Won't hold water	• Pop-up needs adjusting • Bad putty seal on drain

Figure 2.72.2 Troubleshooting problems with lavatories. (*By permission: McGraw-Hill Inc.* 1998 Plumber's Business Planner, *R. Dodge Woodson.*)

TROUBLESHOOTING KITCHEN SINKS

SYMPTOMS	PROBABLE CAUSES
Faucet drips from spout	• Bad washers or cartridge • Bad faucet seats
Faucet leaks at base of spout	• Bad "O" ring
Faucet will not shut off	• Bad washers or cartridge • Bad faucet seats
Poor water pressure	• Partially closed valve • Clogged aerator • Not enough water pressure • Blockage in the faucet • Partially frozen pipe
No water	• Closed valve • Broken pipe • Frozen pipe
Drains slowly	• Partial obstruction in drain or trap
Will not drain	• Blocked drain or trap
Gurgles as it drains	• Partial drainage blockage • Partial blockage in the vent
Won't hold water	• Bad basket strainer • Bad putty seal on drain
Spray attachment will not spray	• Clogged holes in spray head • Kinked spray hose
Spray attachment will not cut off	• Bad spray head

Figure 2.72.3 Troubleshooting problems with kitchen sinks. (*By permission: McGraw-Hill Inc.* 1998 Plumber's Business Planner, *R. Dodge Woodson.*)

TROUBLESHOOTING BATHTUBS

SYMPTOMS	PROBABLE CAUSES
Won't drain	• Clogged drain • Clogged tub waste • Clogged trap
Drains slowly	• Hair in tub waste • Partial drainage blockage
Won't hold water	• Tub waste needs adjusting
Won't release water	• Tub waste needs adjusting
Gurgles as it drains	• Partial drainage blockage • Partial blockage in the vent
Water drips from spout	• Bad faucet washers/cartridge • Bad faucet seats
Water comes out spout and shower at the same time	• Bad diverter washer • Bad diverter seat • Bad diverter
Faucet will not shut off	• Bad washers or cartridge • Bad faucet seats
Poor water pressure	• Partially closed valve • Not enough water pressure • Blockage in the faucet • Partially frozen pipe
No water	• Closed valve • Broken pipe • Frozen pipe

Figure 2.72.4 Troubleshooting problems with bathtubs. (*By permission: McGraw-Hill Inc.* 1998 Plumber's Business Planner, *R. Dodge Woodson.*)

TROUBLESHOOTING SHOWERS

SYMPTOMS	PROBABLE CAUSES
Won't drain	• Clogged drain • Clogged strainer • Clogged trap
Drains slowly	• Hair in strainer • Partial drainage blockage
Gurgles as it drains	• Partial drainage blockage • Partial blockage in the vent
Water drips from shower head	• Bad faucet washers/cartridge • Bad faucet seats
Faucet will not shut off	• Bad washers or cartridge • Bad faucet seats
Poor water pressure	• Partially closed valve • Not enough water pressure • Blockage in the faucet • Partially frozen pipe
No water	• Closed valve • Broken pipe • Frozen pipe

Figure 2.72.5 Troubleshooting problems with showers. (*By permission: McGraw-Hill Inc.* 1998 Plumber's Business Planner, *R. Dodge Woodson.*)

TROUBLESHOOTING GAS WATER HEATERS

SYMPTOMS	PROBABLE CAUSES
Relief valve leaks slowly	• Bad relief valve
Relief valve blows off periodically	• High water temperature • High pressure in tank • Bad relief valve
No hot water	• Out of gas • Pilot light is out • Bad thermostat • Control valve is off • Gas valve closed
Too little hot water	• Bad thermostat • Thermostat needs adjusting
Too much hot water	• Thermostat needs adjusting • Controls are defective • Burner will not shut off
Water leaks from tank	• Hole in tank • Rusted-out fitting in tank

Figure 2.72.6 Troubleshooting problems with residential water heaters (gas). (*By permission: McGraw-Hill Inc.* 1998 Plumber's Business Planner, *R. Dodge Woodson.*)

TROUBLESHOOTING ELECTRIC WATER HEATERS

SYMPTOMS	PROBABLE CAUSES
Relief valve leaks slowly	• Bad relief valve
Relief valve blows off periodically	• High water temperature • High pressure in tank • Bad relief valve
No hot water	• Electrical power is off • Elements are bad • Defective thermostat • Inlet valve is closed
Too little hot water	• An element is bad • Bad thermostat • Thermostat needs adjusting
Too much hot water	• Thermostat needs adjusting • Controls are defective
Water leaks from tank	• Hole in tank • Rusted-out fitting in tank

Figure 2.72.7 Troubleshooting problems with residential water heaters (electric). (*By permission: McGraw-Hill Inc.* 1998 Plumber's Business Planner, *R. Dodge Woodson.*)

TROUBLESHOOTING LAUNDRY TUBS

SYMPTOMS	PROBABLE CAUSES
Faucet drips from spout	• Bad washers or cartridge • Bad faucet seats
Faucet leaks at base of spout	• Bad "O" ring
Faucet will not shut off	• Bad washers or cartridge • Bad faucet seats
Poor water pressure	• Partially closed valve • Clogged aerator • Not enough water pressure • Blockage in the faucet • Partially frozen pipe
No water	• Closed valve • Broken pipe • Frozen pipe
Drains slowly	• Partial obstruction in drain or trap
Will not drain	• Blocked drain or trap
Gurgles as it drains	• Partial drainage blockage • Partial blockage in the vent
Won't hold water	• Bad basket strainer • Bad putty seal on drain

Figure 2.72.8 Troubleshooting problems with laundry tubs. (*By permission: McGraw-Hill Inc.* 1998 Plumber's Business Planner, *R. Dodge Woodson.*)

TROUBLESHOOTING SUBMERSIBLE, POTABLE-WATER PUMPS

SYMPTOMS	PROBABLE CAUSES
Won't start	• No electrical power • Wrong voltage • Bad pressure switch • Bad electrical connection
Starts, but shuts off fast	• Circuit breaker or fuse is inadequate • Wrong voltage • Bad control box • Bad electrical connections • Bad pressure switch • Pipe blockage • Pump is seized • Control box is too hot
Runs, but does not produce water, or only produces a small quantity	• Check valve stuck in closed position • Check valve installed backwards • Bad electrical wiring • Wrong voltage • Pump is sitting above the water in the well • Leak in the piping • Bad pump or motor • Broken pump shaft • Clogged strainer • Jammed impeller
Low water pressure in pressure tank	• Pressure switch needs adjusting • Bad pump • Leak in piping • Wrong voltage
Pump runs too often	• Check valve stuck open • Pressure tank is waterlogged and needs air injected • Pressure switch needs adjusting • Leak in piping • Wrong size pressure tank

Figure 2.72.9 Troubleshooting problems with residential submersible water pumps. (*By permission: McGraw-Hill Inc.* 1998 Plumber's Business Planner, *R. Dodge Woodson.*)

TROUBLESHOOTING JET, POTABLE-WATER PUMPS

SYMPTOMS	PROBABLE CAUSES
Won't start	• No electrical power • Wrong voltage • Bad pressure switch • Bad electrical connection • Bad motor • Motor contacts are open • Motor shaft is seized
Runs, but produces no water	• Needs to be primed • Foot valve is above the water level in the well • Strainer clogged • Suction leak
Starts and stops too often	• Leak in the piping • Bad pressure switch • Bad air control valve • Waterlogged pressure tank • Leak in pressure tank
Low water pressure in pressure tank	• Strainer on foot valve is partially blocked • Leak in piping • Bad air charger • Worn impeller hub • Lift demand is too much for the pump
Pump does not cut off when working pressure is obtained	• Pressure switch is bad • Pressure switch needs adjusting • Blockage in the piping

Figure 2.72.10 Troubleshooting problems with residential jet-type water pumps. (*By permission: McGraw-Hill Inc.* 1998 Plumber's Business Planner, *R. Dodge Woodson.*)

(a) Proportions

(b) display conditions

International Symbol of Accessibility

International TDD Symbol

**International Symbol of
Access for Hearing Loss**

Volume Controlled Telephone

Figure 2.73.0 Approved symbols for handicapped-accessible areas.

Conditions of Use	Short Women		Tall Men	
	in	mm	in	mm
Seated in a wheelchair:				
Manual work:				
Desk or removable armrests	26	660	30	760
Fixed, full-size armrests[2]	32[3]	815	32[3]	815
Light, detailed work:				
Desk or removable armrests	29	735	34	865
Fixed, full-size armrests[2]	32[3]	815	34	865
Seated in a 16-in (405 mm) high chair:				
Manual work	26	660	27	685
Light, detailed work	28	710	31	785

[1] All dimensions are based on a work-surface thickness of 1 1/2 in (38 mm) and a clearance of 1 1/2 in (38 mm) between legs and the underside of a work surface.

[2] This type of wheelchair arm does not interfere with the positioning of a wheelchair under a work surface.

[3] This dimension is limited by the height of the armrests; a lower height would be preferable. Some people in this group prefer lower work surfaces, which require positioning the wheelchair back from the edge of the counter.

Figure 2.74.0 Convenient heights for work surfaces for seated people. (*By permission: Council of American Building Officials (CABO), Falls Church, VA.*)

4.2.3* Wheelchair Turning Space. The space required for a wheelchair to make a 180-degree turn shall be a clear space of 60 in (1525 mm) diameter minimum or a T-shaped space within a 60 in (1525 mm) minimum square with arms 36 in (915 mm) wide minimum and 60 in (1525 mm) long minimum. See Fig. B4.2.3. Wheelchair turning space shall be permitted to include knee and toe clearance in accordance with 4.2.4.3.

4.2.4* Clear Floor or Ground Space for Wheelchairs

4.2.4.1 Size. The clear floor or ground space required to accommodate a single, stationary wheelchair and occupant shall be 30 in by 48 in (760 mm by 1220 mm) minimum. See Fig. B4.2.4.1.

4.2.4.2 Approach. The minimum clear floor or ground space for wheelchairs shall be positioned for either forward or parallel approach to an object. See Fig. B4.2.4.2.

4.2.4.3 Knee and Toe Clearances. Knee clearance shall be 25 in (635 mm) in depth maximum, 30 in (760 mm) wide minimum, and 27 in (685 mm) high minimum. Toe clearance shall be 6 in (150 mm) deep maximum and 9 in (230 mm) high minimum.

4.2.4.4 Relationship of Maneuvering Clearance to Wheelchair Spaces. One full unobstructed side of the clear floor or ground space for a wheelchair shall adjoin or overlap an accessible route or adjoin another wheelchair clear floor space. If a clear floor space is located in an alcove or otherwise confined on all or part of three sides, additional maneuvering clearances shall be provided as follows:

— forward approach. The width of an alcove shall be 36 in (915 mm) minimum when the depth exceeds 10 in (255 mm). See Fig. B4.2.4.4.

— parallel approach. The length of an alcove shall be 60 in (1525 mm) minimum when the depth exceeds 10 in (255 mm). See Fig. B4.2.4.4.

4.2.4.5 Surfaces of Wheelchair Spaces. Clear floor or ground spaces for wheelchairs shall comply with 4.5.

4.2.5 Forward Reach

4.2.5.1 Unobstructed. If the clear floor space allows only forward approach to an object and is unobstructed, the high forward reach permitted shall be 48 in (1220 mm) maximum and the low forward reach shall be 15 in (380 mm) minimum above the floor. See Fig. B4.2.5.1.

4.2.5.2 Obstructed. If the high forward reach is over an obstruction, reach depth and heights shall comply with Table 4.2.5.2. See Fig. B4.2.5.2.

4.2.6* Side Reach

4.2.6.1 Unobstructed. If the clear floor space allows a parallel approach by a person in a wheelchair, the high side reach permitted shall be 54 in (1370 mm) maximum and the low side reach shall be 15 in (380 mm) minimum above the floor. See Fig. B4.2.6.1

4.2.6.2 Obstructed. If the side reach is over an obstruction, the high reach shall be 46 in (1170 mm) maximum providing:

Figure 2.75.0 Wheelchair turning space and reach requirements. (*By permission: Council of American Building Officials (CABO), Falls Church, VA.*)

	in	mm	in	mm
Reach depth	0 – <20	0 – <510	20 – 25	510 – 635
Reach height	48	1220	44	1120

The clear floor space extending under an obstruction shall be equal to or greater than the reach depth for a maximum of 25 in (635 mm).

Figure 2.75.1 Reach limits in inches and millimeters. (*By permission: Council of American Building Officials (CABO), Falls Church, VA.*)

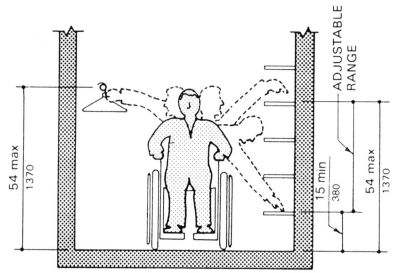

Figure 2.76.0 Reach limits for shelves and closets—illustrated. (*By permission: Council of American Building Officials (CABO), Falls Church, VA.*)

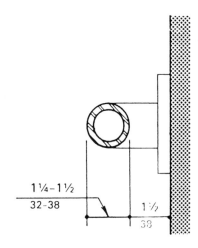

Figure 2.77.0 Size and spacing of grab bars. (*By permission: Council of American Building Officials (CABO), Falls Church, VA.*)

Size and Spacing of Grab Bars

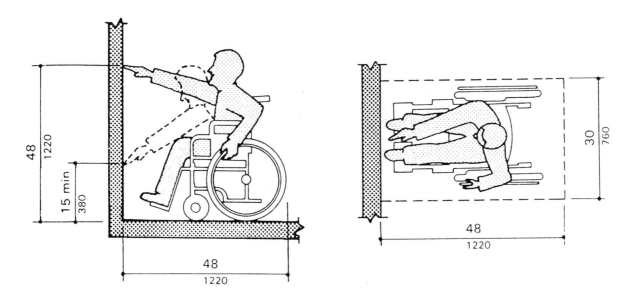

Unobstructed Forward Reach Limit

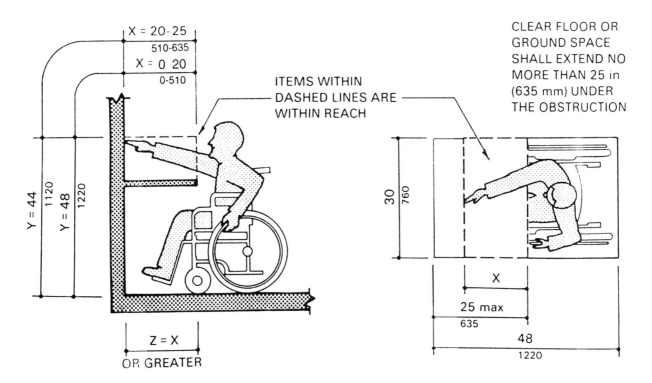

NOTE: x = Reach depth, y = Reach height, z = Clear knee space, z is the clear space below the obstruction, which shall be at least as deep as the reach distance, x.

Forward Reach Over an Obstruction

Figure 2.78.0 Reach limits—forward and over an obstruction—illustrated. (*By permission: Council of American Building Officials (CABO), Falls Church, VA.*)

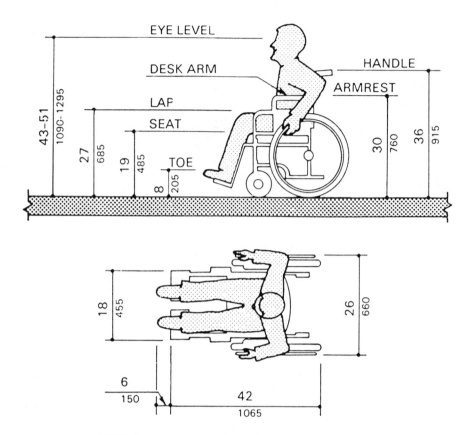

NOTE: Footrests may extend further for very large people.

Figure 2.79.0 Dimensions of adult wheelchairs. (*By permission: Council of American Building Officials (CABO), Falls Church, VA.*)

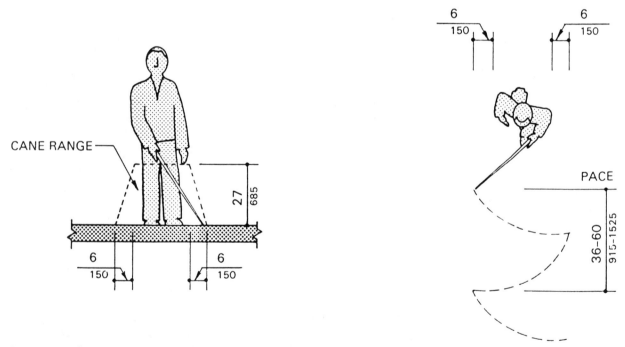

Figure 2.80.0 Cane movement arc employed by handicapped persons. (*By permission: Council of American Building Officials (CABO), Falls Church, VA.*)

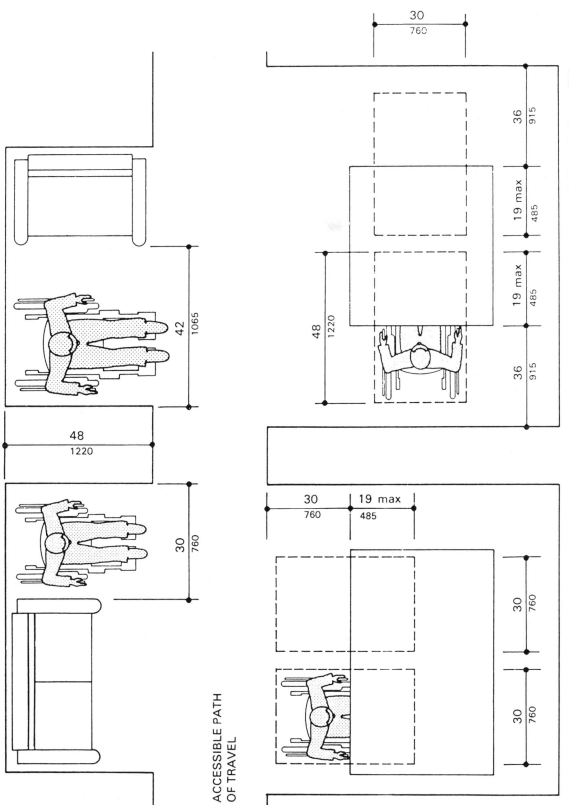

ACCESSIBLE PATH
OF TRAVEL

Figure 2.81.0 Minimum clearances for seating and tables—illustrated. (*By permission: Council of American Building Officials (CABO), Falls Church, VA.*)

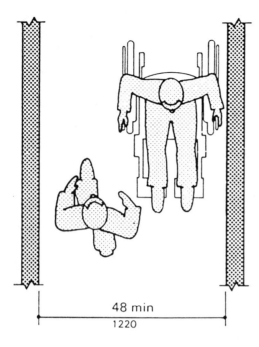

48 min
1220

Figure 2.82.0 Minimum passage for one wheelchair and one ambulatory person. (*By permission: Council of American Building Officials (CABO), Falls Church, VA.*)

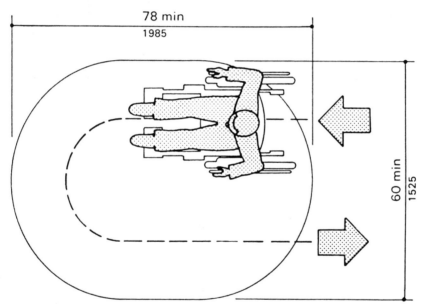

78 min
1985

60 min
1525

Figure 2.82.1 Turning radius of a wheelchair. (*By permission: Council of American Building Officials (CABO), Falls Church, VA.*)

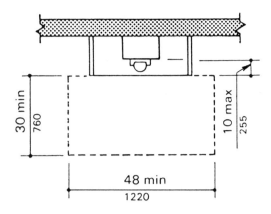

Parallel Approach to Telephone

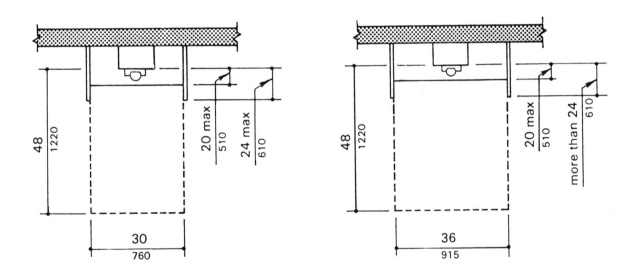

Forward Approach to Telephones

Figure 2.83.0 Forward approach to wall-mounted telephones. (*By permission: Council of American Building Officials (CABO), Falls Church, VA.*)

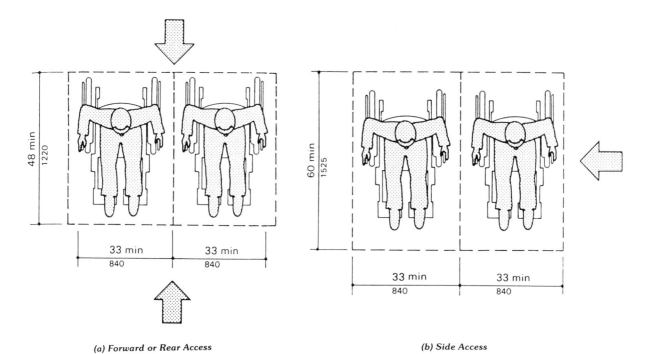

(a) Forward or Rear Access   *(b) Side Access*

Figure 2.84.0   Space requirements for wheelchair seating spaces in a series. (*By permission: Council of American Building Officials (CABO), Falls Church, VA.*)

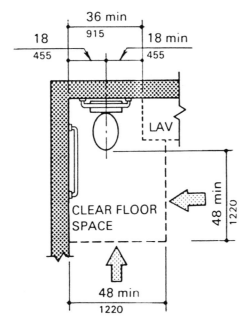

Figure 2.85.0 Clear floor space for water closet in a residence. (*By permission: Council of American Building Officials (CABO), Falls Church, VA.*)

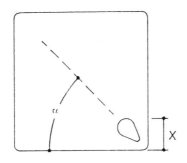

When: x = 3 in   α = 30°max
3 < x < 5 in α = 15°max

Horizontal Angle of Water Stream-Plan View

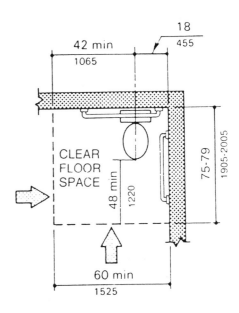

Clear Floor Space at Water Closets

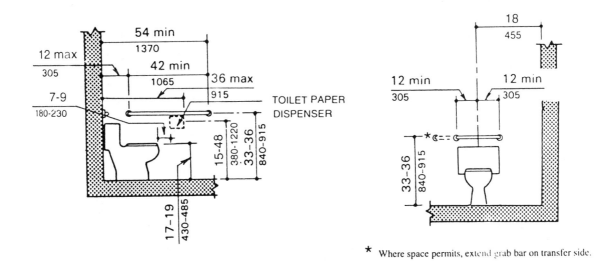

TOILET PAPER
DISPENSER

★ Where space permits, extend grab bar on transfer side.

**Water Closet
Side View**

**Water Closet
Front View**

Figure 2.86.0   Clear space required at water closets. (*By permission: Council of American Building Officials (CABO), Falls Church, VA.*)

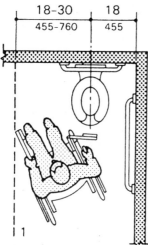

18–30   18
455–760   455

1

TAKES TRANSFER
POSITION, SWINGS
FOOTREST OUT OF
THE WAY, SETS
BRAKES

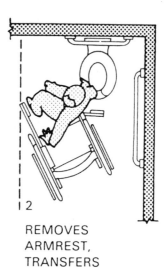

2

REMOVES
ARMREST,
TRANSFERS

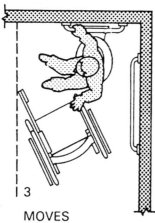

3

MOVES
WHEELCHAIR OUT
OF THE WAY,
CHANGES
POSITION (SOME
PEOPLE FOLD
CHAIR OR PIVOT IT
90° TO THE TOILET)

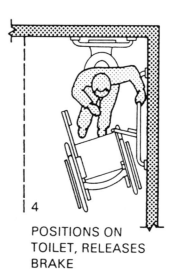

4

POSITIONS ON
TOILET, RELEASES
BRAKE

(a) Diagonal Approach

Figure 2.87.0 Illustrated methods of wheelchair to water closet transfer. *(By permission: Council of American Building Officials (CABO), Falls Church, VA.)*

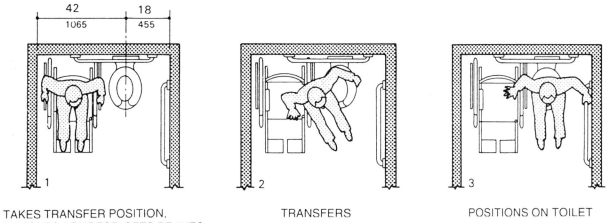

TAKES TRANSFER POSITION,
REMOVES ARMREST, SETS BRAKES

TRANSFERS

POSITIONS ON TOILET

(b) Side Approach

Figure 2.87.1 Wheelchair transfers (continued). *(By permission: Council of American Building Officials (CABO), Falls Church, VA.)*

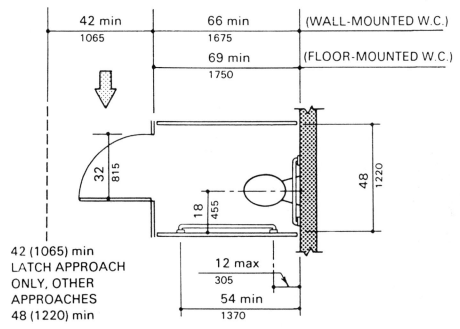

Figure 2.88.0 Alternate wheelchair accessible stall. *(By permission: Council of American Building Officials (CABO), Falls Church, VA.)*

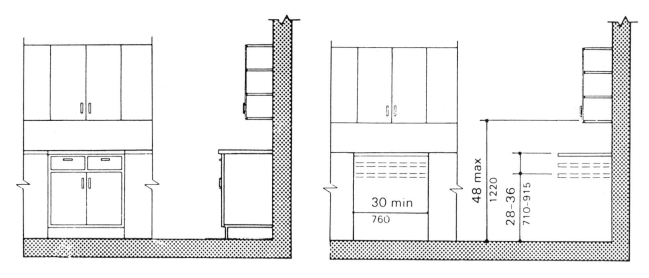

(a) Before Removal of Cabinets and Base   *(b) Cabinets and Base Removed and Height Alternatives*

Figure 2.89.0 Changing kitchen cabinets to handicapped accessible—illustrated. (*By permission: Council of American Building Officials (CABO), Falls Church, VA.*)

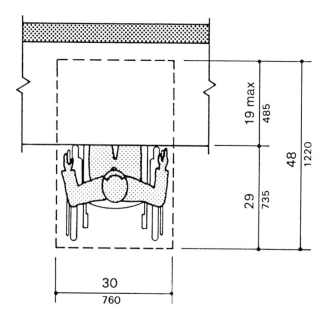

(c) Clear Floor Space under Work Surface

Figure 2.89.1 Counter work surface. (*By permission: Council of American Building Officials (CABO), Falls Church, VA.*)

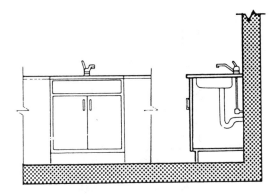

(a) Before Removeral of Cabinets and Base

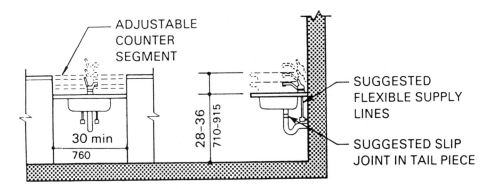

(b) Cabinets and Base Removed and Height Alternatives

Figure 2.89.2 Changing kitchen to meet handicapped needs—illustrated. (*By permission: Council of American Building Officials (CABO), Falls Church, VA.*)

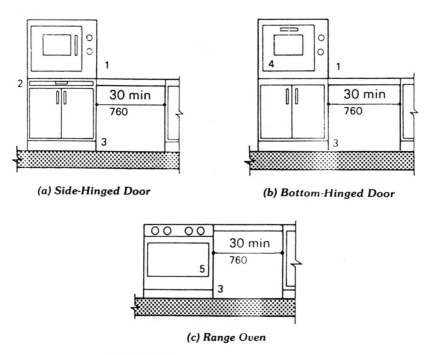

(a) Side-Hinged Door   *(b) Bottom-Hinged Door*

(c) Range Oven

SYMBOL KEY
1. Countertop or wall-mounted oven
2. Pull-out board preferred with side-opening door
3. Clear open space
4. Bottom-hinged door
5. Range oven

Figure 2.89.3 Ovens without self-cleaning feature—access required. (*By permission: Council of American Building Officials (CABO), Falls Church, VA.*)

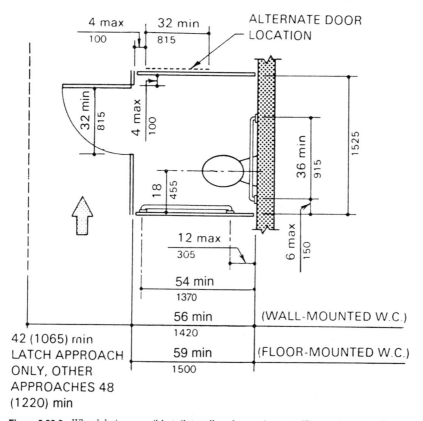

Figure 2.90.0 Wheelchair-accessible toilet stalls—door swing out. (*By permission: Council of American Building Officials (CABO), Falls Church, VA.*)

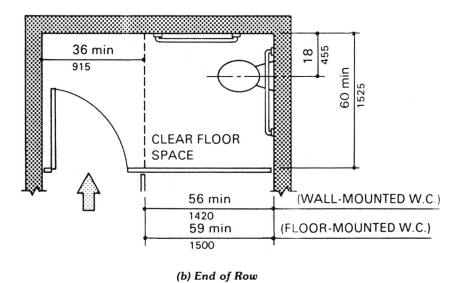

36 min
915

18   455

60 min   1525

CLEAR FLOOR
SPACE

56 min   (WALL-MOUNTED W.C.)
1420
59 min   (FLOOR-MOUNTED W.C.)
1500

(b) End of Row

Figure 2.90.1   Wheelchair-accessible toilet stalls—door swing in. (*By permission: Council of American Building Officials (CABO), Falls Church, VA.*)

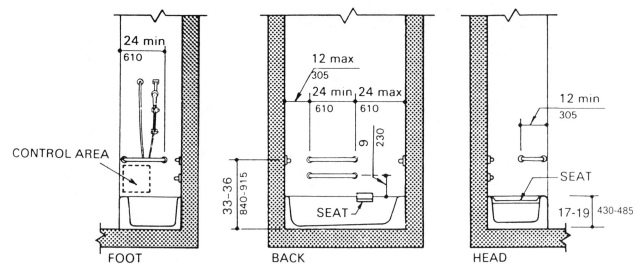

(a) Without Permanent Seat in Tub

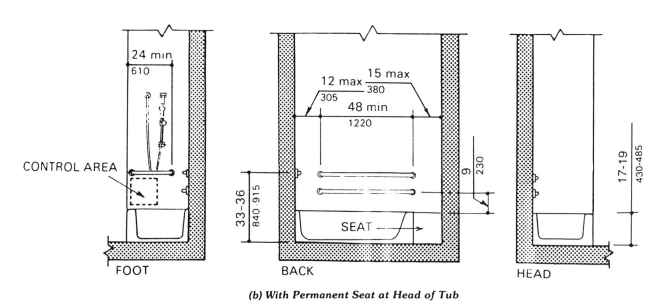

(b) With Permanent Seat at Head of Tub

Figure 2.91.0 Placement of tub seat and other bathtub accessories—illustrated. (*By permission: Council of American Building Officials (CABO), Falls Church, VA.*)

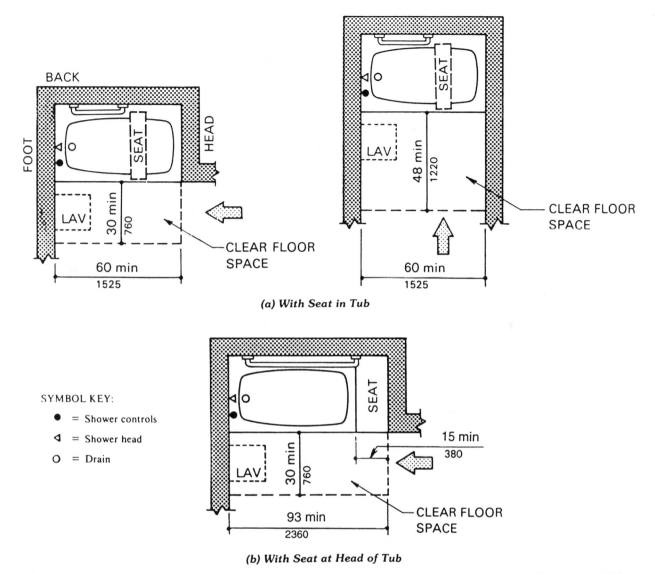

SYMBOL KEY:

● = Shower controls

◁ = Shower head

○ = Drain

(a) With Seat in Tub

(b) With Seat at Head of Tub

Figure 2.91.1 Clear floor space at bathtubs. (*By permission: Council of American Building Officials (CABO), Falls Church, VA.*)

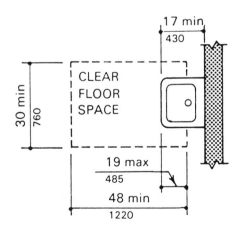

Figure 2.92.0 Clear floor space at lavatories and sinks. (*By permission: Council of American Building Officials (CABO), Falls Church, VA.*)

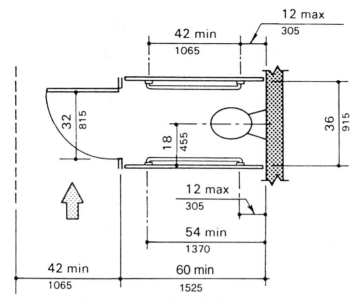

Figure 2.93.0 Ambulatory accessible stall. (*By permission: Council of American Building Officials (CABO), Falls Church, VA.*)

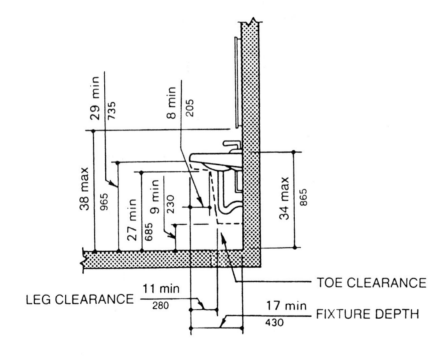

NOTE: Dashed line indicates dimensional clearance of optional under fixture enclosure.

Figure 2.94.0 Leg clearances under a lavatory—illustrated. (*By permission: Council of American Building Officials (CABO), Falls Church, VA.*)

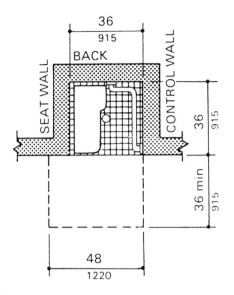

Transfer Type Shower Stall

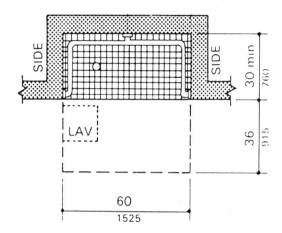

Roll-in Type Shower Stall

Figure 2.95.0 Illustrated types of handicapped shower stalls. (*By permission: Council of American Building Officials (CABO), Falls Church, VA.*)

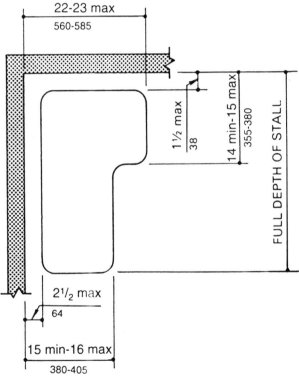

Figure 2.96.0 Shower seat design. (*By permission: Council of American Building Officials (CABO), Falls Church, VA.*)

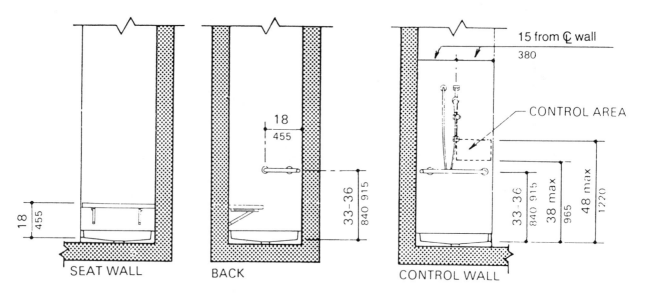

(a) 36-in by 36-in (915-mm by 915-mm) Stall

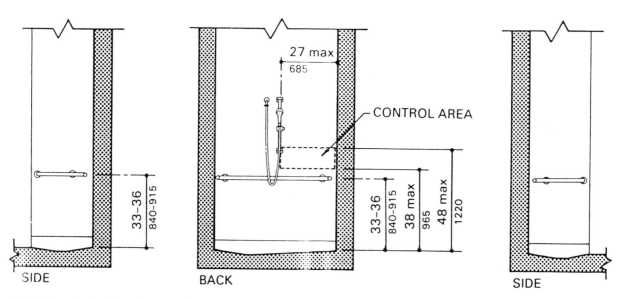

NOTE: Shower head and control area may be on back wall (as shown) or on either side wall.

(b) 30-in by 60-in (760-mm by 1525-mm) Stall

Figure 2.97.0 Placement of shower grab bars. (*By permission: Council of American Building Officials (CABO), Falls Church, VA.*)

City Water Data

State and City	Source of Supply	Maximum Water Temp. F	Hardness PPM
Alabama			
Anniston	W	70	104
Birmingham	S	85	43
Alaska			
Fairbanks	W	46	120
Ketchikan	S	44	4
Arizona			
Phoenix	W	81	210
Tucson	W	80	222
Arkansas			
Little Rock	WS	89	26
California			
Fresno	W	72	87
Los Angeles	WS	79	195
Sacramento	S	83	76
San Francisco	S	66	181
Colorado			
Denver	S	74	123
Pueblo	S	77	279
Connecticut			
Hartford	S	73	12
New Haven	S	76	46
Delaware			
Wilmington	S	83	48
District of Columbia			
Washington	S	84	162
Florida			
Jacksonville	WS	90	305
Miami	W	82	78
Georgia			
Atlanta	S	87	14
Savannah	W	85	120
Hawaii			
Honolulu	S	70	57
Idaho			
Boise	WS	65	71
Illinois			
Chicago	5	73	125
Peoria	W	67	386
Springfield		84	164
Indiana			
Evansville	S	87	140
Fort Wayne	S	84	95
Indianapolis	WS	85	279
Iowa			
Des Moines	S	77	340
Dubuque	W	60	324
Sioux City	W	62	548
Kansas			
Kansas City	S	92	230
Kentucky			
Ashland	S	85	93
Louisville	S	85	104
Louisiana			
New Orleans	S	93	150
Shreveport	S	90	36
Maine			
Portland	S	70	12

City Water Data

State and City	Source of Supply	Maximum Water Temp. F	Hardness PPM
Maryland			
Baltimore	S	75	50
Massachusetts			
Cambridge	S	74	46
Holyoke	S	77	23
Michigan			
Detroit	S	78	100
Muskegon	S	71	153
Minnesota			
Duluth	S	58	54
Minneapolis	S	83	172
Mississippi			
Jackson	S	85	38
Meridian	WS	89	7
Missouri			
Springfield	WS	80	187
St. Louis	S	88	83
Montana			
Butte	WS	54	63
Helena	WS	57	96
Nebraska			
Lincoln	W	63	188
Omaha	S	85	135
New Hampshire			
Berlin	S	69	10
Nashua	W	70	25
Nevada			
Reno	S	63	114
New Jersey			
Atlantic City	WS	73	12
Newark	S	75	29
New Mexico			
Albuquerque	W	72	155
New York			
Albany	S	70	42
Buffalo	S	76	118
New York	WS	73	30
North Carolina			
Asheville	S	79	4
Wilmington	S	89	34
North Dakota			
Bismarck	S	80	172
Ohio			
Cincinnati	S	85	120
Cleveland	S	77	121
Oklahoma			
Oklahoma City	S	83	100
Tulsa	S	85	80
Oregon			
Portland	S	65	10
Pennsylvania			
Philadelphia	S	83	98
Pittsburgh		84	95
Rhode Island			
Providence	S	71	26
South Carolina			
Charleston	S	85	18
Greenville	S	79	4

Figure 2.98.0 Various city water-temperature and hardness figures. (*By permission of The Trane Company, LaCrosse, WI.*)

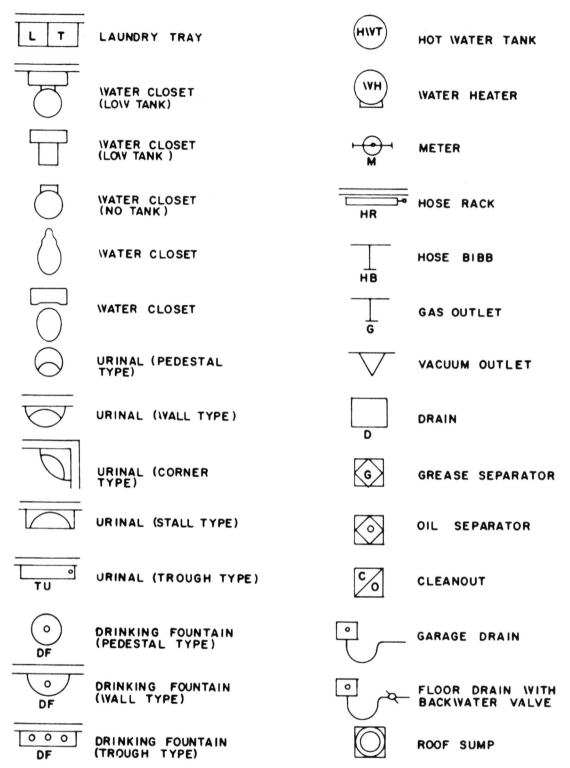

	LAUNDRY TRAY		HOT WATER TANK
	WATER CLOSET (LOW TANK)		WATER HEATER
	WATER CLOSET (LOW TANK)		METER
	WATER CLOSET (NO TANK)		HOSE RACK
	WATER CLOSET		HOSE BIBB
	WATER CLOSET		GAS OUTLET
	URINAL (PEDESTAL TYPE)		VACUUM OUTLET
	URINAL (WALL TYPE)		DRAIN
	URINAL (CORNER TYPE)		GREASE SEPARATOR
	URINAL (STALL TYPE)		OIL SEPARATOR
	URINAL (TROUGH TYPE)		CLEANOUT
	DRINKING FOUNTAIN (PEDESTAL TYPE)		GARAGE DRAIN
	DRINKING FOUNTAIN (WALL TYPE)		FLOOR DRAIN WITH BACKWATER VALVE
	DRINKING FOUNTAIN (TROUGH TYPE)		ROOF SUMP

Figure 2.99.0 Abbreviations, definitions, and symbols that appear on plumbing drawings. (*By permission of Cast Iron Soil Pipe Institute.*)

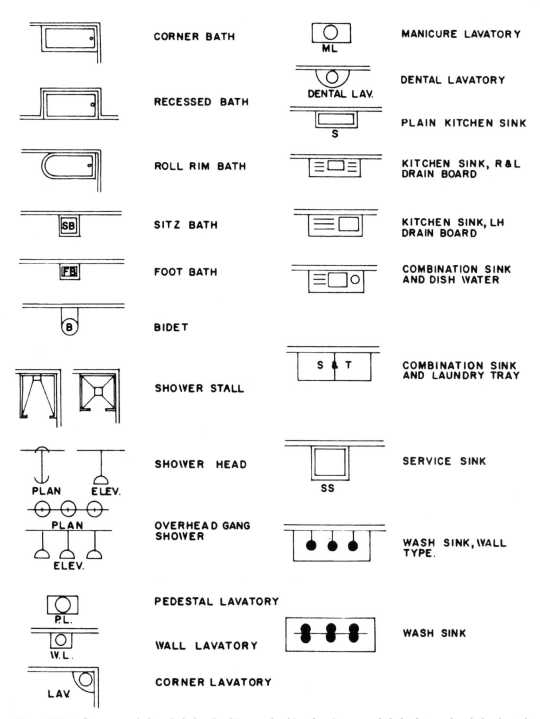

Figure 2.100.0 Recommended symbols for plumbing on plumbing drawings—symbols for fixtures [symbols adopted by the American National Standards Association (ANSI)]. (*By permission of Cast Iron Soil Pipe Institute.*)

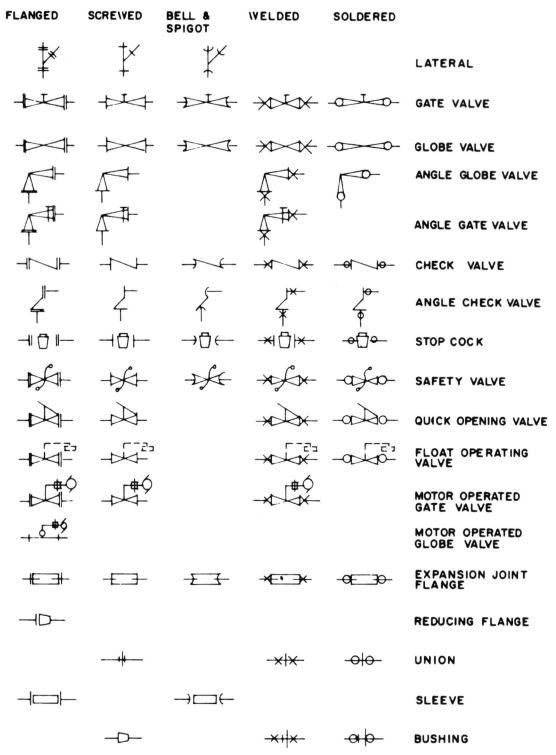

FLANGED	SCREWED	BELL & SPIGOT	WELDED	SOLDERED	
					LATERAL
					GATE VALVE
					GLOBE VALVE
					ANGLE GLOBE VALVE
					ANGLE GATE VALVE
					CHECK VALVE
					ANGLE CHECK VALVE
					STOP COCK
					SAFETY VALVE
					QUICK OPENING VALVE
					FLOAT OPERATING VALVE
					MOTOR OPERATED GATE VALVE
					MOTOR OPERATED GLOBE VALVE
					EXPANSION JOINT FLANGE
					REDUCING FLANGE
					UNION
					SLEEVE
					BUSHING

Figure 2.100.0 (*Continued*)

CHARACTER	PLAN	LINE	OR
CIRC. HOT CITY WATER			
CHILLED DRINK. WATER			
FIRE LINE			F
COLD INDUSTRIAL WATER			
HOT INDUSTRIAL WATER			
CIRC. HOT INDUS. WATER			
AIR			A
GAS			G
OIL			O
VACUUM CLEANER			V
LOCAL OR SURFACE VENT			

CHARACTER	PLAN	LINE
SANITARY SEWAGE		
SOIL STACK	24	
WASTE STACK	17	
VENT STACK	18	
COMBINED SEWAGE		
STORM SEWAGE		
ROOF LEADER		
INDIRECT WASTE		
INDUSTRIAL SEWAGE		
ACID OR CHEMICAL WASTE		
COLD CITY WATER		
HOT CITY WATER		

Figure 2.100.0 (*Continued*)

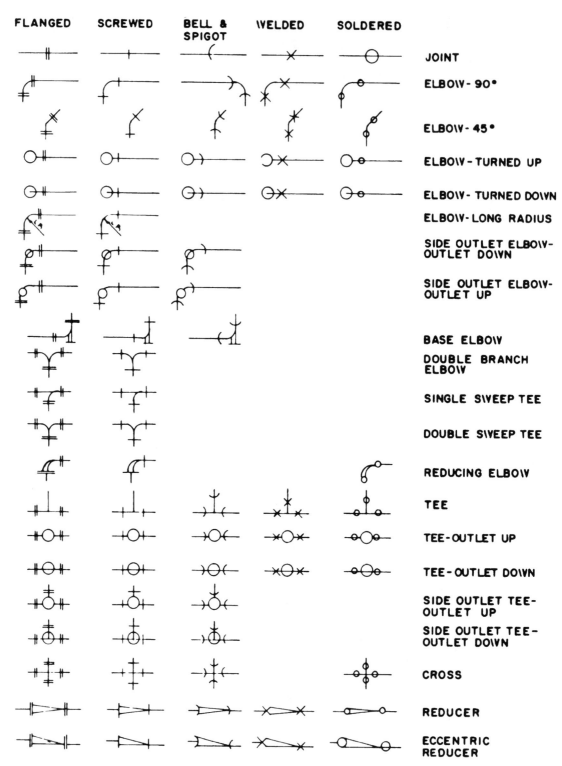

Figure 2.100.0 *(Continued)*

2.101.0   Definitions Used in the Plumbing Trades

Aerobic   Living with air.

Absorption   This term applies to immersion in a fluid for a definite period of time. It is usually expressed as a percentage of the weight of the dry pipe.

Anaerobic   Living without air.

Anchor   Usually pieces of metal used to fasten or secure pipes to the building or structure.

Area of Circle   The square of the radius multiplied by pi (3.1416). Area = π^2 or (r $\times$ r $\times$ 3.1416).

Backfill   That portion of the trench excavation which is replaced after the sewer line has been laid. The material above the pipe up to the original earth line.

Backflow   The flow of water or other liquids, mixture, or substances into the distribution pipe of a potable supply of water from any source other than that intended.

Backflow Preventer   A device or assembly designed to prevent backflow into the potable water system.

Back-siphonage   A term applied to the flow of used water, wastes, and/or contamination into the potable water supply piping, due to vacuums being established in the distribution system, building service, water main, or parts thereof.

Base   The lowest portion or lowest point of a stack of vertical pipe.

Branch   Any part of the piping system other than a main riser or stack.

Cast Iron Soil Pipe   The preferred material for drain, waste, vent, and sewer systems.

Caulking   A method of sealing against water or gas by means of pliable substances such as lead and oakum, etc.

Circumference of a Circle   The diameter of the circle multiplied by pi. Circumference = πD.

Clarified Sewage   A term used for sewage from which suspended matter has been removed.

Code   An ordinance, rule, or regulation that a city or governing body may adopt to control the plumbing work within its jurisdiction.

Coliform Group of Bacteria   Organisms considered in the coli aerogenes group as set forth in the American Water Works Association and the American Public Health Association literature.

Compression   Stress that resists the tendency of two forces acting toward each other.

Conductor   That part of the vertical piping that carries the water from the roof to the storm drain, which starts either 6″ above grade if outside the building, or at the roof sump or gutter if inside the building.

Cross Connection   (or Interconnection) Any physical connection between a city water supply and any waste pipe, soil pipe, sewer, drain, or any private or uncertified water supply. Any potable water supply outlet that is submerged or can be submerged in waste water and/or any other source of contamination.

Crude or Raw Sewage   Untreated sewage.

Dead End   A branch leading from any soil, waste, or vent pipe, building drain, or building sewer, that is terminated at a distance of 2 feet or more by means of a cap, plug, or other fitting not used for admitting water or air to the pipe, except branches serving as cleanout extensions.

Developed Lengths   Length measured along the center line of the pipe and fittings.

Diameter   A straight line that passes through the center of a circle and divides it in half.

Digester and Digestion   That portion of the sewage treatment process in which biochemical decomposition of organic matter takes place, resulting in the formation of simple organic and mineral substances.

Domestic Sewage   Sewage originating principally from dwellings, business buildings, and institutions and usually not containing storm water. In some localities it may include industrial wastes and rainwater from combination sewers.

Drain   Any pipe that carries waste water or waterborne wastes in a building drainage system.

Drain, Building or House   That part of the lowest horizontal piping of a building drainage system that receives and conveys the discharge from soil, waste, and drainage pipes, other than storm drains, from within the walls or footings of any building to the building sewer.

Drains, Combined   That portion of the drainage system within a building that carries storm water and sanitary sewage.

Drains, Storm   Piping and its branches that convey subsoil and/or surface water from areas, courts, roofs, or yards to the building or storm sewer.

Drains, Subsoil   That part of the drainage system that conveys the subsoil, ground, or seepage water from the footings of walls or from under buildings to the building drain, storm water drain, or building sewer.

Dry Weather Flow   Sewage collected during the dry weather that contains little or no groundwater and no storm water.

Ductility   The property of elongation, above the elastic limit, but short of the tensile strength.

Effluent   Sewage, treated or partially treated, flowing from sewage treatment equipment.

Elastic Limit   The greatest stress that a material can withstand without permanent deformation after release of stress.

Erosion   The gradual destruction of metal or other material by the abrasive action of liquids, gases, solids, or mixtures of these materials.

Existing Work   That portion of a plumbing system that has been installed prior to current or contemplated addition, alteration, or correction.

Fixtures, Battery of   Any group of two or more similar adjacent fixtures that discharge into a common horizontal waste or soil branch.

Fixtures, Combination   Any integral unit, such as a kitchen sink and a laundry unit.

Fixtures, Plumbing   Installed receptacles, devices, or appliances that are supplied with water or that receive liquids and/or discharge liquids or liquidborne wastes, either directly or indirectly into drainage system.

Fixture Unit   Amount of fixture discharge equivalent to 7½ gallons or more; 1 cubic foot of water per minute.

Flood Level Rim   The top edge of the receptacle from which water overflows.

Flush Valve   A device located at the bottom of the tank for flushing water closets and similar fixtures.

Flushometer Valve   A device that discharges a predetermined quantity of water to a fixture for flushing purposes; powered by direct water pressure.

Footing   The part of a foundation wall resting on the bearing soil, rock, or piling that transmits the superimposed load to the bearing material.

Fresh Sewage   Sewage of recent origin still containing free dissolved oxygen.

Invert   A line that runs lengthwise along the base of the channel at the lowest point on its wetted perimeter, its slope established when the sewer or drain is installed.

Lateral Sewer   A sewer that does not receive sewage from any other common sewer except house connections.

Leaching Well or Cesspool   Any pit or receptacle having porous walls that permit the contents to seep into the ground.

Leader   The piping from the roof that carries rainwater.

Main Sewer   (Also call the Trunk Sewer) The main stem or principal artery of the sewage system to which branches may be connected.

Master Plumber   A plumber licensed to install and to assume responsibility for contractual agreements pertaining to plumbing and to secure any required permits. The journeyman plumber is licensed to install plumbing under the supervision of a master plumber.

Offset In a line of piping, a combination of pipe, pipes, and/or fittings that join two approximately parallel sections of a line of pipe.

Outfall Sewers Sewers that receive sewage from the collection system and carry it to the point of final discharge or treatment; usually the largest sewer of a system.

Oxidized Sewage Sewage in which the organic matter has been combined with oxygen and has become stable.

Pipe, Horizontal Any pipe installed in a horizontal position or that makes an angle of less than 45° from the horizontal.

Pipe, Indirect Waste Pipe that does not connect directly with the drainage system but conveys liquid wastes into a plumbing fixture or receptacle that is directly connected to the drainage system.

Pipe, Local Ventilating A pipe on the fixture side of the trap through which pipe vapors or foul air can be removed from a room fixture.

Pipe, Soil Any pipe that conveys to the building drain or building sewer the discharge of one or more water closets and/or the discharge of any other fixture receiving fecal matter, with or without the discharge from other fixtures.

Pipe, Special Waste Drain pipe that receives one or more wastes that require treatment before entry into the normal plumbing system; the special waste pipe terminates at the treatment device on the premises.

Pipe, Vertical Any pipe installed in a vertical position or that makes an angle of not more than 45° from the vertical.

Pipe, Waste A pipe that conveys only liquid or liquid-borne waste, free of fecal matter.

Pipe, Water Riser A water supply pipe that extends vertically one full story or more to convey water to branches or fixtures.

Pipe, Water Distribution Pipes that convey water from the service pipe to its points of usage.

Pipes, Water Service That portion of the water piping that supplies one or more structures or premises and that extends from the main to the meter, or if no meter is provided, to the first stop cock or valve inside the premises.

Pitch The amount of slope given to horizontal piping, expressed in inches or vertically projected drop per foot of horizontal pipe.

Plumbing The practice, materials, and fixtures used in the installation, maintenance, extension, and alteration of all piping, fixtures, appliances, and appurtenances in connection with any of the following: sanitary drainage or storm drainage facilities or the venting system and the public or private water-supply systems; also, the practice and materials used in the installation, maintenance, extension, or alteration of water-supply systems and/or the storm water, liquid waste, or sewage system of any premises to their connection with any point of public disposal or other acceptable termina.

Plumbing Inspector Any person who, under the supervision of the authority having jurisdiction, is authorized to inspect plumbing and drainage as defined in the code for the municipality, and complying with the laws of licensing and/or registration of the state, city, or county.

Precipitation The total measurable supply of water received directly from the clouds as snow, rain, hail, and sleet. It is usually expressed in inches per day, month, or year.

Private Use A term that applies to a toilet room or bathroom intended specifically for the use of an individual or family and such visitors as they may permit to use such toilet or bathroom.

Public Use A term that applies to toilet rooms and bathrooms used by employees, occupants, visitors, or patrons, in or about any premises.

Putrefaction Biological decomposition of organic matter with the production of ill-smelling products. It usually takes place where there is a deficiency of oxygen.

Revent (individual vent) That part of a vent pipe line that connects directly with any individual waste pipe or group of wastes, underneath or behind the fixture, and extends to the main or branch vent pipe.

Roughing In   A term concerning the installation of all parts of the plumbing system that should be completed before the installing of the plumbing fixtures. Includes drainage, water supply, vent piping, and necessary fixture connections.

Sanitary Sewer   The conduit of pipe carrying sanitary sewage, storm water, and infiltration of groundwater.

Septic Sewage   Sanitary sewage undergoing putrefaction.

Septic Tank   A receptacle that receives the discharge of a drainage system or part thereof, and is designed and so constructed to separate solids from liquids to discharge into the soil through a system of open-joint or perforated piping, or into a disposal pit.

Sewage   Any liquid waste containing animal, vegetable, or chemical wastes in suspension of solution.

Sewer, Building   (Also called House sewer) That part of the horizontal piping of a drainage system extending from the building drain, storm drain, and/or subsoil drain to its connection into the point of disposal, and carrying the drainage of but one building or part thereof.

Sewer, Building Storm   (Also called House storm sewer) The extension from the building storm drain to the point of disposal.

Sewer, Private   A sewer located on private property that conveys the drainage of one or more buildings to a public sewer or to a privately owned sewage disposal system.

Sewer, Storm   A sewer used to convey rainwater, surface water, condensate, cooling water, or similar water wastes, exclusive of sewage and industrial wastes.

Slick   The thin, oily film that gives the characteristic appearance to the surface of water into which sewage or oily water is discharged.

Sludge   The accumulated suspended solids of sewage deposited in tanks, beds, or basins, mixed with sufficient water to form a semiliquid mass.

Stack   The vertical main of a system of soil, waste, or vent piping.

Stack Vent   The extension of a soil or waste stack above the highest horizontal drain connected to the stack.

Stale Sewage   Sewage that contains little or no oxygen, but is free from putrefaction.

Strain   Change of shape or size produced by stress.

Stress   External forces resisted by reactions within.

Sub-main Sewer   (Also called Branch Sewer) A sewer into which the sewage from two or more lateral sewers is discharged.

Subsoil Drain   A drain that receives the discharge from drains or other wastes, located below the normal grade of the gravity system, that must be emptied by mechanical means.

Sump   A tank or pit that receives the discharge from drains or other wastes, located below the normal grade of the gravity system, that must be emptied by mechanical means.

Tension   That stress that resists the tendency of two forces acting opposite from each other to pull apart two adjoining planes of a body.

Trap   A fitting or device so designed and constructed as to provide, when properly vented, a liquid seal that prevents the back passage of air or sewer gas without materially affecting the flow of sewage or wastewater through it.

Trap Seal   The vertical distance between the crown weir and the top of the dip of the trap

Turbulence   Any deviation from parallel flow.

Underground Piping   Piping in contact with the earth below grade. Pipe in a tunnel or in a watertight trench is not included within the scope of this term.

Vacuum   Any pressure less than that exerted by the atmosphere (may be termed a negative pressure).

Velocity   Time rate of motion in a given direction.

Vent, Circuit A branch vent that serves two or more traps and extends from in front of the last fixture connection of a horizontal branch to the vent stack.

Vent, Common (Also called Dual vent) Vent connecting at the junction of two fixture drains and serving as a vent for both fixtures.

Vent, Continuous A vent that is a continuation of the drain to which it connects. A continuous vent is further defined by the angle that the drain and vent make with the horizontal at the point of connection; for example, vertical continuous waste-and-vent, 45° continuous waste-and-vent, and flat (small angle) continuous waste-and-vent.

Vent, Loop A vent that is connected into the same stack into which the fixtures discharge. If the loop vent serves more than one fixture, it is one type of circuit vent.

Vent Stack A vertical vent pipe installed primarily to provide circulation of air to that part of a venting system to which circuit vents are connected. Branch vents, revents, or individual vents may be led to and connected with a vent stack. The foot of the vent stack may be connected either into a horizontal drainage branch or into a soil or waste stack.

Vent System Pipes installed to provide a flow of air to or from a drainage system or to provide a circulation of air within such system to protect trap seals from siphonage and back pressure.

Vent, Wet A vent that receives the discharge of wastes other than from water closets.

Vent, Yoke A pipe connecting upward from a soil or waste stack to a vent stack for the purpose of preventing pressure changes in stacks.

Venting, Stack A method of venting a fixture through the soil and waste stack.

Vents, Individual Separate vents for each fixture.

Waste The discharge from any fixture, appliance, or appurtenance, in connection with the plumbing system, that does not contain fecal matter. For example, the liquid from a lavatory, tub, sink, or drinking fountain.

Heating, Ventilating, Air Conditioning

Contents

3.0.0 How Things Work

HVAC—heating, ventilation, and air conditioning—involves the cooling and/or heating of air and/or liquids and the movement of the resultant energy to the source where it is required, to warm or cool people or to warm or cool machinery and equipment. The methods used to transfer heat or cooling from its source to the end user varies widely from solar heat to melting ice. Along with controlling temperature and movement of the heating or cooling medium, the amount of moisture (humidity) in the conditioned environment must also be considered, and humidity control remains an essential component of all HVAC systems.

The following "How Things Work" diagrams and related text may familiarize readers with the basics of the more frequently encountered heating and cooling systems and their components. The following diagrams and descriptions are not listed in any special order or priority.

Figure 3.0.1 Elementary Heating Water System

Heating systems producing water in the temperature range of 150° to 250°F are relatively simple. They are composed of a heat source (hot water boiler, in this case), pumps to move the hot water, accessories such as pressure relief valves, air eliminators, mixing valves to adjust water temperature, an expansion tank (to handle the increased volume of water as its temperature rises), and finally, some method to distribute the heat generated to its user.

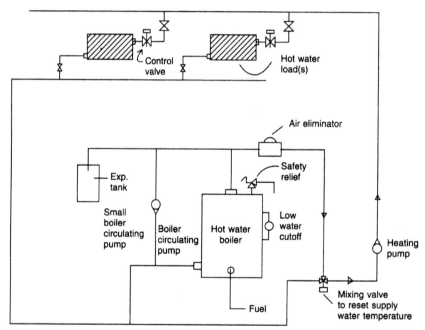

Figure 3.0.1 Elementary heating water system diagram. (*By permission: McGraw Hill* HVAC Systems Design Handbook, *R. W. Haines and C. L. Wilson.*)

Figure 3.0.2   Distribution of Heat in a Hot Water System

Hot water convectors, typically fin tube radiation, are familiar to most contractors, and all such enclosures require a means to allow air to enter and exit the enclosure after being heated by the fin tube element.

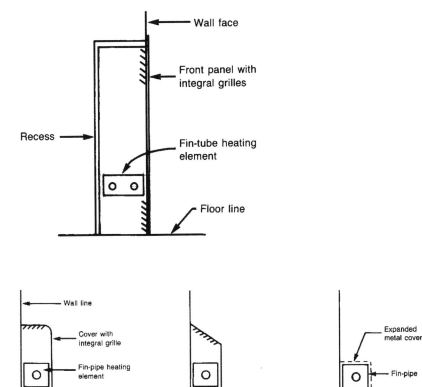

A. Flat top cover   B. Scoping top cover   C. Expanded metal cover

Figure 3.0.2   Typical fin-type enclosures. (*By permission: McGraw Hill* HVAC Systems Design Handbook, *R. W. Haines and C. L. Wilson.*)

Figures 3.0.3 and 3.0.4   Unit Ventilators

These distribution devices contain heating and cooling coils with hot or cold air generated within the unit, or supplied by an exterior source such as a central boiler or chiller. Internal fans distribute the hot or cold air into the area as required, generally by a thermostat within the unit or externally wall mounted.

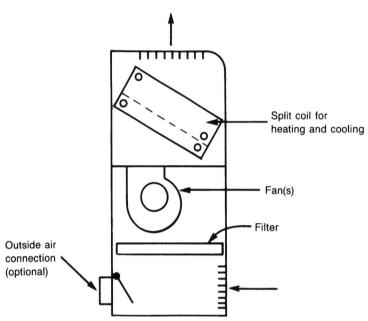

Figure 3.0.3 Fan-coil unit (floor-mounted). (*By permission: McGraw Hill* HVAC Systems Design Handbook, *R. W. Haines and C. L. Wilson.*)

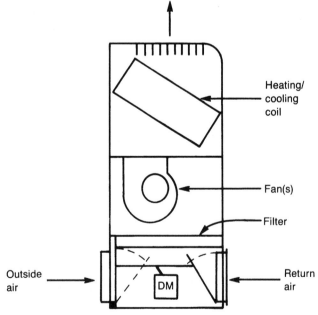

Figure 3.0.4 Unit ventilator. (*By permission: McGraw Hill* HVAC Systems Design Handbook, *R. W. Haines and C. L. Wilson.*)

Figure 3.0.5   Boiler Input and Output

When a boiler or furnace is the heat source, air and fuel create either hot air, hot water, or steam for distribution. The heat source requires a means to purge the products of combustion to the atmosphere via a variety of types of flues or vents.

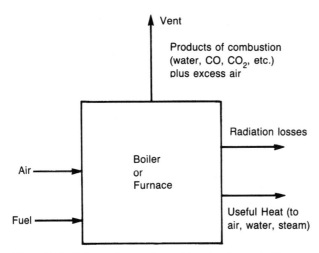

Figure 3.0.5   Boiler or furnace thermal input and output. (*By permission: McGraw Hill* HVAC Systems Design Handbook, *R. W. Haines and C. L. Wilson.*)

Figures 3.0.6 and 3.0.7   Two- and Three-Pipe Hot and Chilled Water Distribution Systems

The two-pipe system has the capability of distributing either hot or chilled water, but not without some form of changeover procedure in which the introduction of heat is shut down and the cooling source fired up to produce cold water for distribution through the single source supply and return system.

The three-pipe system contains a supply pipe for hot water, a supply pipe for cold water, and a common return. This allows a more frequent and potentially easier changeover from heat to air conditioning. Although less expensive to install than a four-pipe system, the operating costs of either the two- or three-pipe systems are not cost-effective.

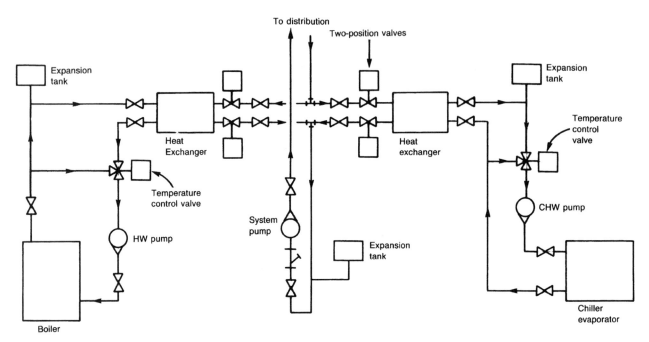

Figure 3.0.6   Central plant serving a two-pipe distribution system. (*By permission: McGraw Hill* HVAC Systems Design Handbook, *R. W. Haines and C. L. Wilson.*)

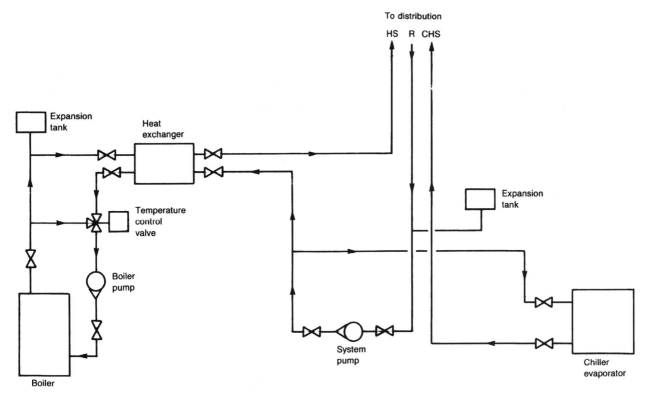

Figure 3.0.7 Central plant serving a three-pipe distribution system. (*By permission: McGraw Hill* HVAC Systems Design Handbook, *R. W. Haines and C. L. Wilson.*)

Figures 3.0.8 and 3.0.9   Multiple Boilers Used as a Heat Source

Multiple boilers are generally piped in parallel and in the case of modular boilers, increased demand allows them to fire up as the load increases and, conversely, shut down individually as demand decreases. When modulars are not used, it is common to size each boiler in a two boiler installation to carry 60% of peak load, alternating operations with each other, or, in case of failure, the active boiler will go "on-line" until the "down" boiler is activated. Pumps can be installed to service one boiler each or to alternate operations when two pumps service both boilers.

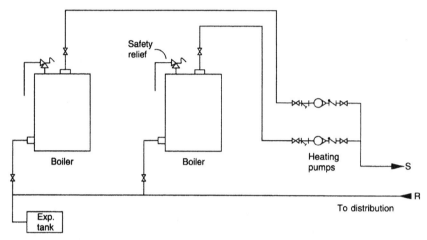

Figure 3.0.8   Central heating plant, multiple boilers/individual pumps. (*By permission: Mc-Graw Hill* HVAC Systems Design Handbook, *R. W. Haines and C. L. Wilson.*)

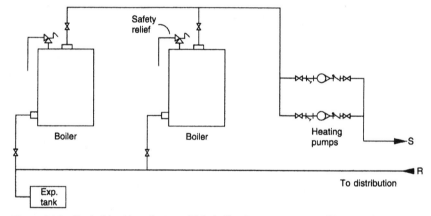

Figure 3.0.9   Central heating plant, multiple boilers/common pumps. (*By permission: Mc-Graw Hill* HVAC Systems Design Handbook, *R. W. Haines and C. L. Wilson.*)

Figure 3.0.10   A Water-to-Air Heat-Pump System Utilizing a Central Plant

When the individual heat-pump units installed through the building call for heat, the return water is cooler because the heat pump removes the heat being supplied to it, thereby requiring the boiler to provide supplemental heat. The reverse is true in the cooling mode, as the heat pump absorbs cool water on their supply side and produces warmer water on its discharge side. The return water is cooled by an evaporative water cooler, such as a cooling tower, which sprays water into the air stream, reducing the temperature of the circulating water.

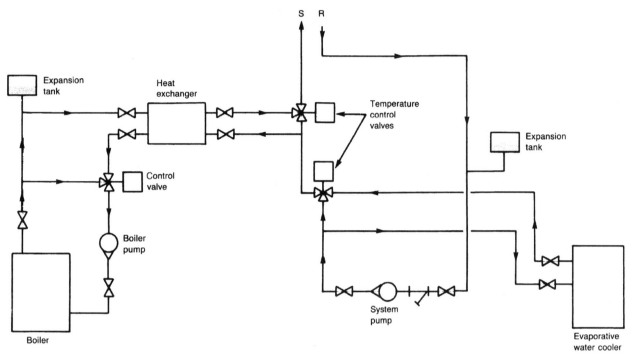

Figure 3.0.10   Central plant serving a water to air heat pump system. (*By permission: McGraw Hill* HVAC Systems Design Handbook, *R. W. Haines and C. L. Wilson.*)

Figure 3.0.11   The Single-Zone Heating/Cooling System Producing Heating and Cooling

Cooling and heating coils placed in the duct system provide the heat or cold air source, which is then distributed throughout the structure by a supply fan.

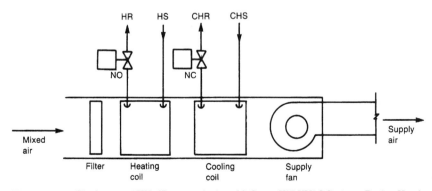

Figure 3.0.11   Single-zone AHU. (*By permission: McGraw Hill* HVAC Systems Design Handbook, *R. W. Haines and C. L. Wilson.*)

Figure 3.0.12   The Single-Zone System with Humidity Control Added

Cooling or heating air without controlling humidity does not produce the most optimum comfort for a building's occupants. Humidification increases the moisture in the air, and the need for this is more apparent in winter, when the drying effect of air is more pronounced.

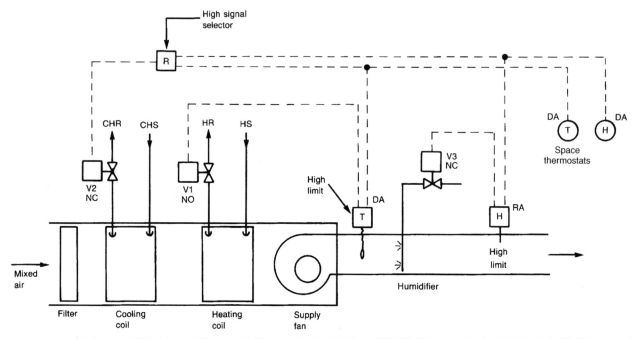

Figure 3.0.12   Single-zone AHU with humidity control. (*By permission: McGraw Hill* HVAC Systems Design Handbook, *R. W. Haines and C. L. Wilson.*)

Figures 3.0.13 and 3.0.14   Expansion Tanks

As water temperature increases, this liquid expands and needs a place to expand to, hence the use of an expansion tank in a water heating system. The closed expansion tank requires draining from time to time as the gas cushion in the tank gradually becomes absorbed and the contents become waterlogged. This situation is similar to residential well water systems utilizing a water storage tank that becomes waterlogged and must be bled from time to time. In both cases, this problem of waterlogging can be eliminated by installing a diaphragm-type tank utilizing a flexible bladder to separate the air from the water.

The open expansion tank is just that: open to the atmosphere and is to be installed at the high point of the piping system. Because this tank is open to the atmosphere, it is subject to corrosion.

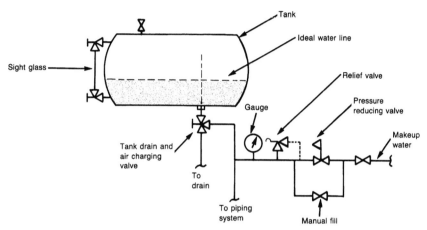

Figure 3.0.13   Closed expansion tank. (*By permission: McGraw Hill* HVAC Systems Design Handbook, *R. W. Haines and C. L. Wilson.*)

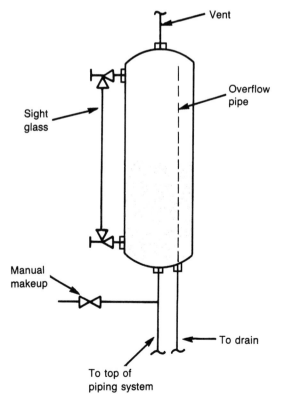

Figure 3.0.14   Open expansion tank. (*By permission: Mc-Graw Hill* HVAC Systems Design Handbook, *R. W. Haines and C. L. Wilson.*)

Figure 3.0.15   A Steam Generating Plant

The basic components of a steam generating system are boilers to generate the steam, feed pumps to keep the proper amount of water flowing to the boiler, a water treatment system to reduce the corrosive effects of the water in the system, and various other accessories such as pumps, low-level water cut-off controls, traps to collect water as the steam condenses, safety relief valves and control valves, and a distribution system.

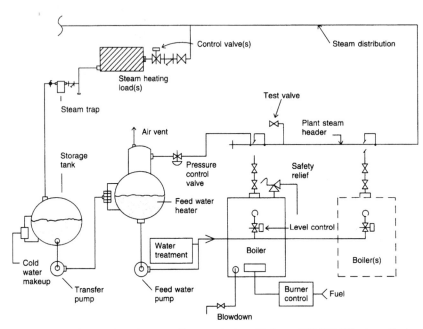

Figure 3.0.15   Steam plant diagram. (*By permission: McGraw Hill* HVAC Systems Design Handbook, *R. W. Haines and C. L. Wilson.*)

Figure 3.0.16   Air Distribution System with an Economizer Control

All AHU systems require outside air intake to mix with recirculating inside air. The degree of indoor air quality determines, in most cases, the amount of outside air to be admitted to the system. Damper motors (DM), controlled by high-limit controls, are used to regulate the amount of outside air admitted to the system. The "economy cycle" causes the outside air dampers to close when the supply fan shuts down. Depending on the outside air temperature, the controller, designated C1 Fig. 3.0.16, adjusts the dampers to maintain the minimum mixed air balance required by the sensor probe in the mixed air stream. As outside air temperatures rise, the damper reacts accordingly and reaches full open.

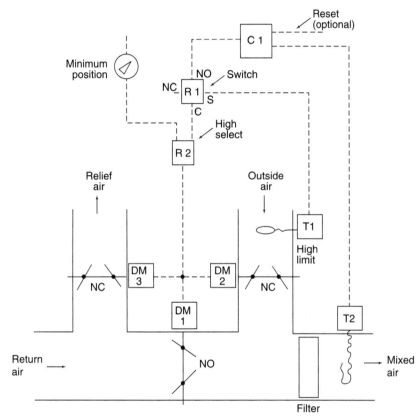

Figure 3.0.16   Classical economy cycle control. (*By permission: McGraw Hill* HVAC Systems Design Handbook, *R. W. Haines and C. L. Wilson.*)

Figure 3.0.17   Outside Air Control When Variable Air Volume (VAV) Distribution Is Used

When VAVs are installed, control is provided by a direct digital control (DDC) system whereby outside air, return air, and relief air dampers are individually regulated to provide the required velocity of air needed in the system, but programming all dampers to affect economy by limiting the amount of outside air being introduced.

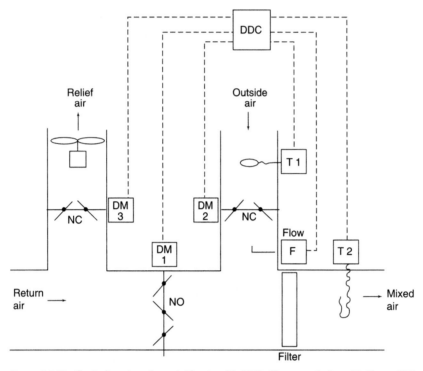

Figure 3.0.17   Control system for outside air with VAV. (*By permission: McGraw Hill* HVAC Systems Design Handbook, *R. W. Haines and C. L. Wilson.*)

Figure 3.0.18   A Multiduct Air Handling Unit

This type of system can provide either heat or cooling in adjacent zones. Fitted with an economy cycle, this system can be cost-effective from an energy conservation point of view. When optimum heating is required, the hot damper is fully opened and other dampers are closed. As a zone requires cooling, the hot damper begins to close and the bypass damper modulates to fully closed, after which the cold damper opens.

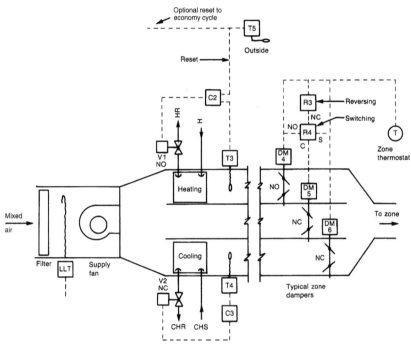

Figure 3.0.18 Three-duct multizone air-handling unit. (*By permission: McGraw Hill* HVAC Systems Design Handbook, *R. W. Haines and C. L. Wilson.*)

Figure 3.0.19   A Constant Volume System (as opposed to a variable volume system)

This system allows a constant flow through a chiller while modulating flow through the air handling unit via a three-way valve that modulates to respond to a decreasing load. When the load on the chiller becomes too small, it shuts down.

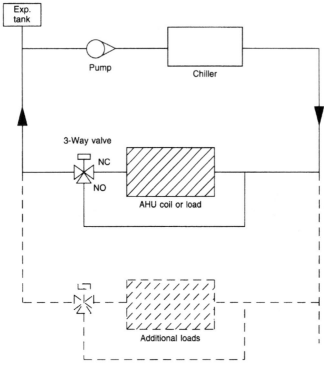

Figure 3.0.19   Constant volume system. (*By permission: McGraw Hill* HVAC Systems Design Handbook, *R. W. Haines and C. L. Wilson.*)

Figure 3.0.20   A Large Central Plant Cooling System

In two buildings in a campus setting, each structure may have its own central plant, with each component designed to act independently within its own system or pumps, condensers and cooling towers may be cross-connected to provide greater system reliability in case of failure of one component or another.

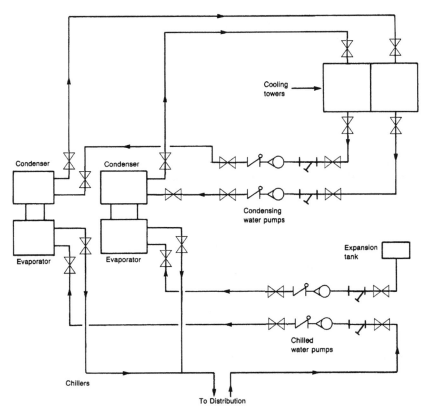

Figure 3.0.20 Central plant, one-to-one arrangement. (*By permission: McGraw Hill* HVAC Systems Design Handbook, *R. W. Haines and C. L. Wilson.*)

Figure 3.0.21   The Closed Circuit Cooling Tower

In a closed circuit cooling tower, cooling water flows through tube coil in the tower and water from a separate coolant water piping system sprays water over the coil and is evaporated. Only the coolant water circuit is open to the atmosphere and requires treatment. Decreased maintenance, however, is offset by lower efficiency of operation. When open circuit cooling towers are used, the exposure to the atmosphere causes water to absorb oxygen and, unless treated, begins to corrode the tower components. The water must also be treated to prevent the growth of algae.

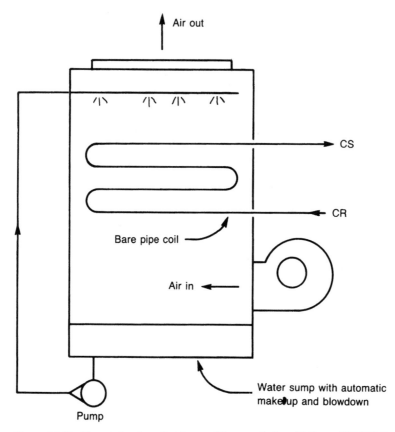

Figure 3.0.21 Closed-circuit cooling tower. (*By permission: McGraw Hill* HVAC *Systems Design Handbook, R. W. Haines and C. L. Wilson.*)

Figures 3.0.22 and 3.0.23

Hot water coils placed in the airstream of a duct need to be protected against winter freeze-ups. When a sensing device controls a three-way valve, sufficient hot water in both temperature and volume is introduced to the coil to prevent freeze-ups. When a three-way valve has not been installed in the system (Fig. 3.0.23), hot water is introduced into the coil directly from the hot water supply when the thermostat dictates.

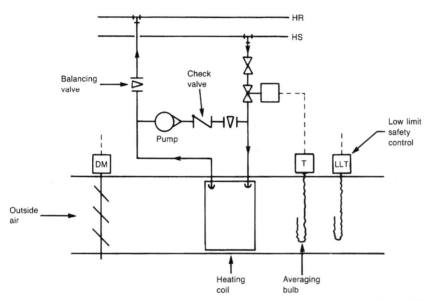

Figure 3.0.22 Freeze protection of hot water heating coil. (*By permission: McGraw Hill* HVAC Systems Design Handbook, *R. W. Haines and C. L. Wilson.*)

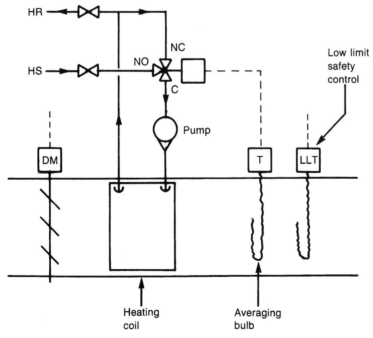

Figure 3.0.23 Freeze protection of hot water heating coil. (*By permission: McGraw Hill* HVAC Systems Design Handbook, *R. W. Haines and C. L. Wilson.*)

Figures 3.0.24 and 3.0.25   Chiller Operations

There are two basic types of chillers, flooded and direct expansion, and there are a few different configurations—shell and tube, shell and coil, plate and double tube—with the shell and tube being the more common type.

The flooded chiller allows refrigerant to flow to the shell and the water flows through the tubes. Because water can be corrosive or contain elements that create scale or buildup in the tubes, even if treated, provisions must be made in these types of chillers for easy removal of the water boxes. The direct expansion (DX) chiller has the refrigerant within the tubes and the water is in the shell. Refrigerant flow is controlled by a thermal expansion valve.

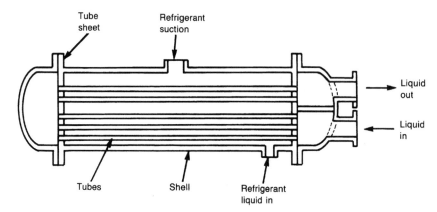

Figure 3.0.24   Flooded liquid chiller. (*By permission: McGraw Hill* HVAC Systems Design Handbook, *R. W. Haines and C. L. Wilson.*)

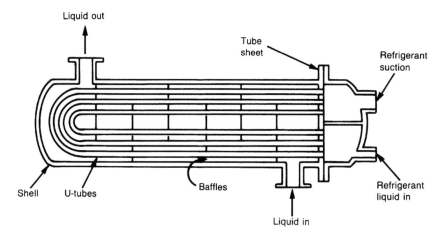

Figure 3.0.25   Direct-expansion chiller (U-tube type). (*By permission: McGraw Hill* HVAC Systems Design Handbook, *R. W. Haines and C. L. Wilson.*)

Figure 3.0.26 Multiple Chiller Installations

In installations involving two or more chillers it becomes necessary to limit chiller cycling for reasons of economy. This can be accomplished by the use of two-way valves and a bypass valve to maintain constant pressure differentials between the supply and return mains.

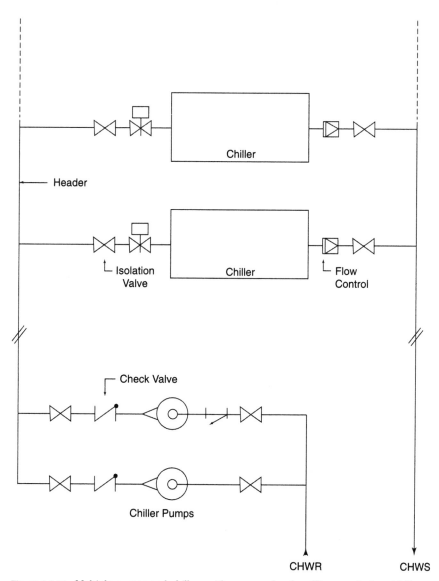

Figure 3.0.26 Multiple pumps and chillers with common header. (*By permission: McGraw Hill* HVAC Systems Design Handbook, *R. W. Haines and C. L. Wilson.*)

Figure 3.0.27   Multiple Chiller Plants with Bypass

With the widespread use of variable speed electric motors, installation using variable speed pumps has gained popularity because it allows the use of different capacity chillers in the same system, and also permits one machine to be used as a standby unit or alternate operation with the primary chiller.

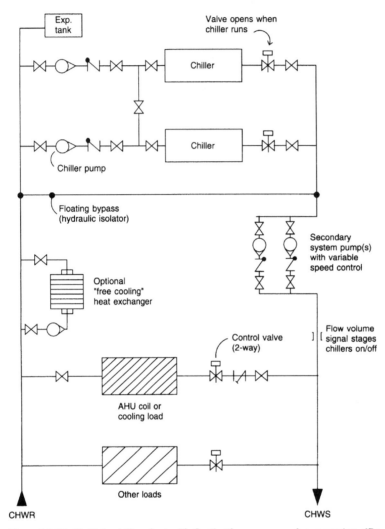

Figure 3.0.27   Multiple chiller plant with floating bypass, secondary pumping. (*By permission: McGraw Hill* HVAC Systems Design Handbook, *R. W. Haines and C. L. Wilson.*)

Figure 3.0.28   Multiple Chillers with Pressure Bypass

Two way valves are able to control the flow through the air handling unit coils efficiently, but to maintain flow through the chillers when reduced loads are placed on the chillers, a bypass valve allows constant differential pressure between the supply and return lines.

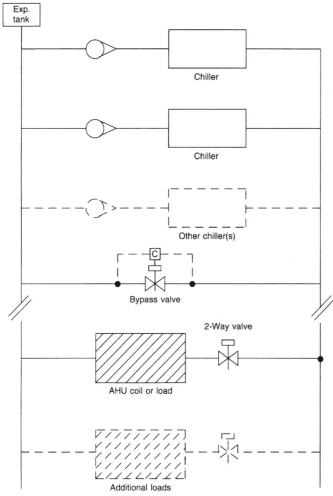

Figure 3.0.28   Multiple chiller plant with pressure bypass. (*By permission: McGraw Hill* HVAC Systems Design Handbook, *R. W. Haines and C. L. Wilson.*)

Figure 3.0.29   Multiple Chillers, Pumps, and Cooling Towers in a Condensing Water System

Installing a header to allow for connecting pumps, chillers, and ccooling towers provides greater flexibility in operating and maintenance costs.

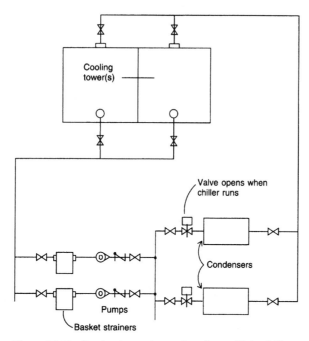

Figure 3.0.29 Condensing water system for multiple chillers, multiple pumps, and multiple cooling towers. (*By permission: McGraw Hill* HVAC Systems Design Handbook, *R. W. Haines and C. L. Wilson.*)

Figure 3.0.30   The Single Zone Air Handling Unit

Sensing devices within the space conveys instructions to the controller (C2) to actuate valves to activate either the heating or cooling coil. Sensor T3 advises the controller of the supply air temperature for further adjustment of the outside air flow.

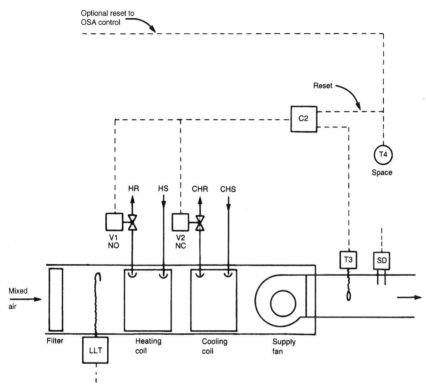

Figure 3.0.30   Single-zone air-handling unit. (*By permission: McGraw Hill* HVAC Systems Design Handbook, *R. W. Haines and C. L. Wilson.*)

Figure 3.0.31   The Multizone Air Handling Unit System

Side-by-side placement of heating and cooling coils in the system's ductwork permits modulated dampers to mix hot and cold air to satisfy the zone requirements. Most of these multizone units are factory assembled and are purchased as such.

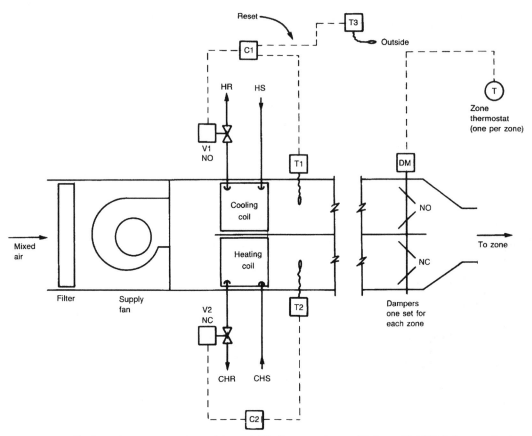

Figure 3.0.31   Traditional arrangement for multizone AHU. (*By permission: McGraw Hill* HVAC Systems Design Handbook, *R. W. Haines and C. L. Wilson.*)

Figure 3.0.32   A Double-Ducted Air Handling Unit Installation

Similar to the multizone unit, this system differs in that the hot and cold air ducts are extended throughout the building rather than restricted to a plenum in the MZ unit. These systems are not economical to operate because they were designed with high velocity–high pressure duct systems requiring electrically inefficient fan operations and adding as much as 10 in. of total pressure across the fan.

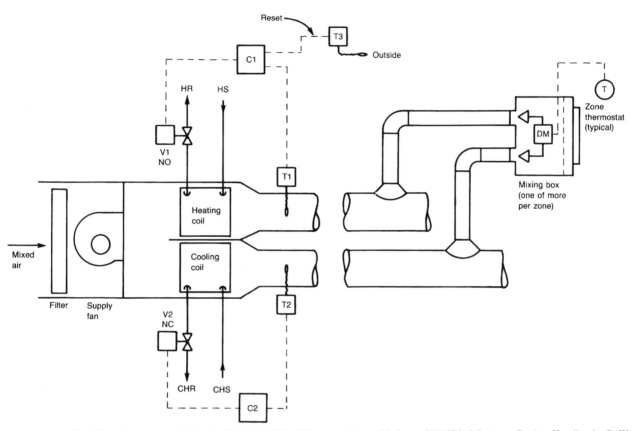

Figure 3.0.32   Traditional arrangement for double-duct AHU. (*By permission: McGraw Hill* HVAC Systems Design Handbook, *R. W. Haines and C. L. Wilson.*)

Figure 3.0.33   A Double-Ducted, Two Fan Air Handling Unit Set-Up

The two fan double-duct system is more efficient, employing one fan for the heating coil and one for the cooling coil, with separate controllers for fans, heating coils, cooling coils, and using a mixing box for air distributed to the exterior space, regulated by a thermostat and a thermostatically controlled duct to the variable air volume diffusers in the interior spaces.

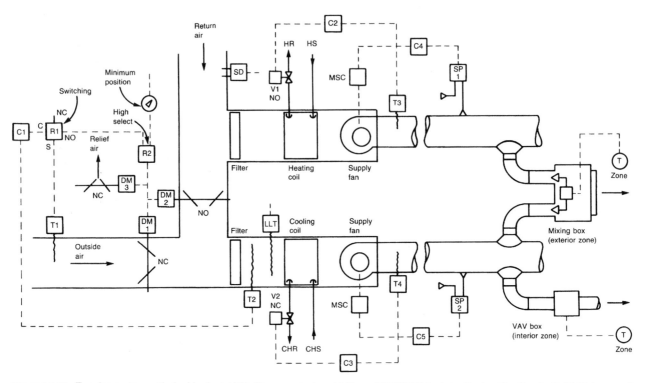

Figure 3.0.33   Two-fan system with double-duct AHU. (*By permission: McGraw Hill* HVAC Systems Design Handbook, *R. W. Haines and C. L. Wilson.*)

Figure 3.0.34   Thermal Storage and the Chiller System

Thermal storage, the use of ice in a cooling system, focuses on utilizing "off-peak" electricity to produce a mass of passive cooling material (ice). The ice is used to lower the water temperature incorporated into the chilled water circulating system.

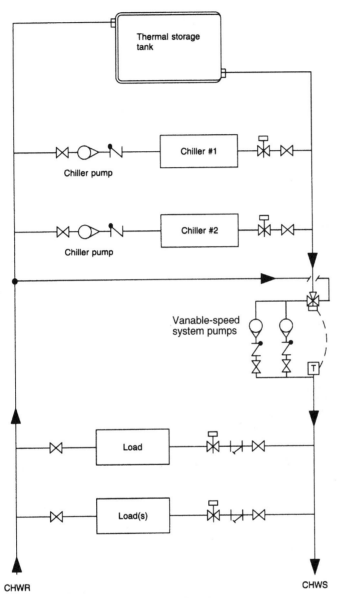

Figure 3.0.34   Chilled water plant with thermal storage. (*By permission: McGraw Hill* HVAC Systems Design Handbook, *R. W. Haines and C. L. Wilson.*)

Figures 3.0.35 and 3.0.36   Heat Exchangers

Heat exchangers transfer heat from one source to another source. One medium—for example, water circulating through the tubes—is heated by another medium—steam in the shell—as shown in Fig. 3.0.35a. Figure 3.0.35b, the steam-to-water heat exchanger, is also known as a steam condenser, because it uses the latent heat of condensation injected into the shell to heat the water that is being passed through the tubes.

Figure 3.0.36 is a water-to-water heat exchanger in which high temperature water enters the tubes and not the shell.

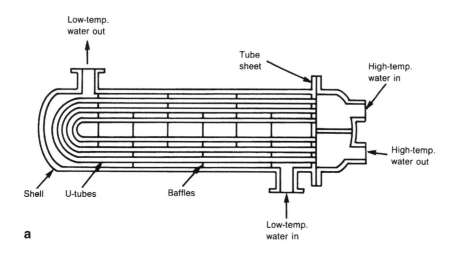

a

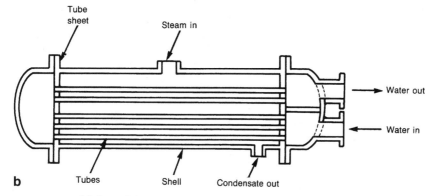

b

Figure 3.0.35 (*a*) Control for steam to hot water heat exchanger. (*b*) Steam-to-water heat exhange. (*By permission: McGraw Hill* HVAC Systems Design Handbook, *R. W. Haines and C. L. Wilson.*)

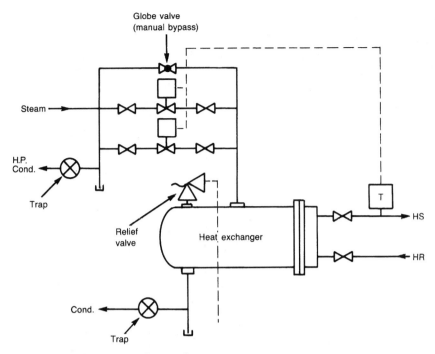

Figure 3.0.36 Water-to-water heat exchanger.

Figures 3.0.37 through 3.0.39   Cogeneration

The deregulation of power generating utilities along with more cost-effective methods to produce electric power on site, independent of the local utility company, has had an accelerating effect on cogeneration. This process involves the use of rejected heat from one energy producing source on site to generate electric power for that user. Various options of power generation range from internal combustion engines to gas- or oil-fired turbines to steam turbines.

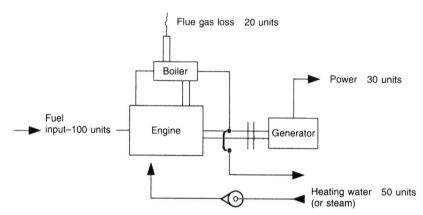

Figure 3.0.37 Engine-driven cogeneration. (*By permission: McGraw Hill* HVAC Systems Design Handbook, *R. W. Haines and C. L. Wilson.*)

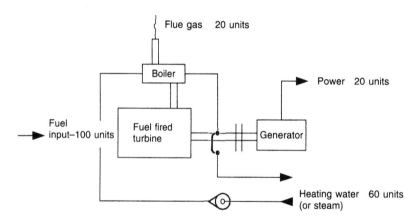

Figure 3.0.38 Turbine-driven cogeneration. (*By permission: McGraw Hill* HVAC Systems Design Handbook, *R. W. Haines and C. L. Wilson.*)

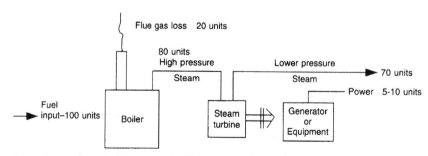

Figure 3.0.39 Steamboiler-topping turbine cogeneration. (*By permission: McGraw Hill* HVAC Systems Design Handbook, *R. W. Haines and C. L. Wilson.*)

Figures 3.0.40–3.0.43   Vibration and Sound Control in Air Distribution Systems

Airborne and structure-borne vibrations are a concern to designers. Sound transmission from a reciprocating machine is transmitted throughout the building as shown in Fig. 3.0.40. Inertia bases under such equipment, merely four-sided steel platforms filled with ready mix concrete (Fig. 3.0.41), reduce vibration transmission, and equipment mounted on an inertia bed with flexible connectors installed in the piping (Fig. 3.0.42) and flexible duct connections (Fig. 3.0.43) decrease vibration transmission even further.

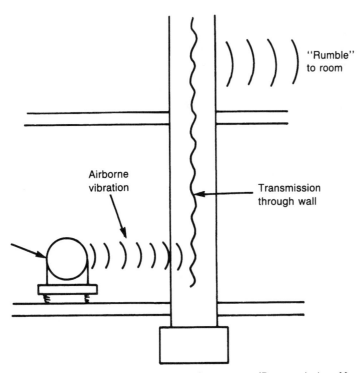

Figure 3.0.40   Sound transmission through structure. (*By permission: McGraw Hill* HVAC Systems Design Handbook, *R. W. Haines and C. L. Wilson.*)

System Design Guidelines

Vibration

System Natural Frequency

The natural frequency of a system is the frequency at which the system prefers to vibrate. It can be calculated by the following equation:

$$f_n = 188 \ (1/d)^{1/2} \text{ (cycles per minute)}$$

The static deflection corresponding to this natural frequency can be calculated by the following equation:

$$d = (188/f_n)^2 \text{ (inches)}$$

By adding vibration isolation, the transmission of vibration can be minimized. A common rule of thumb for selection of vibration isolation is as follows:

Equipment RPM	Static Deflection of Isolation	
	Critical Installation	Non-critical Installation
1200+	1.0 in	0.5 in
600+	1.0 in	1.0 in
400+	2.0 in	1.0 in
300+	3.0 in	2.0 in

Critical installations are upper floor or roof mounted equipment.
Non-critical installations are grade level or basement floor.
Always use total weight of equipment when selecting isolation.
Always consider weight distribution of equipment in selection.

Vibration Severity

When using the Machinery Vibration Severity Chart, the following factors must be taken into consideration:

1. When using displacement measurements only filtered displacement readings (for a specific frequency) should be applied to the chart. Unfiltered or overall velocity readings can be applied since the lines which divide the severity regions are, in fact, constant velocity lines.

2. The chart applies only to measurements taken on the bearings or structure of the machine. The chart does not apply to measurements of shaft vibration.

3. The chart applies primarily to machines which are rigidly mounted or bolted to a fairly rigid foundation. Machines mounted on resilient vibration isolators such as coil springs or rubber pads will generally have higher amplitudes of vibration than those rigidly mounted. A general rule is to allow twice as much vibration for a machine mounted on isolators. However, this rule should not be applied to high frequencies of vibration such as those characteristic of gears and defective rolling-element bearings, as the amplitudes measured at these frequencies are less dependent on the method of machine mounting.

Sound

Sound Power (W) - the amount of power a source converts to sound in watts.

Sound Power Level (LW) - a logarithmic comparison of sound power output by a source to a reference sound source, W_0 (10^{-12} watt).

$$L_W = 10 \ log_{10}(W/W_0) \ dB$$

Sound Pressure (P) - pressure associated with sound output from a source. Sound pressure is what the human ear reacts to.

Sound Pressure Level (Lp) - a logarithmic comparison of sound pressure output by a source to a reference sound source, P_0 (2×10^{-5} Pa).

$$Lp = 20 \ log_{10}(P/P_0) \ dB$$

Even though sound power level and sound pressure level are both expressed in dB, ***THERE IS NO OUTRIGHT CONVERSION BETWEEN SOUND POWER LEVEL AND SOUND PRESSURE LEVEL.*** A constant sound power output will result in significantly different sound pressures and sound pressure levels when the source is placed in different environments.

Rules of Thumb

When specifying sound criteria for HVAC equipment, refer to sound power level, not sound pressure level.

When comparing sound power levels, remember the lowest and highest octave bands are only accurate to about $+/-4$ dB.

Lower freqeuencies are the most difficult to attenuate.

2 × sound pressure (single source) = +3 dB (sound pressure level)
2 × distance from sound source = −6 dB (sound pressure level)
+10 dB (sound pressure level) = 2 × original loudness perception

When trying to calculate the additive effect of two sound sources, use the approximation (logarithms cannot be added directly) on the next page.

Figure 3.0.40.1 Vibration—natural frequencies guidelines. (*By permission: Loren Cook, Springfield, MO.*)

System Design Guidelines

Design Criteria for Room Loudness

Room Type	Sones	Room Type	Sones
Hotels		*Public buildings*	
Lobbies	4.0 to 12	Museums	3 to 9
Banquet rooms	8.0 to 24	Planetariums	2 to 6
Ball rooms	3.0 to 9	Post offices	4 to 12
Individual rooms/suites	2.0 to 6	Courthouses	4 to 12
Kitchens and laundries	7.0 to 12	Public libraries	2 to 6
Halls and corridors	4.0 to 12	Banks	4 to 12
Garages	6.0 to 18	Lobbies and corridors	4 to 12
Residences		*Retail stores*	
Two & three family units	3 to 9	Supermarkets	7 to 21
Apartment houses	3 to 9	Department stores (main floor)	6 to 18
Private homes (urban)	3 to 9	Department stores (upper floor)	4 to 12
Private homes (rural & suburban)	1.3 to 4	Small retail stores	6 to 18
Restaurants		Clothing stores	4 to 12
Restaurants	4 to 12	*Transportation (rail, bus, plane)*	
Cafeterias	6 to 8	Waiting rooms	5 to 12
Cocktail lounges	5 to 15	Ticket sales office	4 to 12
Social clubs	3 to 9	Control rooms & towers	6 to 12
Night clubs	4 to 12	Lounges	5 to 15
Banquet room	8 to 24	Retail shops	6 to 18
Miscellaneous			
Reception rooms	3 to 9		
Washrooms and toilets	5 to 15		
Studios for sound reproduction	1 to 3		
Other studios	4 to 12		

Note: Values shown above are room loudness in sones and are not fan
sone ratings. For additional detail see AMCA publication 302 — Application
of Sone Rating.

Figure 3.0.40.2 Design criteria for room loudness. (*By permission:
Loren Cook, Springfield, MO.*)

Sound Pressure and Sound Pressure Level

Sound Pressure (Pascals)	Sound Pressure Level dB	Typical Environment
200.0	140	30m from military aircraft at take-off
63.0	130	Pneumatic chipping and riveting (operator's position)
20.0	120	Passenger jet takeoff at 100 ft.
6.3	110	Automatic punch press (operator's position)
2.0	100	Automatic lathe shop
0.63	90	Construction site—pneumatic drilling
0.2	80	Computer printout room
0.063	70	Loud radio (in average domestic room)
0.02	60	Restaurant
0.0063	50	Conversational speech at 1m
0.002	40	Whispered conversation at 2m
0.00063	30	
0.0002	20	Background in TV recording studios
0.00002	0	Normal threshold of hearing

Figure 3.0.40.3 Sound pressure and sound pressure levels. (*By permission: Loren Cook, Springfield, MO.*)

System Design Guidelines

Rules of Thumb

Difference between sound pressure levels	dB to add to highest sound pressure level
0	3.0
1	2.5
2	2.1
3	1.8
4	1.5
5	1.2
6	1.0
7	0.8
8	0.6
9	0.5
10+	0

Noise Criteria

Graph sound pressure level for each octave band on NC curve. Highest curve intercepted is NC level of sound source. See **Noise Criteria Curves.**

Figure 3.0.40.4 Differences between sound pressure levels. (*By permission: Loren Cook, Springfield, MO.*)

Sound Power and Sound Power Level

Sound Power (Watts)	Sound Power Level dB	Source
25 to 40,000,000	195	Shuttle Booster rocket
100,000	170	Jet engine with afterburner
10,000	160	Jet aircraft at takeoff
1,000	150	Turboprop at takeoff
100	140	Prop aircraft at takeoff
10	130	Loud rock band
1	120	Small aircraft engine
0.1	110	Blaring radio
0.01	100	Car at highway speed
0.001	90	Axial ventilating fan (2500 m^3h) Voice shouting
0.0001	80	Garbage disposal unit
0.00001	70	Voice—conversational level
0.000001	60	Electronic equipment cooling fan
0.0000001	50	Office air diffuser
0.00000001	40	Small electric clock
0.000000001	30	Voice - very soft whisper

Figure 3.0.40.5 Sound power sources. (*By permission: Loren Cook, Springfield, MO.*)

System Design Guidelines

Vibration Severity

*Use the **Vibration Severity Chart** to determine acceptability of vibration levels measured.*

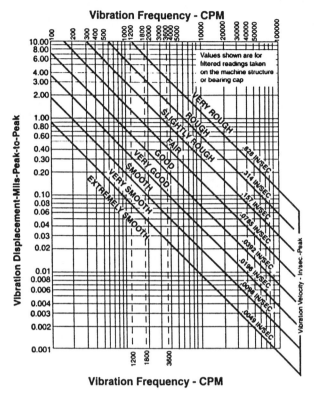

Figure 3.0.40.6 Vibration severity chart. (*By permission: Loren Cook, Springfield, MO.*)

System Design Guidelines

Noise Criteria Curves

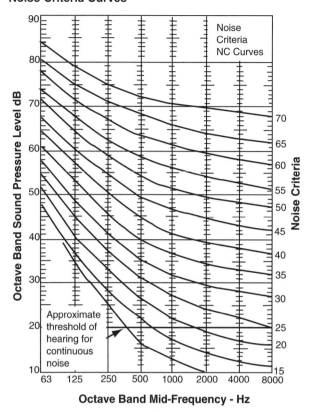

Figure 3.0.40.7 Noise Criteria Curves. (*By permission: Loren Cook, Springfield, MO.*)

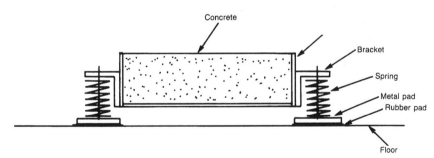

Figure 3.0.41 Inertia base. (*By permission: McGraw Hill* HVAC Systems Design Handbook, *R. W. Haines and C. L. Wilson.*)

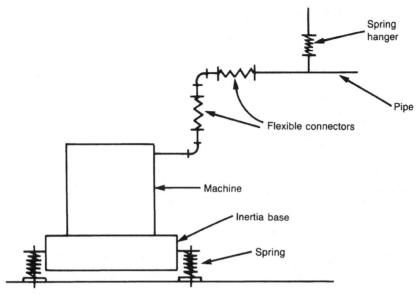

Figure 3.0.42 Piping with flexible connections. (*By permission: McGraw Hill* HVAC Systems Design Handbook, *R. W. Haines and C. L. Wilson.*)

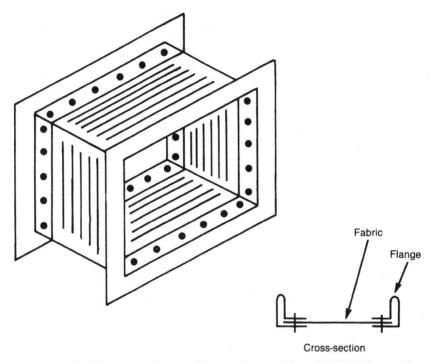

Figure 3.0.43 Flexible connector for duct. (*By permission: McGraw Hill* HVAC Systems Design Handbook, *R. W. Haines and C. L. Wilson.*)

Figures 3.0.44–3.0.46   Air Distribution Ductwork

Figure 3.0.44 contains generally accepted metal gauge thickness for galvanized steel and aluminum ductwork for both residential and nonresidential usage, by duct size. Figure 3.0.45 depicts various types of longitudinal seams used in the fabrication of metal ductwork. Figure 3.0.46 shows the different methods of connections to cross joints of fabricated metal ductwork. Note how direction of airflow is one of the factors in determining the suitability of various types of connections.

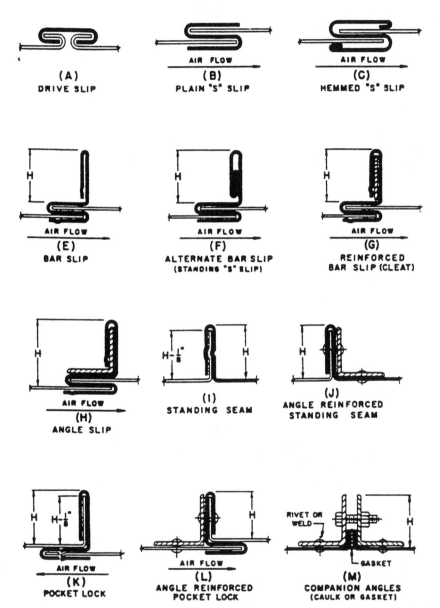

Figure 3.0.44   Typical duct connections cross-joints for sheet metal ductwork (not to scale; H = height referred to in dimensions).

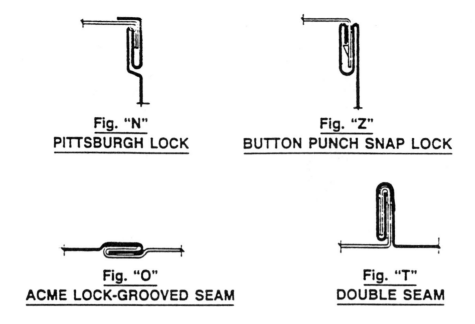

Fig. "N"
PITTSBURGH LOCK

Fig. "Z"
BUTTON PUNCH SNAP LOCK

Fig. "O"
ACME LOCK-GROOVED SEAM

Fig. "T"
DOUBLE SEAM

**Approximately 2" Spacing
Between "Buttons"**

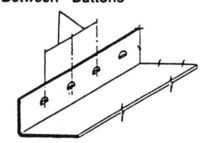

DETAIL NO. 1
MALE PIECE-SNAP LOCK

Figure 3.0.45   Longitudinal seams for sheet metal ductwork.

TYPICAL DUCT CONSTRUCTION SHEET METAL GAGES
IN ONE- AND TWO-FAMILY DWELLINGS

Metal Gauges (duct not enclosed in partitions)		
ROUND DUCTS		
Diameter, inches	Minimum thickness galvanized sheet gage	Minimum thickness aluminum B & S gage
Less than 12	30	26
12-14	28	26
15-18	26	24
Over 18	24	22
RECTANGULAR DUCTS		
Width, inches	Minimum thickness galvanized sheet gage	Minimum thickness aluminum B & S gage
Less than 14	28	24
14-24	26	22
25-30	24	22
Over 30	22	20
Metal Gauges (duct enclosed in partitions)		
Width, inches	Minimum thickness galvanized sheet gage	Minimum thickness aluminum B & S gage
14 or less	30	26
Over 14	28	24

TYPICAL DUCT CONSTRUCTION SHEET METAL GAGES
(All uses except 1- and 2-family dwellings)

RECTANGULAR DUCTS		
Maximum side inches	Steel min. Galv. Sheet Gage	Aluminum Min. B&S Gage
Through 12	26 (0.022 in.)	24 (0.020 in.)
13 through 30	24 (0.028 in.)	22 (0.025 in.)
31 through 54	22 (0.034 in.)	20 (0.032 in.)
55 through 84	20 (0.040 in.)	18 (0.040 in.)
Over 84	18 (0.052 in.)	16 (0.051 in.)

ROUND DUCTS			
	Spiral seam duct	Longitudinal seam duct	Fittings
Diameter inches	Steel min. Galv. Sht. Gage	Steel min. Galv. Sht. Gage	Steel min. Galv. Sht. Gage
Through 12	28.(0.019 in.)	26 (0.022 in.)	26 (0.022 in.)
13 through 18	26 (0.022 in.)	24 (0.028 in.)	24 (0.028 in.)
19 through 28	24 (0.028 in.)	22 (0.034 in.)	22 (0.034 in.)
29 through 36	22 (0.034 in.)	20 (0.040 in.)	20 (0.040 in.)
37 through 52	20 (0.040 in.)	18 (0.052 in.)	18 (0.052 in.)

Figure 3.0.46 Typical sheet metal gauges used in residential and nonresidential construction.

Figures 3.0.47–3.0.57   Ductwork Dampers and Mixing Boxes

The distribution ductwork cannot function adequately without other components. Dampers regulate the flow through all or certain portions of the ductwork. Gate dampers (Fig. 3.0.47) are usually of the fully open or fully closed variety. Shutter-type dampers (Fig. 3.0.48) are generally of the fire/smoke type and remain open until fully closing in response to the presence of heat or smoke. Splitter dampers (Fig. 3.0.49) and butterfly dampers (Fig. 3.0.50) are used to balance the air flow. Multiblade dampers such as the ones shown in Figs. 3.0.51 and 3.0.52 are used to balance airflow and are generally activated by the system's automatic controls; however, they can also be manually adjusted.

Mixing boxes in a VAV system regulate the amount of hot or cold air that is to be introduced into the steady flow of air streaming through the distribution device. The fan-powered mixing box (Fig. 3.0.53) mixes return air with conditioned supply air at a constant volume. The bypass VAV box (Fig. 3.0.54) receives air from the constant volume supply source but has the ability to divert excess air back to the return air plenum. The double-ducted VAV mix boxes in Fig. 3.0.55 allow hot and cold dampers to operate independently. When maximum heat is required, the hot damper opens fully, and as this load decreases, the hot damper modulates to close. At a predetermined point the cooling damper opens allowing air volume to remain stable as the hot damper travels to full close and the cooling damper opens further.

Air velocity is affected by restrictions in flow; in particular, various configurations of elbows. In straight ductwork, air flows evenly (Fig. 3.0.56); however, the introduction of an elbow creates turbulence in the flow, as shown in Fig. 3.0.57.

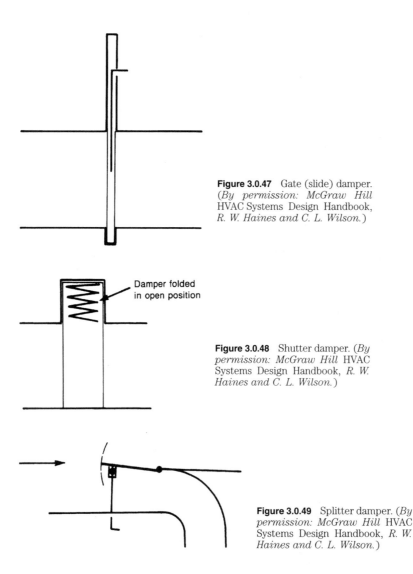

Figure 3.0.47   Gate (slide) damper. (*By permission: McGraw Hill* HVAC Systems Design Handbook, *R. W. Haines and C. L. Wilson.*)

Damper folded in open position

Figure 3.0.48   Shutter damper. (*By permission: McGraw Hill* HVAC Systems Design Handbook, *R. W. Haines and C. L. Wilson.*)

Figure 3.0.49   Splitter damper. (*By permission: McGraw Hill* HVAC Systems Design Handbook, *R. W. Haines and C. L. Wilson.*)

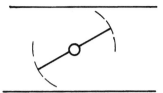

Figure 3.0.50 Butterfly damper. (*By permission: McGraw Hill* HVAC Systems Design Handbook, *R. W. Haines and C. L. Wilson.*)

Figure 3.0.51 Parallel-blade damper. (*By permission: McGraw Hill* HVAC Systems Design Handbook, *R. W. Haines and C. L. Wilson.*)

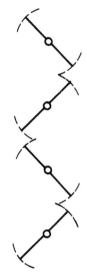

Figure **3.0.52** Opposed-blade damper, cross-sectional view. (*By permission: McGraw Hill* HVAC Systems Design Handbook, *R. W. Haines and C. L. Wilson.*)

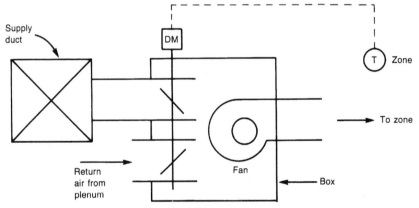

Figure 3.0.53 Fan-powered mixing box. (*By permission: McGraw Hill* HVAC Systems Design Handbook, *R. W. Haines and C. L. Wilson.*)

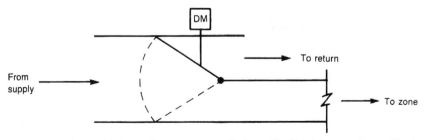

Figure 3.0.54 Bypass VAV box. (*By permission: McGraw Hill* HVAC Systems Design Handbook, *R. W. Haines and C. L. Wilson.*)

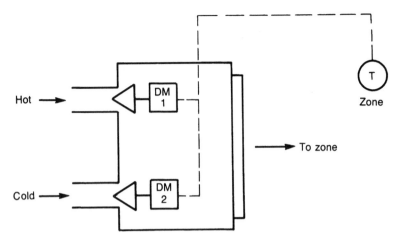

A. Mixing box arrangement.

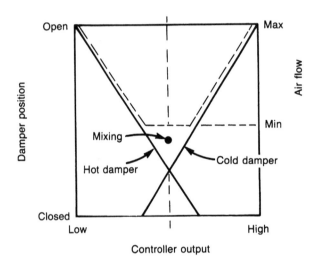

B. Damper operating sequence.

Figure 3.0.55 Double-duct VAV mixing box. (*By permission: McGraw Hill* HVAC Systems Design Handbook, *R. W. Haines and C. L. Wilson.*)

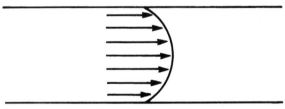

Figure 3.0.56 Air velocity pattern in straight duct. (*By permission: McGraw Hill* HVAC Systems Design Handbook, *R. W. Haines and C. L. Wilson.*)

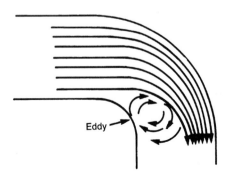

Figure 3.0.57 Air velocity pattern at a duct elbow. (*By permission: McGraw Hill* HVAC Systems Design Handbook, *R. W. Haines and C. L. Wilson.*)

Figure 3.0.58   The Air-Cooled Direct Expansion Packaged Unit

A cut-away section of a packaged direct expansion packaged unit reveals each component in the system.

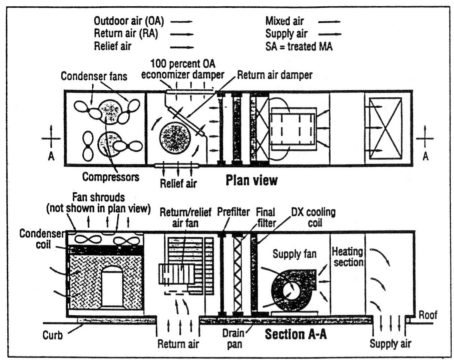

Figure 3.0.58   The air-cooled DX packaged unit (not to scale). (*By permission: HPAC magazine—Heating/Piping/Air Conditioning—December 1998.*)

Figure 3.0.59   A Heating Coil Piping Diagram

Hot water generated by a furnace or boiler is pumped into the air flow via a set of circulation pumps, often alternating in operation—a typical fan coil arrangement.

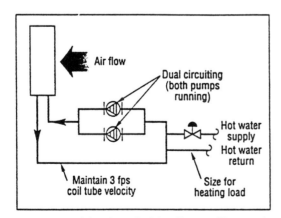

Figure 3.0.59   A heating coil piping diagram. (*By permission: HPAC magazine—Heating/Piping/Air Conditioning—December 1998.*)

Figure 3.0.60   A Simplified Schematic of an Open Recirculating and Closed Loop System Typical of a Commercial Office Building Installation

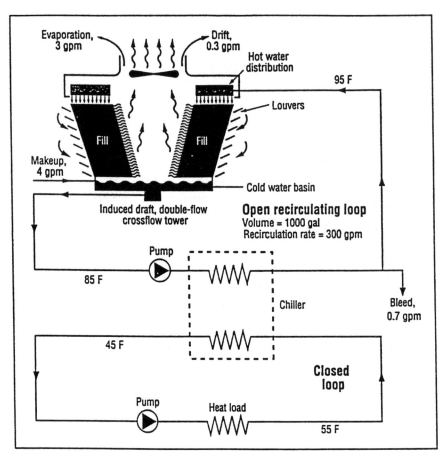

Figure 3.0.60   Simplified open recirculating/closed loop installation. (*By permission: HPAC Magazine—Heating/Piping/Air Conditioning—Jan. 1999.*)

Figure 3.0.61   How a Firetube Boiler Operation Differs from a Water-Tube Boiler

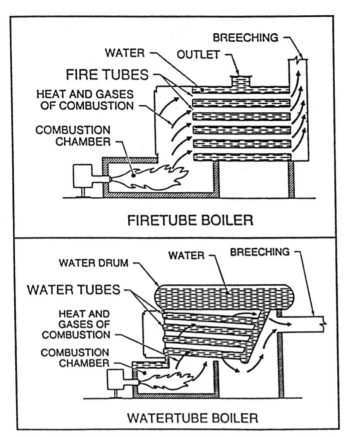

Figure 3.0.61   Firetube vs. watertube boilers. (*By permission: HPAC Magazine—Heating/Piping/Air Conditioning—Jan. 1999.*)

3.1.0 Common Boiler Types

	CAST IRON	MEMBRANE WATERTUBE	ELECTRIC	FIREBOX	FIRETUBE	FLEXIBLE WATERTUBE	INDUSTRIAL WATERTUBE	VERTICAL FIRETUBE
Efficiency	Low	Medium	High	Medium	High	Medium	Medium	Low/Medium
Floor Space Required	Low	Very Low	Very Low	Medium	Medium/ High	Low	High	Very Low
Maintenance	Medium/High	Medium	Medium/High	Low	Low	Medium	High	Low
Initial Cost	Medium	Low/Medium	High	Low	Medium/ High	Low/Medium	High	Low
No. of Options Available	Low	Medium	Medium	Low/Medium	High	Medium	High	Low
Pressure Range	HW/LPS	HW/LPS HPS to 600 psig	HW/LPS HPS to 900 PSIG	HW/LPS	HW/LPS HPS to 350 psig	HW/LPS	High Temp HW HPS to 900 psig	HW/LPS HPS to 150 psig
Typical Sizes	To 200 hp	To 250 hp	To 300 hp	To 300 hp	To 1500 hp	To 250 hp		To 100 hp
Typical Applications	Heating/ Process	Heating/ Process	Heating/ Process	Heating	Heating/ Process	Heating	Process	Heating/ Process
Comments	Field Erectable					Field Erectable		

Scotch Marine - The Classic Firetube Boiler

The Scotch Marine style of boiler has become so popular in the last 40 years that it frequently is referred to simply as "a fire-tube boiler." Firetube boilers are available for low or high pressure steam, or for hot water applications. Firetube boilers are typically used for applications ranging from 15 to 1500 horsepower. A firetube boiler is a cylindrical vessel, with the flame in the furnace and the combustion gases inside the tubes. The furnace and tubes are within a larger vessel, which contains the water and steam.

The firetube construction provides some characteristics that differentiate it from other boiler types. Because of its vessel size, the firetube contains a large amount of water, allowing it to respond to load changes with minimum variation in steam pressure.

Steam pressure in a firetube boiler is generally limited to approximately 350 psig. To achieve higher pressure, it would be necessary to use a very thick shell and tube sheet material. For this reason, a watertube boiler is generally used if pressure above 350 psig desgn is needed.

Firetube boilers are usually built similar to a shell and tube heat exchanger. A large quantity of tubes results in more heating surface per boiler horsepower, which greatly improves heat transfer and efficiency.

Firetube boilers are rated in boiler horsepower (BHP), which should not be confused with other horsepower measurements.

The furnace and the banks of tubes are used to transfer heat to the water. Combustion occurs within the furnace and the flue gases are routed through the tubes to the stack outlet. Firetube boilers are available in two, three and four pass designs. A single "pass" is defined as the area where combustion gases travel the length of the boiler. Generally, boiler efficiencies increase with the number of passes.

Firetube boilers are available in either dryback or wetback design. In the dryback boiler, a refractory-lined chamber, outside of the vessel, is used to direct the combustion gases from the furnace to the tube banks. Easy access to all internal areas of the boiler including tubes, burner, furnace, and refractory, is available from either end of the boiler. This makes maintenance easier and reduces associated costs.

The wetback boiler design has a water cooled turn around chamber used to direct the flue gases from the furnace to the tube banks. The wetback design requires less refractory maintenance; however, internal pressure vessel maintenance, such as cleaning, is more difficult and costly. In addition, the wetback design is more prone to water side sludge buildup, because of the restricted flow areas near the turn around chamber.

(By permission of Cleaver Brooks, Milwaukee, Wisconsin)

3.2.0 Hot-Water Boiler (Schematic)

A. Heavy steel boiler frame, built and stamped in accordance with the appropriate ASME Boiler Code.

B. Large volume water leg downcomers promote rapid internal circulation and temperature equalization.

C. Bryan bent water tubes are flexible, individually replaceable without welding or rolling.

D. Internal water-cooled furnace with low heat release rate.

E. Water side interior accessible for cleanout and inspection, front and rear openings, upper and lower drums.

F. Boiler tube and furnace area access panels: heavy gauge steel-lined with high temperature ceramic fiber and insulation, bolted and tightly sealed to boiler frame.

G. Combustion chamber and burner head are completely accessible via manway in end of combustion chamber.

H. Heavy gauge steel boiler jacket with rust-resistant zinc coating and enamel finish. Insulated with fiberglass to insure exceptionally cool outer surface.

I. Rear flame observation port in access door at rear of boiler.

J. Minimum sized flue vent.

K. Forced draft, flame retention head-type burner. Efficient combustion of oil or gas, quiet operation.

L. Control panel: all controls installed with connections to terminal strip.

(By permission of Bryan Steam Boilers, Peru, Indiana)

3.2.1 Exploded View of Hot-Water Boiler

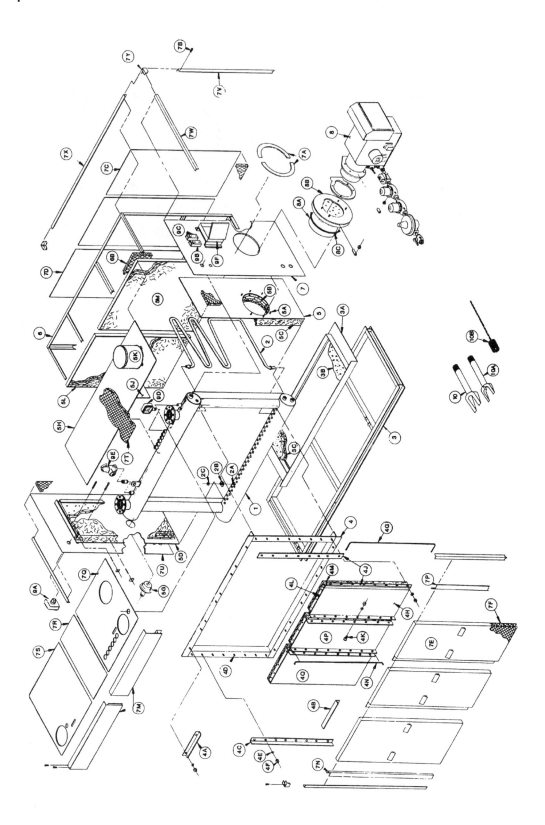

3.2.2 Hot-Water Boiler (Parts List)

ITEM NO.	DESCRIPTION
	BOILER FRAME ASSEMBLY(Less Tubes)
	BOILER TUBE ASSEMBLY
	"A" Outside Tubes
	"B" Inside Tubes
	Tube Studs
	Tube Clamps
	Tube Nuts
	BOILER BASE ASSEMBLY
	Boiler Floor Pan Assembly
	Floor Pan Insul.(Min. Fib.)Layers
	Floor Pan Refr.(Castable) Bags
	TUBE ACCESS PANEL ASSEMBLY
	Tube Access Panel Frame
	Panel Clamp, Top
	Panel Clamp, Bottom
	Panel Clamp, End
	Panel Nuts - 3/8"-16
	Rope Gasket (Ft.)
	FRONT (HINGED) PANEL ASSEMBLY
	Panel
	Hinge and Shim
	Bolts - 3/8"-16 x 1"
	Washers - 3/8"
	FRONT PANEL INSULATION ASSEMBLY
	Insul.(Mineral Fib.)19" x 60-9/16"
	Refractory(Cer.Fib.) 29" x 60-9/16"
	Rope Gasket (Ft.)
	CENTER PANEL ASSEMBLY
	Panel
	Bolts - 3/8"-16 x 1"
	Nuts - 3/8"-16
	CENTER PANEL INSULATION ASSEMBLY
	Insul.(Min. Fib)17-5/8"x 60-9/16"
	Refr.(Cer. Fib)30-1/8"x 60-9/16"
	Rope Gasket (Ft.)
	END PANEL ASSEMBLY "A"
	Panel "A" - 20-15/16" Wide
	Bolts - 3/8"-16 x 1"
	Nuts - 3/8"-16
	END PANEL INSULATION ASSEMBLY "A"
	Insul.(Min. Fib)16-15/16"X60-9/16"
	Refr.(Cer. Fib)29-7/16"X 60-9/16"
	Rope Gasket (Ft.)
	END PANEL ASSEMBLY "B"
	Panel "B" 24-1/8" Wide
	Bolts 3/8"-16
	Washers - 3/8"
	Nuts - 3/8"
	END PANEL INSULATION ASSEMBLY "B"
	Insul.(Min. Fib)20-1/8"x 60-9/16"
	Refr.(Cer. Fib.)30-5/8"x 60-9/16"
	Rope Gasket (Ft.)
o	END PANEL ASSEMBLY "C"
	Panel "C"- 33-13/16" Wide
	Bolts - 3/8"-16 x 1"
	Washers - 3/8"
	Nuts - 3/8"
	END PANEL INSULATION ASSEMBLY "C"
	Insul.(Min. Fib)29-13/16"X 60-9/16
	Refr.(Cer. Fib.)43-5/16"X 60-9/16"
	Rope Gasket (Ft.)
5	FLUE COLLECTOR ASSEMBLY
5A	FLUE COLLECTOR FRONT PLATE ASS'Y
	Angle Iron 14 Ft.
	Weld Studs- 3/8"-16 x 2-1/2" Long
	FRONT PLATE INSULATION ASSEMBLY
5B	Insul.(Min. Fib.)28-1/2"x 70-5/8"
5C	Refr.(Cer. Fib.)40-1/2"x 70-5/8"
5D	FLUE COLLECTOR REAR PLATE ASS'Y
	Angle Iron 14 Ft.
	Weld Studs-3/8"-16 x 1-1/2" Long
	REAR PLATE INSULATION ASS'Y
5E	Insul.(Min. Fib.)28-1/2"x 70-5/8"
5F	Refr.(Cer. Fib.)40-1/2"x 70-5/8"
5G	Rear Access Plug/Site Port
5H	FLUE COLL. TOP PLATE ASSEMBLY
5J	VERTICAL TUBE BAFFLE ASSEMBLY
5K	FLUE EXTENSION/CONNECTION
5L	FLUE COLLECTOR SIDE ASSEMBLY "A"
	Side Panel "A" - 22-9/16" Wide
	SIDE PANEL "A" INSULATION ASS'Y
	Insul.(Cer. Fib.)22-9/16"x 70-3/4"
5M	Insul.(Cer. Fib.)23-1/16"x 70-3/4"
o	FLUE COLLECTOR SIDE ASSEMBLY "B"
	Side Panel "B" 44-1/4" Wide
	SIDE PANEL "B" INSUL. ASSEMBLY
	Insul. (Cer. Fib.)43-3/4"x 70-3/4"
o	FLUE COLLECTOR SIDE ASSEMBLY "C"
	Side Panel "C" 45-1/16" Wide
	SIDE PANEL "C" INSUL. ASSEMBLY
	Insul.(Cer. Fib.)45-1/16"X 70-3/4"
	Insul.(Cer. Fib.)45-9/16"X 70-3/4"
o	FLUE COLLECTOR SIDE ASSEMBLY "D"
	Side Panel "D" 53-7/8" Wide
	SIDE PANEL "D" INSUL. ASSEMBLY
	Flue Coll. Side Assembly (Cont'd)
	Insul.(Car. Fib.)53-7/8"x 70-3/4"
	Insul.(Car. Fib.)54-3/8"x 70-3/4"
o	FLUE COLLECTOR SIDE ASSEMBLY "E"
	Side Panel "E" 57-1/8" Wide
	SIDE PANEL "E" INSULATION ASS'Y
	Insul.(Cer. Fib.)57-1/8"x 70-3/4"
	Insul.(Cer. Fib.)57-5/8"x 70-3/4"
6	JACKET FRAME ASSEMBLY
6A	JACKET INSULATION ASSEMBLY
	Insulation (Sq. Ft.)
6B	Insul.(Min. Fib.)29-13/16"X 60-9/16
7	JACKET FRONT PANEL ASSEMBLY
7A	Jacket Front
	Front Burner Plug Cover
7B	Jacket Screws
7C	JACKET SIDE ASSEMBLY
	38" Panel
7D	Modular Panel
	Jacket Screws
7E	JACKET ACCESS PANEL ASSEMBLY "A"
	Panel "A" Front-30-9/16" x 74-1/2"
7F	PANEL "A" INSULATION ASSEMBLY
	Insul.(Fib'gls W/F)30-1/2" x 72"
o	ACCESS PANEL ASSEMBLY "B"
	Panel "B" 17-1/2 x 74-1/2"
	PANEL "B" INSULATION ASSEMBLY
	Insul. (Fib'gls W/F) 17-1/2" x 72"
	ACCESS PANEL ASSEMBLY "C"
	Panel "C" 22-5/16" x 74-1/2"
	PANEL "C" INSULATION ASSEMBLY
	Insul.(Fib'gls W/F)17-1/2" x 72"
o	ACCESS PANEL "D"
	Panel "D" 27-1/8" x 72"
	PANEL "D" INSULATION ASSEMBLY
	Insulation (Fib'gls W/F)27" x 72"
	ACCESS PANEL ASSEMBLY "E"
	Panel "E" 30-5/16" x 74-1/2"
	PANEL "E" INSULATION ASSEMBLY
	Insul.(Fib'gls W/ F)30-1/4" x 72"
7M	JACKET FILLER PANEL ASSEMBLY
	Filler Panel
7N	JACKET FILLER STRIP ASSEMBLY
	Filler Strip Left
7P	Filler Strip Right
7Q	JACKET TOP PANEL ASSEMBLY
	Top Panel - Front
	Jacket Screws
7R	Top Panel - Center
	Jacket Screws
7S	Top Panel - Rear
	Jacket Screws
7T	JACKET TOP INSULATION (Sq. Ft.)
	JACKET REAR PANEL ASSEMBLY
	Rear Panel
	Rear Panel Access Plug Cover
	Jacket Screws
7U	JACKET REAR INSUL. ASSEMBLY
	Insulation (Fib'gls W/F)
7V	JACKET TRIM ASSEMBLY
	Vertical Moldings
7W	Horizontal Moldings (Ends)
7X	Horizontal Moldings (Sides)
7Y	Corner Moldings
	Jacket Screws
8	FORCED DRAFT BURNER ASSEMBLY
8A	Forced Draft Burner Assembly
8B	Forced Draft Burner Plug
8C	Forced Draft Burner Spacer
	Forced Draft Burner Gasket
9	WATER TRIM ASSEMBLY
9A	Low Water Cut-off
9B	High Limit
9C	Operator
9D	Pressure/Temperature Control
9E	Relief Valve
9F	Control Panel

(By permission of Bryan Boilers, Peru, Indiana)

3.3.0 Typical Steam Boiler System

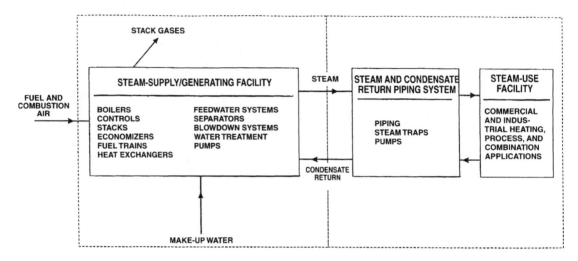

Typical Steam System

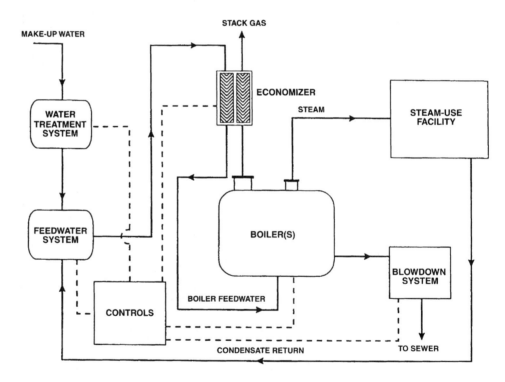

Schematic Diagram of a Generic Steam-Generating Facility

(By permission of Cleaver Brooks, Milwaukee, Wisconsin)

Steam is an invisible gas generated by adding heat energy to water in a boiler. Enough energy must be added to raise the temperature of the water to the boiling point. Then additional energy—without any further increase in temperature—changes the water to steam.

Steam is a very efficient and easily controlled heat transfer medium. It is most often used for transporting energy from a central location (the boiler) to any number of locations in the plant where it is used to heat air, water or process applications.

As noted, additional Btu are required to make boiling water change to steam. These Btu are not lost but stored in the steam ready to be released to heat air, cook tomatoes, press pants or dry a roll of paper.

The heat required to change boiling water into steam is called the heat of vaporization or latent heat. The quantity is different for every pressure/temperature combination, as shown in the steam tables.

Steam at Work...
How the Heat of Steam is Utilized

Heat flows from a higher temperature level to a lower temperature level in a process known as heat transfer. Starting in the combustion chamber of the boiler, heat flows through the boiler tubes to the water. When the higher pressure in the boiler pushes steam out, it heats the pipes of the distribution system. Heat flows from the steam through the walls of the pipes into the cooler surrounding air. This heat transfer changes some of the steam back into water. That's why distribution lines are usually insulated to minimize this wasteful and undesirable heat transfer.

When steam reaches the heat exchangers in the system, the story is different. Here the transfer of heat from the steam is desirable. Heat flows to the air in an air heater, to the water in a water heater or to food in a cooking kettle. Nothing should interfere with this heat transfer.

Condensate Drainage...
Why It's Necessary

Condensate is the by-product of heat transfer in a steam system. It forms in the distribution system due to unavoidable radiation. It also forms in heating and process equipment as a result of desirable heat transfer from the steam to the substance heated. Once the steam has condensed and given up its valuable latent heat, the hot condensate must be removed immediately. Although the available heat in a pound of condensate is negligible as compared to a pound of steam, condensate is still valuable hot water and should be returned to the boiler.

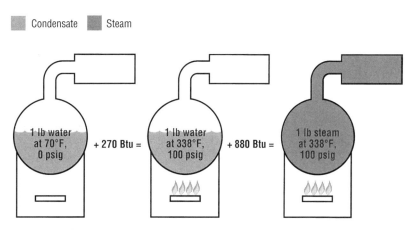

These drawings show how much heat is required to generate one pound of steam at 100 pounds per square inch pressure. Note the extra heat and higher temperature required to make water boil at 100 pounds pressure than at atmospheric pressure. Note, too, the lesser amount of heat required to change water to steam at the higher temperature.

These drawings show how much heat is required to generate one pound of steam at atmospheric pressure. Note that it takes 1 Btu for every 1° increase in temperature up to the boiling point, but that it takes more Btu to change water at 212°F to steam at 212°F.

Definitions

- **The Btu.** A Btu—British thermal unit—is the amount of heat energy required to raise the temperature of one pound of cold water by 1°F. Or, a Btu is the amount of heat energy given off by one pound of water in cooling, say, from 70°F to 69°F.
- **Temperature.** The degree of hotness with no implication of the amount of heat energy available.
- **Heat.** A measure of energy available with no implication of temperature. To illustrate, the one Btu which raises one pound of water from 39°F to 40°F could come from the surrounding air at a temperature of 70°F or from a flame at a temperature of 1,000°F.

Figure 3.4.0 Basic steam concepts. *(By permission: Associated Steam Specialty Company, Drexel Hill, PA.)*

The need to drain the distribution system. Condensate lying in the bottom of steam lines can be the cause of one kind of water hammer. Steam traveling at up to 100 miles per hour makes "waves" as it passes over this condensate (Fig. 5-2). If enough condensate forms, high-speed steam pushes it along, creating a dangerous slug which grows larger and larger as it picks up liquid in front of it. Anything which changes the direction—pipe fittings, regulating valves, tees, elbows, blind flanges—can be destroyed. In addition to damage from this "battering ram," high-velocity water may erode fittings by chipping away at metal surfaces.

The need to drain the heat transfer unit. When steam comes in contact with condensate cooled below the temperature of steam, it can produce another kind of water hammer known as *thermal shock*. Steam occupies a much greater volume than condensate, and when it collapses suddenly, it can send shock waves throughout the system. This form of water hammer can damage equipment, and it signals that condensate is not being drained from the system.

Obviously, condensate in the heat transfer unit takes up space and reduces the physical size and capacity of the equipment. Removing it quickly keeps the unit full of steam (Fig. 5-3). As steam condenses, it forms a film of water on the inside of the heat exchanger. Non-condensable gases do not change into a liquid and flow away by gravity. Instead, they accumulate as a thin film on the surface of the heat exchanger—along with dirt and scale. All are potential barriers to heat transfer (Fig. 5-1).

The need to remove air and CO_2. Air is always present during equipment start-up and in the boiler feedwater. Feedwater may also contain dissolved carbonates which release carbon dioxide gas. The steam velocity pushes the gases to the walls of the heat exchangers where they may block heat transfer. This compounds the condensate drainage problem because these gases must be removed along with the condensate.

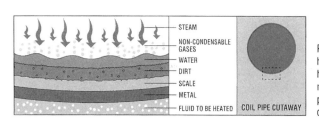

Potential barriers to heat transfer: steam heat and temperature must penetrate these potential barriers to do their work.

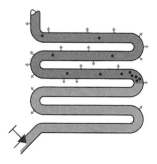

Coil half full of condensate can't work at full capacity.

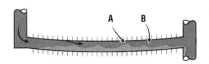

Condensate allowed to collect in pipes or tubes is blown into waves by steam passing over it until it blocks steam flow at point **A**. Condensate in area **B** causes a pressure differential that allows steam pressure to push the slug of condensate along like a battering ram.

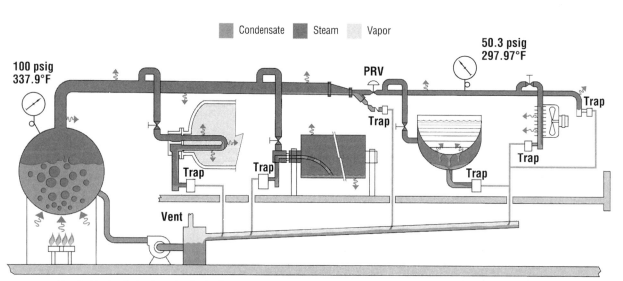

Note that heat radiation from the distribution system causes condensate to form and, therefore, requires steam traps at natural low points or ahead of control valves. In the heat exchangers, traps perform the vital function of removing the condensate before it becomes a barrier to heat transfer. Hot condensate is returned through the traps to the boiler for reuse.

Figure 3.4.0 *(Continued)*

Benefits of Balancing a Circuit

- An installed hydronic circuit never matches design flow. Actual flow has favored and unfavored circuits to or from pumps, chillers, condensers, and other mechanical equipment. Balancing restores the flow so that it is as close as possible to design values, supplying desired or required flow to all parts of the building.

A balanced circuit saves energy because:

- heating is distributed in the optimum way throughout the building, allowing occupants to be comfortable even with lower average building temperatures.
- cooling is distributed in the optimum way throughout the building, allowing occupants to be comfortable even with higher average building temperatures.
- pumps operate against the lowest possible load and use less electricity.
- small temperature drops (short circuits) are eliminated and therefore unnecessary cycling of the system is eliminated.
- Capital costs are reduced because maintenance crews will not order unnecessary equipment (such as additional pumps, chillers, or boilers) due to unbalanced flows in the circuits.
- Maintenance costs are reduced in a balanced circuit because mechanical equipment shares the load evenly, as opposed to certain pieces of equipment being overworked. MTBF (mean time between failures) and equipment life are prolonged.
- Intent of design can better be maintained because maintenance crews do not have to blindly throttle valves throughout the system.
- Occupant comfort is enhanced during heating and cooling seasons; complaints are reduced.
- Circuit performance is optimized, producing desired heating/cooling values for minimum energy expense and service costs.
- Designer and contractor are praised, not faulted.

Typical Applications

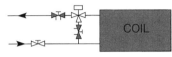

Coil – CW or HW

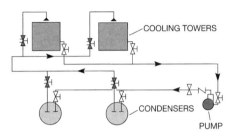

Cooling Tower & Chiller Connections

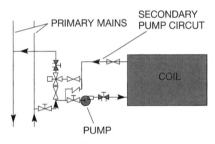

Primary-Secondary Connections

 – Indicates ABV

Sizing and Installation of Balancing Valves ½ – 12″

Sizing

Minimum flow requires: (a) a 1-foot pressure drop across the valve to obtain an accurate meter reading; and (b) that the valve should be from 50% to 100% open at set point for greatest accuracy.

Maximum flow is determined by:

(a) the highest velocity acceptable; and (b) the largest allowable pressure drop.

Installation

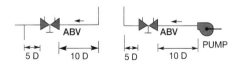

To assure flow accuracy, large turbulences must be avoided in the inlet of the balancing valve. To obtain an accurate flow reading at any metering station, good, straight, air-free flow is required.

Figure 3.4.1   Benefits of balancing a circuit. (*By permission: Associated Steam Specialty Company, Drexel Hill, PA.*)

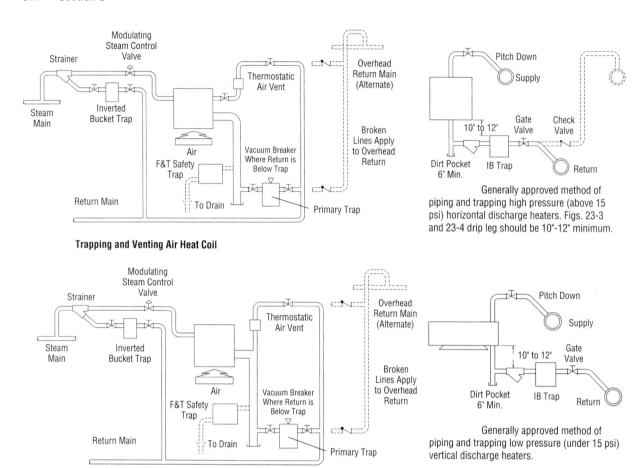

Trapping and Venting Air Heat Coil

Generally approved method of piping and trapping high pressure (above 15 psi) horizontal discharge heaters. Figs. 23-3 and 23-4 drip leg should be 10"-12" minimum.

Generally approved method of piping and trapping low pressure (under 15 psi) vertical discharge heaters.

Figure 3.4.2 Trapping and venting coils. (*By permission: Associated Steam Specialty Company, Drexel Hill, PA.*)

Trap Selection and Safety Factors

This chart provides recommendations for traps likely to be most effective in various applications. The recommended safety factors ensure proper operation under varying conditions. For more specific information on recommended traps and safety factors, contact your Armstrong representative.

Application	1st Choice	2nd Choice	Safety Factor
Boiler Header	IBLV	F&T	1.5:1
(Superheat)	IBCV Burnished	Wafer	Start-up Load
Steam Mains & Branch Lines (Non-Freezing)	IB (CV if pressure varies)	F&T	2:1, 3:1 if @ end of main, ahead of valve, or on branch
(Freezing)	IB	Thermostatic or Disc	Same as above
Steam Separator	IBLV	DC	3:1
Steam quality 90% or less	DC		3:1
Tracer Lines	IB	Thermostatic or Disc	2:1
Unit Heaters & Air Handlers (Constant Pressure)	IBLV	F&T	3:1
(0-15 Variable Pressure)	F&T	IBLV	2:1 @ ½ psi Differential
(16-30 Variable Pressure)	F&T	IBLV	2:1 @ 2 psi Differential
(>30 Variable Pressure)	F&T	IBLV	3:1 @ ½ Max. Pressure Differential
Finned Radiation & Pipe Coils (Constant Pressure)	IB	Thermostatic	3:1 for quick heating 2:1 normally
(Variable Pressure)	F&T	IB	3:1 for quick heating 2:1 normally
Process Air Heaters (Constant Pressure)	IB	F&T	2:1
(Variable Pressure)	F&T	IBLV	3:1 @ ½ Max. Pressure Differential
Steam Absorption Machine (Chiller)	F&T	IB Ext. air vent	2:1 @ ½ psi Differential
Shell & Tube Heat Exchangers, Pipe & Embossed Coils (Constant Pressure)	IB	DC or F&T	2:1
(Variable Pressure)	F&T	DC or IBT (If >30 PSI IBLV)	<15 psi 2:1 @ ½ psi, 16-30 psi 2:1 @ 2 psi, >30 psi 3:1 @ ½ Max. Pressure Differential
Evaporator Single Effect & Multiple Effect	DC	IBLV or F&T	2:1, If Load 50,000 lbs/hr use 3:1
Jacketed Kettles (Gravity Drain)	IBLV	F&T or Thermostatic	3:1
(Syphon Drain)	DC	IBLV	3:1
Rotating Dryers	DC	IBLV	3:1 for DC, 8:1 for IB constant pressure, 10:1 for IB variable pressure
Flash Tanks	IBLV	DC or F&T	3:1

IBLV = Inverted Bucket Large Vent
IBCV = Inverted Bucket Internal Check Valve
IBT = Inverted Bucket Thermic Vent
F&T = Float & Thermostatic
DC = Differential Condensate Controller
Thermo. = Thermostatic

Use an IB with external air vent above the F&T pressure limitations or if the steam is dirty. All safety factors are at the operating pressure differential unless otherwise noted.

Figure 3.4.3 Trap selection and safety factors. (*By permission: Associated Steam Specialty Company, Drexel Hill, PA.*)

How to Drain Air Distribution Systems

Air distribution systems make up the vital link between compressors and the vast amount of air-utilizing equipment. They represent the method by which air is actually transported to all parts of the plant to perform specific functions.

The three primary components of air distribution systems are air mains, air branch lines, and air distribution manifolds. They each fill certain requirements of the system and together with separators and traps contribute to efficient air utilization. Common to all air distribution systems is the need for drip legs at various intervals. These drip legs are provided to:

1) Let liquid escape by gravity from the fast moving air.
2) Store the liquid until the pressure differential can discharge it through the drain trap.
3) Serve as dirt pockets for the inevitable dirt and grit that will accumulate in the distribution system.

Air mains are one of the most common applications for drain traps.

These lines need to be kept free of liquid to keep the supplied equipment operating properly. Inadequately trapped air mains often result in water hammer and slugs of liquid which can damage control valves and other equipment. There is also a freeze potential wherever water is allowed to accumulate. In areas where air is moving slowly, the accumulation of water can effectively reduce the pipe size, thereby increasing the pressure drop and wasting energy.

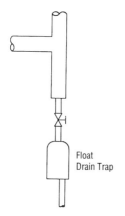

Drain trap installed straight under a low point.

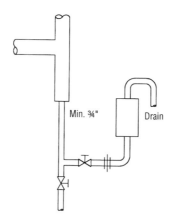

Inverted bucket drain trap installed on compressed air line contaminated by oil.

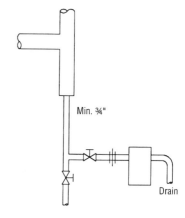

Inverted bucket drain trap installed on compressed air line contaminated by oil.

Recommendation Chart (See Chart on Gatefold B for "FEATURE CODE" References)

Equipment Being Drained	1st Choice and Feature Code	Alternate Choice and Feature Code
Air Mains	FF B, C, D, J, M	FP*

* IB is a good alternative where heavy oil carryover is likely.

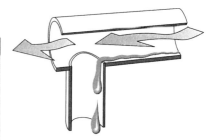

Drip leg length should be at least 1½ times the diameter of the main and never less than 10". Drip leg diameter should be the same size as the main, up to 4" pipe size and at least ½ of the diameter of the main above that, but never less than 4".

Figure 3.4.4 How to drain air distribution systems. (*By permission: Associated Steam Specialty Company, Drexel Hill, PA.*)

Steam Coil Typical Arrangements

- Standard Coils (Type S)

- Tandem Coils (Type T)

- For Vertical or
  Horizontal Air Flow

- With Headers
  Inside the Casing

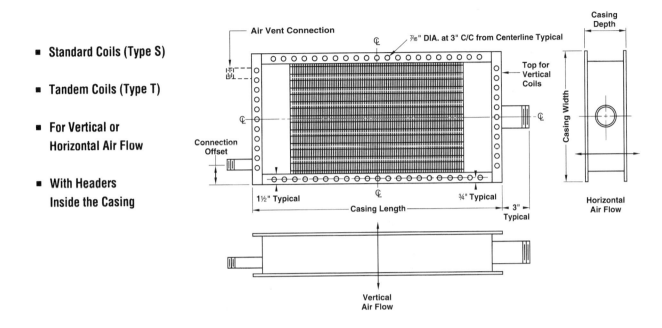

- Standard Coils (Type S)

- Tandem Coils (Type T)

- For Vertical or
  Horizontal Air Flow

- With Headers
  Outside the Casing

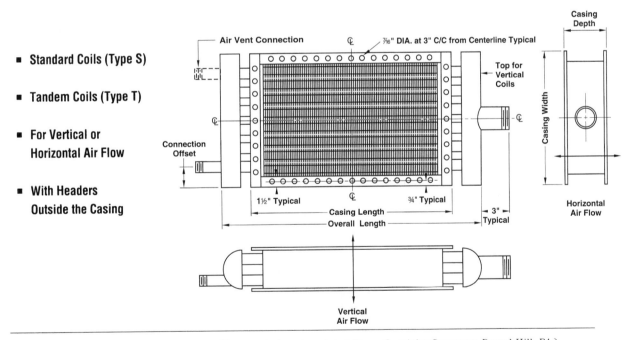

Figure 3.4.5 Steam coil typical arrangements. *(By permission: Associated Steam Specialty Company, Drexel Hill, PA.)*

3.5.0   Summary of Federal EPA Rules for Boilers Built/Modified after June 9, 1989

RULES FOR SULFUR DIOXIDE (SO$_2$) EMISSIONS

1. Coal Firing

1.2 lb SO$_2$/MMBtu Limit all 10-100 MMBtu.
90% SO$_2$ reduction required if > 75 MMBtu and > 55% annual coal capacity.
Initial performance testing required within 180 days of start-up.
30 day rolling average used in calculations.
Continuous Emission Monitoring System (CEMS) required except:
Fuel analysis may be used (before cleanup equipment).
Units < 30 MMBtu may use supplier certificate for compliance.

2. Residual Oil Firing

Limit of 0.5 lb SO$_2$/MMBtu or 0.5% sulfur in fuel.
CEMS required to meet SO$_2$ limit except fuel analysis can be used as fired condition before cleanup equipment.
Fuel sulfur limit compliance can be:
Daily as fired fuel analysis.
As delivered (before used) fuel analysis.
Fuel supplier certificate for units < 30 MMBtu.
Initial performance testing and 30 day rolling average required except for supplier certificate.

3. Distillate Oil Firing (ASTM grades 1 and 2)

Limit 0.5% sulfur in fuel (required in ASTM standard).
Compliance by fuel supplier certificate.
No monitoring or initial testing required.

RULES FOR PARTICULATE MATTER (PM) EMISSIONS

1. General

Limits established only for units between 30-100 MMBtu.
All coal, wood and residual oil fired units > 30 MMBtu must meet opacity limit of 20%, except one 6 minute/hour opacity of 27%. CEMS required to monitor opacity.

2. Coal Firing

0.05 lb/MMBtu limit if > 30 MMBtu and > 90% annual coal capacity.
0.10 lb/MMBtu limit if > 30 MMBtu and < 90% annual coal capacity.
20% opacity (CEMS) and initial performance tests on both PM limit and opacity.

3. Wood Firing

0.10 lb/MMBtu limit if > 30 MMBtu and > 30% annual wood capacity.
0.30 lb/MMBtu limit if > 30 MMBtu and < 30% annual wood capacity.
Opacity limits and initial testing per above.

4. Oil Firing

All units > 30 MMBtu subject to opacity limit, only residual oil firing must use CEMS.
Initial performance testing required.

REPORTING REQUIREMENTS

Owners or operators of all affected units must submit information to the administrator, even if they are not subject to any emission limits or testing. Required reports include:
Information on unit size, fuels, start-up dates and other equipment information.
Initial performance test results, CEMS performance evaluation.
Quarterly reports on SO$_2$ and/or PM emission results, including variations from limits and corrective action taken.
For fuel supplies certificate, information on supplies and details of sampling and testing for coal and residual oil.
Records must be maintained for two years.

(By permission of Cleaver Brooks, Milwaukee, Wisconsin)

3.6.0   Boiler Feedback Systems (Illustrated)

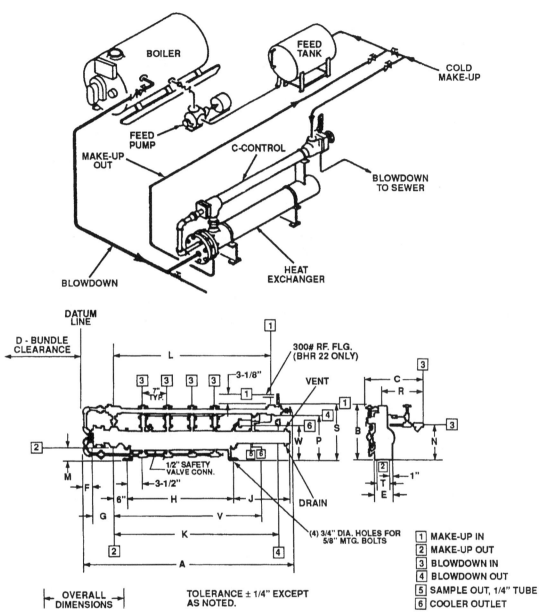

(*By permission of Cleaver Brooks, Milwaukee, Wisconsin*)

3.7.0   Typical Firetube Boiler Fuel Consumption for No. 2 and No. 6 Oil

Typical Firetube Boiler Fuel Consumption Rates - No. 6 Oil (gal/hr) [A]

AVERAGE OUTPUT	BOILER EFFICIENCY					
	86%	84%	82%	80%	78%	76%
BHP						
100	26	27	27	28	29	29
200	52	53	54	56	57	59
300	78	80	82	84	86	88
400	104	106	109	112	114	117
500	130	133	136	140	143	147
600	156	159	163	168	172	176
700	182	186	191	196	200	206
800	208	213	218	224	229	235
900	234	239	245	252	257	264
1000	260	266	272	280	286	294

A. Based on 150,000 Btu/gallon.

Typical Firetube Boiler Fuel Consumption Rates - No. 2 Oil (gal/hr) [A]

AVERAGE OUTPUT	BOILER EFFICIENCY					
	86%	84%	82%	80%	78%	76%
BHP						
100	28	28	29	30	31	31
200	56	57	58	60	61	63
300	83	85	87	90	92	94
400	111	114	117	120	123	126
500	139	142	146	149	153	157
600	167	171	175	179	184	189
700	195	199	204	209	215	220
800	222	228	233	239	245	252
900	250	256	262	269	276	283
1000	278	285	292	299	307	315

A. Based on 140,000 Btu/gallon.

(By permission of Cleaver Brooks, Milwaukee, Wisconsin)

3.8.0   Boiler Economizer Features and Schematic

Reduces Fuel Use and Cost:

- Recovers heat from flue gases that would otherwise be wasted.
- Heat is used to raise boiler feedwater temperature prior to entering the boiler.

Load Changes:

- Rapid changes in load demands can be met faster due to higher feedwater temperature.

Emissions:

- Reduced fuel-firing rates for any given steam output means reduced NOx emissions.

ASME Construction:

- Ensures high quality design and manufacturing standards.
- Provides safety and reliability.

High Efficiency Heat Exchanger:

- Provides continuous, high-frequency resistance welding.
- Provides uniform fin-to-tube contact for maximum heat transfer.
- Fin tubing offers up to 12 times the heat exchange surface of bare tubing of the same diameter.

Self-Drainng Design:

- Suitable for outdoor installation.

Low Pressure Drop:

- Provides gas side pressure drops of 0.8" WC or less.
- Permits use of smaller forced draft fans.
- Permits use of existing fans in almost all installations.

Gas Tight Combustion Stack:

- Provides inner casing of carbon steel.
- Provides outer casing of weather resistant, corrugated, galvanized carbon steel.
- Compact dimensions provide for easy installation.

Feedwater Preheating System:

- Controls cold end corrosion through all flow rates.
- Prevents the forming of corrosive acids in the economizer.
- Prevents stack corrosion.

(By permission of Cleaver Brooks, Milwaukee, Wisconsin)

3.8.0   Boiler Economizer Features and Schematic (Continued)

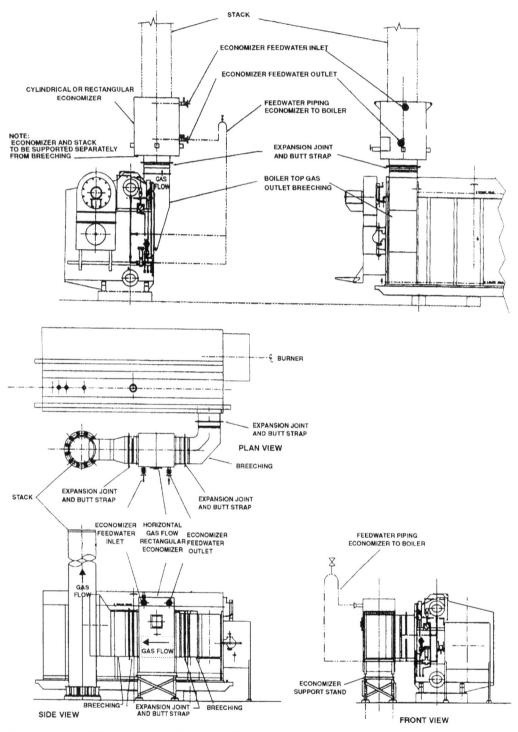

(By permission of Cleaver Brooks, Milwaukee, Wisconsin)

3.9.0 Boiler Stack Options

Stack Product Offering and Application Information

MODEL NO.	AMERI-VENT	CBS-I	CBS-II	CBS-III	ICBS
Description	Type "B" Gas Vent	Single Wall Stainless Steel	Double Wall Air Insulated	Triple Wall Air Insulated	Double wall either 1", 2" or 4" Material Insulated
Applications	AGA Listed Gas Appliances	Air/Product Containment Breeching Systems, Grease Duct	Boiler and Breeching Systems, Engine/Turbine Exhausts, Grease/Oven Exhausts, Air/Particle Containment		
Fuel Types	Natural or LP Gas	LP; Natural Gas; #2, #4[A], #5[A] or #6[A] Fuel Oil; Wood; Coal[A]; Grease Vapors; Caustic Fumes; Particles			
Exhaust Pressures	Neutral or Negative	Positive, Neutral or Negative			
Exhaust Temp. Continuous/Intermittent	400°F Plus Ambient	Air Product Containment or 2000°F for Grease Duct	100 °F Continuous, 1400 °F Intermittent 1400 °Continuous. 1800 °F Intermittent Grease Duct 500 °F Continuous		
Diameters	3" through 30"	6" through 48"			
Materials	Inner: .012" Aluminum 3" to 6" .014" Aluminum 7" to 18" .018" Aluminum 20" to 30"	Inner: Standard 304SS .035" All Diameters (Optional .035" 316SS Available)	Inner: .035" 20-ga 304SS Standard (Optional .035" 20-ga 316SS Available)	Inner: .035" 20-ga 304SS Standard (Optional .035" 20-ga 316SS Available)	Inner: .035" 20-ga 304SS Standard (Optional .035" 20-ga 316SS Available)
	Outer: 28-ga Galvanized Steel 3" to 30"	Outer: N/A	Outer: .025" 24-ga Aluminum Coated Steel 6" to 24" .034" 20-ga Aluminum Coated Steel 26" to 48" Optional 304 or 316SS Available)	Center: .025" 24-ga Aluminum Coated Steel 6" to 24" .034" 20-ga Aluminum Coated Steel 26" to 48" (Optional 304 or 316SS Available)	Insulation Material: Eleven Pound Fiber Insulation of 1", 2" or 4" thickness
				Outer: .025" 24-ga Aluminum Coated Steel 6" to 24" .034" 20-ga Aluminum Coated Steel 26" to 48" (Optional 304 or 316SS Available)	Outer: .025" 24-ga Aluminum Coated Steel 6" to 24" .034" 20-ga Aluminum Coated Steel 26" to 48" (Optional 304 or 316SS Available)
Insulation	1/4" Air 3" through 6" 1/2" Air 7" through 30"	N/A	1" Air	Innerwall to Center = 1" Air Center to Outerwall = 1/2" Air	1" Material 2" Material or 4" Material
Application Ref. & Listings	Complies with one or more of the following: AGA; HUD; NBC; UBC; UMC; NMC; SMC; SBCCI; ICBO; BOCA; UL-103, 710, 411; ULC-S604; NFPA-85, A, B, D, 31, 34, 37, 54, 96, 211.				

A. Recommended 316 Stainless Steel.

(By permission of Cleaver Brooks, Milwaukee, Wisconsin)

3.10.0   Typical Stack Construction

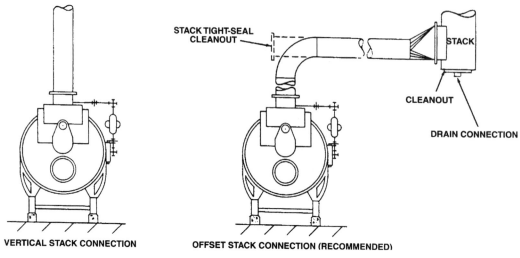

VERTICAL STACK CONNECTION   OFFSET STACK CONNECTION (RECOMMENDED)

Typical Stack Locations

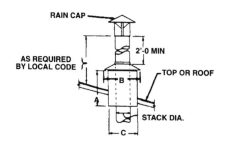

BOILER HP	STACK DIAMETER (IN.)	A (IN.)	B (IN.)	C (IN.)
15-20	6	15	15	12
25-40, 50A	8	20	20	16
50-60	10	25	25	20
70-100A, 125A	12	30	30	24
125-200	16	40	40	32
250-350	20	50	50	40
400-800	24	60	60	48

Typical Stack Construction

(By permission of Cleaver Brooks, Milwaukee, Wisconsin)

3.11.0   Stack Expansion/Contraction and Installation Concerns

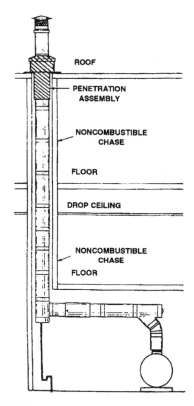

ROOF

PENETRATION ASSEMBLY

NONCOMBUSTIBLE CHASE

FLOOR

DROP CEILING

NONCOMBUSTIBLE CHASE

FLOOR

SYSTEMS EXTENDING THROUGH ANY STORY ABOVE THE BOILER ROOM REQUIRE A NONCOMBUSTIBLE CHASE ENCLOSURE FROM THE BOILER ROOM TO THE ROOF AND A PENETRATION ASSEMBLY OR ROOF SUPPORT ASSEMBLY AT THE ROOF LEVEL

DETAIL: OUTER WALL REMOVED FOR CLARITY

WALL SUPPORT ASSEMBLY

12.5 FT.

EXPANSION JOINT (SEE DETAIL)

INNER WALL

37.5 FT.

WALL GUIDE ASSEMBLY

1-1/2 IN.

COLLAR

1/2 IN.

FIXED POINT AT BOILER

THERMAL EXPANSION OCCURS BETWEEN ANY TWO FIXED POINTS IN THE STACK SYSTEM. NOTE: DISTANCES SHOWN ON ILLUSTRATION ARE FOR EXAMPLE ONLY.

When stack systems are exposed to the heating and cooling of normal operation, the components will expand and contract. The systems are designed to adjust to this movement, provided the amount and direction of expansion is accurately calculated, and the system is correctly installed.

The amount of thermal expansion that will occur depends on the length of breeching, height of the stack, temperature of the flue gas, and arrangement of the system. Therefore, the following must be considered.

Thermal Expansion

The CBS/ICBS systems use two different parts to compensate for thermal expansion between two fixed points in the system.

(By permission of Cleaver Brooks, Milwaukee, Wisconsin)

They are: expansion joints and bellows joints.

Each type of joint is:

- Designed to compensate for linear expansion only.

- Never used to correct for misalignment between components.

- Not load bearing. Therefore, these systems are usually between support or guide assemblies.

Expansion Direction

To determine expansion direction, it is necessary to understand how the expansion or bellows joint works. The expansion or bellows joint itself does not expand. In fact, the opposite happens. It compresses to absorb the movement of the parts expanding around it. This expansion movement occurs from fixed points in the system toward the expansion or bellows joint.

3.12.0 Schematic of a Typical Custom-Built HVAC Unit

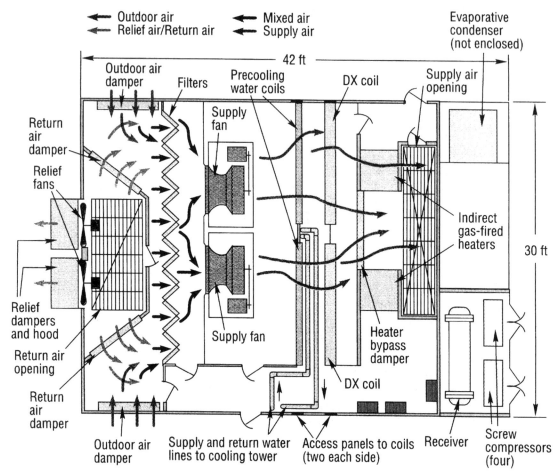

← Outdoor air   ← Mixed air
← Relief air/Return air   ← Supply air

Evaporative condenser (not enclosed)

Outdoor air damper   Filters   Precooling water coils   DX coil   Supply air opening

Return air damper

Relief fans

Supply fan

Relief dampers and hood

Return air opening

Return air damper

Supply fan

Indirect gas-fired heaters

Heater bypass damper

DX coil

42 ft

30 ft

Outdoor air damper   Supply and return water lines to cooling tower   Access panels to coils (two each side)   Receiver   Screw compressors (four)

(Reprinted by permission from Heating/Piping/Air Conditioning *magazine, December 1996)*

3.13.0   Schematic of Indirect Evaporative Precooling System

Components include:

1. Stand along cooling tower

2. Water pump and piping

3. Water cooling coils

4. Centrifugal separator

5. Chemical treatment system

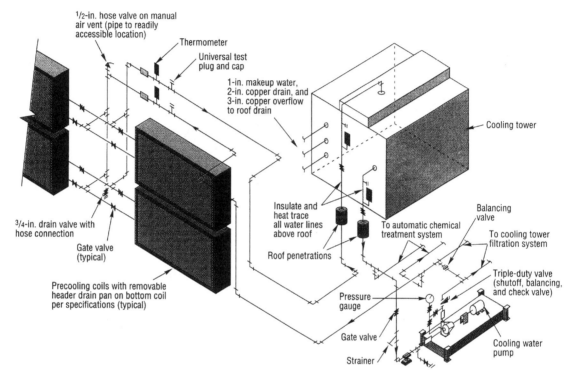

(Reprinted by permission from Heating/Piping/Air Conditioning magazine, December 1996)

3.14.0   Heat-Pump Operation Schematics

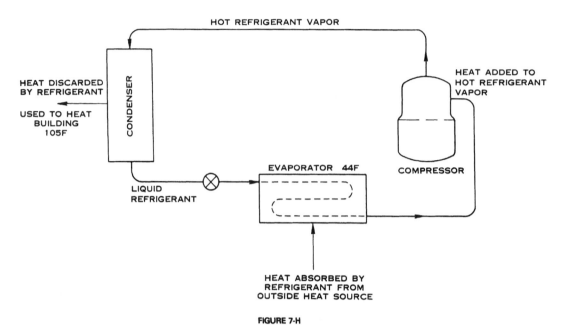

FIGURE 7-H

HEAT PUMP ON HEATING CYCLE

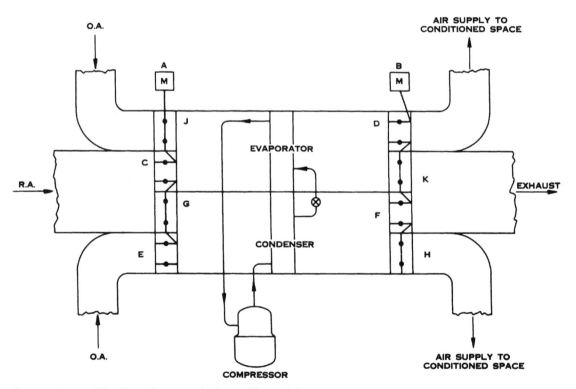

(By permission of The Trane Company, LaCrosse, Wisconsin)

3.15.0   Air-Cooled Condenser and Subcooling System (Illustrated)

System Piping Suggestions

If an air conditioning system with an air-cooled condenser will operate only when the outdoor temperature is above 40 F, a simple fan cycling or multilouvered damper control is usually adequate. The shutter control will follow the system load variations closely enough so there should be neither head pressure nor starting problems. The system piping can be simple, as illustrated in the piping diagram.

As will be noted, this system does not employ the conventional liquid receiver. The air condenser has sufficient volume to hold the charge on a system where the components are reasonably close together. Since the accumulator between the condensing circuit and the subcooler of the air condenser can handle a small variation in liquid volume, this would not be considered a critically charged system.

Compressor Unit Piping

Evaporator Piping — Direct Expansion Coil

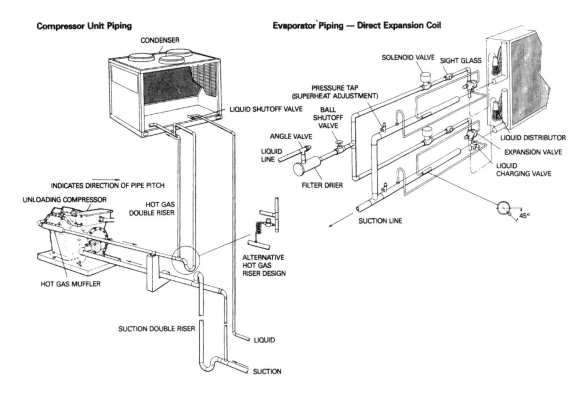

TYPICAL PIPING ARRANGEMENT OF SYSTEM WITH AIR-COOLED CONDENSER AND SUBCOOLING. NO HEAD PRESSURE CONTROL, OR HEAD PRESSURE CONTROL MAY BE WITH SHUTTERS.

(By permission of The Trane Company, LaCrosse, Wisconsin)

3.16.0   Variable Air Volume (VAV) Systems Diagrammed

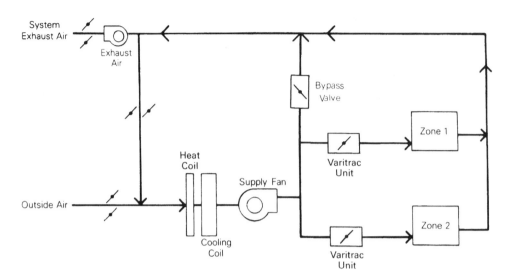

CHANGEOVER-BYPASS VARIABLE AIR VOLUME SYSTEM

The changeover-bypass variable air volume system offers perhaps the least expensive temperature control for a large number of zones when compared to other variable air volume systems. It is a flexible system in that it is relatively easy and inexpensive to subdivide a building into additional new zones should it become necessary after the initial building and system design have been completed. Operating and first cost savings are both possible through the ability to use building load diversity to not only reduce the installed system equipment size but also to reduce its energy use through more efficient operation of smaller pieces of equipment at part-load conditions.

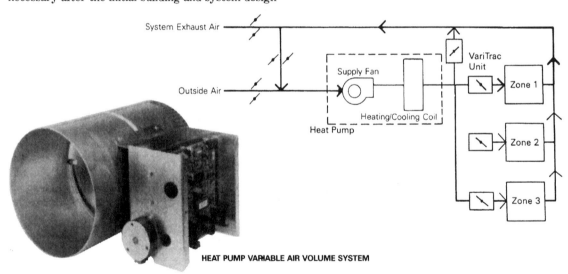

HEAT PUMP VARIABLE AIR VOLUME SYSTEM

(By permission of The Trane Company, LaCrosse, Wisconsin)

3.16.1 Variable Air Volume (VAV) Diagrams Showing Radiation Heating, Reheat and Fan-Powered Systems

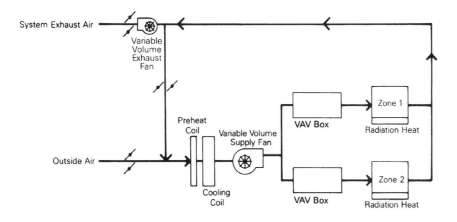

VARIABLE AIR VOLUME COOLING WITH PERIMETER RADIATION HEATING

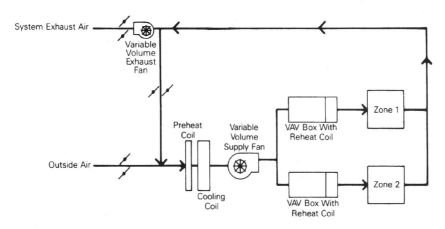

VARIABLE AIR VOLUME REHEAT SYSTEM

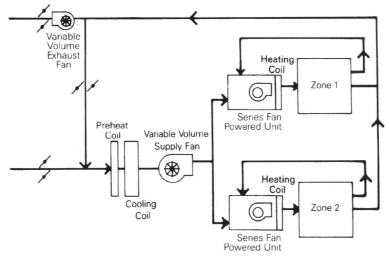

SERIES FAN POWERED VARIABLE AIR VOLUME SYSTEM

(By permission of The Trane Company, LaCrosse, Wisconsin)

VAV Troubleshooting guidelines

Diffuser Dumps Cold A
Airflow too low (velocity too slow).
Check to determine if box is reducing too far.
Evaluate box minimum setting.
Diffuser is too large; check installation.

Conditioned Space Is Too Cold
Supply air temperature is too cold.
Too much supply air.
Diffuser pattern or throw is incorrect causing drafts.
Temperature sensor is located incorrectly or needs calibration.

Conditioned Space Is Too Warm.
Supply air temperature setting is too warm.
Not enough supply air.
Refrigeration system not operating properly.
Fan-coil evaporator is iced over because of low airflow.
Temperature sensor is located incorrectly or needs calibration.
Low pressure duct leaking.
Low pressure duct not insulated.
Cold air from diffuser isn't mixing properly with room air; increase volume or velocity, change or retrofit diffuser.

Noise.
Too much air in low pressure duct; check box maximum setting.
Static pressure in the system is too high.
Diffuser is too small.
Diffuser is dampered at face; always damper at takeoff.
Pattern controllers loose; tighten or remove.

Not Enough Air
Box not operating properly; check minimum setting; reset as necessary.
Not enough static pressure at box inlet for proper operation.
Damper in VAV box is closed; may be loose on shaft or frozen.
Low pressure damper closed.
Restrictions in low pressure duct.
Remove pattern controllers in diffusers.
Low pressure duct is leaking, disconnected or twisted.
Install VAV diffusers where applicable; these diffusers close down at the face

producing higher air velocities as they reduce in size.
Install fan powered boxes.

Box Not Operating Properly
Not enough static pressure at the inlet.
Too much static pressure at the inlet.
Static pressure sensor is defective, clogged, or located incorrectly.
Static pressure setting on controller is incorrect.
Static pressure controller needs calibrating.
Fan speed not correct.
Inlet duct leaking or disconnected.
Main duct work improperly designed.
Not enough straight duct on the inlet of the box.
Diversity is incorrect.
Box is wrong size or wrong nameplate.
Check installation. Leak test if necessary.
Damper loose on shaft.
Linkage from actuator to damper is incorrect or binding.
Actuator is defective.
Controls are defective, need calibration or are set incorrectly.
Volume controller not set properly for normally open or normally closed operation.
Damper linked incorrectly: NO for NC operation or vice versa. Pneumatic tubing to controller is piped incorrect, leaking, or pinched. Restrictor in pneumatic tubing is missing, broken, placed incorrectly, wrong size, or clogged.
Oil or water in pneumatic lines.
No power to controls.
Wired incorrectly.
PC board defective.

Fan Not Operating Properly
Inlet vanes or centrifugal fans not operating properly.
Pitch on vaneaxial fans not adjusted correctly.
Fan rotating backwards.
Return air fan not tracking with supply fan. Check static pressure sensors, airflow measuring stations and move, clean or calibrate. Consider replacing return fan with a relief fan.

Negative Pressure in The Building
(Office buildings should be maintained at +0.03 to +0.05 inches of water).
1. Caused by stack effect or improper return air control. Seal building properly. Balance return system, and install manual balancing dampers as needed to control OA, RA and EA at the unit. Get return fan to track with supply fan; consider replacing return fan with relief fan.
Check that static pressure sensors are properly located and working.
Install pressure controlled return air dampers in return air shafts from ceiling plenums.
2. Caused by the supply fan reducing air volume and a reduction of outside air for (1) the constant volume exhaust fans and (2) exfiltration.
Increase minimum outside air by:
a. Operating manual volume damper in outside air duct and/or increasing outside air duct size.
b. Control OA from supply fan. As fan slows, outside air damper opens.
c. Control OA damper from flow monitor in OA duct to maintain a constant minimum OA volume.

Inadequate Amount of Outside Air for Proper Ventilation
Caused by the supply fan reducing air volume and a reduction of outside air for code or building requirements.
Increase minimum outside air by:
1. Opening manual volume damper in outside air duct and/or increasing outside air duct size.
2. Control OA from supply fan. As fan slows, outside air damper opens.
3. Control OA damper from flow monitor in OA duct to maintain a constant minimum OA volume.

Figure 3.16.2 VAV troubleshooting tips. (*By permission: Airflow Technical Products, Inc., Netcong, NJ.*)

COMMON TERMS

Absolute Pressure: The total of the indicated gage pressure plus the atmospheric pressure. Abbreviated "psia" for pounds per square inch absolute.

Atmospheric Pressure: The pressure exerted upon the earth's surface by the air because of the gravitational attraction of the earth. Standard atmosphere pressure at sea level is 14.7 pounds per square inch (psi). Measured with a barometer.

Barometer: An instrument for measuring atmospheric pressure.

Calibration: Determining or correcting the error of an existing scale.

CEFAPP: Close enough for all practical purposes.

Differential Pressure Gage: An instrument that reads the difference between two pressures directly and therefore, eliminates the need to take two separate pressures and then calculate the difference.

Electronic Instruments: Most of the mechanical analog instruments now have electronic digital counterparts. All instruments, analog or digital, should be checked against a sheltered set before each balancing project. Pressure measuring instruments should be checked against a standard liquid-filled manometer.

Gage: An instrument for measuring pressure.

Gage Pressure: The pressure that's indicated on the gage.

Harmonics: For a strobe light tachometer, harmonics are frequencies of light flashed that are a multiple or submultiple of the actual rotating speed. For example, if the light frequency is either exactly two times or exactly one-half the actual speed of the rotating equipment, the part will appear stationary but the image won't be as sharp as when the rpm is correct.

Manometer: An instrument for measuring pressures. Essentially a U-tube partly filled with a liquid, usually water, mercury or a light oil. The pressure exerted on the liquid is indicated by the liquid displaced. A manometer can be used as a differential pressure gage.

Meniscus: The curved surface of the liquid column in a manometer. In manometers that measure air pressures, the liquid is either water or a light oil. In a manometers that measure water pressures, the liquid is mercury.

Operating Load Point: Actual system operating capacity when an instrument reading is taken.

Parallax: A false reading that happens when the eye of the reader isn't exactly perpendicular to the lines on the instrument scale.

Pitot Tube: A sensing device used to measure total pressures in a fluid stream. It was invented by a French physicist, Henri Pitot, in the 1700's.

Sheltered Set: A sheltered set of instruments is a group of instruments used only to check the calibration of field instruments.

Sensitivity: A measure of the smallest incremental change to which an instrument can respond.

AIR INSTRUMENTS

Air Differential Pressure Gage: Magnehelic "R" and Capsuhelic "R" are two brands of air differential gages. These gages contain no liquid, but instead work on a diaphragm and pointer system which move within a certain pressure range. Although absolutely level mounting isn't necessary, if the position of the gage is changed, resetting of the zero adjustment may be required for proper gage reading as specified by the manufacturer. These gages have two sets of tubing connector ports for different permanent mounting positions; however, only one set of ports is used for readings and the other set is capped off. The ports are stamped on the gages as "high" pressure and "low" pressure and when used with a Pitot tube, static tip or other sensing device to measure total pressure, static pressure, or velocity pressure, the tubing is connected from the sensing device to the tubing connector ports in the same manner as the connection to the inclined-vertical manometer. These gages are generally used more for static pressure readings than for velocity pressures readings. They should be checked frequently against a manometer.

Anemometer: An instrument used to measure air velocities.

Capture Hood: An instrument which captures the air of a supply, return or exhaust terminal and guides it over a flow-measuring device. It measures airflow directly in cubic feet per minute. Calibration by the manufacturer should be done every 6 months, especially if the instrument isn't checked periodically against a sheltered instrument.

Compound Gages: Compound gages measure pressures both above and below atmospheric. They read in pounds per square inch above atmospheric and inches of mercury below atmospheric. Compound gages are calibrated to read zero at atmospheric pressure.

Cubic Feet Per Minute (CFM): A unit of measurement. The volume or rate of airflow.

Deflecting Vane Anemometer: The deflecting vane anemometer gives instantaneous, direct readings in feet per minute. It's used most often for determining air velocity through supply, return and exhaust air grilles, registers or diffusers. It may also have attachments for measuring low velocities in an open space or at the face of a fume hood. With other attachments, Pitot traverses and static pressures can be taken. To use this instrument refer to the manufacturer's recommendation for usage, proper attachment selection and sensor placement. A correction (Ak) factor is also needed when measuring grilles, registers or diffusers. Calibration by the manufacturer should be done every 6 months, especially if the instrument isn't checked periodically against a sheltered instrument.

Feet per Minute (FPM): A unit of measurement. The velocity of the air.

Hot Wire Anemometer: This instrument measures instantaneous air velocity in feet per minute using an electrically heated wire. As air passes over the wire, the wire's resistance is changed and this change is shown as velocity on the instrument's scale. This instrument is very position sensitive when used to measure air velocities. Therefore, it's important to ensure that the probe is held at right angles to the airflow. The hot wire anemometer is most often used to measure low velocities such as found at the face of fume hoods; however, some instruments can also measure temperatures and static pressures. Calibration by the manufacturer should be done every 6 months, especially if the instrument isn't checked periodically against a sheltered instrument.

Inches of Water Gage or Column (In. WG or In. WC): A unit of air pressure measurement equal to the pressure exerted by a column of water 1 inch high.

Inclined Manometer, Inclined-Vertical Manometer: The inclined manometer has an inclined scale which reads in inches of water gage in various ranges such as 0 to 0.25 in.wg, 0 to 0.50 in.wg, or 0 to 1.0 in.wg. The inclined-vertical manometer has both an inclined scale that reads 0 - 1.0 in.wg and a vertical scale for reading greater pressures such as 1.0 to 10 in.wg.
The inclined and inclined-vertical manometer have a left and right tube connection for attaching tubing from a Pitot tube or other sensing device to the manometer.

Figure 3.16.2.1 Air balancing glossary of terms. (*By permission: Airflow Technical Products, Inc., Netcong, NJ.*)

3.17.0   Single- and Two-Pipe Cooling System Diagrams

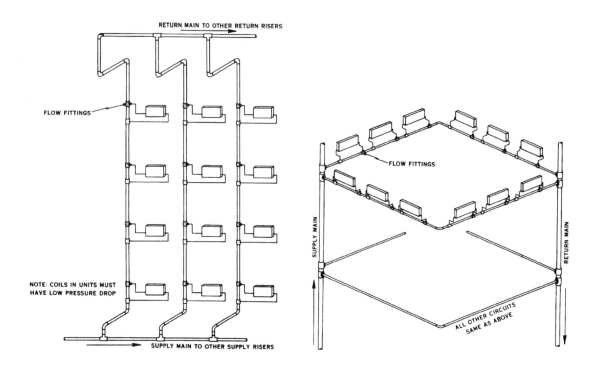

SINGLE-SUPPLY, SINGLE-RETURN RISER

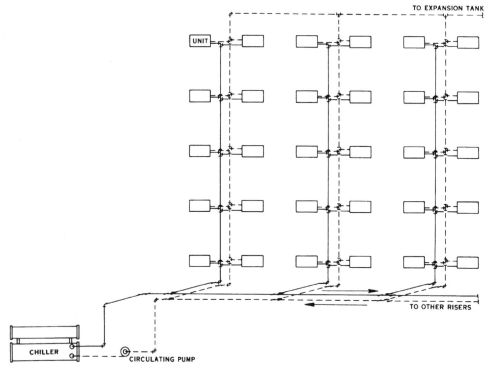

TWO-PIPE DIRECT RETURN MAINS AND RISERS

(By permission of The Trane Company, LaCrosse, Wisconsin)

3.18.0 Two-Pipe Reverse Main and Three-Pipe Heating/Cooling Piping Diagrams

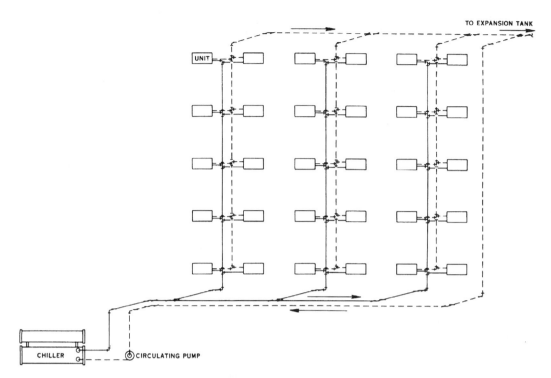

TWO-PIPE REVERSE RETURN MAINS AND RISERS

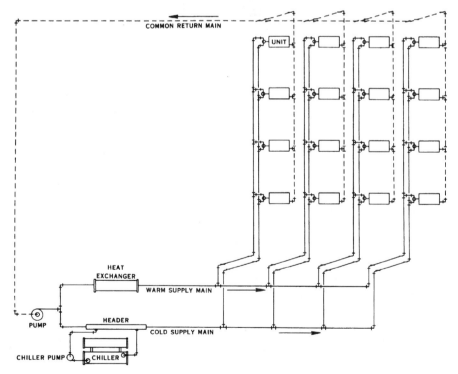

SIMPLE THREE-PIPE WATER DISTRIBUTION

(By permission of The Trane Company, LaCrosse, Wisconsin)

3.19.0   Four-Pipe Systems with One- and Two-Coil Piping Diagrams

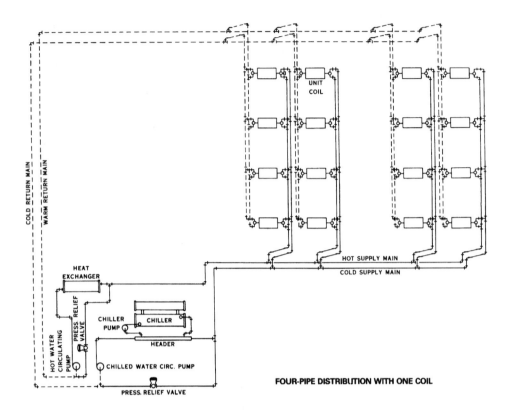

FOUR-PIPE DISTRIBUTION WITH ONE COIL

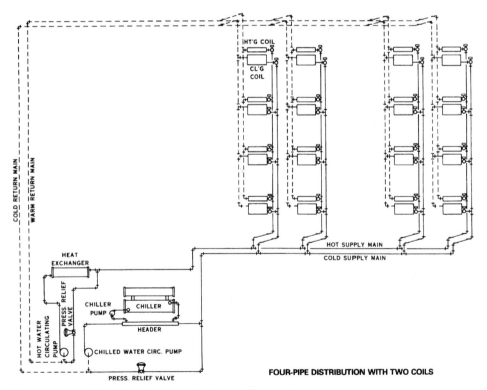

FOUR-PIPE DISTRIBUTION WITH TWO COILS

(By permission of The Trane Company, LaCrosse, Wisconsin)

3.20.0   Shell and Coil, and Shell and Tube Condensers (Illustrated and Described)

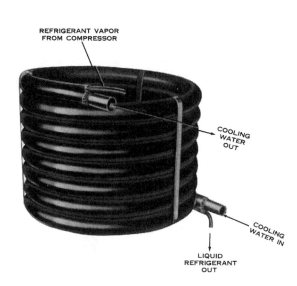

DOUBLE-TUBE CONDENSER

Shell-and-Coil Condenser

A shell-and-coil condenser is simply a continuous copper coil mounted inside a steel shell. Water flows through the coil and refrigerant vapor from the compressor is discharged inside the shell to condense on the outside of the cold tubes. In many designs, the shell also serves as a liquid receiver.

The shell-and-coil condenser has a low manufacturing cost but this is offset by the disadvantage that this type condenser is difficult to service in the field. If a leak develops in the coil, the head from the shell must be removed and the entire coil pulled from the shell in order to find and repair the leak. A continuous coil is a nuisance to clean whereas straight tubes are easy to clean with mechanical tube cleaners. Thus, with some types of fouled cooling water, it may be difficult to maintain a high rate of heat transfer with a shell-and-coil condenser.

Shell-and-Tube Condenser

The shell-and-tube condenser, permits a large amount of condensing surface to be installed in a comparatively small space. The condenser consists of a large number of ¾ or ⅝-inch tubes installed inside a steel shell. The water flows inside the tubes while the vapor flows outside, around the nest of tubes. The vapor condenses on the outside surface of the tubes and drips to the bottom of the condenser, which may be used as a receiver for the storage of liquid refrigerant. Shell-and-tube condensers are used for practically all water-cooled refrigeration systems.

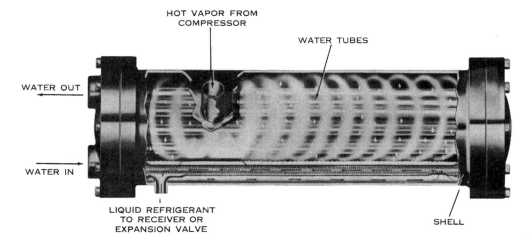

SHELL-AND-TUBE CONDENSER

(By permission of The Trane Company, LaCrosse, Wisconsin)

3.21.0   Shell and Tube Evaporator (Diagram and Description)

Shell-and-Tube Evaporators

There are two common types of shell-and-tube evaporators used to provide chilled water for air conditioning systems. There are the same two types as discussed previously with fin and tube coil evaporators; the **flooded** type and the **direct expansion** (dry) type. In the flooded type, the shell contains a tube bundle through which water to be chilled is pumped. Half to three-fourths of the tube bundle is immersed in liquid refrigerant, which boils because of the heat received from the water being cooled.

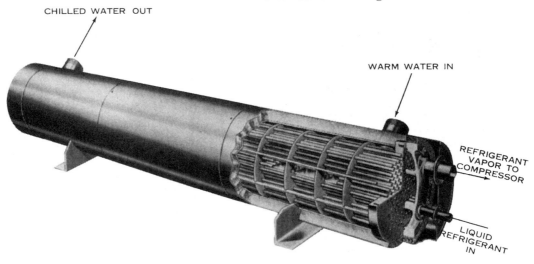

CHILLED WATER OUT

WARM WATER IN

REFRIGERANT VAPOR TO COMPRESSOR

LIQUID REFRIGERANT IN

A DIRECT EXPANSION SHELL-AND-TUBE EVAPORATOR

(By permission of The Trane Company, LaCrosse, Wisconsin)

3.22.0   Evaporative Condenser (Diagram and Description)

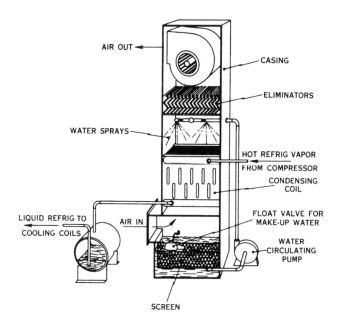

EVAPORATIVE CONDENSER

The evaporative condenser is a form of water-cooled condenser that offers a means of conserving water by combining the condenser and the cooling tower into one piece of equipment.

(By permission of The Trane Company, LaCrosse, Wisconsin)

3.23.0   Heating with a Chiller (Diagram and Description)

Freezing water to ice in the ice storage system removes 144 Btu per pound of ice generated by the chiller. This is why the ice storage system designed with a heat-recovery loop can also make the chiller into a water-source heat pump for cold-weather heating.

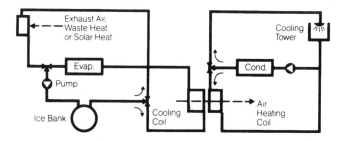

ICE STORAGE SYSTEM SCHEMATIC WITH HEAT RECOVERY LOOP

(By permission of The Trane Company, LaCrosse, Wisconsin)

3.24.0   Typical Evaporative Cooler (Diagram and Description)

A typical evaporative cooler is a metal housing with three sides containing porous material kept saturated with water. A pump lifts water from the sump in the bottom of the unit and delivers it to perforated troughs at the top of the unit. The fan draws outside air through the saturated material and discharges it directly into the conditioned space or into a duct system for distribution into several rooms. The porous material is generally spun-glass fibers, aspen excelsior pads, or tinsel made of copper or aluminum. The discharge line from the pump is usually plastic tubing, although copper tubing or iron pipe is sometimes used. A float valve is normally provided to replenish the water evaporated into the air passing throught the unit.

Generally, this valve is set to waste a fixed amount of water at all times. This ensures there is a continual dilution of the natural minerals in the water that are left behind due to evaporation. This is commonly called "blowdown" and provides protection against a sticking float valve.

Variations in this design are offered by several manufacturers for applications, primarily in dry climates with a low design wet-bulb temperature.

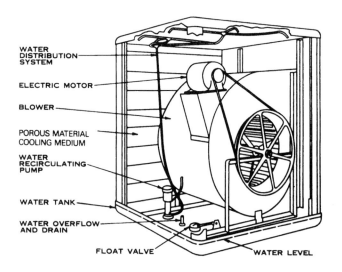

TYPICAL EVAPORATIVE COOLER

(By permission of The Trane Company, LaCrosse, Wisconsin)

3.25.0   Typical Flow Diagram for an Ice Storage System

It is important to note that while making ice at night, the chiller must cool the water–glycol solution down to 26°F, rather than producing 44°F water required for conventional air-conditioning systems.

This has the net effect of "derating" the nominal chiller capacity by a substantial amount (typically 25–30 percent). The compressor efficiency at this time is only slightly reduced because the lower nighttime outdoor ambient wet bulb temperatures result in cooler condenser water from the cooling tower, which lowers the condensing temperature to keep the chiller operating efficiently. Similarly, chillers with air-cooled condensing also benefit from cooler outdoor ambient dry-bulb temperatures to lower the system condensing temperature at night.

The temperature modulating valve in the bypass loop has the added advantage of providing excellent capacity control. During mild temperature days, typically in the spring and fall, the chiller is often capable of providing all the necessary cooling capacity for the building without the use of cooling capacity from the ice storage system. When the building's actual cooling load is equal to or less than the chiller capacity at the time, all of the system coolant flows through the bypass loop.

It is important that the coolant chosen be an ethylene glycol–based industrial coolant, such as Dowthern SR-1 or UCAR Thermofluid 17, which is specially formulated for low viscosity and good heat-transfer properties. Either of these fluids contain a multicomponent corrosion inhibitor, which

3.25.0 Typical Flow Diagram for an Ice Storage System (Continued)

is effective with most materials of construction including aluminum, copper, silver solder, and plastics. Further, they contain no antileak agents and produce no films to interfere with heat transfer efficiency. They also permit use of standard pumps, seals, and air handling coils. It should be noted, however, that because of the slight difference in heat-transfer properties between water and the mild glycol solution, the cooling coil capacities must be increased by approximately 5 percent. It is also important that the water and glycol solution be thoroughly mixed before the solution is placed into the system.

The use of ice storage system technology opens new doors to other economic opportunities in system design. These offer significant potential for not only first-cost savings, but also operating cost savings that should be evaluated on a life-cycle cost basis using a computerized economic analysis program such as the Trane Air Conditioning Economics (TRACE®) program.

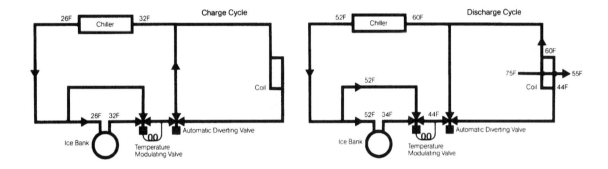

TYPICAL FLOW DIAGRAMS FOR A PARTIAL ICE STORAGE SYSTEM

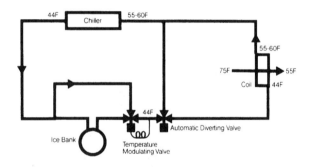

TYPICAL FLOW DIAGRAM FOR A PARTIAL ICE STORAGE SYSTEM — WITH ALL COOLANT THROUGH THE BYPASS LOOP

(By permission of The Trane Company, LaCrosse, Wisconsin)

Why Humidification is Important

Relative Humidity(RH):
The ratio of the vapor pressure (or mole fraction) of water vapor in the air to the vapor pressure (or mole fraction) of saturated air at the same dry-bulb temperature and pressure.

Sensible Heat:
Heat that when added to or taken away from a substance causes a change in temperature or, in other words, is "sensed" by a thermometer. Measured in Btu.

Latent Heat:
Heat that when added to or taken away from a substance causes or accompanies a phase change for that substance. This heat does not register on a thermometer, hence its name "latent" or hidden. Measured in Btu.

Dew Point:
The temperature at which condensation occurs (100%RH) when air is cooled at a constant pressure without adding or taking away water vapor.

Evaporative Cooling:
A process in which liquid water is evaporated into air. The liquid absorbs the heat necessary for the evaporation process from the air, thus, there is a reduction in air temperature and an increase in the actual water vapor content of the air.

Enthalpy:
Also called heat content, this is the sum of the internal energy and the product of the volume times the pressure. Measured in Btu/lb.

Hygroscopic Materials:
Materials capable of absorbing or giving up moisture.

Phase:
The states of existence for a substance, solid, liquid, or gas (vapor).

Humidification is simply the addition of water to air. However, humidity exerts a powerful influence on environmental and physiological factors. Improper humidity levels (either too high or too low) can cause discomfort for people, and can damage many kinds of equipment and materials. Conversely, the proper type of humidification equipment and controls can help you achieve effective, economical, and trouble-free control of humidity.

As we consider the importance of humidity among other environmental factors— temperature, cleanliness, air movement, and thermal radiation—it is important to remember that humidity is perhaps the least evident to human perception. Most of us will recognize and react more quickly to temperature changes, odors or heavy dust in the air, drafts, or radiant heat. Since relative humidity interrelates with these variables, it becomes a vital ingredient in total environmental control.

Humidity and Temperature

Humidity is water vapor or moisture content always present in the air. Humidity is definable as an absolute measure: the amount of water vapor in a unit of air. But this measure of humidity does not indicate how dry or damp the air is. This can only be done by computing the ratio of the actual partial vapor pressure to the saturated partial vapor pressure at the same temperature. This is relative humidity, expressed by the formula:

$$RH = \dfrac{vp_a}{vp_s} \bigg|_t$$

vp_a = actual vapor pressure
vp_s = vapor pressure at saturation
t = dry-bulb temperature

For practical purposes, at temperatures and pressures normally encountered in building systems, relative humidity is considered as the amount of water vapor in the air compared to the amount the air can hold at a given temperature.

"At a given temperature" is the key to understanding relative humidity. Warm air has the capacity to hold more moisture than cold air. For example, 10,000 cubic feet of 70°F air can hold 80,550 grains of moisture. The same 10,000 cubic feet of air at 10°F can hold only 7,760 grains of moisture.

70°F
80,550 Grains

10°F
7,760 Grains

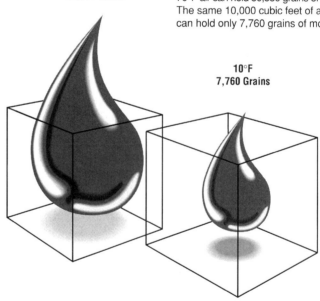

If the 10,000 cubic feet of 10°F air held 5,820 grains of moisture, its relative humidity would be 75%. If your heating system raises the temperature of this air to 70°F with no moisture added, it will still contain 5,820 grains of moisture. However, at 70°F, 10,000 cubic feet of air can hold 80,550 grains of moisture. So the 5,820 grains it actually holds give it a relative humidity of slightly more than 7%. That's very dry...drier than the Sahara Desert.

Figure 3.26.0 Why humidification is important. (*By permission: Associated Steam Specialty Company, Drexel Hill, PA.*)

How Humidifiers Work

Steam Humidification

Unlike other humidification methods, steam humidifiers have a minimal effect on dry-bulb (DB) temperatures. The steam humidifier discharges ready-made water vapor. This water vapor does not require any additional heat as it mixes with the air and increases relative humidity. Steam is pure water vapor existing at 212°F (100°C). This high temperature creates a perception that steam, when discharged into the air, will actually increase air temperature. This is a common misconception. In truth, as the humidifier discharges steam into the air, a steam/air mixture is established. In this mixture steam temperature will rapidly decrease to essentially the air temperature.

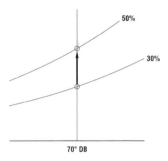

The psychrometric chart helps illustrate that steam humidification is a constant DB process. Starting from a point on any DB temperature line, steam humidification will cause movement straight up along the constant DB line. The example illustrates that 70°F DB is constant as we increase RH from 30%-50%. This is true because steam contains the necessary heat (enthalpy) to add moisture without increasing or decreasing DB temperature. Actual results utilizing high pressure steam or large RH increases (more than 50%) increase DB by 1° to 2°F. As a result, no additional heating or air conditioning load occurs.

Direct Steam Injection Humidifiers

The most common form of steam humidifier is the direct steam injection type. From a maintenance point of view, direct steam humidification systems require very little upkeep. The steam supply itself acts as a cleaning agent to keep system components free of mineral deposits that can clog water spray and evaporative pan systems.

Response to control and pinpoint control of output are two other advantages of the direct steam humidification method. Since steam is ready-made water vapor, it needs only to be mixed with air to satisfy the demands of the system. In addition, direct steam humidifiers can meter output by means of a modulating control valve. As the system responds to control, it can position the valve anywhere from closed to fully open. As a result, direct steam humidifiers can respond more quickly and precisely to fluctuating demand.

The high temperatures inherent in steam humidification make it virtually a sterile medium. Assuming boiler makeup water is of satisfactory quality and there is no condensation, dripping or spitting in the ducts, no bacteria or odors will be disseminated with steam humidification.

Corrosion is rarely a concern with a properly installed steam system. Scale and sediment—whether formed in the unit or entrained in the supply steam—are drained from the humidifier through the steam trap.

Steam-to-Steam Humidifiers

Steam-to-steam humidifiers use a heat exchanger and the heat of treated steam to create a secondary steam for humidification from untreated water. The secondary steam is typically at atmospheric pressure, placing increased importance on equipment location.

Maintenance of steam-to-steam humidifiers is dependent on water quality. Impurities such as calcium, magnesium and iron can deposit as scale, requiring frequent cleaning. Response to control is slower than with direct steam because of the time required to boil the water.

Electronic Steam Humidifiers (Electrode)

Electronic steam humidifiers are used when a source of steam is not available. Electricity and water create steam at atmospheric pressure. Electrode-type units pass electrical current through water to provide proportional output. Use with pure demineralized, deionized or distilled water *alone* will generally not provide sufficient conductivity for electrode units.

Water quality affects the operation and maintenance of electrode-type humidifiers. Use with hard water requires more frequent cleaning, and pure softened water can shorten electrode life. Microprocessor-based diagnostics assist with troubleshooting.

Direct Steam Humidification

Steam-to-Steam Humidification

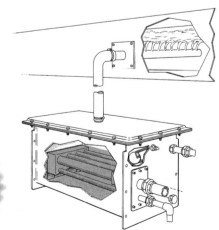

Figure 3.26.1 How humidifiers work. (*By permission: Associated Steam Specialty Company, Drexel Hill, PA.*)

Electrode units are easily adaptable to different control signals and offer full modulated output. However, the need to boil the water means control will not compare with direct-injection units.

Electronic Steam Humidifiers (Ionic Bed)

Ionic bed electronic humidifiers typically use immersed resistance heating elements to boil water. Since current does not pass through water, conductivity is not a concern. Therefore, ionic bed technology makes the humidifier versatile enough to accommodate various water qualities. These units work by using ionic bed cartridges containing a fibrous media to attract solids from water as its temperature rises, minimizing the buildup of solids inside the humidifier. Water quality does not affect operation, and maintenance typically consists of simply replacing the cartridges.

Ionic bed humidifiers are adaptable to different control signals and offer full modulated output. Control is affected by the need to boil the water, however.

Water Spray

The water spray process can create potential temperature control problems. In order to become water vapor or humidity, water requires approximately 1,000 Btu per pound to vaporize. This heat must be drawn from the air, where it will hopefully vaporize. If not enough heat is available quickly enough, the water remains a liquid. This unvaporized water can result in overhumidification, and the water can "plate out" on surfaces, creating a sanitation hazard.

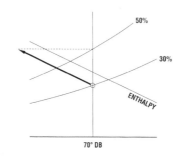

Water spray contains virtually none of the heat of vaporization required to increase the RH of the air to desired conditions. For this reason, water spray humidification is a virtually constant enthalpy process. However, as the psychrometric example illustrates, DB temperature changes as we increase RH from 30%-50%. The result of this loss of DB temperature is an increased heating load to maintain 70°F.

Response of water spray humidifiers to control is slow due to the need for evaporation to take place before humidified air can be circulated. On/off control of output means imprecise response to system demand and continual danger of saturation. Water spray systems can distribute large amounts of bacteria, and unevaporated water discharge can collect in ducts, around drains and drip pans, and on eliminator plates, encouraging the growth of algae and bacteria. Corrosion is another ongoing problem with water spray humidification. Scale and sediment can collect on nozzles, ductwork, eliminator plates, etc., leading to corrosion and high maintenance costs.

Evaporative Pan

The evaporative pan method uses steam, hot water or electricity to provide energy for heating coils which in turn heat water and create water vapor. This method is most effective when installed in smaller capacity environments either in the air handling system or individually within the area(s) to be humidified.

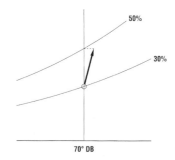

Evaporative pan humidification can increase dry-bulb temperature as measured on the psychrometric chart. This unwanted temperature change may occur as air is forced across the warmed water in the pan. The increase in DB can cause damaging results in process applications and increase the need for humidity control. The psychrometric chart helps illustrate that evaporative pan humidification is not a constant DB process. This example shows DB temperature increasing as we move from 30%-50% RH. To maintain a constant DB of 70°F some cooling load (air conditioning) is required.

Electronic Steam Humidification
Figure 15-1.

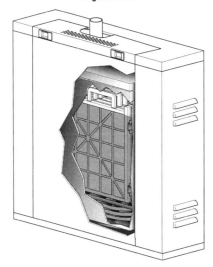

Water Spray
Figure 15-2.

Figure 3.26.1 *(Continued)*

3.26.2   Types of Humidifiers (Illustrated and Described)

Hot Element Humidifier

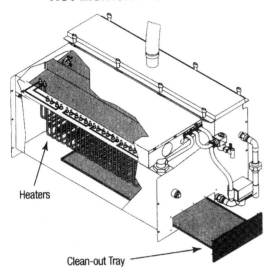

Resistance heating elements are submerged in an evaporating chamber full of water.

Atomizing Humidifier

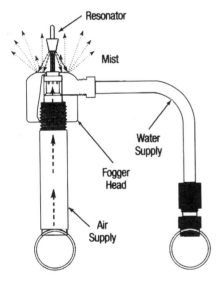

Compressed air is used to provide an ultrasonic shock wave to atomize the water into a mist, which is absorbed into the airstream.

Steam-to-Steam Humidifier

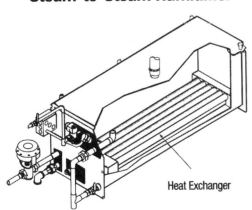

Humidification steam output is modulated to match load conditions by the use of a valve, controlling the flow to the heat exchanger.

(Reprinted by permission from Heating/Piping/Air Conditioning *magazine, December 1996)*

3.26.2   Types of Humidifiers (Illustrated and Described) (Continued)

Replaceable Cylinder Humidifier

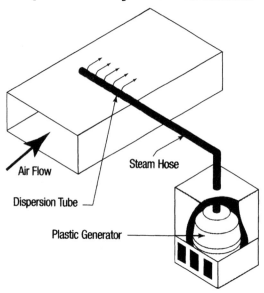

Consists of a replaceable plastic evaporator mounted in a cabinet along with a fill and drain, valve, and an electronic water-level detector. Electronic plates inside the plastic evaporator allow electric current to pass through the water, causing it to boil and create steam.

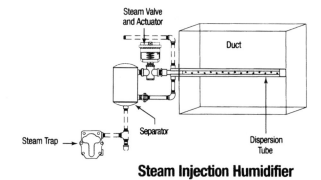

Direct steam injection is used in buildings with a boiler where the introduction of boiler chemicals into the air is not objectionable.

Steam Injection Humidifier

(Reprinted by permission from Heating/Piping/Air Conditioning *magazine, December 1996)*

Steam: Recommended for virtually all commercial, institutional and industrial applications. Where steam is not available, small capacity needs up to 50-75 lbs/hr can be met best using ionic bed type self-contained steam generating units. Above this capacity range, central system steam humidifiers are most effective and economical. Steam should be specified with caution where humidification is used in small, confined areas to add large amounts of moisture to hygroscopic materials. We recommend that you consult your Armstrong Representative regarding applications where these conditions exist.

Evaporative Pan: Recommended only as an alternative to self-contained steam generating unit humidifiers for small load commercial or institutional applications. Generally not recommended where load requirements exceed 50-75 lbs/hr.

Water Spray: Recommended for industrial applications where evaporative cooling is required; typical application is summertime humidification of textile mills in the southern U.S.

steam is the best natural medium for humidification. It provides ready-made vapor produced in the most efficient evaporator possible, the boiler. There is no mineral dust deposited, and because there is no liquid moisture present, steam creates no sanitation problems, will not support the growth of algae or bacteria, has no odor and creates no corrosion or residual mineral scale. Unless there are particular requirements to an application that can only be met with evaporative pan or water spray methods, steam humidification will meet system needs most effectively and economically.

specify steam boilers and generators solely for humidification when the building to be humidified does not have a steam supply. The minimum humidification load where this becomes economically feasible falls in the range of 50-75 lbs/hr. Steam generator capacity is generally specified 50% greater than maximum humidification load, depending on the amount of piping and number of humidifiers and distribution manifolds that must be heated. Typical piping for boiler-humidifier installations is shown in Fig. 17-1.

Typical Piping for Boiler-Humidifier Installation

Design Guidelines—Boiler-Humidifier Combinations

1. Boiler gross output capacity should be at least 1.5 times the total humidification load.

2. Water softeners should be used on boiler feedwater.

3. Condensate return system is not necessary (unless required by circumstances).

4. Boiler pressure should be at 15 psig or less.

5. An automatic blowdown system is desirable.

6. All steam supply piping should be insulated.

7. No limit to size or number of humidifiers from one boiler.

Figure 3.26.3 Recommended applications. (*By permission: Associated Steam Specialty Company, Drexel Hill, PA.*)

Determining Humidity Requirements of Materials

No single level of relative humidity provides adequate moisture content in all hygroscopic materials. Moisture content requirements vary greatly from one material to the next. We will discuss typical hygroscopic materials which require specific RH levels to avoid moisture loss and materials deterioration and/or production problems that result.

Table 10-1. Recommended Relative Humidities

PROCESS OR PRODUCT	Temp.°F	%RH	PROCESS OR PRODUCT	Temp.°F	%RH	PROCESS OR PRODUCT	Temp.°F	%RH
Residences	70-72	30	Switchgear:			**Tea**		
			Fuse & cutout assembly	73	50	Packaging	65	65
Libraries & Museums			Capacitor winding	73	50			
Archival	55-65	35	Paper storage	73	50	**Tobacco**		
Art storage	60-72	50	Conductor wrapping with yarn	75	65-70	Cigar & cigarette making	70-75	55-65
Stuffed fur animals	40-50	50	Lightning arrester assembly	68	20-40	Softening	90	85-88
			Thermal circuit breakers			Stemming & stripping	75-85	70-75
Communication Centers			assembly & test	75	30-60	Packing & shipping	73-75	65
Telephone terminals	72-78	40-50	High-voltage transformer repair	79	55	Filler tobacco casing		
Radio & TV studios	74-78	30-40	Water wheel generators:			& conditioning	75	75
			Thrust runner lapping	70	30-50	Filter tobacco storage		
General Commercial & Public Buildings			Rectifiers:			& preparation	77	70
	70-74	20-30	Processing selenium &			Wrapper tobacco storage		
(including cafeterias, restaurants, airport terminals, office buildings, & bowling centers)			copper oxide plates	73	30-40	& conditioning	75	75
Hospitals & Health Facilities			**Fur**			**Pharmaceuticals**		
General clinical areas	72	30-60	Storage	40-50	55-65	Powder storage (prior to mfg) *		*
Surgical area						Manufactured powder storage		
Operating rooms	68-76	50-60	**Gum**			& packing areas	75	35
Recovery rooms	75	50-60	Manufacturing	77	33	Milling room	75	35
Obstetrical			Rolling	68	63	Tablet compressing	75	35
Full-term nursery	75	30-60	Stripping	72	53	Tablet coating room	75	35
Special care nursery	75-80	30-60	Breaking	73	47	Effervescent tablets & powders	75	20
			Wrapping	73	58	Hypodermic tablets	75	30
Industrial Hygroscopic Materials						Colloids	75	30-50
Abrasive			**Leather**			Cough drops	75	40
Manufacture	79	50	Drying	68-125	75	Glandular products	75	5-10
Ceramics			Storage, winter room temp.	50-60	40-60	Ampoule manufacturing	75	35-50
Refractory	110-150	50-90				Gelatin capsules	75	35
Molding room	80	60-70	**Lenses (Optical)**			Capsule storage	75	35
Clay storage	60-80	35-65	Fusing	75	45	Microanalysis	75	50
Decalcomania production	75-80	48	Grinding	80	80	Biological manufacturing	75	35
Decorating room	75-80	48				Liver extracts	75	35
			Matches			Serums	75	50
Cereal			Manufacture	72-73	50	Animal rooms	75-80	50
Packaging	75-80	45-50	Drying	70-75	60	Small animal rooms	75-78	50
			Storage	60-63	50	* Store in sealed plastic containers in sealed drums.		
Distilling								
Storage			**Mushrooms**			**Photographic Processing**		
Grain	6	35-40	Spawn added	60-72	nearly sat.	Photo Studio		
Liquid yeast	32-33		Growing period	50-60	80	Dressing room	72-74	40-50
General manufacturing	60-75	45-60	Storage	32-35	80-85	Studio (camera room)	72-74	40-50
Aging	65-72	50-60				Film darkroom	70-72	45-55
			Paint Application			Print darkroom	70-72	45-55
Electrical Products			Oils, paints: Paint spraying	60-90	80	Drying room	90-100	35-45
Electronics & X-ray:						Finishing room	72-75	40-55
Coil & transformer winding	72	15	**Plastics**			Storage room		
Semi conductor			Manufacturing areas:			b/w film & paper	72-75	40-60
assembly	68	40-50	Thermosetting molding			color film & paper	40-50	40-50
Electrical instruments:			compounds	80	25-30	Motion picture studio	72	40-55
Manufacture			Cellophane wrapping	75-80	45-65			
& laboratory	70	50-55				**Static Electricity Control**		
Thermostat assembly			**Plywood**			Textiles, paper, explosive control		> 55
& calibration	75	50-55	Hot pressing (resin)	90	60			
Humidistat assembly			Cold pressing	90	15-25	**Clean Rooms & Spaces**		45
& calibration	75	50-55						
Small mechanisms:			**Rubber-Dipped Goods**			**Data Processing**	72	45-50
Close tolerance assembly	72	40-45	Cementing	80	25-30*			
Meter assembly & test	75	60-63	Dipping surgical articles	75-80	25-30*	**Paper Processing**		
			Storage prior to manufacture	60-75	40-50*	Finishing area	70-75	40-45
			Laboratory (ASTM Standard)	73.4	50*	Test laboratory	73	50
			* Dew point of air must be below evaporation temperature of solvent					

Abstracted from the ASHRAE Handbook Systems and Applications

Figure 3.26.4   Determining humidity requirements. (*By permission: Associated Steam Specialty Company, Drexel Hill, PA.*)

Understanding Condensation

Causes

Condensation on windows and other cool surfaces in the home can be both annoying and possibly injurious to your home. Because the most often visible condensation is seen on windows, it is easy to blame condensation as being a window fault. This is not true in most cases. Any cool surface will cause excess humidity to condense on it. If there is condensation on windows, you may be assured there is condensation on walls. This is more serious since that can penetrate the walls and cause internal problems.

The cause of condensation is air saturated with too much humidity or water. When this happens, air cannot hold the excess humidity. It gets rid of it by condensing it on the most convenient cool surface.

Where Does Humidity Come From?

- Normal breathing and perspiration by a family of four adds a half pint of water to the air each hour.
- Cooking can add up to four or five pints of water per day.
- A shower can add another half pint.
- Dishwashers, washing machines, and dryers can add several pints of water to the air.
- Humidifiers which are adding too much humidity.
- Poorly insulated crawl spaces which allow humidity to invade the home.
- New homes will often emit excess humidity for the normal drying out of the building products. This is normal and will usually adjust itself within a year or less.

In other words, if condensation is to be reduced, the source and amount of humidity in the air needs to be determined.

How Much Humidity Is Too Much and How Much Can Air Hold?

Warmer air holds more moisture than cool or cold air. This is illustrated on a humid, hot summer day when condensation appeared on a cold glass. This means that the amount of moisture in the air has reached its maximum and can't hold any more. Therefore, it gets rid of it by condensing it on the nearest cool or cold surface.

As air cools, it can't hold as much moisture and therefore, condensation will appear more quickly.

So what is the ideal amount of relative humidity in the air? Based on keeping an indoor temperature of 70° F, it will vary with the outdoor temperature. But as a guide, the following relationship should help.

Outside Air Temp.	Inside Relative Humidity
-20° F or Colder	15% Maximum
-20° F	20% Maximum
-10° F	25% Maximum
0° F	30% Maximum
10° F	35% Maximum

Figure 3.26.5   Understanding condensation

20° F   40% Maximum

If your relative humidity is above these levels, you probably will have condensation on any cool surface.

Is Condensation More Prevalent Today Than It Used to Be?
In some cases this may be true. In older houses, the insulation and weather-stripping and other house tightening factors allowed the house to breathe and exchange drier air with inside more humid air.

Of course, windows also were not so air tight and caused colder air to enter the house and can also cause the surface of the window to be colder.

Today, because we are all energy conscious, houses and windows are far more energy efficient. This make us all more comfortable, but may trap humid air inside the home.

So, What Do You Do Now?
The obvious answer is to reduce the humidity and decrease the number of cool surfaces in your home.

Your first step is to find what the humidity level in your home is. This will need to be monitored regularly as the temperature outside varies; Devices which measure humidity are called hygrometers. They can be purchased at most reliable hardware and home center stores.

Here are a few things you can do to control humidity:

- Make sure your humidifier is working correctly. Turn it down as the weather becomes colder.
- Vent all appliances and vents to the outside.
- Vent attic and crawl spaces.
- Cover the earth in your crawl space with a vapor barrier.
- Run exhaust fans while cooking or bathing.
- If you have a forced air furnace, make sure your home is properly ventilated by installing a fresh intake.
- Don't store firewood inside.
- As a temporary solution, you may want to try opening your windows a little each day to allow the exchange of colder' drier air with warmer more humid air. This should not effect your energy bill in any substantial manner.
- Install energy efficient windows.

What is an Energy Efficient Window and Why Will It Help?

Windows, doors, and skylights have become an important part of the energy saving plan. They do not allow cold air to enter around a window, thus cooling the surface. Spacers between glazing in double or triple glazed windows are more energy efficient and do not allow cold air to migrate through them causing the glazing to cool.

Special metallic coatings have been developed (known as Low E or low emissivity) which reflects radiant heat and restricts its flow through glass. During cold weather, it will keep heat inside. In hot weather, it keeps heat outside.

Using energy efficient windows will keep the interior glass surfaces warmer and thus

Figure 3.26.5 *(Continued)*

reduce the interior cool surfaces on which moisture can condense.

Summary

- In order to reduce condensation:
- Reduce the amount of moisture in the air as the outside temperature gets colder (see chart page 4)
- Make sure your home is properly ventilated.
- Use exhaust fans.
- Use vapor barriers on the earth in your crawl space.
- Use a hygrometer to measure and regulate your humidity level.
- Use energy efficient windows (Installing a storm window in and older home may help.)
- If building a new home, make sure your builder is using only kiln dried lumber and make sure he places heat vents beneath patio doors.

Remember, there is always a possibility that in very cold, unusual circumstances, you may still have some temporary condensation. But if the humidity level is proper and the home correctly vented, this will be short lived.

Choose Windows Which Are Certified

Windows, doors, and skylights, which are Hallmark, certified by the National Wood Window and Door Association have undergone rigorous structural, air, and water infiltration testing.

Windows, which are certified by the National Fenestration Rating Council, have undergone intensive evaluation for energy efficiency.

Figure 3.26.5 *(Continued)*

Checklist for Condensation and Potential Mold Growth

Key
- ● Likely remedial measure
- ○ Possible remedial measure

Remedial measures and Action by Householder (● likely, ○ possible)

Measure		Symptom 1: Dampness/mold growth in cold weather only	Symptom 2: Dampness/mold growth in rooms with large areas of exposed walls, roofs, or floors	Symptom 3: Dampness/mold growth in bedrooms	Symptom 4: Dampness/mold growth in or behind cupboards or in upper corners of rooms	Symptom 5: Dampness/mold growth at times of rain or snow	Symptom 6: Moisture staining, tidemarks on ceilings, walls, etc.
Action by Householder	Mold Treatment	●	○	○	○		
	Reduce Use of LPG, etc.	○	○	○	○		
	Adequate Ventilation	●	○	○	●		
	Adequate Heat Input	●	●	●	○		
Remedial Measures	Mold Treatment	●	○	○	○		
Building	Alter Fittings				●		
	Double Glazing		○	○			
	Insulate Floor	○	●				
	Insulate Walls	○	●	○			
	Insulate Roof	○	○	○			
Ventilation	Cupboard Ventilators			○	○		
	Window Vent (Slot Type)	○	●	●	●		
	Exhaust Vent, Bathroom	○	○	○	○		
	Exhaust Vent, Kitchen	○	○		○		
Heating System	Supplement, Hall, Ldg.		●	○			
	Supplement, Bedroom		●	●			
	Supplement, Livingroom		●				
	Maintain	○	○			Repair, overhaul, replace as necessary	Repair, overhaul, replace as necessary

Possible Causes or Influences

Symptom	Heating System	Ventilation	Building	Householder
Dampness/mold growth in cold weather only	Low or intermittent heat input leading to inadequate background heating	Heating and ventilation not matched; usually too little ventilation due to absence of natural paths of ventilation	Inappropriate construction for intermittent heating, i.e., slow response heavy weight	Heat input too low, particularly in bedrooms; intermittent heating; drying clothes indoors; use of LPG or wood
Dampness/mold growth in rooms with large areas of exposed walls, roofs, or floors	Low or intermittent heat input leading to inadequate background heating	Heating and ventilation not matched	Exposed or projecting areas of building that are poorly insulated (including stairwells); downstairs bedrooms; large windows, esp. full-height windows	Heat input too low, particularly in bedrooms; uninsulated water pipes; intermittent heating; use of LPG or wood
Dampness/mold growth in bedrooms	Unheated bedrooms; lack of background heat in dwelling generally	Poor ventilation not matched with heating; moist air entering from kitchens and bathrooms	Exposed or projecting bedrooms, poorly insulated, esp. north-facing unheated bedrooms downstairs; bedrooms over unheated spaces, e.g., walkways, garages	Heat input too low, particularly at night; leaving kitchen and bathroom doors open; use of LPG or wood
Dampness/mold growth in or behind cupboards or in upper corners of rooms		Stagnant air pockets; lack of air movement behind furniture; lack of air movement in cupboards, etc.	Built-in cupboards or wardrobes on outside walls	Bad positioning of furniture; overfilling of cupboards; putting clothes away damp
Dampness/mold growth at times of rain or snow, especially with recurrence in summer during periods of rain			Dampness from water penetration; building cracks; defective flashing or pointing; blocked gutters; cracked gutters/rainwater pipes; exposure to driving rain	
Moisture staining, tidemarks on ceilings, walls, etc.	Radiator, boiler, pipe leaks		Dampness from plumbing leaks; cracked water pipes, tanks, etc.	

Figure 3.26.6   Checklist for condensation and potential mold growth.

Air Movement and Humidity

Another variable, air movement in the form of infiltration and exfiltration from the building, influences the relationship between temperature and relative humidity. Typically, one to three times every hour (and many more times with forced air make-up or exhaust) cold outdoor air replaces your indoor air. Your heating system heats this cold, moist outdoor air, producing warm, dry indoor air.

Evaporative Cooling

We've discussed the effects of changing temperature on relative humidity. Altering RH can also cause temperature to change. For every pound of moisture evaporated by the air, the heat of vaporization reduces the sensible heat in the air by about 1,000 Btu. This can be moisture absorbed from people or from wood, paper, textiles, and other hygroscopic materials in the building. Conversely, if hygroscopic materials absorb moisture from humid air, the heat of vaporization can be released to the air, raising the sensible heat.

Dew Point

Condensation will form on windows whenever the temperature of the glass surface is below the dew point of the air. Table 5-2, from data presented in the ASHRAE Handbook & Product Directory, indicates combinations of indoor relative humidity and outside temperature at which condensation will form. Induction units, commonly used below windows in modern buildings to blow heated air across the glass, permit carrying higher relative humidities without visible condensation.

Table 5-1. Grains of Water per Cubic Foot of Saturated Air and per Pound of Dry Air at Various Temperatures. (Abstracted from ASHRAE Handbook)

°F	Per cu ft	Per lb Dry Air	°F	Per cu ft	Per lb Dry Air	°F	Per cu ft	Per lb Dry Air	°F	Per cu ft	Per lb Dry Air
-10	0.28466	3.2186	50	4.106	53.38	78	10.38	145.3	106	23.60	364.0
- 5	0.36917	4.2210	51	4.255	55.45	79	10.71	150.3	107	24.26	375.8
0	0.47500	5.5000	52	4.407	57.58	80	11.04	155.5	108	24.93	387.9
5	0.609	7.12	53	4.561	59.74	81	11.39	160.9	109	25.62	400.3
10	0.776	9.18	54	4.722	61.99	82	11.75	166.4	110	26.34	413.3
15	0.984	11.77	55	4.889	64.34	83	12.11	172.1	111	27.07	426.4
20	1.242	15.01	56	5.060	66.75	84	12.49	178.0	112	27.81	440.4
25	1.558	19.05	57	5.234	69.23	85	12.87	184.0	113	28.57	454.5
30	1.946	24.07	58	5.415	71.82	86	13.27	190.3	114	29.34	469.0
31	2.033	25.21	59	5.602	74.48	87	13.67	196.7	115	30.13	483.9
32	2.124	26.40	60	5.795	77.21	88	14.08	203.3	120	34.38	566.5
33	2.203	27.52	61	5.993	80.08	89	14.51	210.1	125	39.13	662.6
34	2.288	28.66	62	6.196	83.02	90	14.94	217.1	130	44.41	774.9
35	2.376	29.83	63	6.407	86.03	91	15.39	224.4	135	50.30	907.9
36	2.469	31.07	64	6.622	89.18	92	15.84	231.8	140	56.81	1064.7
37	2.563	32.33	65	6.845	92.40	93	16.31	239.5	145	64.04	1250.9
38	2.660	33.62	66	7.074	95.76	94	16.79	247.5	150	71.99	1473.5
39	2.760	34.97	67	7.308	99.19	95	17.28	255.6	155	80.77	1743.0
40	2.863	36.36	68	7.571	102.8	96	17.80	264.0	160	90.43	2072.7
41	2.970	37.80	69	7.798	106.4	97	18.31	272.7	165	101.0	2480.8
42	3.081	39.31	70	8.055	110.2	98	18.85	281.7	170	112.6	2996.0
43	3.196	40.88	71	8.319	114.2	99	19.39	290.9	175	125.4	3664.5
44	3.315	42.48	72	8.588	118.2	100	19.95	300.5	180	139.2	4550.7
45	3.436	44.14	73	8.867	122.4	101	20.52	310.3	185	154.3	5780.6
46	3.562	45.87	74	9.153	126.6	102	21.11	320.4	190	170.7	7581.0
47	3.692	47.66	75	9.448	131.1	103	21.71	330.8	195	188.6	10493.0
48	3.826	49.50	76	9.749	135.7	104	22.32	341.5	200	207.9	15827.0
49	3.964	51.42	77	10.06	140.4	105	22.95	352.6			

Table 5-2. Relative Humidities at Which Condensation Will Appear on Windows at 70°F When Glass Surface is Unheated

Outdoor Temperature	Single Glass	Double Glass (Storm Windows or Thermal Glass)
-10	11%	38%
0	16%	42%
+10	21%	49%
+20	28%	56%
+30	37%	63%
+40	48%	71%

Figure 3.27.0 Air movement and humidity, evaporative cooling, and dew point. (*By permission: Associated Steam Specialty Company, Drexel Hill, PA.*)

Estimated Dew Point of Compressed Air

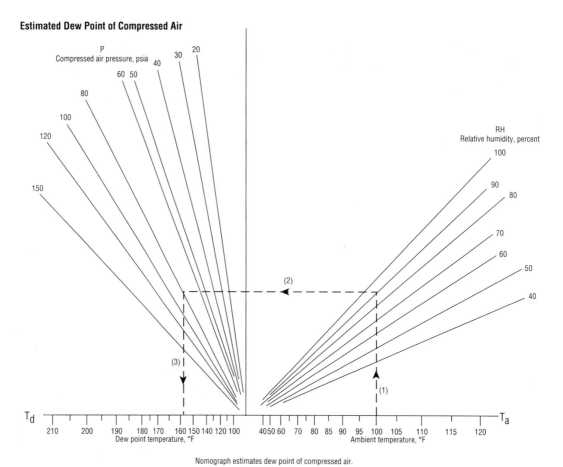

Nomograph estimates dew point of compressed air.

Figure 3.27.1 Estimating dew point of compressed air. (*By permission: Associated Steam Specialty Company, Drexel Hill, PA.*)

Direct Acting

For Steam, Air, and Non-Corrosive Gas Service

The simplest of pressure reducing valves, the direct acting type operates with either a flat diaphragm or convoluted bellows. Since it is self-contained, it does not need an external sensing line downstream to operate. It is the smallest and most economical of the three types and is designed for low to moderate flows. Accuracy of direct acting PRVs is typically ±10%.

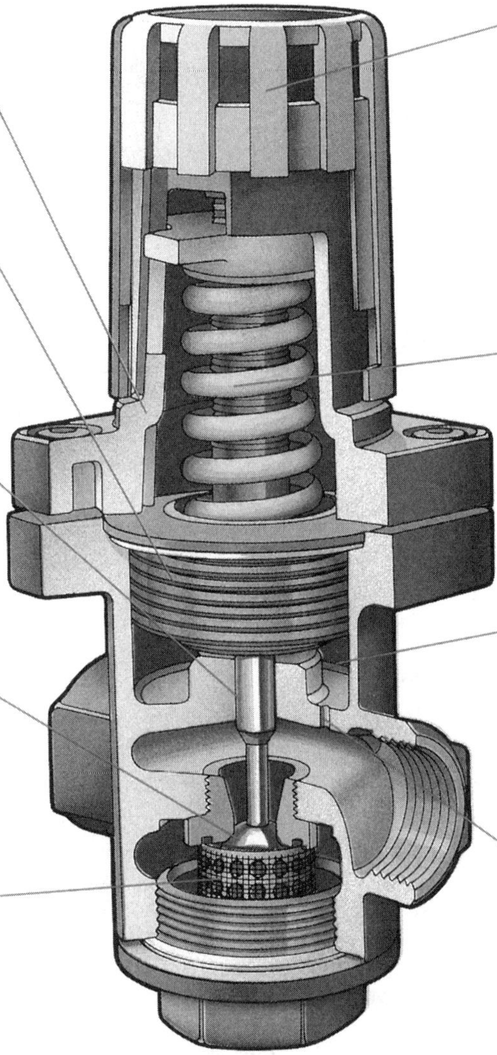

Bronze body for corrosion resistance.

Convoluted phosphor bronze bellows allows smaller body size than conventional direct-acting diaphragm PRVs.

Stainless steel stem and bronze body minimize the sticking common with conventional valves.

Hardened stainless steel valve and seat with spring return to assure tight shutoff.

Integral mesh strainer eliminates fouling due to dirt.

Plastic cap doesn't absorb heat so it won't burn an operator's hands. Permits changing of set pressure without tools. Simply lift cap and turn to desired setting.

Interchangeable range spring is easily changed to different pressure by removing four (4) cap screws.

Downstream pressure sensed outside the bellows vs. inside, which assures bellows integrity and longer bellows life.

Integral sensing port to eliminate piping of external sensing line.

Figure 3.28.0 Pressure reducing valve—direct acting. (*By permission: Associated Steam Specialty Company, Drexel Hill, PA.*)

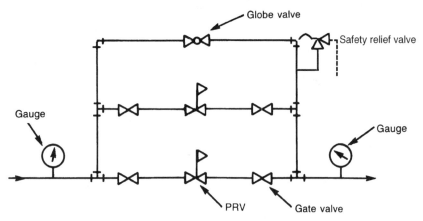

Figure 3.28.1 Pressure-reducing station. (*By permission: McGraw Hill* HVAC Systems Design Handbook, *R. W. Haines and C. L. Wilson.*)

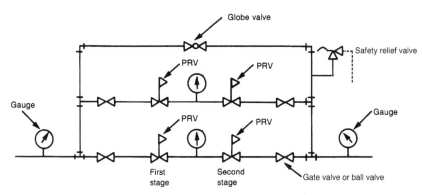

Figure 3.28.2 Two-stage pressure-reducing station. (*By permission: McGraw Hill* HVAC Systems Design Handbook, *R. W. Haines and C. L. Wilson.*)

3.29.0 Mechanical Draft Towers (Illustrated and Described)

Mechanical Draft

Mechanical draft towers use either single or multiple fans to prove flow of a known volume of air through the tower. Thus their thermal performance is considered to be more stable and is affected by fewer psychometric variables than that of the natural draft atmospheric towers. The presence of fans also provides a means of regulating air flow to compensate for changing atmospherc and load conditions through fan capacity modulation of speed and/or cycling.

Mechanical draft towers are categorized as either "induced draft," wherein a fan located in the exiting air stream draws air through the tower or "forced draft," in which the fan is located in the ambient air stream entering the tower, and the air is blown through the tower.

Induced Draft

An induced draft cooling tower is provided with a top-mounted fan that induces atmospheric air to flow up through the tower, as warm water falls downward. An induced draft tower may have only spray nozzles for water breakup or it may be filled with various slat and deck arrangements. There are several types of induced draft cooling towers.

In a counterflow induced draft tower, a top-mounted fan induces air to enter the sides of the tower and flow vertically upward as the water cascades through the tower. The counterflow tower is particularly well adapted to a restricted space as the discharge air is directed vertically upward, and the sides require only a minimum clearance for air intake area. The primary breakup of water may be by either pressure spray or by gravity from pressure-filled fumes.

A doubleflow induced draft tower has a top-mounted fan to induce air to flow across the fill material. The air is then turned vertically in the center of the tower. The distinguishing characteristics of

3.29.0   Mechanical Draft Towers (Illustrated and Described)

a doubleflow induced draft tower are the two air intakes on opposite sides of the tower and the horizontal flow of air through the fill sections.

Comparing counterflow and doubleflow induced draft towers of equal capacity, the doubleflow tower would be somewhat wider but the height would be much less. Cooling towers must be braced against the wind. From a structural standpoint, therefore, it is much easier to design a doubleflow than a counterflow tower as the low silhouette of the doubleflow type offers much less resistance to the force of the winds.

Mechanical equipment for counterflow and doubleflow towers is mounted on top of the tower and is readily accessible for inspection and maintenance. The water distributing systems are completely open on top of the tower and can be inspected during operation. This makes it possible to adjust the float valves and clean stopped-up nozzles while the towers are operating.

The crossflow induced draft tower is a modified version of the doubleflow induced draft tower. The fan in a crossflow cooling tower draws air through a horizontal opening and discharges the air vertically.

In some situations, an indoor location for the cooling tower may be desirable. An induced draft tower of the counterflow or crossflow design is generally selected for indoor installation. Two connections to the outside are usually required: one for drawing outdoor air into the tower, and the other for discharging it back to the outside. A centrifugal blower is often necessary for this application to overcome the static pressure of the ductwork. Many options are possible as to point of air entrance and air discharge. This flexibility is often important in designing an indoor installation. Primary water breakup is by pressure spray and fill of various types.

An indoor installation of an induced draft counterflow cooling tower is shown. In this particular case, air required for operation of the tower is being taken from the basement. As the cooling tower fan is therefore exhausting air from the building, the quantity of air exhausted must be included in sizing the outside air intake for the air conditioning system.

The induced draft cooling tower, for indoor installation, is usually a completely assembled packaged unit, but is so designed that it can be partially disassembled to permit passage through limited entrances. Indoor installations of cooling towers are becoming more popular. External space restrictions; architectural compatibility (including aesthetics), and convenience for observation, diagnostics, and maintenance all combine to favor an indoor location. The installation cost is somewhat higher than an outdoor location. Packaged towers are generally available in capacities to serve the cooling requirements of refrigeration plants in the 5 to 100 ton range.

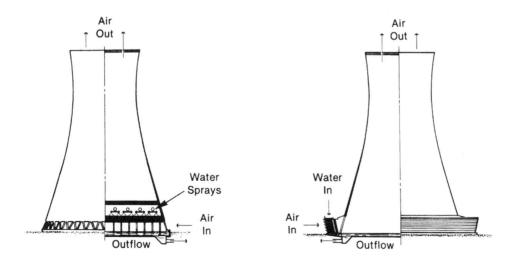

COUNTERFLOW NATURAL DRAFT TOWER   CROSSFLOW NATURAL DRAFT TOWER

(By permission of The Trane Company, LaCrosse, Wisconsin)

3.29.0   Mechanical Draft Towers (Illustrated and Described) (Continued)

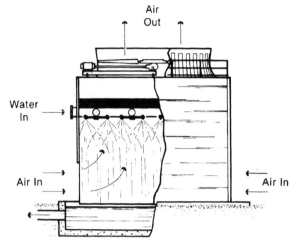

(A) SPRAY-FILLED, COUNTERFLOW COOLING TOWER

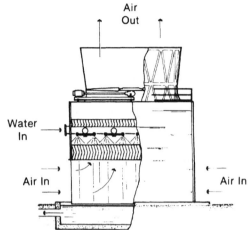

(B) COUNTERFLOW COOLING TOWER WITH FILL BANK

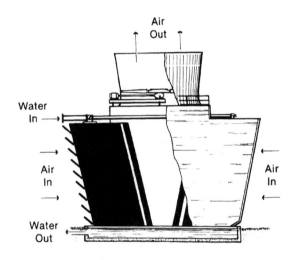

A DOUBLEFLOW INDUCED DRAFT COOLING TOWER

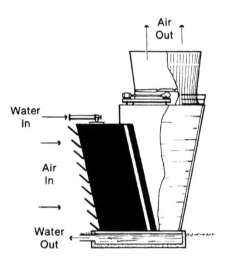

A SINGLE FLOW CROSS-FLOW INDUCED DRAFT COOLING TOWER

(By permission of The Trane Company, LaCrosse, Wisconsin)

To Avoid Them

Oil. A critical drainage problem exists at points where oil may be present in the compressed air (principally at intercoolers. aftercoolers and receivers).

Two facts create this problem:
1) Oil is lighter than water and will float on top of water.
2) Compressor oil when cooled tends to become thick and viscous.

The beaker simulates any drain trap that has its discharge valve at the bottom. Fig. 4-1. Like the beaker. the trap will fill with heavy oil that may be thick and viscous.

Compare with Fig. 4-2 which shows an identical beaker except that the discharge valve is at the same level as the oil. Oil will

that for every 19 drops of water and one of oil that enter the beaker, exactly 19 drops of water and one drop of oil will leave. The beaker always will be filled with water.

The conclusion is obvious. When there is an oil-water mixture to be drained from an air separator or receiver. use a trap with the discharge valve at the top.

Dirt and Grit. While scale and sediment is seldom a problem between the compressor and receiver. it is encountered in the air distribution system. particularly when the piping is old. In this situation. scale will be carried to a drain trap along with the water. If the drain trap is not designed to handle dirt and grit. the trap may fail to drain water and oil. or the trap valve may not close.

Air Loss. Often in compressed air systems. the solution to one problem may also cause another problem. For example. a common method of draining unwanted moisture is to

also creates a leak. The immediate problem is solved. but the "solution" has an obvious. and usually underestimated. cost of continual air loss.

How much air is lost depends on orifice size and line pressure (see Table 5-1). The overall result is a decrease in line pressure. the loss of up to a third of the system's compressed air. and the cost of compressing it.

Leak control involves:
Looking for leaks during shutdown with an ultrasonic leak detector.
Determining total leakage by observing how fast pressure drops with the compressor off. both before and after a leak survey.
Fixing leaks at joints. valves. and similar points.
Replacing cracked open valves with drain traps.
Checking the system regularly.

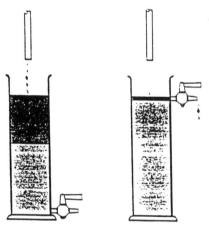

If a beaker collecting oil and water is drained from the bottom at the same rate that oil and water enter. it will eventually fill entirely with oil because oil floats on water.

If a beaker collecting oil and water is drained from the top at the same rate that oil and water enter, it soon will be entirely filled with water because the oil floats on the water.

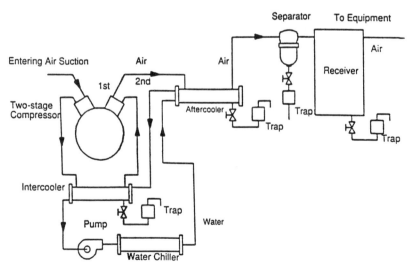

Figure 3.30.0 Drainage problems in compressed air systems and how to avoid them. (*By permission: Associated Steam Specialty Company, Drexel Hill, PA.*)

Drainage Methods
Manual
Liquid may be discharged continuously through cracked open valves, or periodically by opening manually operated drain valves.

Open drains are a continuous waste of air or gas—and the energy to produce it. A valve manually opened will be left open until air blows freely. Frequently, however, the operator will delay or forget to close the valve and precious air or gas is lost.

Automatic
Automatic drainage equipment which is adequate for the system is seldom a part of the original system. However, subsequent installation of automatic drain traps will significantly reduce energy and maintenance costs.

Drain Traps
Water collected in separators and drip legs must be removed continuously without wasting costly air or gas. In instances where drain traps are not part of the system design, manual drain valves are usually opened periodically or left cracked open to drain constantly. In either case, the valves are opened far enough that some air and gas are lost along with the liquid. To eliminate this problem, a drain trap should be installed at appropriate points to remove liquid continuously and automatically without wasting air or gas.

The job of the drain trap is to get liquid and oil out of the compressed air/gas system. In addition, for overall efficiency and economy, the trap must also provide:

Operation that is relatively trouble-free with minimal need for adjustment or maintenance.
Reliable operation even though dirt, grit and oil are present in the line.
Long operating life.
Minimal air loss.
Ease of repair.

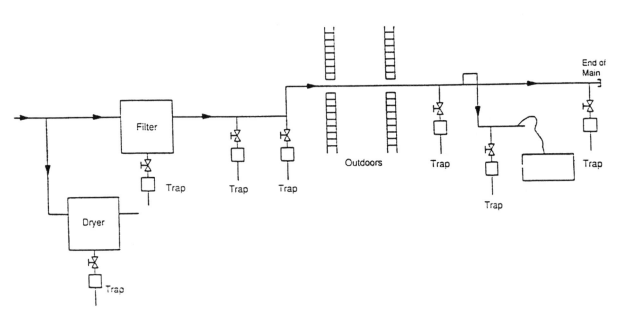

Drain trap locations in a compressed air system.
The use of drain traps is an effective way to remove water that collects in many places in a compressed air system. Each trap location must be considered individually.

Figure 3.30.2 Drainage methods—manual, automatic, and drain traps. (*By permission: Associated Steam Specialty Company, Drexel Hill, PA.*)

Drain Trap Selection

Do-It-Yourself Sizing is required at times. Fortunately, drain trap sizing is simple when you know or can figure:
1. Liquid loads in lbs/hr
2. Pressure differential
3. Maximum allowable pressure

1. Liquid Load. Each "How To" section of this handbook contains formulas and useful information on proper sizing procedures and safety factors.

2. Pressure Differential. *Maximum differential* is the difference between main pressure, or the downstream pressure of a PRV, and return line pressure. See Fig. 10-1. The drain trap must be able to open against this pressure differential.

Operating differential. When the plant is operating at capacity, the pressure at the trap inlet may be lower than main pressure. And the pressure in the return header may go above atmospheric.

If the operating differential is at least 80% of the maximum differential, it is safe to use maximum differential in selecting traps.

IMPORTANT: Be sure to read the discussion on page 11 which deals with less common, but important, reductions in pressure differential.

3. Maximum Allowable Pressure. The trap must be able to withstand the maximum allowable pressure of the system, or design pressure. It may not have to operate at this pressure, but it must be able to contain it. As an example, the maximum inlet pressure is 150 psig and the return line pressure is 15 psig. This results in a differential pressure of 135 psi, however, the trap must be able to withstand 150 psig maximum allowable pressure. See Fig. 10-1.

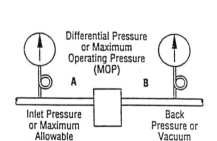

Differential Pressure or Maximum Operating Pressure (MOP)

A   B

Inlet Pressure or Maximum Allowable Pressure (MAP)

Back Pressure or Vacuum

"A" minus "B" is Pressure Differential: If "B" is back pressure, subtract it from "A". If "B" is vacuum, add it to "A".

Figure 3.30.3 Drain trap selection. (*By permission: Associated Steam Specialty Company, Drexel Hill, PA.*)

For Heavy Oil/Water Service

BVSW inverted bucket drain traps are designed for systems with heavy oil or water services.

An inverted bucket is used because the discharge valve is at the top, so oil is discharged first and the trap body is almost completely filled with water at all times.

BVSW stands for Bucket Vent Scrubbing Wire. This 1/16" dia. wire swings freely from the trap cap and extends through the bucket vent. Its function is to prevent reduction of vent size by buildup of solids or heavy oil in the vent itself. The up and down motion of the bucket relative to the vent scrubbing wire keeps the vent clean and full size.

Operation of Inverted Bucket Drain Traps

1. Since there is seldom sufficient accumulation of water to float the bucket and close the valve, the trap must be primed on initial start-up or after draining for cleaning. Step 1 shows "after operating" primed condition with oil in the top of bucket and a very thin layer of oil on top of water in the trap body.

2. When valve in line to trap is opened, air enters bucket, displacing liquid. When bucket is two-thirds full of air it becomes buoyant and floats. This closes the discharge valve. As bucket rises, the vent scrubbing wire removes oil and any dirt from bucket vent.

Both liquid and air in the trap are at full line pressure so no more liquid or air can enter trap until some liquid or air escapes through discharge valve. Static head "H" forces air through bucket vent. The air rises to top of trap and displaces water which enters bucket at bottom to replace the air that passes through the vent. Just as soon as the bucket is less than two-thirds full of air it loses buoyancy and starts to pull on valve lever as shown in Step 3.

Operation of the BVSW Inverted Bucket Drain Trap

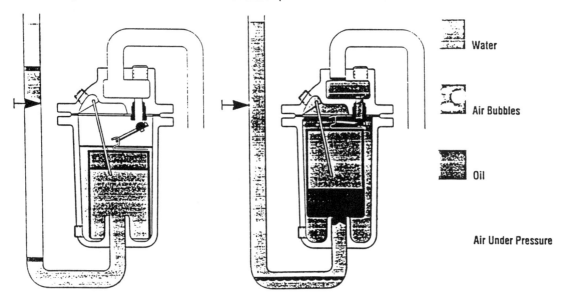

☐ Water

☐ Air Bubbles

■ Oil

Air Under Pressure

1. Trap primed, air off. bucket down, trap valve open.

2. Trap in service, bucket floating. Air passes through bucket vent and collects at top of trap.

Figure 3.30.4   Inverted bucket traps for heavy oil and water service. (*By permission: Associated Steam Specialty Company, Drexel Hill, PA.*)

3. Note that liquid level at top of trap has dropped and the liquid level in the bucket has risen. The volume of water displaced by air exactly equals the volume of water that entered the bucket. During this valve-closed part of the operating cycle—Steps 2 and 3—water and oil are collecting in the horizontal line ahead of the trap. When bucket is about two-thirds full of liquid it exerts enough pull on lever to crack open the discharge valve.

4. Two things happen simultaneously. a) The accumulated air at top of trap is discharged immediately, followed by oil and any water that enters the trap while the valve is cracked. b) Pressure in trap body is lowered slightly allowing accumulated liquid in horizontal line to enter the trap. Air displaces liquid from the bucket until it floats and closes the discharge valve restoring the condition shown in Step 2.

5. When full buoyancy is restored, trap bucket is two-thirds full of air. Oil that has entered while trap was open flows under bottom of bucket and rises to top of water in trap body. The trap normally discharges small quantities of air several times per minute.

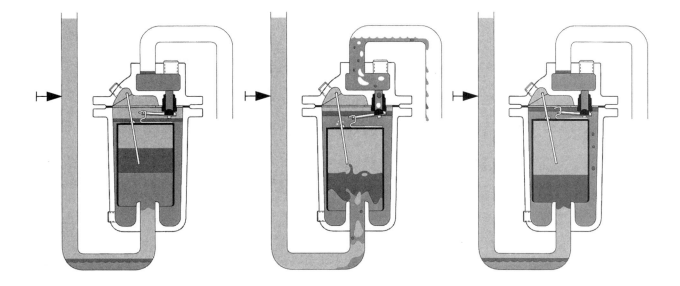

3. Water enters bucket to replace air passing through bucket vent. This increases weight of bucket until...

4. Pull on lever cracks valve. Air at top of trap escapes followed by oil and water. Liquid in pipe ahead of trap enters bucket followed by air.

5. Air displaces liquid and excess oil from bucket restoring condition shown in Step 2.

Figure 3.30.4 *(Continued)*

Float Type Drain Traps

Closed Float

Hollow thin wall metal floats are attached through linkages to valves at the trap bottom, and a seat with an appropriately sized orifice is inserted at the trap outlet. Floats are selected to provide adequate buoyancy to open the valve against the pressure difference. Discharge usually is to atmosphere, so the pressure drop is equal to the system air pressure. The float and linkage are made of stainless steel, and the valve and seat are hardened stainless steel for wear resistance and long life. The body is cast iron, stainless steel, or cast or forged steel depending on gas pressure. Bodies may be made of stainless steel to resist corrosive gas mixtures.

Entering liquid drops to the bottom of the body. As liquid level rises, the ball is floated upward thereby causing the valve to open sufficiently that outlet flow balances inlet flow. Subsequent change of incoming flow raises or lowers water level further opening or throttling the valve. Thus discharge is proportionally modulated to drain liquid completely and continuously. However, gas flow may be constant or it may abruptly change depending on system demand characteristics. Liquid formation may be sporadic, or the nature of flow generation may cause surges. At times, flow will be very low requiring operation to throttle the flow or even tight shut-off. Tightness of closure, gas leakage, and trap cost will depend on the design of linkage and valve.

Free Floating Lever

The discharge from the No. 1-LD is continuous. The opening of the valve is just wide enough to remove the liquid as fast as it comes to the trap. Thus, at times, the valve is barely cracked from its seat.

Operation of the No. 1-LD Free Floating Lever Drain Trap. As water begins to fill the body of the trap, the float rises opening the discharge valve. Motion of the free floating valve lever is guided to provide precise closure.

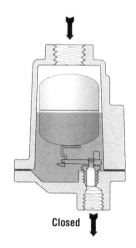

Closed

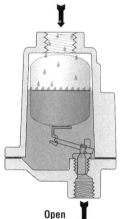

Open

Free Floating Linkage

A hemispherical ball-shaped valve is attached to linkage which is suspended freely on two guide pins. There is no fixed pivot or rigid guides, therefore the attachment is loose. There are no critical alignments, and the lever and valve may move in all directions. Consequently, the lever may move the valve to the seat in any alignment. As the valve approaches the seat, the pressure pushes the round valve into the square edge orifice of the seat affecting a line seal to attain bubble-tight closure.

Free Floating Linkage

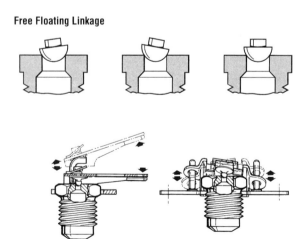

Figure 3.30.5 Closed float traps. (*By permission: Associated Steam Specialty Company, Drexel Hill, PA.*)

Fixed Pivot Comcal Valve

A conically shaped valve is attached to a fixed pivot leverage system. The fixed pivot does not allow the valve to move freely to conform to the seat for tight closure. Thus, they may not seal tightly, and some loss of air or gas may be expected.

Operation of No. 21 Fixed Pivot Drain Trap

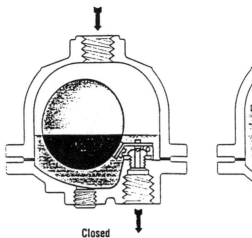

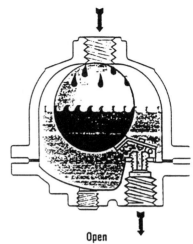

Closed

Open

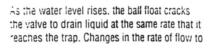

As the water level rises. the ball float cracks the valve to drain liquid at the same rate that it reaches the trap. Changes in the rate of flow to

the trap adjust the float level and the degree of opening of the valve.

Snap Action

Because of the sporadic liquid flow. much of the time the valve is only slightly opened. If there is fine dirt or grit in the gas. particles may accumulate and clog the partially open valve. or they may lodge between the valve and seat preventing closure.

To overcome this. a special toggle spring-operated valve is used.

A flat spring attached to the leverage system holds the valve closed until liquid level is high enough for the buoyancy to exceed the spring force.

Then the valve is snapped open. and the accumulated dirt and grit can be flushed through the wide open valve. When the body is near empty, buoyancy is reduced enough to permit the spring to snap the valve closed.

Operation of the No. 71-A Snap Action Drain Trap

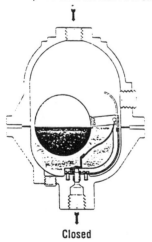

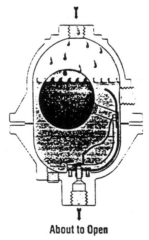

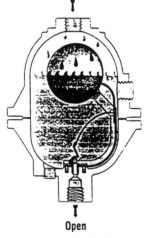

Closed

About to Open

Open

Filling Cycle. Trap valve has just closed. Spring bowed to right. Float rides high in water because no force is exerted on spring. As water enters, float rises. storing energy in spring. This increases submergence of float.

Float now is more than half submerged and spring has assumed a "handlebar mustache" shape. Energy stored in spring is due to increased displacement of water. A very slight rise in water level causes spring to snap to the left....

Instantly the valve opens wide. This releases energy from spring and float again rides high in water. As water level drops. weight of float bends spring to right causing snap closing of valve before all the water has been discharged

Figure 3.30.6   Fixed, pivot, and snap action traps. (*By permission: Associated Steam Specialty Company, Drexel Hill, PA.*)

3.31.0 Equivalent Rectangular Duct Dimension Tables

Duct Diameter, in.	Rectangular Size, in.	Aspect Ratio														
		1.00	1.25	1.50	1.75	2.00	2.25	2.50	2.75	3.00	3.50	4.00	5.00	6.00	7.00	8.00
6	Width	—	6													
	Height	—	5													
7	Width	6	8													
	Height	6	6													
8	Width	7	9	9	11											
	Height	7	7	6	6											
9	Width	8	9	11	11	12	14									
	Height	8	7	7	6	6	6									
10	Width	9	10	12	12	14	14	15	17							
	Height	9	8	8	7	7	6	6	6							
11	Width	10	11	12	14	14	16	18	17	18	21					
	Height	10	9	8	8	7	7	7	6	6	6					
12	Width	11	13	14	14	16	16	18	19	21	21	24				
	Height	11	10	9	8	8	7	7	7	6	6	6				
13	Width	12	14	15	16	18	18	20	19	21	25	24	30			
	Height	12	11	10	9	9	8	8	7	7	7	6	6			
14	Width	13	14	17	18	18	20	20	22	24	25	28	30	36		
	Height	13	11	11	10	9	9	8	8	8	7	7	6	6		
15	Width	14	15	17	18	20	20	23	25	24	28	28	35	36	42	
	Height	14	12	11	10	10	9	9	9	8	8	7	7	6	6	
16	Width	15	16	18	19	20	23	23	25	27	28	32	35	42	42	48
	Height	15	13	12	11	10	10	9	9	9	8	8	7	7	6	6
17	Width	16	18	20	21	22	25	25	28	27	32	32	35	42	49	48
	Height	16	14	13	12	11	11	10	10	9	9	8	7	7	7	6
18	Width	16	19	21	23	24	25	28	28	30	32	36	40	42	49	56
	Height	16	15	14	13	12	11	11	10	10	9	9	8	7	7	7
19	Width	17	20	21	23	24	27	28	30	30	35	36	40	48	49	56
	Height	17	16	14	13	12	12	11	11	10	10	9	8	8	7	7
20	Width	18	20	23	25	26	27	30	30	33	35	40	45	48	56	56
	Height	18	16	15	14	13	12	12	11	11	10	10	9	8	8	7
21	Width	19	21	24	26	28	29	30	33	33	39	40	45	54	56	64
	Height	19	17	16	15	14	13	12	12	11	11	10	9	9	8	8
22	Width	20	23	26	26	28	32	33	36	36	39	44	50	54	56	64
	Height	20	18	17	15	14	14	13	13	12	11	11	10	9	8	8
23	Width	21	24	26	28	30	32	35	36	39	42	44	50	54	63	64
	Height	21	19	17	16	15	14	14	13	13	12	11	10	9	9	8
24	Width	22	25	27	30	32	34	35	39	39	42	48	55	60	63	72
	Height	22	20	18	17	16	15	14	14	13	12	12	11	10	9	9
25	Width	23	25	29	30	32	36	38	39	42	46	48	55	60	70	72
	Height	23	20	19	17	16	16	15	14	14	13	12	11	10	10	9
26	Width	24	26	30	32	34	36	38	41	42	46	52	55	66	70	72
	Height	24	21	20	18	17	16	15	15	14	13	13	11	11	10	9
27	Width	25	28	30	33	36	38	40	41	45	49	52	60	66	70	80
	Height	25	22	20	19	18	17	16	15	15	14	13	12	11	10	10
28	Width	26	29	32	35	36	38	43	44	45	49	56	60	66	77	80
	Height	26	23	21	20	18	17	17	16	15	14	14	12	11	11	10
29	Width	27	30	33	35	38	41	43	44	48	53	56	65	72	77	88
	Height	27	24	22	20	19	18	17	16	16	15	14	13	12	11	11
30	Width	27	31	35	37	40	43	45	47	48	53	60	65	72	77	88
	Height	27	25	23	21	20	19	18	17	16	15	15	13	12	11	11
31	Width	28	31	35	39	40	43	45	50	51	56	60	70	78	84	88
	Height	28	25	23	22	20	19	18	18	17	16	15	14	13	12	11
32	Width	29	33	36	39	42	45	48	50	54	56	60	70	78	84	96
	Height	29	26	24	22	21	20	19	18	18	16	15	14	13	12	12
33	Width	30	34	38	40	44	47	50	52	54	60	64	75	78	91	96
	Height	30	27	25	23	22	21	20	19	18	17	16	15	13	13	12
34	Width	31	35	39	42	44	47	50	52	57	60	64	75	84	91	96
	Height	31	28	26	24	22	21	20	19	19	17	16	15	14	13	12
35	Width	32	36	39	42	46	50	53	55	57	63	68	75	84	91	104
	Height	32	29	26	24	23	22	21	20	19	18	17	15	14	13	13
36	Width	33	36	41	44	48	50	53	55	60	63	68	80	90	98	104
	Height	33	29	27	25	24	22	21	20	20	18	17	16	15	14	13
38	Width	35	39	44	47	50	54	58	61	63	67	72	85	96	105	112
	Height	35	31	29	27	25	24	23	22	21	19	18	17	16	15	14

*Shaded area not recommended.

(By permission of American Society of Heating, Refrigerating and Air-Conditioning Engineers, Inc. Atlanta, Georgia, from their 1993 ASHRAE Fundamentals Handbook)

3.31.0 Equivalent Rectangular Duct Dimension Tables (Continued)

Duct Diameter, in.	Rectangular Size, in.	Aspect Ratio														
		1.00	1.25	1.50	1.75	2.00	2.25	2.50	2.75	3.00	3.50	4.00	5.00	6.00	7.00	8.00
40	Width	37	41	45	49	52	56	60	63	66	70	76	90	96	105	120
	Height	37	33	30	28	26	25	24	23	22	20	19	18	16	15	15
42	Width	38	43	48	51	56	59	63	66	69	74	80	90	102	112	120
	Height	38	34	32	29	28	26	25	24	23	21	20	18	17	16	15
44	Width	40	45	50	54	58	61	65	69	72	81	84	95	108	119	128
	Height	40	36	33	31	29	27	26	25	24	23	21	19	18	17	16
46	Width	42	48	53	56	60	65	68	72	75	84	88	100	114	126	136
	Height	42	38	35	32	30	29	27	26	25	24	22	20	19	18	17
48	Width	44	49	54	60	62	68	70	74	78	88	92	105	120	126	136
	Height	44	39	36	34	31	30	28	27	26	25	23	21	20	18	17
50	Width	46	51	57	61	66	70	75	77	81	91	96	110	120	133	144
	Height	46	41	38	35	33	31	30	28	27	26	24	22	20	19	18
52	Width	48	54	59	63	68	72	78	83	84	95	100	115	126	140	152
	Height	48	43	39	36	34	32	31	30	28	27	25	23	21	20	19
54	Width	49	55	62	67	70	77	80	85	90	98	104	120	132	147	160
	Height	49	44	41	38	35	34	32	31	30	28	26	24	22	21	20
56	Width	51	58	63	68	74	79	83	88	93	102	108	125	138	147	160
	Height	51	46	42	39	37	35	33	32	31	29	27	25	23	21	20
58	Width	53	60	66	70	76	81	85	91	96	105	112	130	144	154	168
	Height	53	48	44	40	38	36	34	33	32	30	28	26	24	22	21
60	Width	55	61	68	74	78	83	90	94	99	109	116	130	144	161	
	Height	55	49	45	42	39	37	36	34	33	31	29	26	24	23	
62	Width	57	64	71	75	82	88	93	96	102	112	120	135	150	168	
	Height	57	51	47	43	41	39	37	35	34	32	30	27	25	24	
64	Width	59	65	72	79	84	90	95	99	105	116	124	140	156		
	Height	59	52	48	45	42	40	38	36	35	33	31	28	26		
66	Width	60	68	75	81	86	92	98	105	108	119	128	145	162		
	Height	60	54	50	46	43	41	39	38	36	34	32	29	27		
68	Width	62	70	77	82	90	95	100	107	111	123	132	150	168		
	Height	62	56	51	47	45	42	40	39	37	35	33	30	28		
70	Width	64	71	80	86	92	99	105	110	114	126	136	155			
	Height	64	57	53	49	46	44	42	40	38	36	34	31			
72	Width	66	74	81	88	94	101	108	113	117	130	140	160			
	Height	66	59	54	50	47	45	43	41	39	37	35	32			
74	Width	68	76	84	91	98	104	110	116	123	133	144	165			
	Height	68	61	56	52	49	46	44	42	41	38	36	33			
76	Width	70	78	86	93	100	106	113	118	126	137	148	165			
	Height	70	62	57	53	50	47	45	43	42	39	37	33			
78	Width	71	80	89	95	102	110	115	121	129	140	152				
	Height	71	64	59	54	51	49	46	44	43	40	38				
80	Width	73	83	90	98	104	113	118	124	132	144	156				
	Height	73	66	60	56	52	50	47	45	44	41	39				
82	Width	75	84	93	100	108	115	123	129	135	147	160				
	Height	75	67	62	57	54	51	49	47	45	42	40				
84	Width	77	86	95	103	110	117	125	132	138	151	164				
	Height	77	69	63	59	55	52	50	48	46	43	41				
86	Width	79	88	98	105	112	119	128	135	141	154	168				
	Height	79	70	65	60	56	53	51	49	47	44	42				
88	Width	80	90	99	107	116	124	130	138	144	158					
	Height	80	72	66	61	58	55	52	50	48	45					
90	Width	82	93	102	110	118	126	133	140	147	161					
	Height	82	74	68	63	59	56	53	51	49	46					
92	Width	84	94	104	112	120	128	138	143	150	165					
	Height	84	75	69	64	60	57	55	52	50	47					
94	Width	86	96	107	116	124	131	140	146	153	168					
	Height	86	77	71	66	62	58	56	53	51	48					
96	Width	88	99	108	117	126	135	143	151	159						
	Height	88	79	72	67	63	60	57	55	53						
98	Width	90	100	111	119	128	137	145	154	162						
	Height	90	80	74	68	64	61	58	56	54						
100	Width	91	103	113	123	132	140	148	157	165						
	Height	91	82	75	70	66	62	59	57	55						
102	Width	93	105	116	124	134	142	153	160	168						
	Height	93	84	77	71	67	63	61	58	56						
104	Width	95	106	117	128	136	146	155	162							
	Height	95	85	78	73	68	65	62	59							

*Shaded area not recommended.

(By permission of American Society of Heating, Refrigerating and Air-Conditioning Engineers, Inc. Atlanta, Georgia, from their 1993 ASHRAE Fundamentals Handbook)

3.31.0 Equivalent Rectangular Duct Dimension Tables (Continued)

Duct Diameter, in.	Rectangular Size, in.	Aspect Ratio														
		1.00	1.25	1.50	1.75	2.00	2.25	2.50	2.75	3.00	3.50	4.00	5.00	6.00	7.00	8.00
106	Width	97	109	120	130	140	149	158	165							
	Height	97	87	80	74	70	66	63	60							
108	Width	99	110	122	131	142	151	160	168							
	Height	99	88	81	75	71	67	64	61							
110	Width	101	113	125	135	144	153	163								
	Height	101	90	83	77	72	68	65								
112	Width	102	115	126	137	146	158	165								
	Height	102	92	84	78	73	70	66								
114	Width	104	116	129	140	150	160									
	Height	104	93	86	80	75	71									
116	Width	106	119	131	142	152	162									
	Height	106	95	87	81	76	72									
118	Width	108	121	134	144	154	164									
	Height	108	97	89	82	77	73									
120	Width	110	123	135	147	158										
	Height	110	98	90	84	79										

*Shaded area not recommended.

(By permission of American Society of Heating, Refrigerating and Air-Conditioning Engineers, Inc. Atlanta, Georgia, from their 1993 ASHRAE Fundamentals Handbook)

3.32.0 Equivalent Spiral, Flat, Oval Duct Dimensions

Duct Diameter, in.	Major Axis (a), in. / Minor Axis (b), in.														
	3	4	5	6	7	8	9	10	11	12	14	16	18	20	22	24
5	8															
5.5	9	7														
6	11	9														
6.5	12	10	8													
7	15	12	10	8												
7.5	19	13	—	9												
8	22	15	11	—												
8.5		18	13	11	10											
9		20	14	12	—	10										
9.5		21	18	14	12	—										
10			19	15	13	11										
10.5			21	17	15	13	12									
11				19	16	14	—	12								
11.5				20	18	16	14	—								
12				23	20	17	15	13								
12.5					25	21	—	—	15	14						
13					28	23	19	17	16	—	14					
13.5					30	—	21	18	—	16	—					
14					33	—	22	20	18	17	15					
14.5					36	—	24	22	19	—	17					
15						39	—	27	23	21	19	18				
16						45	—	30	—	24	22	20	17			
17						52	—	35	—	27	24	21	19			
18						59	—	39	—	30	—	25	22	19		
19						46	—	34	—	28	23	21				
20						50	—	38	—	31	27	24	21			
21						58	—	43	—	34	28	25	23			
22						65	—	48	—	37	31	29	26			
23						71	—	52	—	42	34	30	27			
24						77	—	57	—	45	38	33	29	26		
25							63	—	50	41	36	32	29			
26							70	—	56	45	38	34	31			
27							76	—	59	49	41	37	34			
28									65	52	46	40	36			
29										72	58	49	43	39	35	
30										78	61	54	46	40	38	
31										81	67	57	49	44	39	37
32											71	60	53	47	42	40
33											77	66	56	51	46	41
34												69	59	55	47	44
35												76	65	58	50	46
36												79	68	61	53	49
37													71	64	57	52
38													78	67	60	55
40														77	69	62
42															75	68
44															82	74

(By permission of American Society of Heating, Refrigerating and Air-Conditioning Engineers, Inc. Atlanta, Georgia, from their 1993 ASHRAE Fundamentals Handbook)

Duct Design

Velocity and Velocity Pressure Relationships

Velocity (fpm)	Velocity Pressure (in wg)	Velocity (fpm)	Velocity Pressure (in wg)
300	0.0056	3500	0.7637
400	0.0097	3600	0.8079
500	0.0155	3700	0.8534
600	0.0224	3800	0.9002
700	0.0305	3900	0.9482
800	0.0399	4000	0.9975
900	0.0504	4100	1.0480
1000	0.0623	4200	1.0997
1100	0.0754	4300	1.1527
1200	0.0897	4400	1.2069
1300	0.1053	4500	1.2624
1400	0.1221	4600	1.3191
1500	0.1402	4700	1.3771
1600	0.1596	4800	1.4364
1700	0.1801	4900	1.4968
1800	0.2019	5000	1.5586
1900	0.2250	5100	1.6215
2000	0.2493	5200	1.6857
2100	0.2749	5300	1.7512
2200	0.3017	5400	1.8179
2300	0.3297	5500	1.8859
2400	0.3591	5600	1.9551
2500	0.3896	5700	2.0256
2600	0.4214	5800	2.0972
2700	0.4544	5900	2.1701
2800	0.4887	6000	2.2443
2900	0.5243	6100	2.3198
3000	0.5610	6200	2.3965
3100	0.5991	6300	2.4744
3200	0.6384	6400	2.5536
3300	0.6789	6500	2.6340
3400	0.7206	6600	2.7157

For calculation of velocity pressures at velocities other than those listed above: $P_v = (V/4005)^2$

For calculation of velocities when velocity pressures are known:

$$V = 4005 \sqrt{(V_p)}$$

Figure 3.32.1 Duct velocity and velocity pressure relationships. (*By permission: Loren Cook, Springfield, MO.*)

Duct Design

U.S. Sheet Metal Gauges

Gauge No.	Steel (Manuf. Std. Ga.)		Galvanized (Manuf. Std. Ga.)	
	Thick. in.	Lb./ft.^{2}	Thick.in.	Lb./ft.^{2}
26	.0179	.750	.0217	.906
24	.0239	1.00	.0276	1.156
22	.0299	1.25	.0336	1.406
20	.0359	1.50	.0396	1.656
18	.0478	2.00	.0516	2.156
16	.0598	2.50	.0635	2.656
14	.0747	3.125	.0785	3.281
12	.1046	4.375	.1084	4.531
10	.1345	5.625	.1382	5.781
8	.1644	6.875	.1681	7.031
7	.1793	7.50	—	—

Gauge No.	Mill Std. Thick Aluminum*		Stainless Steel (U.S. Standard Gauge)	
	Thick. in.	Lb./ft.^{2}	Thick.in.	Lb./ft.^{2}
26	.020	.282	.0188	.7875
24	.025	.353	.0250	1.050
22	.032	.452	.0312	1.313
20	.040	.564	.0375	1.575
18	.050	.706	.050	2.100
16	.064	.889	.062	2.625
14	.080	1.13	.078	3.281
12	.100	1.41	.109	4.594
10	.125	1.76	.141	5.906
8	.160	2.26	.172	7.218
7	.190	2.68	.188	7.752

Aluminum is specified and purchased by material thickness rather than gauge.

Figure 3.32.2 U.S. sheet metal gauges—steel, galvanized, and aluminum. (*By permission: Loren Cook, Springfield, MO.*)

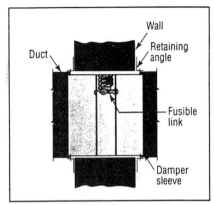

Fig. 1   Type A has the blade stack and channels in the air stream and is meant for transfer grille/louver low velocity type applications.

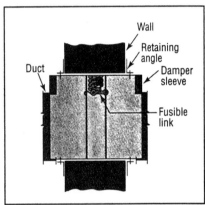

Fig. 2   Type B has the blade stack out of the air stream but the channel is in it.

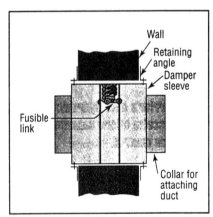

Fig. 3   Type C has both the blade stack and channels clear of the air stream.

Figure 3.33.0   Fire dampers—typical installations. Why fire dampers? When an air duct or transfer grill results in the penetration of a fire-rated wall or floor (2 hr or more for a duct or any rating for a transfer grill), a fire damper is installed to restrict the passage of flame. The damper uses a fusible that causes damper blades to close or a curtain blade stack to drop on heating above the link's rating. (*By permission: HPAC Magazine— Heating/Piping/Air Conditioning—December 1998.*)

3.34.0 Typical Fan Configurations

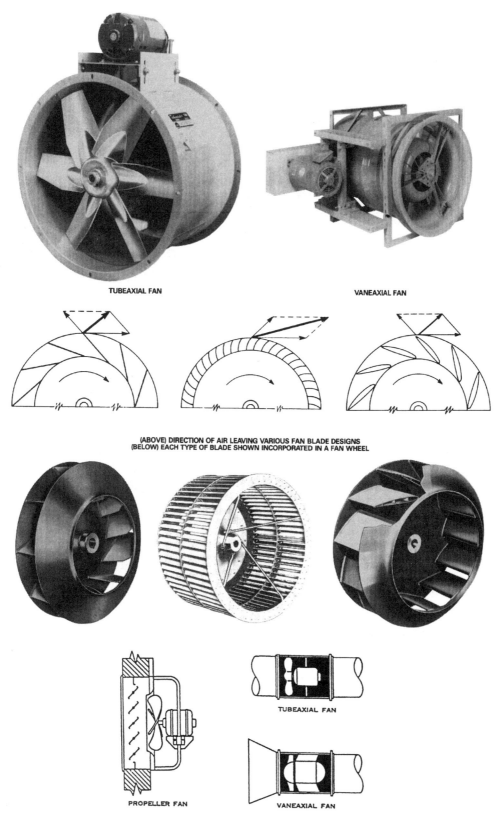

TUBEAXIAL FAN

VANEAXIAL FAN

(ABOVE) DIRECTION OF AIR LEAVING VARIOUS FAN BLADE DESIGNS
(BELOW) EACH TYPE OF BLADE SHOWN INCORPORATED IN A FAN WHEEL

PROPELLER FAN

TUBEAXIAL FAN

VANEAXIAL FAN

(By permission of The Trane Company, LaCrosse, Wisconsin)

Fan Laws

The simplified form of the most commonly used fan laws include.

- **CFM varies directly with RPM**
  $CFM_1/CFM_2 = RPM_1/RPM_2$

- **SP varies with the square of the RPM**
  $SP_1/SP_2 = (RPM_1/RPM_2)^2$

- **HP varies with the cube of the RPM**
  $HP_1/HP_2 = (RPM_1/RPM_2)^3$

Fan Performance Tables and Curves

Performance tables provide a simple method of fan selection. However, it is critical to evaluate fan performance curves in the fan selection process as *the margin for error is very slim when selecting a fan near the limits of tabular data.* The performance curve also is a valuable tool when evaluating fan performance in the field.

Fan performance tables and curves are based on standard air density of 0.075 lb/ft^3. When altitude and temperature differ significantly from standard conditions (sea level and 70° F) performance modification factors must be taken into account to ensure proper performance.

For further information refer to *Use of Air Density Factors— An Example,* page 3.

Fan Testing—Laboratory, Field

Fans are tested and performance certified under ideal laboratory conditions. When fan performance is measured in field conditions, the difference between the ideal laboratory condition and the actual field installation must be considered. Consideration must also be given to fan inlet and discharge connections as they will dramatically affect fan performance in the field. If possible, readings must be taken in straight runs of ductwork in order to ensure validity. If this cannot be accomplished, motor amperage and fan RPM should be used along with performance curves to estimate fan performance.

Air Density Factors for Altitude and Temperature

Altitude (ft.)	Temperature							
	70	100	200	300	400	500	600	700
0	1.000	.946	.803	.697	.616	.552	.500	.457
1000	.964	.912	.774	.672	.594	.532	.482	.441
2000	.930	.880	.747	.648	.573	.513	.465	.425
3000	.896	.848	.720	.624	.552	.495	.448	.410
4000	.864	.818	.694	.604	.532	.477	.432	.395
5000	.832	.787	.668	.580	.513	.459	.416	.380
6000	.801	.758	.643	.558	.493	.442	.400	.366
7000	.772	.730	.620	.538	.476	.426	.386	.353
8000	.743	.703	.596	.518	.458	.410	.372	.340
9000	.714	.676	.573	.498	.440	.394	.352	.326
10000	.688	.651	.552	.480	.424	.380	.344	.315
15000	.564	.534	.453	.393	.347	.311	.282	.258
20000	.460	.435	.369	.321	.283	.254	.230	.210

Use of Air Density Factors—An Example

A fan is selected to deliver 7500 CFM at 1-1/2 inch SP at altitude of 6000 feet above sea level and an operating temperature of 200° F. From the table above, **Air Density Factors for Altitude and Temperature,** the air density correction factor is determined to be .643 by using the fan's operating altitude and temperature. Divide the design SP by the air density correction factor.

1.5" SP/.643 = 2.33" SP

Referring to the fan's performance rating table, it is determined that the fan must operate at 976 RPM to develop the desired 7500 CFM at 6000 foot above sea level and at an operating temperature of 200° F.

The BHP (Brake Horsepower) is determined from the fan's performance table to be 3.53. This is corrected to conditions at altitude by multiplying the BHP by the air density correction factor.

3.53 BHP × .643 = 2.27 BHP

The final operating conditions are determined to be 7500 CFM 1-1/2" SP, 976 RPM, and 2.27 BHP.

Fan Types

Axial Fan—An axial fan discharges air parallel to the axis of the impeller rotation. As a general rule, axial fans are preferred for high volume, low pressure, and non-ducted systems.

Axial Fan Types
Propeller, Tube Axial and Vane Axial.

Centrifugal Fan—Centrifugal fans discharge air perpendicular to the axis of the impeller rotation. As a general rule, centrifugal fans are preferred for higher pressure ducted systems.

Centrifugal Fan Types
Backward Inclined, Airfoil, Forward Curved, and Radial Tip.

Fan Selection Criteria

Before selecting a fan, the following information is needed.

- Air volume required—CFM
- System resistance—SP
- Air density (Altitude and Temperature)
- Type of service
  - Environment type
  - Materials/vapors to be exhausted
  - Operation temperature
- Space limitations
- Fan type
- Drive type (Direct or Belt)
- Noise criteria
- Number of fans
- Discharge
- Rotation
- Motor position
- Expected fan life in years

Figure 3.34.1 Fan basics. (*By permission: Loren Cook, Springfield, MO.*)

System Design Guidelines

General Ventilation

- Locate intake and exhaust fans to make use of prevailing winds.
- Locate fans and intake ventilators for maximum sweeping effect over the working area.
- If filters are used on gravity intake, size intake ventilator to keep intake losses below 1/8" SP.
- Avoid fans blowing opposite each other. When necessary, separate by at least 6 fan diameters.
- Use Class B insulated motors where ambient temperatures are expected to be high for air-over motor conditions.
- If air moving over motors contains hazardous chemicals or particles, use explosion-proof motors mounted in or out of the airstream, depending on job requirements.
- For hazardous atmosphere applications use fans of non-sparking construction.*

Process Ventilation

- Collect fumes and heat as near the source of generation as possible.
- Make all runs of ducts as short and direct as possible.
- Keep duct velocity as low as practical considering capture for fumes or particles being collected.
- When turns are required in the duct system use long radius elbows to keep the resistance to a minimum (preferably 2 duct diameters).
- After calculating duct resistance, select the fan having reserve capacity beyond the static pressure determined.
- Use same rationale regarding intake ventilators and motors as in General Ventilation guidelines above.
- Install the exhaust fan at a location to eliminate any recirculation into other parts of the plant.
- When hoods are used, they should be sufficient to collect all contaminating fumes or particles created by the process.

* Refer to AMCA Standard 99; See page 4.

Figure 3.34.2 General ventilation guidelines. (*By permission: Loren Cook, Springfield, MO.*)

Air Quality Method

Designing for acceptable indoor air quality requires that we address:
- Outdoor air quality
- Design of the ventilation systems
- Sources of contaminants
- Proper air filtration
- System operation and maintenance

Determine the number of people occupying the respective building spaces. Find the CFM/person requirements in Ventilation Rates for Acceptable Indoor Air Quality, page 42. Calculate the required outdoor air volume as follows:

$$People = Occupancy/1000 \times Floor\ Area\ (ft^2)$$
$$CFM = People \times Outdoor\ Air\ Requirement\ (CFM/person)$$

Outdoor air quantities can be reduced to lower levels if proper particulate and gaseous air filtration equipment is utilized.

Air Change Method

Find total volume of space to be ventilated. Determine the required number of air changes per hour.

$$CFM = Bldg.\ Volume\ (ft^3)\ /\ Air\ Change\ Frequency$$

Consult local codes for air change requirements or, in absence of code, refer to "Suggested Air Changes", page 41.

Heat Removal Method

When the temperature of a space is higher than the ambient outdoor temperature, general ventilation may be utilized to provide "free cooling". Knowing the desired indoor and the design outdoor dry bulb temperatures, and the amount of heat removal required (BTU/Hr):

$$CFM = Heat\ Removal\ (BTU/Hr)\ /\ (1.10 \times Temp\ diff)$$

Suggested Air Changes

Type of Space	Air Change Frequency (minutes)
Assembly Halls	3-10
Auditoriums	4-15
Bakeries	1-3
Boiler Rooms	2-4
Bowling Alleys	2-8
Dry Cleaners	1-5
Engine Rooms	1-1.5
Factories (General)	1-5
Forges	1-2
Foundries	1-4
Garages	2-10
Generating Rooms	2-5
Glass Plants	1-2
Gymnasiums	2-10
Heat Treat Rooms	0.5-1
Kitchens	1-3
Laundries	2-5
Locker Rooms	2-5
Machine Shops	3-5
Mills (Paper)	2-3
Mills (Textile)	5-15
Packing Houses	2-15
Recreation Rooms	2-8
Residences	2-5
Restaurants	5-10
Retail Stores	3-10
Shops (General)	3-10
Theaters	3-8
Toilets	2-5
Transformer Rooms	1-5
Turbine Rooms	2-6
Warehouses	2-10

Figure 3.34.3 Acceptable indoor air qualities and suggested air changes by type of space. (*By permission: Loren Cook, Springfield, MO.*)

System Design Guidelines

Kitchen Ventilation
Hoods and Ducts
- Duct velocity should be between 1500 and 4000 fpm
- Hood velocities (not less than 50 fpm over face area between hood and cooking surface)
  - Wall Type - 80 CFM/ft2
  - Island Type - 125 CFM/ft2
- Extend hood beyond cook surface 0.4 x distance between hood and cooking surface

Filters
- Select filter velocity between 100 - 400 fpm
- Determine number of filters required from a manufacturer's data (usually 2 cfm exhaust for each sq. in. of filter area maximum)
- Install at 45 - 60° to horizontal, never horizontal
- Shield filters from direct radiant heat
- Filter mounting height:
  - No exposed cooking flame—1-1/2' minimum to filter
  - Charcoal and similar fires—4' minimum to filter
- Provide removable grease drip pan
- Establish a schedule for cleaning drip pan and filters and follow it diligently

Fans
- Use upblast discharge fan
- Select design CFM based on hood design and duct velocity
- Select SP based on design CFM and resistance of filters and duct system
- Adjust fan specification for expected exhaust air temperature

System Design Guidelines

Design Criteria for Room Loudness

Room Type	Sones	Room Type	Sones
Auditoriums		**Indoor sports activities**	
Concert and opera halls	1.0 to 3	Gymnasiums	4 to 12
Stage theaters	1.5 to 5	Coliseums	3 to 9
Movie theaters	2.0 to 6	Swimming pools	7 to 21
Semi-outdoor amphi-theaters	2.0 to 6	Bowling alleys	4 to 12
Lecture halls	2.0 to 6	Gambling casinos	4 to 12
Multi-purpose	1.5 to 5	**Manufacturing areas**	
Courtrooms	3.0 to 9	Heavy machinery	25 to 60
Auditorium lobbies	4.0 to 12	Foundries	20 to 60
TV audience studios	2.0 to 6	Light machinery	12 to 36
Churches and schools		Assembly lines	12 to 36
Sanctuaries	1.7 to 5	Machine shops	15 to 50
Schools & classrooms	2.5 to 8	Plating shops	20 to 50
Recreation halls	4.0 to 12	Punch press shops	50 to 60
Kitchens	6.0 to 18	Tool maintenance	7 to 21
Libraries	2.0 to 6	Foreman's office	5 to 15
Laboratories	4.0 to 12	General storage	10 to 30
Corridors and halls	5.0 to 15	**Offices**	
Hospitals and clinics		Executive	2 to 6
Private rooms	1.7 to 5	Supervisor	3 to 9
Wards	2.5 to 8	General open offices	4 to 12
Laboratories	4.0 to 12	Tabulation/computation	6 to 18
Operating rooms	2.5 to 8	Drafting	4 to 12
Lobbies & waiting rooms	4.0 to 12	Professional offices	3 to 9
Halls and corridors	4.0 to 12	Conference rooms	1.7 to 5
		Board of Directors	1 to 3
		Halls and corridors	5 to 15

Note: Values showns above are room loudness in sones and are not fan sone ratings. For additional detail see AMCA publication 302 - Application of Sone Rating.

Figure 3.34.4   Kitchen ventilation guidelines/design criteria for room loudness. (*By permission: Loren Cook, Springfield, MO.*)

Fan Basics

Fan Installation Guidelines
Centrifugal Fan Conditions
Typical Inlet Conditions

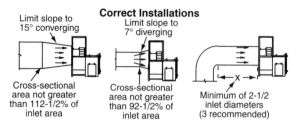

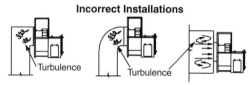

Typical Outlet Conditions

Figure 3.34.5 Fan installation guidelines. (*By permission: Loren Cook, Springfield, MO.*)

Fan Basics

Fan Troubleshooting Guide

Low Capacity or Pressure

- Incorrect direction of rotation – Make sure the fan rotates in same direction as the arrows on the motor or belt drive assembly.
- Poor fan inlet conditions –There should be a straight, clear duct at the inlet.
- Improper wheel alignment.

Excessive Vibration and Noise

- Damaged or unbalanced wheel.
- Belts too loose; worn or oily belts.
- Speed too high.
- Incorrect direction of rotation. Make sure the fan rotates in same direction as the arrows on the motor or belt drive assembly.
- Bearings need lubrication or replacement.
- Fan surge.

Overheated Motor

- Motor improperly wired.
- Incorrect direction of rotation. Make sure the fan rotates in same direction as the arrows on the motor or belt drive assembly.
- Cooling air diverted or blocked.
- Improper inlet clearance.
- Incorrect fan RPM.
- Incorrect voltage.

Overheated Bearings

- Improper bearing lubrication.
- Excessive belt tension.

Figure 3.34.6 Troubleshooting guide. (*By permission: Loren Cook, Springfield, MO.*)

Fan Basics

Classifications for Spark Resistant Construction†

Fan applications may involve the handling of potentially explosive or flammable particles, fumes or vapors. Such applications require careful consideration of all system components to insure the safe handling of such gas streams. This AMCA Standard deals only with the fan unit installed in that system. The Standard contains guidelines which are to be used by both the manufacturer and user as a means of establishing general methods of construction. The exact method of construction and choice of alloys is the responsibility of the manufacturer; however, the customer must accept both the type and design with full recognition of the potential hazard and the degree of protection required.

Construction Type

A. All parts of the fan in contact with the air or gas being handled shall be made of nonferrous material. Steps must also be taken to assure that the impeller, bearings, and shaft are adequately attached and/or restrained to prevent a lateral or axial shift in these components.

B. The fan shall have a nonferrous impeller and nonferrous ring about the opening through which the shaft passes. Ferrous hubs, shafts, and hardware are allowed provided construction is such that a shift of impeller or shaft will not permit two ferrous parts of the fan to rub or strike. Steps must also be taken to assure the impeller, bearings, and shaft are adequately attached and/or restrained to prevent a lateral or axial shift in these components.

C. The fan shall be so constructed that a shift of the impeller or shaft will not permit two ferrous parts of the fan to rub or strike.

Notes

1. No bearings, drive components or electrical devices shall be placed in the air or gas stream unless they are constructed or enclosed in such a manner that failure of that component cannot ignite the surrounding gas stream.

2. The user shall electrically ground all fan parts.

3. For this Standard, nonferrous material shall be a material with less than 5% iron or any other material with demonstrated ability to be spark resistant.

Classifications for Spark Resistant Construction (cont.)

4. The use of aluminum or aluminum alloys in the presence of steel which has been allowed to rust requires special consideration. Research by the U.S. Bureau of Mines and others has shown that aluminum impellers rubbing on rusty steel may cause high intensity sparking.

The use of the above Standard in no way implies a guarantee of safety for any level of spark resistance. "Spark resistant construction also does not protect against ignition of explosive gases caused by catastrophic failure or from any airstream material that may be present in a system."

Standard Applications

- Centrifugal Fans
- Axial and Propeller Fans
- Power Roof Ventilators

This standard applies to ferrous and nonferrous metals. The potential questions which may be associated with fans constructed of FRP, PVC, or any other plastic compound were not addressed.

Figure 3.34.7 Fan spark resistance construction. (*By permission: Loren Cook, Springfield, MO.*)

Fan Basics

Impeller Designs - Axial

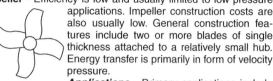

Propeller - Efficiency is low and usually limited to low pressure applications. Impeller construction costs are also usually low. General construction features include two or more blades of single thickness attached to a relatively small hub. Energy transfer is primarily in form of velocity pressure.

Applications - Primary applications include low pressure, high volume air moving applications such as air circulation within a space or ventilation through a wall without attached duct work. Used for replacement air applications.

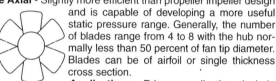

Tube Axial - Slightly more efficient than propeller impeller design and is capable of developing a more useful static pressure range. Generally, the number of blades range from 4 to 8 with the hub normally less than 50 percent of fan tip diameter. Blades can be of airfoil or single thickness cross section.

Applications - Primary applications include low and medium pressure ducted heating, ventilating, and air conditioning applications where air distribution on the downstream side is not critical. Also used in some industrial applications such as drying ovens, paint spray booths, and fume exhaust systems.

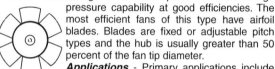

Vane Axial - Solid design of the blades permits medium to high pressure capability at good efficiencies. The most efficient fans of this type have airfoil blades. Blades are fixed or adjustable pitch types and the hub is usually greater than 50 percent of the fan tip diameter.

Applications - Primary applications include general heating, ventilating, and air conditioning systems in low, medium, and high pressure applications. Advantage where straight through flow and compact installation are required. Air distribution on downstream side is good. Also used in some industrial applications such as drying ovens, paint spray booths, and fume exhaust systems. Relatively more compact than comparable centrifugal type fans for the same duty.

Impeller Designs - Centrifugal

Airfoil - Has the highest efficiency of all of the centrifugal impeller designs with 9 to 16 blades of airfoil contour curved away from the direction of rotation. Air leaves the impeller at a velocity less than its tip speed. Relatively deep blades provide for efficient expansion with the blade passages. For the given duty, the airfoil impeller design will provide for the highest speed of the centrifugal fan designs.

Applications - Primary applications include general heating systems, and ventilating and air conditioning systems. Used in larger sizes for clean air industrial applications providing significant power savings.

Figure 3.34.8 Fan impeller design. (*By permission: Loren Cook, Springfield, MO.*)

Impeller Designs - Centrifugal (cont.)

Backward Inclined, Backward Curved - Efficiency is slightly less than that of the airfoil design. Backward inclined or backward curved blades are single thickness with 9 to 16 blades curved or inclined away from the direction of rotation. Air leaves the impeller at a velocity less than its tip speed. Relatively deep blades provide efficient expansion with the blade passages.

Applications - Primary applications include general heating systems, and ventilating and air conditioning systems. Also used in some industrial applications where the airfoil blade is not acceptable because of a corrosive and/or erosive environment.

Radial - Simplest of all centrifugal impellers and least efficient. Has high mechanical strength and the impeller is easily repaired. For a given point of rating, this impeller requires medium speed. Classification includes radial blades and modified radial blades), usually with 6 to 10 blades.

Applications - Used primarily for material handling applications in industrial plants. Impeller can be of rugged construction and is simple to repair in the field. Impeller is sometimes coated with special material. This design also is used for high pressure industrial requirements and is not commonly found in HVAC applications.

Forward Curved - Efficiency is less than airfoil and backward curved bladed impellers. Usually fabricated at low cost and of lightweight construction. Has 24 to 64 shallow blades with both the heel and tip curved forward. Air leaves the impeller at velocities greater than the impeller tip speed. Tip speed and primary energy transferred to the air is the result of high impeller velocities. For the given duty, the wheel is the smallest of all of the centrifugal types and operates most efficiently at lowest speed.

Applications - Primary applications include low pressure heating, ventilating, and air conditioning applications such as domestic furnaces, central station units, and packaged air conditioning equipment from room type to roof top units.

Fan Basics

Terminology for Centrifugal Fan Components

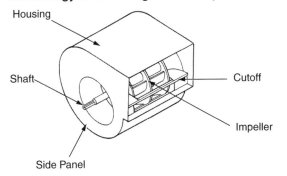

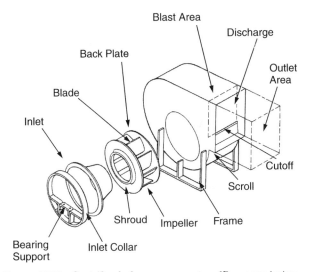

Figure 3.34.9 Centrifugal fan components. (*By permission: Loren Cook, Springfield, MO.*)

Fan Basics

Motor Positions for Belt Drive Centrifugal Fans†

To determine the location of the motor, face the drive side of the fan and pick the proper motor position designated by the letters W, X, Y or Z as shown in the drawing below.

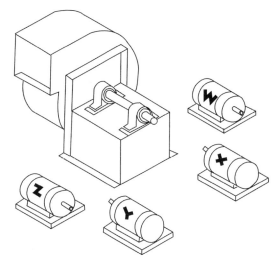

Figure 3.34.10 Motor positions for belt drive centrifugal fans. (*By permission: Loren Cook, Springfield, MO.*)

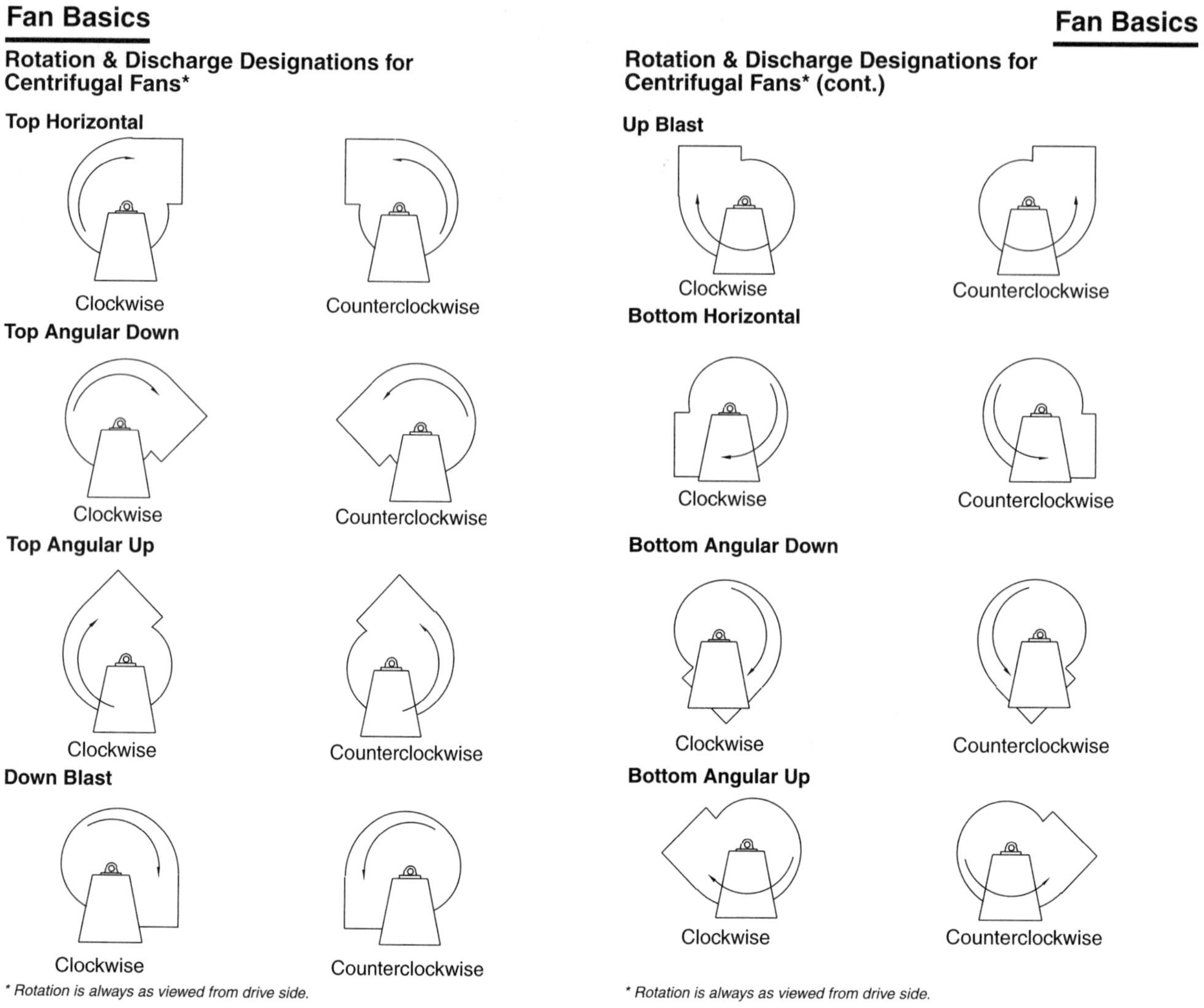

Fan Basics

Rotation & Discharge Designations for Centrifugal Fans*

Top Horizontal

Clockwise   Counterclockwise

Top Angular Down

Clockwise   Counterclockwise

Top Angular Up

Clockwise   Counterclockwise

Down Blast

Clockwise   Counterclockwise

* Rotation is always as viewed from drive side.

Rotation & Discharge Designations for Centrifugal Fans* (cont.)

Up Blast

Clockwise   Counterclockwise

Bottom Horizontal

Clockwise   Counterclockwise

Bottom Angular Down

Clockwise   Counterclockwise

Bottom Angular Up

Clockwise   Counterclockwise

* Rotation is always as viewed from drive side.

Figure 3.34.11   Rotation and discharge designations for centrifugal fans. (*By permission: Loren Cook, Springfield, MO.*)

Fan Basics

Drive Arrangements for Centrifugal Fans†

SW - Single Width, **SI** - Single Inlet
DW - Double Width, **DI** - Double Inlet

Arr. 1 SWSI - For belt drive or direct drive connection. Impeller over-hung. Two bearings on base.

Arr. 2 SWSI - For belt drive or direct drive connection. Impeller over-hung. Bearings in bracket supported by fan housing.

Arr. 3 SWSI - For belt drive or direct drive connection. One bearing on each side supported by fan housing.

Arr. 3 DWDI - For belt drive or direct connection. One bearing on each side and supported by fan housing.

Fan Basics

Drive Arrangements for Centrifugal Fans (cont.)

SW - Single Width, **SI** - Single Inlet
DW - Double Width, **DI** - Double Inlet

Arr. 4 SWSI - For direct drive. Impeller over-hung on prime mover shaft. No bearings on fan. Prime mover base mounted or integrally directly connected.

Arr. 7 SWSI - For belt drive or direct connection. Arrangement 3 plus base for prime mover.

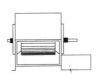

Arr. 7 DWDI - For belt drive or direct connection. Arrangement 3 plus base for prime mover.

Arr. 8 SWSI - For belt drive or direct connection. Arrangement 1 plus extended base for prime mover.

Arr. 9 SWSI - For belt drive. Impeller overhung, two bearings, with prime mover outside base.

Arr. 10 SWSI - For belt drive. Impeller overhung, two bearings, with prime mover inside base.

†Adapted from AMCA Standard 99-2404-78

Figure 3.34.12 Drive arrangements for centrifugal fans. (*By permission: Loren Cook, Springfield, MO.*)

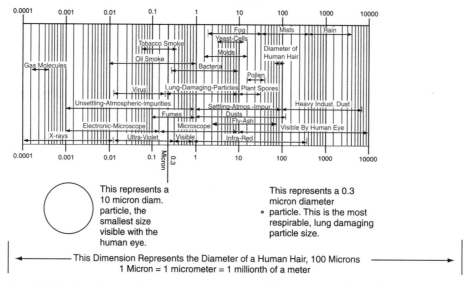

Figure 3.34.13 Relative size of common air contaminants. (*By permission: Loren Cook, Springfield, MO.*)

Filter Comparison

Filter Type	ASHRAE Arrestance Efficiency	ASHRAE Atmospheric Dust Spot Efficiency	Initial Pressure Drop (IN.WG)	Final Pressure Drop (IN.WG)
Permanent	60-80%	8-12%	0.07	.5
Fiberglass Pad	70-85%	15-20%	0.17	.5
Polyester Pad	82-90%	15-20%	0.20	.5
2" Throw Away	70-85%	15-20%	0.17	.5
2" Pleated Media	88-92%	25-30%	0.25	.5-.8
60% Cartridge	97%	60-65%	0.3	1.0
80% Cartridge	98%	80-85%	0.4	1.0
90% Cartridge	99%	90-95%	0.5	1.0
HEPA	100%	99.97%	1.0	2.0

Figure 3.34.14 Various filter comparisons. (*By permission: Loren Cook, Springfield, MO.*)

3.35.0   Rate of Heat Gain from Selected Office Equipment

Appliance	Size	Maximum Input Rating, Btu/h	Standby Input Rating, Btu/h	Recommended Rate of Heat Gain, Btu/h
Check processing workstation	12 pockets	16400	8410	8410
Computer devices				
Card puncher		2730 to 6140	2200 to 4800	2200 to 4800
Card reader		7510	5200	5200
Communication/transmission		6140 to 15700	5600 to 9600	5600 to 9600
Disk drives/mass storage		3410 to 34100	3412 to 22420	3412 to 22420
Magnetic ink reader		3280 to 16000	2600 to 14400	2600 to 14400
Microcomputer	16 to 640 KByte[a]	340 to 2050	300 to 1800	300 to 1800
Minicomputer		7500 to 15000	7500 to 15000	7500 to 15000
Optical reader		10240 to 20470	8000 to 17000	8000 to 17000
Plotters		256	128	214
Printers				
Letter quality	30 to 45 char/min	1200	600	1000
Line, high speed	5000 or more lines/min	4300 to 18100	2160 to 9040	2500 to 13000
Line, low speed	300 to 600 lines/min	1540	770	1280
Tape drives		4090 to 22200	3500 to 15000	3500 to 15000
Terminal		310 to 680	270 to 600	270 to 600
Copiers/Duplicators				
Blue print		3930 to 42700	1710 to 17100	3930 to 42700
Copiers (large)	30 to 67[a] copies/min	5800 to 22500	3070	5800 to 22500
Copiers (small)	6 to 30[a] copies/min	1570 to 5800	1020 to 3070	1570 to 5800
Feeder		100	—	100
Microfilm printer		1540	—	1540
Sorter/collator		200 to 2050	—	200 to 2050
Electronic equipment				
Cassette recorders/players		200	—	200
Receiver/tuner		340	—	340
Signal analyzer		90 to 2220	—	90 to 2220
Mailprocessing				
Folding machine		430	—	270
Inserting machine	3600 to 6800 pieces/h	2050 to 11300	—	1330 to 7340
Labeling machine	1500 to 30000 pieces/h	2050 to 22500	—	1330 to 14700
Postage meter		780	—	510
Wordprocessors/Typewriters				
Letter quality printer	30 to 45 char/min	1200	600	1000
Phototypesetter		5890	—	5180
Typewriter		270	—	230
Wordprocessor		340 to 2050	—	300 to 1800
Vending machines				
Cigarette		250	51 to 85	250
Cold food/beverage		3920 to 6550	—	1960 to 3280
Hot beverage		5890	—	2940
Snack		820 to 940	—	820 to 940
Miscellaneous				
Barcode printer		1500	—	1260
Cash registers		200	—	160
Coffee maker	10 cups	5120	—	3580 sensible 1540 latent
Microfiche reader		290	—	290
Microfilm reader		1770	—	1770
Microfilm reader/printer		3920	—	3920
Microwave oven	1 ft^3	2050	—	1360
Paper shredder		850 to 10240	—	680 to 8250
Water cooler	32 qt/h	2390	—	5970

[a]Input is not proportional to capacity.

(By permission of American Society of Heating, Refrigerating and Air-Conditioning Engineers, Inc. Atlanta, Georgia, from their 1993 ASHRAE Fundamentals Handbook)

Heating & Refrigeration

Heat Loss Estimates

The following will give quick estimates of heat requirements in a building knowing the cu.ft. volume of the building and design conditions.

Type of Structure		Masonry Wall			Insulated Steel Wall		
		Indoor Temp (F)					
		60°	65°	70°	60°	65°	70°
		BTU/Cubic Foot			BTU/Cubic Foot		
Single Story 4 Walls Exposed		3.4	3.7	4.0	2.2	2.4	2.6
Single Story One Heated Wall		2.9	3.1	3.4	1.9	2.0	2.2
Single Floor One Heated Wall Heated Space Above		1.9	2.0	2.2	1.3	1.4	1.5
Single Floor Two Heated Walls Heated Space Above		1.4	1.5	1.6	0.9	1.0	1.1
Single Floor Two Heated Walls		2.4	2.6	2.8	1.6	1.7	1.8
Multi-Story	2 Story	2.9	3.1	3.4	1.9	2.1	2.2
	3 Story	2.8	3.0	3.2	1.8	2.0	2.1
	4 Story	2.7	2.9	3.1	—	—	—
	5 Story	2.6	2.8	3.0	—	—	—

The following correction factors must be used and multiplied by the answer obtained above.

Corrections for Outdoor Design		Corrections for "R" Factor (Steel Wall)	
Temperature	Multiplier	"R" Factor	Multiplier
+50	.23	8	1.0
+40	.36	10	.97
+30	.53	12	95
+20	.69	14	.93
+10	.84	16	.92
+ 0	1.0	19	.91
-10	1.15		
-20	1.2		
-30	1.46		

Heating & Refrigeration

Heat Loss Estimates (cont.)

Considerations Used for Corrected Values

1—0°F Outdoor Design (See Corrections)
2—Slab Construction—If Basement is involved multiply final BTUH by 1.7.
3—Flat Roof
4—Window Area is 5% of Wall Area
5—Air Change is .5 Per Hour.

Fuel Comparisons**

This provides equivalent BTU Data for Various Fuels.

Natural Gas	1,000,000 BTU = 10 Therms or 1,000,000 BTU = (1000 Cu. Ft.)
Propane Gas	1,000,000 BTU = 46 Lb. or 1,000,000 BTU = 10.88 Gallon
No. 2 Fuel Oil	1,000,000 BTU = 7.14 Gallon
Electrical Resistance	1,000,000 BTU = 293 KW (Kilowatts)
Municipal Steam	1,000,000 BTU = 1000 Lbs. Condensate
Sewage Gas	1,000,000 BTU = 1538 Cu.Ft. to 2380 Cu.Ft.
LP/Air Gas	1,000,000 BTU = 46 Lb. Propane or 1,000,000 BTU = 10.88 Gallon Propane or 1,000,000 BTU = 690 Cu.Ft. Gas/Air Mix

Fuel Gas Characteristics

Natural Gas	925 to 1125 BTU/Cu.Ft.	.6 to .66 Specific Gravity
Propane Gas	2550 BTU/Cu.Ft.	1.52 Specific Gravity
*Sewage Gas	420 to 650 BTU/Cu.Ft.	.55 to .85 Specific Gravity
*Coal Gas	400 to 500 BTU/Cu.Ft.	.5 to .6 Specific Gravity
*LP/Air Mix	1425 BTU/Cu.Ft.	1.29 Specific Gravity

* Before attempting to operate units on these fuels, contact manufacturer.
** Chemical Rubber Publishing Co., Handbook of Chemistry and Physics.

Figure 3.35.1   Heat loss estimates for various structures/fuel comparisons. (*By permission: Loren Cook, Springfield, MO.*)

Coefficients of Transmission (U) of Masonry Walls

Coefficients are expressed in Btu per (hour) (square foot) (Fahrenheit degree difference in temperature between the air on the two sides), and are based on an outside wind velocity of 15 mph

Exterior Facing Material	R	AV R	Backing Material	R	None U	Plas. 5/8 In. On Wall (Sand Agg.) U	Plas. 5/8 In. On Wall (Lt. Wt. Agg.) U	Metal Lath and 3/4 In. Plas. On Furring (Sand Agg.) U	Metal Lath and 3/4 In. Plas. On Furring (Lt. Wt. Agg.) U	Gypsum Lath (3/8 In.) and 1/2 In. Plas. On Furring No. Plas. U	Gypsum Lath (Sand Agg.) U	Gypsum Lath (Lt. Wt. Agg.) U	Insul. Bd. Lath (1/2 In.) & 1/2 In. Plas. On Furring No. Plas. U	Insul. Bd. Lath (Sand Agg.) U	Wood Lath and 1/2 In. Plas. (Sand Agg.) U	No.
Resistance →						0.11	0.39	0.13	0.47	0.32	0.41	0.64	1.43	1.52	0.40	
					A	B	C	D	E	F	G	H	I	J	K	
Face Brick 4 In.	0.44		Concrete Block (Cinder Agg.) (4 In.)	1.11	0.41	0.39	0.35	0.28	0.26	0.27	0.26	0.25	0.21	0.20	0.26	1
Stone 4 In.	0.32		(8 In.)	1.72	0.33	0.32	0.29	0.24	0.22	0.23	0.23	0.21	0.18	0.18	0.23	2
		.39	(12 In.)	1.89	0.31	0.30	0.28	0.23	0.21	0.22	0.22	0.21	0.18	0.17	0.22	3
Precast Concrete (Sand Agg.) 4 In.	0.32		(Lt. Wt. Agg.) (4 In.)	1.50	0.35	0.34	0.31	0.25	0.23	0.24	0.24	0.22	0.19	0.19	0.24	4
6 In.	0.48		(8 In.)	2.00	0.30	0.29	0.27	0.23	0.21	0.22	0.21	0.20	0.17	0.17	0.21	5
			(12 In.)	2.27	0.28	0.27	0.25	0.21	0.20	0.20	0.20	0.19	0.17	0.16	0.20	6
			(Sand Agg.) (4 In.)	0.71	0.49	0.46	0.41	0.32	0.29	0.30	0.29	0.27	0.22	0.22	0.29	7
			(8 In.)	1.11	0.41	0.39	0.35	0.28	0.26	0.27	0.26	0.25	0.21	0.20	0.26	8
			(12 In.)	1.28	0.38	0.37	0.33	0.27	0.25	0.26	0.25	0.24	0.20	0.20	0.25	9
			Hollow Clay Tile (4 In.)	1.11	0.41	0.39	0.35	0.28	0.26	0.27	0.26	0.25	0.21	0.20	0.26	10
			(8 In.)	1.85	0.31	0.30	0.28	0.23	0.22	0.22	0.22	0.21	0.18	0.18	0.22	11
			(12 In.)	2.50	0.26	0.25	0.24	0.20	0.19	0.19	0.19	0.18	0.16	0.16	0.19	12
			Concrete (Sand Agg.) (4 In.)	0.32	0.60	0.56	0.49	0.36	0.32	0.34	0.33	0.31	0.25	0.24	0.33	13
			(6 In.)	0.48	0.55	0.52	0.45	0.34	0.31	0.32	0.31	0.29	0.24	0.23	0.31	14
			(8 In.)	0.64	0.51	0.48	0.42	0.32	0.29	0.31	0.30	0.28	0.23	0.22	0.30	15
Common Brick 4 In.	0.80		Concrete Block (Cinder Agg.) (4 In.)	1.11	0.36	0.35	0.32	0.26	0.24	0.25	0.24	0.23	0.19	0.19	0.24	16
			(8 In.)	1.72	0.29	0.29	0.26	0.22	0.21	0.21	0.21	0.20	0.17	0.17	0.21	17
		.72	(12 In.)	1.89	0.28	0.27	0.25	0.21	0.20	0.21	0.20	0.19	0.17	0.17	0.20	18
			(Lt. Wt. Agg.) (4 In.)	1.50	0.32	0.30	0.28	0.23	0.22	0.22	0.22	0.21	0.18	0.18	0.22	19
Precast Concrete (Sand Agg.) 8 In.	0.64		(8 In.)	2.00	0.27	0.26	0.25	0.21	0.20	0.20	0.20	0.19	0.16	0.16	0.20	20
			(12 In.)	2.27	0.25	0.25	0.23	0.20	0.19	0.19	0.19	0.18	0.16	0.16	0.19	21
			(Sand Agg.) (4 In.)	0.71	0.42	0.40	0.36	0.29	0.26	0.27	0.27	0.25	0.21	0.21	0.27	22
			(8 In)	1.11	0.36	0.35	0.32	0.26	0.24	0.25	0.24	0.23	0.19	0.19	0.24	23
			(12 In.)	1.28	0.34	0.33	0.30	0.25	0.23	0.24	0.23	0.22	0.19	0.18	0.23	24
			Hollow Clay Tile (4 In.)	1.11	0.36	0.35	0.32	0.26	0.24	0.25	0.24	0.23	0.19	0.19	0.24	25
			(8 In.)	1.85	0.28	0.28	0.26	0.22	0.20	0.21	0.20	0.19	0.17	0.17	0.20	26
			(12 In.)	2.50	0.24	0.23	0.22	0.19	0.18	0.18	0.18	0.17	0.15	0.15	0.18	27
			Concrete (Sand Agg.) (4 In.)	0.32	0.50	0.48	0.42	0.32	0.29	0.30	0.30	0.28	0.23	0.22	0.30	28
			(6 In.)	0.48	0.47	0.44	0.39	0.31	0.28	0.29	0.28	0.27	0.22	0.22	0.28	29
			(8 In.)	0.64	0.43	0.41	0.37	0.29	0.27	0.28	0.27	0.26	0.21	0.21	0.27	30

Figure 3.35.2 Coefficients of transmission (U) through masonry walls.

Coefficients of Transmission (U) of Solid Masonry Walls

Coefficients are expressed in Btu per (hour) (square foot) (Fahrenheit degree difference in temperature between the air on the two sides), and are based on an outside wind velocity of 15 mph

Exterior Construction[3]		None	Plas. 5/8 In. On Wall		Metal Lath and 3/4 In. Plas. On Furring		Gypsum Lath (3/8 In.) and 1/2 In. Plas. On Furring			Insul. Bd. Lath (1/2 In.) and 1/2 In. Plas. On Furring		Wood Lath and 1/2 In. Plas.	
			(Sand Agg.)	(Lt. Wt. Agg.)	(Sand Agg.)	(Lt. Wt. Agg.)	No. Plas.	(Sand Agg.)	(Lt. Wt. Agg.)	No. Plas.	(Sand Agg.)	(Sand Agg.)	
Resistance ↓ →			0.11	0.39	0.13	0.47	0.32	0.41	0.64	1.43	1.52	0.40	
		U	U	U	U	U	U	U	U	U	U	U	
Material	R	A	B	C	D	E	F	G	H	I	J	K	No.
Brick (Face And Common) [4]													
(6 In.)	0.61	0.68	0.64	0.54	0.39	0.34	0.36	0.35	0.33	0.26	0.25	0.35	1
(8 In.)	1.24	0.48	0.45	0.41	0.31	0.28	0.30	0.29	0.27	0.22	0.22	0.29	2
(12 In.)	2.04	0.35	0.33	0.30	0.25	0.23	0.24	0.23	0.22	0.19	0.19	0.23	3
(16 In.)	2.84	0.27	0.26	0.25	0.21	0.19	0.20	0.20	0.19	0.16	0.16	0.20	4
Brick (Common Only)													
(8 In.)	1.60	0.41	0.39	0.35	0.28	0.26	0.27	0.26	0.25	0.21	0.20	0.26	5
(12 In.)	2.40	0.31	0.30	0.27	0.23	0.21	0.22	0.22	0.21	0.18	0.17	0.22	6
(16 In.)	3.20	0.25	0.24	0.23	0.19	0.18	0.19	0.18	0.18	0.16	0.15	0.18	7
Stone (Lime And Sand)													
(8 In.)	0.64	0.67	0.63	0.53	0.39	0.34	0.36	0.35	0.32	0.26	0.25	0.35	8
(12 In.)	0.96	0.55	0.52	0.45	0.34	0.31	0.32	0.31	0.29	0.24	0.23	0.31	9
(16 In.)	1.28	0.47	0.45	0.40	0.31	0.28	0.29	0.28	0.27	0.22	0.22	0.29	10
(24 In.)	1.92	0.36	0.35	0.32	0.26	0.24	0.25	0.24	0.23	0.19	0.19	0.24	11
Hollow Clay Tile													
(8 In.)	1.85	0.36	0.36	0.32	0.26	0.24	0.25	0.25	0.23	0.20	0.19	0.25	12
(12 In.)	2.22	0.33	0.31	0.29	0.24	0.22	0.23	0.22	0.21	0.18	0.18	0.23	13
(12 In.)	2.50	0.30	0.29	0.27	0.22	0.21	0.22	0.21	0.20	0.17	0.17	0.21	14
Poured Concrete													
30 Lb. Per Cu Ft													
(4 In.)	4.44	0.19	0.19	0.18	0.16	0.15	0.15	0.15	0.14	0.13	0.13	0.15	15
(6 In.)	6.66	0.13	0.13	0.13	0.12	0.11	0.11	0.11	0.11	0.10	0.10	0.11	16
(8 In.)	8.88	0.10	0.10	0.10	0.09	0.09	0.09	0.09	0.09	0.08	0.08	0.09	17
(10 In.)	11.10	0.08	0.08	0.08	0.08	0.07	0.08	0.08	0.07	0.07	0.07	0.08	18
80 Lb. Per Cu Ft													
(6 In.)	2.40	0.31	0.30	0.27	0.23	0.21	0.22	0.22	0.21	0.18	0.17	0.22	19
(8 In.)	3.20	0.25	0.24	0.23	0.19	0.18	0.19	0.18	0.18	0.16	0.15	0.18	20
(10 In.)	4.00	0.21	0.20	0.19	0.17	0.16	0.16	0.16	0.15	0.14	0.14	0.16	21
(12 In.)	4.80	0.18	0.17	0.17	0.15	0.14	0.14	0.14	0.14	0.12	0.12	0.14	22
140 Lb. Per Cu Ft													
(6 In.)	0.48	0.75	0.69	0.58	0.41	0.36	0.38	0.37	0.34	0.27	0.26	0.37	23
(8 In.)	0.64	0.67	0.63	0.53	0.39	0.34	0.36	0.35	0.32	0.26	0.25	0.35	24
(10 In.)	0.80	0.61	0.57	0.49	0.36	0.32	0.34	0.33	0.31	0.25	0.24	0.33	25
(12 In.)	0.96	0.55	0.52	0.45	0.34	0.31	0.32	0.31	0.29	0.24	0.23	0.31	26
Concrete Block													
(Gravel Agg.) (8 In.)	1.11	0.52	0.48	0.43	0.33	0.29	0.31	0.30	0.28	0.23	0.22	0.30	27
(12 In.)	1.28	0.47	0.45	0.40	0.31	0.28	0.29	0.28	0.27	0.22	0.22	0.29	28
(Cinder Agg.) (8 In.)	1.72	0.39	0.37	0.34	0.27	0.25	0.26	0.25	0.24	0.20	0.20	0.25	29
(12 In.)	1.89	0.36	0.35	0.32	0.26	0.24	0.25	0.24	0.23	0.19	0.19	0.24	30
(Lt. Wt. Agg.) (8 In.)	2.00	0.35	0.34	0.31	0.26	0.23	0.24	0.24	0.22	0.19	0.19	0.24	31
(12 In.)	2.27	0.32	0.31	0.28	0.24	0.22	0.23	0.22	0.21	0.18	0.18	0.22	32

[3] If stucco or structural glass is applied to the exterior, the additional resistance value of 0.10 would have a negligible effect on the U value.
[4] Brick, 6 in. (5½ in. actual) is assumed to have no backing. Walls 8, 12 and 16 in. have 4 in. of face brick and balance of common brick.

Figure 3.35.2 *(Continued)*

Coefficients of Transmission (U) of Masonry Cavity Walls

Coefficients are expressed in Btu per (hour) (square foot) (Fahrenheit degree difference in temperature between the air on the two sides), and are based on an outside wind velocity of 15 mph

Exterior Construction			Inner Section		None	Plas. 5/8 In. On Wall		Metal Lath and 3/4 In. Plas. On Furring		Gypsum Lath (3/8 In.) and 1/2 In. Plas. On Furring			Insul. Bd. Lath (1/2 In.) & 1/2 In. Plas. On Furring		Wood Lath and 1/2 In. Plas.	
						(Sand Agg.)	(Lt. Wt. Agg.)	(Sand Agg.)	(Lt. Wt. Agg.)	No. Plas.	(Sand Agg.)	(Lt. Wt. Agg.)	No. Plas.	(Sand Agg.)	(Sand Agg.)	
			Resistance ↓→			0.11	0.39	0.13	0.47	0.32	0.41	0.64	1.43	1.52	0.40	
		AV			U	U	U	U	U	U	U	U	U	U	U	
Material	R	R	Material	R	A	B	C	D	E	F	G	H	I	J	K	No.
Face Brick (4 In.)		0.44	Concrete Block (4 In.)													
			(Gravel Agg.)	0.71	0.34	0.32	0.30	0.25	0.23	0.23	0.23	0.22	0.19	0.18	0.23	1
			(Cinder Agg.)	1.11	0.30	0.29	0.27	0.22	0.21	0.21	0.21	0.20	0.17	0.17	0.21	2
			(Lt. Wt. Agg.)	1.50	0.27	0.26	0.24	0.21	0.19	0.20	0.19	0.19	0.16	0.16	0.19	3
			Common Brick (4 In.)	0.80	0.33	0.32	0.29	0.24	0.22	0.23	0.23	0.21	0.18	0.18	0.23	4
			Clay Tile (4 In.)	1.11	0.30	0.29	0.27	0.22	0.21	0.21	0.21	0.20	0.17	0.17	0.21	5
Common Brick (4 In.) Concrete Block (Gravel Agg.)	0.80 0.71	0.76	Concrete Block (4 In.)													
			(Gravel Agg.)	0.71	0.30	0.29	0.27	0.23	0.21	0.22	0.21	0.20	0.18	0.17	0.21	6
			(Cinder Agg.)	1.11	0.27	0.26	0.25	0.21	0.19	0.20	0.20	0.19	0.16	0.16	0.20	7
			(Lt. Wt. Agg.)	1.50	0.25	0.24	0.22	0.19	0.18	0.19	0.18	0.18	0.1 S	0.15	0.18	8
			Common Brick (4 In.)	0.80	0.30	0.29	0.27	0.22	0.21	0.21	0.21	0.20	0.17	0.17	0.21	9
			Clay Tile (4 In.)	1.11	0.27	0.26	0.25	0.21	0.19	0.20	0.20	0.19	0.16	0.16	0.20	10
Concrete Block (Cinder Agg.) (4 In.)		1.11	Concrete Block													
			(Gravel Agg.)	0.71	0.27	0.27	0.25	0.21	0.20	0.20	0.20	0.19	0.17	0.16	0.20	11
			(Cinder Agg.)	1.11	0.25	0.24	0.23	0.19	0.18	0.19	0.18	0.18	0.16	0.15	0.18	12
			(Lt. Wt. Agg.)	1.50	0.23	0.22	0.21	0.18	0.17	0.17	0.17	0.17	0.15	0.14	0.17	13
			Common Brick (4 In.)	0.80	0.27	0.26	0.24	0.21	0.19	0.20	0.20	0.19	0.16	0.16	0.20	14
			Clay Tile (4 In.)	1.11	0.25	0.24	0.23	0.19	0.18	0.19	0.18	0.18	0.16	0.15	0.18	15

Figure 3.35.2 (*Continued*)

Coefficients of Transmission (U) of Frame Walls

These Coefficients are expressed in Btu per (hour) (square foot) (Fahrenheit degree difference in temperature between the air on the two sides), ond are based on an outside wind velocity of 15 mph

Exterior[1]			Interior Finish		Type of Sheathing[3]						
					None, Bldg. Paper	Gypsum Board ½"	Ply-wood 5/16"	Wood 25/32" & Bldg. Paper	Insulation Board Sheathing ½"	Insulation Board Sheathing 25/32"	
			Resistance ↓ →		0.06	0.45	0.39	1.04	1.32	2.06	
Material	R	Av. R	Material	R	U A	U B	U C	U D	U E	U F	No.
			None	-	0.57	0.47	0.48	0.36	0.33	0.27	1
			Gypsum Bd. (⅜")	0.32	0.33	0.29	0.30	0.25	0.23	0.20	2
			Gypsum Lath (⅜") and ½" Plas. (Lt. Wt. Agg.)	0.64	0.30	0.27	0.27	0.23	0.22	0.20	3
Wood Siding			Gypsum Lath (⅜") and ½" Plas. (Sand Agg.)	0.41	0.32	0.28	0.29	0.24	0.23	0.19	4
Drop-(1" x 8")	0.79		Metal Lath and ¾" Plas. (Lt. Wt. Agg.)	0.47	0.31	0.28	0.28	0.24	0.22	0.19	5
Bevel (½" x 8")	0.81	0.85[2]	Metal Lath and ¾" Plas. (Sand Agg.)	0.13	0.35	0.31	0.31	0.26	0.24	0.21	6
Wood Shingles 7½" Exposure	0.87		Insul. Bd. (½")	1.43	0.24	0.22	0.22	0.19	0.18	0.16	7
Wood Panels (¾")	0.94		Insul. Bd. Lath (½") and ½" Plas. (Sand Agg.)	1.52	0.24	0.22	0.22	0.19	0.18	0.16	8
			Plywood (¼")	0.31	0.33	0.29	0.30	0.25	0.23	0.20	9
			Wood Panels (¾")	0.94	0.27	0.25	0.25	0.22	0.20	0.18	10
			Wood Lath and ½" Plas. (Sand Agg.)	0.40	0.32	0.28	0.29	0.24	0.23	0.19	11
			None	—	0.73	0.56	0.58	0.42	0.38	0.30	12
			Gypsum Bd. (⅜")	0.32	0.37	0.33	0.33	0.27	0.25	0.21	13
			Gypsum Lath (⅜") and ½" Plas. (Lt. Wt. Agg.)	0.64	0.33	0.30	0.30	0.25	0.24	0.20	14
			Gypsum Lath (⅜") and ½" Plas. (Sand Agg.)	0.41	0.36	0.32	0.32	0.27	0.25	0.21	15
Face-Brick Veneer[4]	0.44	0.45[2]	Metal Lath and ¾" Plas. (Lt. Wt. Agg.)	0.47	0.35	0.31	0.32	0.26	0.25	0.21	16
Plywood (⅝")	0.47		Metal Lath and ¾" Plas. (Sand Agg.)	0.13	0.40	0.35	0.36	0.29	0.27	0.22	17
			Insul. Bd. (½")	1.43	0.26	0.24	0.24	0.21	0.20	0.17	18
			Insul. Bd. Lath (½") and ½" Plas. (Sand Agg.)	1.52	0.26	0.23	0.24	0.21	0.19	0.17	19
			Plywood (¼")	0.31	0.38	0.33	0.33	0.27	0.26	0.21	20
			Wood Panels (¾")	0.94	0.30	0.27	0.28	0.23	0.22	0.19	21
			Wood Lath and ½" Plas. (Sand Agg.)	0.40	0.36	0.32	0.32	0.27	0.25	0.21	22
			None	—	0.43	0.37	0.38	0.30	0.28	0.23	23
			Gypsum Bd. (⅜")	0.32	0.28	0.25	0.25	0.22	0.20	0.18	24
Wood Shingles Over			Gypsum Lath (⅜") and ½" Plas. (Lt. Wt. Agc.)	0.64	0.25	0.23	0.23	0.20	0.19	0.17	25
Insul.: Backer			Gypsum Lath (⅜") and ½" Plas. (Sand Agg.)	0.41	0.27	0.24	0.25	0.21	0.20	0.18	26
Bd. (5/16")	1.40	1.42[2]	Metal Lath and ¾" Plas. (Lt. Wt. Agg.)	0.47	0.27	0.24	0.24	0.21	0.20	0.17	27
Asphalt Insul. Siding	1.45		Metal Lath and ¾" Plas. (Sand Agg.)	0.13	0.29	0.26	0.27	0.23	0.21	0.18	28
			Insul. Bd. (½")	1.43	0.21	0.20	0.20	0.18	0.17	0.15	29
			Insul. Bd. Lath (½") and ½" Plas. (Sand Agg.)	1.52	0.21	0.19	0.19	0.17	0.16	0.15	30
			Plywood (¼")	0.31	0.28	0.25	0.25	0.22	0.20	0.18	31
			Wood Panels (¾")	0.94	0.24	0.22	0.22	0.19	0.18	0.16	32
			Wood Lath and ½" Plas. (Sand Agg.)	0.40	0.27	0.24	0.25	0.21	0.20	0.18	33
			None	—	0.91	0.67	0.70	0.48	0.42	0.32	34
			Gypsum Bd. (⅜")	0.32	0.42	0.36	0.37	0.30	0.27	0.23	35
			Gypsum Lath (⅜") and ½" Plas. (Lt. Wt. Agg.)	0.64	0.37	0.32	0.33	0.27	0.25	0.21	36
Asbestos-Cement			Gypsum Lath (⅜") and ½" Plas. (Sand Agg.)	0.41	0.40	0.35	0.36	0.29	0.27	0.22	37
Siding	0.21		Metal Lath and ¾" Plas. (Lt. Wt. Agg.)	0.47	0.39	0.34	0.35	0.28	0.26	0.22	38
Stucco[5] 1 In	0.20	0.19[2]	Metal Lath and ¾" Plas. (Sand Agg.	0.13	0.45	0.39	0.40	0.31	0.29	0.24	39
Asphalt Roll			Insul. Bd. (½")	1.43	0.29	0.26	0.26	0.22	0.21	0.18	40
Siding	0.15		Insul. Bd. Lath (½") and ½" Plas. (Sand Agg.)	1.52	0.28	0.25	0.26	0.22	0.21	0.18	41
			Plywood (¼")	0.31	0.42	0.36	0.37	0.30	0.27	0.23	42
			Wood Panels (¾")	0.94	0.33	0.29	0.30	0.25	0.23	0.20	43
			Wood Lath and ½" Plas. (Sand Agg.)	0.40	0.40	0.35	0.36	0.29	0.27	0.22	44

[1] Note that although several types of exterior finish may be grouped because they have approximately the same thermal resistance value, it is not implied that all types may be suitable for application over all types of sheathing listed.
[2] Average resistance of items listed. This average was used in computation of U values shown.
[3] Building paper is not included except where noted.
[4] Small air space between building paper and brick veneer neglected.
[5] Where stucco is applied over insulating board or gypsum sheathing, building paper is generally required, but the change in U value is negligible.

Figure 3.35.2 *(Continued)*

Design Temperature Differences

Item No.	Item	Temperature Difference* (F)
1	Walls, Exterior	17
2	Glass In Exterior Walls	17
3	Glass In Partitions	10
4	Store Show Windows Having A Large Lighting Load	30
5	Partitions	10
6	Partitions, Or Glass In Partitions, Adjacent To Laundries, Kitchens, Or Boiler Rooms	25
7	Floors Above Unconditioned Rooms	10
8	Floors On Ground	0
9	Floors Above Basements	0
10	Floors Above Rooms Or Basements Used As Laundries, Kitchens, Or Boiler Rooms	35
11	Floors Above Vented Spaces	17
12	Floors Above Unvented Spaces	
13	Ceilings With Unconditioned Rooms Above	10
14	Ceilings With Rooms Above Used As Laundries, Kitchens, Etc.	20
15	Ceiling With Roof Directly Above (No Attic)	17
16	Ceiling With Totally Enclosed Attic Above	17
17	Ceiling With Cross-Ventilated Attic Above	17

*These temperature differences are based on the assumption that the air conditioning system is being designed to maintain an inside temperature 17 F lower than the outdoor temperature. For air conditioning systems designed to maintain a greater temperature difference than 17 F between the inside and outside, add to the values in the above table, the difference between the assumed design temperature difference and 17 F.

Figure 3.35.3 Design temperatures differences. (*By permission: Loren Cook, Springfield, MO.*)

Overall Heat Transfer Coefficients for Glass and Glass Blocks

Description	Outdoor Exposure
Single-Glass Windows	1.06
Double-Glass Windows	0.64
Triple-Glass Windows	0.34
Glass Block	0.56
Glass Block	0.48

Figure 3.35.4 Heat transfer through glass and glass blocks. (*By permission: Loren Cook, Springfield, MO.*)

3.36.0 Thermal Properties of Common Building Materials

Description	Density, lb/ft³	Conductivity[b] (k), Btu·in / h·ft²·°F	Conductance (C), Btu / h·ft²·°F	Resistance [c](R) Per Inch Thickness (1/k), °F·ft²·h / Btu·in	Resistance [c](R) For Thickness Listed (1/C), °F·ft²·h / Btu	Specific Heat, Btu / lb·°F
Brick, fired clay *continued*	100	4.2–5.1	—	0.24–0.20	—	—
	90	3.6–4.3	—	0.28–0.24	—	—
	80	3.0–3.7	—	0.33–0.27	—	—
	70	2.5–3.1	—	0.40–0.33	—	—
Clay tile, hollow						
1 cell deep ...... 3 in.	—	—	1.25	—	0.80	0.21
1 cell deep ...... 4 in.	—	—	0.90	—	1.11	—
2 cells deep ...... 6 in.	—	—	0.66	—	1.52	—
2 cells deep ...... 8 in.	—	—	0.54	—	1.85	—
2 cells deep ...... 10 in.	—	—	0.45	—	2.22	—
3 cells deep ...... 12 in.	—	—	0.40	—	2.50	—
Concrete blocks[f]						
Limestone aggregate						
8 in., 36 lb, 138 lb/ft³ concrete, 2 cores	—	—	—	—	—	—
Same with perlite filled cores	—	—	0.48	—	2.1	—
12 in., 55 lb, 138 lb/ft³ concrete, 2 cores	—	—	—	—	—	—
Same with perlite filled cores	—	—	0.27	—	3.7	—
Normal weight aggregate (sand and gravel)						
8 in., 33-36 lb, 126-136 lb/ft³ concrete, 2 or 3 cores	—	—	0.90–1.03	—	1.11–0.97	0.22
Same with perlite filled cores	—	—	0.50	—	2.0	—
Same with verm. filled cores	—	—	0.52–0.73	—	1.92–1.37	—
12 in., 50 lb, 125 lb/ft³ concrete, 2 cores	—	—	0.81	—	1.23	0.22
Medium weight aggregate (combinations of normal weight and lightweight aggregate)						
8 in., 26-29 lb, 97-112 lb/ft³ concrete, 2 or 3 cores..	—	—	0.58–0.78	—	1.71–1.28	—
Same with perlite filled cores	—	—	0.27–0.44	—	3.7–2.3	—
Same with verm. filled cores	—	—	0.30	—	3.3	—
Same with molded EPS (beads) filled cores	—	—	0.32	—	3.2	—
Same with molded EPS inserts in cores	—	—	0.37	—	2.7	—
Lightweight aggregate (expanded shale, clay, slate or slag, pumice)						
6 in., 16-17 lb 85-87 lb/ft³ concrete, 2 or 3 cores...	—	—	0.52–0.61	—	1.93–1.65	—
Same with perlite filled cores	—	—	0.24	—	4.2	—
Same with verm. filled cores	—	—	0.33	—	3.0	—
8 in., 19-22 lbs, 72-86 lb/ft³ concrete,	—	—	0.32–0.54	—	3.2–1.90	0.21
Same with perlite filled cores	—	—	0.15–0.23	—	6.8–4.4	—
Same with verm. filled cores	—	—	0.19–0.26	—	5.3–3.9	—
Same with molded EPS (beads) filled cores	—	—	0.21	—	4.8	—
Same with UF foam filled cores	—	—	0.22	—	4.5	—
Same with molded EPS inserts in cores	—	—	0.29	—	3.5	—
12 in., 32-36 lb, 80-90 lb/ft³ concrete, 2 or 3 cores...	—	—	0.38–0.44	—	2.6–2.3	—
Same with perlite filled cores	—	—	0.11–0.16	—	9.2–6.3	—
Same with verm. filled cores	—	—	0.17	—	5.8	—
Stone, lime, or sand						
Quartzitic and sandstone	180	72	—	0.01	—	—
	160	43	—	0.02	—	—
	140	24	—	0.04	—	—
	120	13	—	0.08	—	0.19
Calcitic, dolomitic, limestone, marble, and granite..	180	30	—	0.03	—	—
	160	22	—	0.05	—	—
	140	16	—	0.06	—	—
	120	11	—	0.09	—	0.19
	100	8	—	0.13	—	—
Gypsum partition tile						
3 by 12 by 30 in., solid	—	—	0.79	—	1.26	0.19
3 by 12 by 30 in., 4 cells	—	—	0.74	—	1.35	—
4 by 12 by 30 in., 3 cells	—	—	0.60	—	1.67	—
Concretes						
Sand and gravel or stone aggregate concretes (concretes with more than 50% quartz or quartzite sand have conductivities in the higher end of the range) ..	150	10.0–20.0	—	0.10–0.05	—	—
	140	9.0–18.0	—	0.11–0.06	—	0.19–0.24
	130	7.0–13.0	—	0.14–0.08	—	—
Limestone concretes	140	11.1	—	0.09	—	—
	120	7.9	—	0.13	—	—
	100	5.5	—	0.18	—	—
Gypsum-fiber concrete (87.5% gypsum, 12.5% wood chips)	51	1.66	—	0.60	—	0.21
Cement/lime, mortar, and stucco	120	9.7	—	0.10	—	—
	100	6.7	—	0.15	—	—
	80	4.5	—	0.22	—	—
Lightweight aggregate concretes						
Expanded shale, clay, or slate; expanded slags; cinders; pumice (with density up to 100 lb/ft³); and scoria (sanded concretes have conductivities in the higher end of the range)	120	6.4–9.1	—	0.16–0.11	—	—
	100	4.7–6.2	—	0.21–0.16	—	0.20
	80	3.3–4.1	—	0.30–0.24	—	0.20
	60	2.1–2.5	—	0.48–0.40	—	—
	40	1.3	—	0.78	—	—

(By permission of American Society of Heating, Refrigerating and Air-Conditioning Engineers, Inc. Atlanta, Georgia, from their 1993 ASHRAE Fundamentals Handbook)

3.36.0   Thermal Properties of Common Building Materials (Continued)

Description	Density, lb/ft³	Conductivity[b] (k), Btu·in h·ft²·°F	Conductance (C), Btu h·ft²·°F	Resistance [c](R) Per Inch Thickness (1/k), °F·ft²·h Btu·in	Resistance [c](R) For Thickness Listed (1/C), °F·ft²·h Btu	Specific Heat, Btu lb·°F
Expanded polystyrene, molded beads	1.0	0.26	—	3.85	—	—
	1.25	0.25	—	4.00	—	—
	1.5	0.24	—	4.17	—	—
	1.75	0.24	—	4.17	—	—
	2.0	0.23	—	4.35	—	—
Cellular polyurethane/polyisocyanurate[i] (CFC-11 exp.) (unfaced)	1.5	0.16–0.18	—	6.25–5.56	—	0.38
Cellular polyisocyanurate[i] (CFC-11 exp.)(gas-permeable facers).............	1.5–2.5	0.16–0.18	—	6.25–5.56	—	0.22
Cellular polyisocyanurate[j] (CFC-11 exp.)(gas-impermeable facers)...........	2.0	0.14	—	7.04	—	0.22
Cellular phenolic (closed cell)(CFC-11, CFC-113 exp.)	3.0	0.12	—	8.20	—	—
Cellular phenolic (open cell)	1.8–2.2	0.23	—	4.40	—	—
Mineral fiber with resin binder	15.0	0.29	—	3.45	—	0.17
Mineral fiberboard, wet felted						
Core or roof insulation.........................	16–17	0.34	—	2.94	—	—
Acoustical tile	18.0	0.35	—	2.86	—	0.19
Acoustical tile	21.0	0.37	—	2.70	—	—
Mineral fiberboard, wet molded						
Acoustical tile[k]	23.0	0.42	—	2.38	—	0.14
Wood or cane fiberboard						
Acoustical tile[k] ....0.5 in.	—	—	0.80	—	1.25	0.31
Acoustical tile[k] ....0.75 in.	—	—	0.53	—	1.89	—
Interior finish (plank, tile)	15.0	0.35	—	2.86	—	0.32
Cement fiber slabs (shredded wood with Portland cement binder)	25–27.0	0.50–0.53	—	2.0–1.89	—	—
Cement fiber slabs (shredded wood with magnesia oxysulfide binder)	22.0	0.57	—	1.75	—	0.31
Loose Fill						
Cellulosic insulation (milled paper or wood pulp)	2.3–3.2	0.27–0.32	—	3.70–3.13	—	0.33
Perlite, expanded	2.0–4.1	0.27–0.31	—	3.7–3.3	—	0.26
	4.1–7.4	0.31–0.36	—	3.3–2.8	—	—
	7.4–11.0	0.36–0.42	—	2.8–2.4	—	—
Mineral fiber (rock, slag, or glass)[g]						
approx. 3.75–5 in............................	0.6–2.0	—	—	—	11.0	0.17
approx. 6.5–8.75 in.	0.6–2.0	—	—	—	19.0	—
approx. 7.5–10 in.	0.6–2.0	—	—	—	22.0	—
approx. 10.25–13.75 in.	0.6–2.0	—	—	—	30.0	—
Mineral fiber (rock, slag, or glass)[g]						
approx. 3.5 in. (closed sidewall application)	2.0–3.5	—	—	—	12.0–14.0	—
Vermiculite, exfoliated..........................	7.0–8.2	0.47	—	2.13	—	0.32
	4.0–6.0	0.44	—	2.27	—	—
Spray Applied						
Polyurethane foam	1.5–2.5	0.16–0.18	—	6.25–5.56	—	—
Ureaformaldehyde foam	0.7–1.6	0.22–0.28	—	4.55–3.57	—	—
Cellulosic fiber	3.5–6.0	0.29–0.34	—	3.45–2.94	—	—
Glass fiber	3.5–4.5	0.26–0.27	—	3.85–3.70	—	—
METALS (See Chapter 36, Table 3)						
ROOFING						
Asbestos-cement shingles	120	—	4.76	—	0.21	0.24
Asphalt roll roofing	70	—	6.50	—	0.15	0.36
Asphalt shingles	70	—	2.27	—	0.44	0.30
Built-up roofing ....0.375 in.	70	—	3.00	—	0.33	0.35
Slate...................................0.5 in.	—	—	20.00	—	0.05	0.30
Wood shingles, plain and plastic film faced	—	—	1.06	—	0.94	0.31
PLASTERING MATERIALS						
Cement plaster, sand aggregate....................	116	5.0	—	0.20	—	0.20
Sand aggregate ....0.375 in.	—	—	13.3	—	0.08	0.20
Sand aggregate ....0.75 in.	—	—	6.66	—	0.15	0.20
Gypsum plaster:						
Lightweight aggregate ....0.5 in.	45	—	3.12	—	0.32	—
Lightweight aggregate ....0.625 in.	45	—	2.67	—	0.39	—
Lightweight aggregate on metal lath ....0.75 in.	—	—	2.13	—	0.47	—
Perlite aggregate	45	1.5	—	0.67	—	0.32
Sand aggregate	105	5.6	—	0.18	—	0.20
Sand aggregate ....0.5 in.	105	—	11.10	—	0.09	—
Sand aggregate ....0.625 in.	105	—	9.10	—	0.11	—
Sand aggregate on metal lath ....0.75 in.	—	—	7.70	—	0.13	—
Vermiculite aggregate..........................	45	1.7	—	0.59	—	—
MASONRY MATERIALS *Masonry Units*						
Brick, fired clay...............................	150	8.4–10.2	—	0.12–0.10	—	—
	140	7.4–9.0	—	0.14–0.11	—	—
	130	6.4–7.8	—	0.16–0.12	—	—
	120	5.6–6.8	—	0.18–0.15	—	0.19
	110	4.9–5.9	—	0.20–0.17	—	—

(By permission of American Society of Heating, Refrigerating and Air-Conditioning Engineers, Inc. Atlanta, Georgia, from their 1993 ASHRAE Fundamentals Handbook)

3.36.0   Thermal Properties of Common Building Materials (Continued)

Description	Density, lb/ft³	Conductivity[b] (k), Btu·in / h·ft²·°F	Conductance (C), Btu / h·ft²·°F	Resistance [c](R) Per Inch Thickness (1/k), °F·ft²·h / Btu·in	Resistance [c](R) For Thickness Listed (1/C), °F·ft²·h / Btu	Specific Heat, Btu / lb·°F
BUILDING BOARD						
Asbestos-cement board	120	4.0	—	0.25	—	0.24
Asbestos-cement board ....0.125 in.	120	—	33.00	—	0.03	
Asbestos-cement board ....0.25 in.	120	—	16.50	—	0.06	
Gypsum or plaster board................0.375 in.	50	—	3.10	—	0.32	0.26
Gypsum or plaster board................0.5 in.	50	—	2.22	—	0.45	
Gypsum or plaster board................0.625 in.	50	—	1.78	—	0.56	
Plywood (Douglas Fir)[d]	34	0.80	—	1.25	—	0.29
Plywood (Douglas Fir) ....0.25 in.	34	—	3.20	—	0.31	
Plywood (Douglas Fir) ....0.375 in.	34	—	2.13	—	0.47	
Plywood (Douglas Fir) ....0.5 in.	34	—	1.60	—	0.62	
Plywood (Douglas Fir) ....0.625 in.	34	—	1.29	—	0.77	
Plywood or wood panels....................0.75 in.	34	—	1.07	—	0.93	0.29
Vegetable fiber board						
Sheathing, regular density[e] ....0.5 in.	18	—	0.76	—	1.32	0.31
....0.78125 in.	18	—	0.49	—	2.06	
Sheathing intermediate density[e] ....0.5 in.	22	—	0.92	—	1.09	0.31
Nail-base sheathing[e].......................0.5 in.	25	—	0.94	—	1.06	0.31
Shingle backer ....0.375 in.	18	—	1.06	—	0.94	0.31
Shingle backer ....0.3125 in.	18	—	1.28	—	0.78	
Sound deadening board....................0.5 in.	15	—	0.74	—	1.35	0.30
Tile and lay-in panels, plain or acoustic	18	0.40	—	2.50	—	0.14
.........0.5 in.	18	—	0.80	—	¹.25	
.........0.75 in.	18	—	0.53	—	1.89	
Laminated paperboard...........................	30	0.50	—	2.00	—	0.33
Homogeneous board from repulped paper	30	0.50	—	2.00	—	0.28
Hardboard[e]						
Medium density..............................	50	0.73	—	1.37	—	0.31
High density, service-tempered grade and service grade..	55	0.82	—	1.22	—	0.32
High density, standard-tempered grade	63	1.00	—	1.00	—	0.32
Particleboard[e]						
Low density	37	0.71	—	1.41	—	0.31
Medium density..............................	50	0.94	—	1.06	—	0.31
High density................................	62.5	1.18	—	0.85	—	0.31
Underlayment........................0.625 in.	40	—	1.22	—	0.82	0.29
Waferboard	37	0.63	—	1.59	—	—
Wood subfloor..........................0.75 in.	—	—	1.06	—	0.94	0.33
BUILDING MEMBRANE						
Vapor—permeable felt...........................	—	—	16.70	—	0.06	
Vapor—seal, 2 layers of mopped 15-lb felt	—	—	8.35	—	0.12	
Vapor—seal, plastic film	—	—	—	—	Negl.	
FINISH FLOORING MATERIALS						
Carpet and fibrous pad	—	—	0.48	—	2.08	0.34
Carpet and rubber pad	—	—	0.81	—	1.23	0.33
Cork tile ....0.125 in.	—	—	3.60	—	0.28	0.48
Terrazzo ....1 in.	—	—	12.50	—	0.08	0.19
Tile—asphalt, linoleum, vinyl, rubber	—	—	20.00	—	0.05	0.30
vinyl asbestos................................						0.24
ceramic....................................						0.19
Wood, hardwood finish ....0.75 in.	—	—	1.47	—	0.68	
INSULATING MATERIALS						
Blanket and Batt[f,g]						
Mineral fiber, fibrous form processed from rock, slag, or glass						
approx. 3–4 in.	0.4–2.0	—	0.091	—	11	
approx. 3.5 in.	0.4–2.0	—	0.077	—	13	
approx. 3.5 in.	1.2–1.6	—	0.067	—	15	
approx. 5.5–6.5 in.	0.4–2.0	—	0.053	—	19	
approx. 5.5 in.	0.6–1.0	—	0.048	—	21	
approx. 6–7.5 in.	0.4–2.0	—	0.045	—	22	
approx. 8.25–10 in.	0.4–2.0	—	0.033	—	30	
approx. 10–13 in.	0.4–2.0	—	0.026	—	38	
Board and Slabs						
Cellular glass	8.0	0.33	—	3.03	—	0.18
Glass fiber, organic bonded......................	4.0–9.0	0.25	—	4.00	—	0.23
Expanded perlite, organic bonded	1.0	0.36	—	2.78	—	0.30
Expanded rubber (rigid)	4.5	0.22	—	4.55	—	0.40
Expanded polystyrene, extruded (smooth skin surface) (CFC-12 exp.)	1.8–3.5	0.20	—	5.00	—	0.29
Expanded polystyrene, extruded (smooth skin surface) (HCFC-142b exp.)[h].......................	1.8–3.5	0.20	—	5.00	—	0.29

(By permission of American Society of Heating, Refrigerating and Air-Conditioning Engineers, Inc. Atlanta, Georgia, from their 1993 ASHRAE Fundamentals Handbook)

Conversion Factors

Power

Multiply	By	To Get
Boiler hp	33,472	Btu/hr
Boiler hp	34.5	lbs H_2O evap. at 212°F
Horsepower	2,540	Btu/hr
Horsepower	550	ft-lbs/sec
Horsepower	33,000	ft-lbs/min
Horsepower	42.42	Btu/min
Horsepower	0.7457	Kilowatts
Kilowatts	3,415	Btu/hr
Kilowatts	56.92	Btu/min
Watts	44.26	ft-lbs/min
Watts	0.7378	ft-lbs/sec
Watts	0.05692	Btu/min
Tons refrig.	12,000	Btu/hr
Tons refrig.	200	Btu/min
Btu/hr	0.00002986	Boiler hp
lbs H_2O evap. at 212°F	0.0290	Boiler hp
Btu/hr	0.000393	Horsepower
ft-lbs/sec	0.00182	Horsepower
ft-lbs/min	0.0000303	Horsepower
Btu/min	0.0236	Horsepower
Kilowatts	1.341	Horsepower
Btu/hr	0.000293	Kilowatts
Btu/min	0.01757	Kilowatts
ft-lbs/min	0.02259	Watts
ft-lbs/sec	1.355	Watts
Btu/min	1.757	Watts
Btu/hr	0.0000833	Tons refrig.
Btu/min	0.005	Tons refrig.

Energy

Multiply	By	To Get
Btu	778	ft-lbs
Btu	0.000393	hp-hrs
Btu	0.000293	kw-hrs
Btu	0.0010307	{lbs H_2O evap. at 212°F
Btu	0.293	Watt-hrs
ft-lbs	0.3765	Watt-hrs
Latent heat) of ice	143.33	Btu/lb H_2O
lbs H_2O evap.) at 212°F	0.284	kw-hrs
lbs H_2O evap.) at 212°F	0.381	hp-hrs
ft-lbs	0.001287	Btu
hp-hrs	2,540	Btu
kw-hrs	3,415	Btu
lbs H_2O evap.) at 212°F	970.4	Btu
Watt-hrs	3.415	Btu
Watt-hrs	2,656	ft-lbs
Btu/lb H_2O	0.006977	{Latent heat of ice
kw-hrs	3.52	{lbs H_2O evap. at 212°F
hp-hrs	2.63	{lbs. H_2O evap. at 212°F

Pressure

Multiply	By	To Get
atmospheres	29.92	{in Mercury (at 62°F)
atmospheres	406.8	{in H_2O (at 62°F)
atmospheres	33.90	{ft. H_2O (at 62°F)
atmospheres	14.70	lbs/in²
atmospheres	1.058	ton/ft²
in. H_2O} (at 62°F)	0.0737	{in. Mercury (at 62°F)
ft H_2O} (at 62°F)	0.881	{in. Mercury (at 62°F)
ft H_2O} (at 62°F)	0.4335	lbs/in²
ft H_2O} (at 62°F)	62.37	lbs/ft²
in. Mercury} (at 62°F)	70.73	lbs/ft²
in. Mercury} (at 62°F)	0.4912	lbs/in²
in. Mercury} (at 62°F)	0.03342	atmospheres
in. H_2O} (at 62°F)	0.002458	atmospheres
ft. H_2O} (at 62°F)	0.0295	atmospheres
lbs/in²	0.0680	atmospheres
ton/ft²	0.945	atmospheres
in. Mercury} (at 62°F)	13.57	{in. H_2O (at 62°F)
in. Mercury} (at 62°F)	1.131	{ft H_2O (at 62°F)
lbs/in²	2.309	{ft H_2O (at 62°F)
lbs/ft²	0.01603	{ft H_2O (at 62°F)
lbs/ft²	0.014138	{in. Mercury (at 62°F)
lbs/in²	2.042	{in. Mercury (at 62°F)
lbs/in²	0.0689	Bar
lbs/in²	0.0703	kg/cm²

Velocity of Flow

Multiply	By	To Get
ft/min	0.01139	miles/hr
ft/min	0.01667	ft/sec
cu ft/min	0.1247	gal/sec
cu ft/sec	448.8	gal/min
miles/hr	88	ft/min
ft/sec	60	ft/min
gal/sec	8.02	cu ft/min
gal/min	0.002228	cu ft/sec

Heat Transmission

Multiply	By	To Get
Btu/in} /sq ft /hr/°F	0.0833	{Btu/ft /sq ft /hr/°F
Btu/ft} /sq ft /hr /°F	12	{Btu/in /sq ft /hr/ °F

Weight

Multiply	By	To Get
lbs	7,000	grains
lbs H_2O (60°F)	0.01602	cu ft H_2O
lbs H_2O (60°F)	0.1198	gal H_2O
tons (long)	2,240	lbs
tons (short)	2,000	lbs
grains	0.000143	lbs
cu ft H_2O	62.37	lbs H_2O (60°F)
gal H_2O	8.3453	lbs H_2O (60°F)
lbs	0.000446	tons (long)
lbs	0.000500	tons (short)

Circular Measure

Multiply	By	To Get
Degrees	0.01745	Radians
Minutes	0.00029	Radians
Diameter	3.142	Circumference
Radians	57.3	Degrees
Radians	3,438	Minutes
Circumference	0.3183	Diameter

Volume

Multiply	By	To Get
Barrels (oil)	42	gal (oil)
cu ft	1,728	cu in
cu ft	7.48	gal
cu in	0.00433	gal
gal (oil)	0.0238	barrels (oil)
cu in	0.000579	cu ft
gal	0.1337	cu ft
gal	231	cu in

Temperature

F = (°C x 1.8) + 32
C = (°F - 32) ÷ 1.8

Fractions and Decimals

Multiply	By	To Get
Sixty-fourths	0.015625	Decimal
Thirty-seconds	0.03125	Decimal
Sixteenths	0.0625	Decimal
Eighths	0.125	Decimal
Fourths	0.250	Decimal
Halves	0.500	Decimal
Decimal	64	Sixty-fourths
Decimal	32	Thirty-seconds
Decimal	16	Sixteenths
Decimal	8	Eighths
Decimal	4	Fourths
Decimal	2	Halves

Gallons shown are U.S. standard.

Figure 3.37.0   Conversion factors. (*By permission: Associated Steam Specialty Company, Drexel Hill, PA.*)

Useful Engineering Tables

Schedule 40 Pipe, Standard Dimensions

Size (in)	Diameters External (in)	Diameters Approximate Internal (in)	Nominal Thickness (in)	Circumference External (in)	Circumference Internal (in)	Transverse Areas External (sq in)	Transverse Areas Internal (sq in)	Transverse Areas Metal (sq in)	Length of Pipe per sq ft External Surface Feet	Length of Pipe per sq ft Internal Surface Feet	Length of Pipe Containing One Cubic Foot Feet	Nominal Weight per foot Plain Ends	Nominal Weight per foot Threaded and Coupled	Number Threads per Inch of Screw
⅛	0.405	0.269	0.068	1.272	0.845	0.129	0.057	0.072	9.431	14.199	2533.775	0.244	0.245	27
¼	0.540	0.364	0.088	1.696	1.114	0.229	0.104	0.125	7.073	10.493	1383.789	0.424	0.425	18
⅜	0.675	0.493	0.091	2.121	1.549	0.358	0.191	0.167	5.658	7.747	754.360	0.567	0.568	18
½	0.840	0.622	0.109	2.639	1.954	0.554	0.304	0.250	4.547	6.141	473.906	0.850	0.852	14
¾	1.050	0.824	0.113	3.299	2.589	0.866	0.533	0.333	3.637	4.635	270.034	1.130	1.134	14
1	1.315	1.049	0.133	4.131	3.296	1.358	0.864	0.494	2.904	3.641	166.618	1.678	1.684	11½
1¼	1.660	1.380	0.140	5.215	4.335	2.164	1.495	0.669	2.301	2.767	96.275	2.272	2.281	11½
1½	1.900	1.610	0.145	5.969	5.058	2.835	2.036	0.799	2.010	2.372	70.733	2.717	2.731	11½
2	2.375	2.067	0.154	7.461	6.494	4.430	3.355	1.075	1.608	1.847	42.913	3.652	3.678	11½
2½	2.875	2.469	0.203	9.032	7.757	6.492	4.788	1.704	1.328	1.547	30.077	5.793	5.819	8
3	3.500	3.068	0.216	10.996	9.638	9.621	7.393	2.228	1.091	1.245	19.479	7.575	7.616	8
3½	4.000	3.548	0.226	12.566	11.146	12.566	9.886	2.680	0.954	1.076	14.565	9.109	9.202	8
4	4.500	4.026	0.237	14.137	12.648	15.904	12.730	3.174	0.848	0.948	11.312	10.790	10.889	8
5	5.563	5.047	0.258	17.477	15.856	24.306	20.006	4.300	0.686	0.756	7.198	14.617	14.810	8
6	6.625	6.065	0.280	20.813	19.054	34.472	28.891	5.581	0.576	0.629	4.984	18.974	19.185	8
8	8.625	7.981	0.322	27.096	25.073	58.426	50.027	8.399	0.442	0.478	2.878	28.554	28.809	8
10	10.750	10.020	0.365	33.772	31.479	90.763	78.855	11.908	0.355	0.381	1.826	40.483	41.132	8
12	12.750	11.938	0.406	40.055	37.699	127.640	111.900	15.740	0.299	0.318	1.288	53.600	—	—
14	14.000	13.125	0.437	43.982	41.217	153.940	135.300	18.640	0.272	0.280	1.069	63.000	—	—
16	16.000	15.000	0.500	50.265	47.123	201.050	176.700	24.350	0.238	0.254	0.817	78.000	—	—
18	18.000	16.874	0.563	56.548	52.998	254.850	224.000	30.850	0.212	0.226	0.643	105.000	—	—
20	20.000	18.814	0.593	62.831	59.093	314.150	278.000	36.150	0.191	0.203	0.519	123.000	—	—
24	24.000	22.626	0.687	75.398	71.063	452.400	402.100	50.300	0.159	0.169	0.358	171.000	—	—

Diameters and Areas of Circles and Drill Sizes

Drill Size	Dia.	Area	Drill Size	Dia.	Area	Drill Size	Dia.	Area	Drill Size	Dia.	Area
3/64	.0469	.00173	27	.1440	.01629	C	.2420	.04600	27/64	.4219	.13920
55	.0520	.00212	26	.1470	.01697	D	.2460	.04753	7/16	.4375	.15033
54	.0550	.00238	25	.1495	.01705	1/4	.2500	.04909	29/64	.4531	.16117
53	.0595	.00278	24	.1520	.01815	E	.2500	.04909	15/32	.4688	.17257
1/16	.0625	.00307	23	.1540	.01863	F	.2570	.05187	31/64	.4844	.18398
52	.0635	.00317	5/32	.1562	.01917	G	.2610	.05350	1/2	.500	.19635
51	.0670	.00353	22	.1570	.01936	17/64	.2656	.05515	33/64	.5156	.20831
50	.0700	.00385	21	.1590	.01986	H	.2660	.05557	17/32	.5312	.22166
49	.0730	.00419	20	.1610	.02036	I	.2720	.05811	9/16	.5625	.24850
48	.0760	.00454	19	.1660	.02164	J	.2770	.06026	19/32	.5937	.27688
5/64	.0781	.00479	18	.1695	.02256	K	.2810	.06202	5/8	.6250	.30680
47	.0785	.00484	11/64	.1719	.02320	9/32	.2812	.06213	21/32	.6562	.33824
46	.0810	.00515	17	.1730	.02351	L	.2900	.06605	11/16	.6875	.37122
45	.0820	.00528	16	.1770	.02461	M	.2950	.06835	23/32	.7187	.40574
44	.0860	.00581	15	.1800	.02545	19/64	.2969	.06881	3/4	.7500	.44179
43	.0890	.00622	14	.1820	.02602	N	.3020	.07163	25/32	.7812	.47937
42	.0935	.00687	13	.1850	.02688	5/16	.3125	.07670	13/16	.8125	.51849
3/32	.0938	.00690	3/16	.1875	.02761	O	.3160	.07843	27/32	.8437	.55914
41	.0960	.00724	12	.1890	.02806	P	.3230	.08194	7/8	.8750	.60132
40	.0980	.00754	11	.1910	.02865	21/64	.3281	.08449	29/32	.9062	.64504
39	.0995	.00778	10	.1935	.02941	Q	.3320	.08657	15/16	.9375	.69029
38	.1015	.00809	9	.1960	.03017	R	.3390	.09026	31/32	.9687	.73708
37	.1040	.00850	8	.1990	.03110	11/32	.3438	.09281	1	1.0000	.78540
36	.1065	.00891	7	.2010	.03173	S	.3480	.09511	1-1/16	1.0625	.88664
7/64	.1094	.00940	13/64	.2031	.03241	T	.3580	.10066	1-1/8	1.1250	.99402
35	.1100	.00950	6	.2040	.03268	23/64	.3594	.10122	1-3/16	1.1875	1.1075
34	.1110	.00968	5	.2055	.03317	U	.3680	.10636	1-1/4	1.2500	1.2272
33	.1130	.01003	4	.2090	.03431	3/8	.3750	.11045	1-5/16	1.3125	1.3530
32	.1160	.01039	3	.2130	.03563	V	.3770	.11163	1-3/8	1.3750	1.4849
31	.1200	.01131	7/32	.2188	.03758	W	.3860	.11702	1-7/16	1.4375	1.6230
1/8	.1250	.01227	2	.2210	.03836	25/64	.3906	.11946	1-1/2	1.5000	1.7671
30	.1285	.01242	1	.2280	.04083	X	.3970	.12379	1-5/8	1.6250	2.0739
29	.1360	.01453	A	.2340	.04301	Y	.4040	.12819	1-3/4	1.7500	2.4053
28	.1405	.01550	15/64	.2344	.04314	13/32	.4062	.12962	1-7/8	1.8750	2.7612
9/64	.1406	.01553	B	.2380	.04449	Z	.4130	.13396		2.0000	3.1416

Figure 3.38.0   Useful engineering tables. (*By permission: Associated Steam Specialty Company, Drexel Hill, PA.*)

Formulas & Conversion Factors

Miscellaneous Formulas (cont.)

Speed—A-C Machinery

$$\text{Synchronous RPM} = \frac{\text{Hertz x 120}}{\text{Poles}}$$

$$\text{Percent Slip} = \frac{\text{Synchronous RPM - Full-Load RPM}}{\text{Synchronous RPM}} \text{ x 100}$$

Motor Application

$$\text{Torque (lb.-ft.)} = \frac{\text{Horsepower x 5250}}{\text{RPM}}$$

$$\text{Horsepower} = \frac{\text{Torque (lb.-ft.) x RPM}}{5250}$$

Time for Motor to Reach Operating Speed (seconds)

$$\text{Seconds} = \frac{\text{WK}^2 \text{ x Speed Change}}{308 \text{ x Avg. Accelerating Torque}}$$

$$\text{Average Accelerating Torque} = \frac{[(\text{FLT} + \text{BDT})/2] + \text{BDT} + \text{LR1}}{3}$$

WK^2 = Inertia of Rotor + Inertia of Load (lb.-ft.^{2})
FLT = Full-Load Torque   BDT = Breakdown Torque
LRT = Locked Rotor Torque

$$\text{Load WK}^2 \text{ (at motor shaft)} = \frac{\text{WK}^2 \text{ (Load) x Load RPM}^2}{\text{Motor RPM}^2}$$

Shaft Stress (P.S.I.) $= \dfrac{\text{HP x 321,000}}{\text{RPM x Shaft Dia.}^3}$

Change in Resistance Due to Change in Temperature

$$R_C = R_H \text{ x } \frac{(K + T_C)}{(K + T_H)}$$

$$R_H = R_C \text{ x } \frac{(K + T_H)}{(K + T_C)}$$

K   = 234.5 - Copper
    = 236 - Aluminum
    = 180 - Iron
    = 218 - Steel
R_C = Cold Resistance (OHMS)
R_H = Hot Resistance (OHMS)
T_C = Cold Temperature (°C)
T_H = Hot Temperature (°C)

Figure 3.39.0 Formulas and conversion factors for AC machinery. (*By permission: Loren Cook, Springfield, MO.*)

Formulas & Conversion Factors

Miscellaneous Formulas

OHMS Law

Ohms = Volts/Amperes (R = E/I)
Amperes = Volts/Ohms (I = ER)
Volts = Amperes × Ohms (E = IR)

Power—A-C Circuits

$$\text{Efficiency} = \frac{746 \times \text{Output Horsepower}}{\text{Input Watts}}$$

$$\text{Three-Phase Kilowatts} = \frac{\text{Volts} \times \text{Amperes} \times \text{Power Factor} \times 1.732}{1000}$$

$$\text{Three-Phase Volt-Amperes} = \text{Volts} \times \text{Amperes} \times 1.732$$

$$\text{Three-Phase Amperes} = \frac{746 \times \text{Horsepower}}{1.732 \times \text{Volts} \times \text{Efficiency} \times \text{Power Factor}}$$

$$\text{Three Phase Efficiency} = \frac{746 \times \text{Horsepower}}{\text{Volts} \times \text{Amperes} \times \text{Power Factor} \times 1.732}$$

$$\text{Three-Phase Power Factor} = \frac{\text{Input Watts}}{\text{Volts} \times \text{Amperes} \times 1.732}$$

$$\text{Single-Phase Kilowatts} = \frac{\text{Volts} \times \text{Amperes} \times \text{Power Factor}}{1000}$$

$$\text{Single-Phase Amperes} = \frac{746 \times \text{Horsepower}}{\text{Volts} \times \text{Efficiency} \times \text{Power Factor}}$$

$$\text{Single-Phase Efficiency} = \frac{746 \times \text{Horsepower}}{\text{Volts} \times \text{Amperes} \times \text{Power Factor}}$$

$$\text{Single-Phase Power Factor} = \frac{\text{Input Watts}}{\text{Volts} \times \text{Amperes}}$$

$$\text{Horsepower (3 Ph)} = \frac{\text{Volts} \times \text{Amperes} \times 1.732 \times \text{Efficiency} \times \text{Power Factor}}{746}$$

$$\text{Horsepower (1 Ph)} = \frac{\text{Volts} \times \text{Amperes} \times \text{Efficiency} \times \text{Power Factor}}{746}$$

Power—D-C Circuits

$$\text{Watts} = \text{Volts} \times \text{Amperes} \ (W = EI)$$

$$\text{Amperes} = \frac{\text{Watts}}{\text{Volts}} \ (I = W/E)$$

$$\text{Horsepower} = \frac{\text{Volts} \times \text{Amperes} \times \text{Efficiency}}{746}$$

Figure 3.40.0 Ohms law and AC and DC circuit formulas. (*By permission: Loren Cook, Springfield, MO.*)

Formulas & Conversion Factors

Miscellaneous Formulas (cont.)

Where:

BHP = Brake Horsepower
GPM = Gallons per Minute
FT = Feet
PSI = Pounds per Square Inch
PSIG = Pounds per Square Inch Gauge
PSF = Pounds per Square Foot
PIW = Inches of Water Gauge

Area and Circumference of Circles

Diameter (inches)	Area (sq.in.)	Area (sq. ft.)	Circumference (feet)
1	0.7854	0.0054	0.2618
2	3.142	0.0218	0.5236
3	7.069	0.0491	0.7854
4	12.57	0.0873	1.047
5	19.63	0.1364	1.309
6	28.27	0.1964	1.571
7	38.48	0.2673	1.833
8	50.27	0.3491	2.094
9	63.62	0.4418	2.356
10	78.54	0.5454	2.618
11	95.03	0.6600	2.880
12	113.1	0.7854	3.142
13	132.7	0.9218	3.403
14	153.9	1.069	3.665
15	176.7	1.227	3.927
16	201.0	1.396	4.189
17	227.0	1.576	4.451
18	254.7	1.767	4.712
19	283.5	1.969	4.974
20	314.2	2.182	5.236
21	346.3	2.405	5.498
22	380.1	2.640	5.760
23	415.5	2.885	6.021
24	452.4	3.142	6.283

Formulas & Conversion Factors

Area and Circumference of Circles (cont.)

Diameter (inches)	Area (sq.in.)	Area (sq. ft.)	Circumference (feet)
25	490.9	3.409	6.545
26	530.9	3.687	6.807
27	572.5	3.976	7.069
28	615.7	4.276	7.330
29	660.5	4.587	7.592
30	706.8	4.909	7.854
31	754.7	5.241	8.116
32	804.2	5.585	8.378
33	855.3	5.940	8.639
34	907.9	6.305	8.901
35	962.1	6.681	9.163
36	1017.8	7.069	9.425
37	1075.2	7.467	9.686
38	1134.1	7.876	9.948
39	1194.5	8.296	10.21
40	1256.6	8.727	10.47
41	1320.2	9.168	10.73
42	1385.4	9.621	10.99
43	1452.2	10.08	11.26
44	1520.5	10.56	11.52
45	1590.4	11.04	11.78
46	1661.9	11.54	12.04
47	1734.9	12.05	12.30
48	1809.5	12.57	12.57
49	1885.7	13.09	12.83
50	1963.5	13.64	13.09
51	2043	14.19	13.35
52	2124	14.75	13.61
53	2206	15.32	13.88
54	2290	15.90	14.14
55	2376	16.50	14.40
56	2463	17.10	14.66
57	2552	17.72	14.92

Figure 3.41.0 Formulas and conversion factors—areas and circumferences. (*By permission: Loren Cook, Springfield, MO.*)

Heating & Refrigeration

Water Flow and Piping (cont.)
Example: If design values were 200 gpm and 40 ft head and actual flow were changed to 100 gpm, the new head would be:

$$h_2 = 40 \left(\frac{100}{200}\right)^2 = 10 \text{ ft}$$

$$\text{Pump hp} = \frac{\text{gpm x ft head x sp gr}}{3960 \text{ x \% efficiency}}$$

Typical single suction pump efficiencies, %:

1/12 to 1/2 hp	40 to 55
3/4 to 2	45 to 60
3 to 10	50 to 65

double suction pumps:

20 to 50	60 to 80

Friction Loss for Water Flow
Average value—new pipe. Used pipe add 50%
Feet loss / 100 ft—schedule 40 pipe

US Gpm	1/2 in. v Fps	1/2 in. h_F FtHd	3/4 in. v Fps	3/4 in. h_F FtHd	1 in. v Fps	1 in. h_F FtHd	1-1/4 in. v Fps	1-1/4 in. h_F FtHd
2.0	2.11	5.5						
2.5	2.64	8.2						
3.0	3.17	11.2						
3.5	3.70	15.3						
4	4.22	19.7	2.41	4.8				
5	5.28	29.7	3.01	7.3				
6			3.61	10.2	2.23	3.1		
8			4.81	17.3	2.97	5.2		
10			6.02	26.4	3.71	7.9		
12					4.45	11.1	2.57	2.9
14					5.20	14.0	3.00	3.8
16					5.94	19.0	3.43	4.8

Heating & Refrigeration

Friction Loss for Water Flow (cont.)

US Gpm	1-1/2 in. v Fps	1-1/2 in. h_F FtHd	2 in. v Fps	2 in. h_F FtHd	2-1/2 in. v Fps	2-1/2 in. h_F FtHd	1-1/4 in. v Fps	1-1/4 in. h_F FtHd
18	2.84	2.8					3.86	6.0
20	3.15	3.4					4.29	7.3
22	3.47	4.1					4.72	8.7
24	3.78	4.8					5.15	10.3
26	4.10	5.5					5.58	11.9
28	4.41	6.3					6.01	13.7
30	4.73	7.2					6.44	15.6
35	5.51	9.6					7.51	20.9
40	6.30	12.4	3.82	3.6				
45	7.04	15.5	4.30	4.4				
50			4.78	5.4				
60			5.74	7.6	4.02	3.1		
70			6.69	10.2	4.69	4.2	3 in. v Fps	3 in. h_F FtHd
80			7.65	13.1	5.36	5.4		
100					6.70	8.2	5.21	3.9
120					8.04	11.5	5.21	3.9
140					9.38	15.5	6.08	5.2
160							6.94	6.7
180							7.81	8.4
200							8.68	10.2

Adapted from "Numbers", Bill Holladay and Cy Otterholm, 1985

Figure 3.42.0   Water flow and frictional loss for water flow. (*By permission: Loren Cook, Springfield, MO.*)

Classification	Occupancy Sq. Ft/Person		Lights Watts/Sq.Ft.		Refrigeration Sq.Ft/Ton‡		Air Quantities CFM/Sq.Ft.					
							East-South-West		North		Internal	
	Lo	Hi	Lo	Hi	Lo	Hi	Lo	Hi	Lo	Hi	Lo	Hi
Beauty & Barber Shops	45	25	3.0*	9.0*	240	105	1.5	4.2	1.1	2.6	0.9	2.0
Dept. Stores-Basement	30	20	2.0	4.0	340	225	—	—	—	—	0.7	1.2
Main Floor	45	16	3.5	9.0†	350	150	—	—	—	—	0.9	2.0
Upper Floors	75	40	2.0	3.5†	400	280	—	—	—	—	0.8	1.2
Clothing Stores	50	30	1.0	4.0	345	185	0.9	1.6	0.7	1.4	0.6	1.1
Drug Stores	35	17	1.0	3.0	180	110	1.8	3.0	1.0	1.8	0.7	1.3
Discount Stores	35	15	1.5	5.0	345	120	0.7	2.0	0.6	1.6	0.5	1.1
Shoe Stores	50	20	1.0	3.0	300	150	1.2	2.1	1.0	1.8	0.8	1.2
Malls	100	50	1.0	2.0	365	160	—	—	—	—	1.1	2.5
Refrigeration for Central Heating and Cooling Plant												
Urban Districts						285						
College Campuses						240						
Commercial Centers						200						
Residential Centers						375						

Refrigeration and air quantities for applications listed in this table of cooling load check figures are based on all-air system and normal outdoor air quantities for ventilation except as noted.

Notes: ‡Refrigeration loads are for entire application.

†Includes other loads expressed in Watts sq.ft.

♦Air quantities for heavy manufacturing areas are based on supplementary means to remove excessive heat.

*Air quantities for hospital patient rooms and office buildings (except internal areas) are based on induction (air-water) system.

Figure 3.43.0   Cooling load check figures. (*By permission: Loren Cook, Springfield, MO.*)

Screwed fittings, turbulent flow only, equipment length in feet.

Fittings	Pipe Size							
	1/2	3/4	1	1-1/4	1-1/2	2	2-1/2	3
Standard 90° Ell	3.6	4.4	5.2	6.6	7.4	8.5	9.3	11
Long rad. 90° Ell	2.2	2.3	2.7	3.2	3.4	3.6	3.6	4.0
Standard 45° Ell	.71	.92	1.3	1.7	2.1	2.7	3.2	3.9
Tee Line flow	1.7	2.4	3.2	4.6	5.6	7.7	9.3	12
Tee Br flow	4.2	5.3	6.6	8.7	9.9	12	13	17
180° Ret bend	3.6	4.4	5.2	6.6	7.4	8.5	9.3	11
Globe Valve	22	24	29	37	42	54	62	79
Gate Valve	.56	.67	.84	1.1	1.2	1.5	1.7	1.9
Angle Valve	15	15	17	18	18	18	18	18
Swing Check	8.0	8.8	11	13	15	19	22	27
Union or Coupling	.21	.24	.29	.36	.39	.45	.47	.53
Bellmouth inlet	.10	.13	.18	.26	.31	.43	.52	.67
Sq mouth inlet	.96	1.3	1.8	2.6	3.1	4.3	5.2	6.7
Reentrant pipe	1.9	2.6	3.6	5.1	6.2	8.5	10	13
Sudden enlargement	Feet of liquid loss = $\dfrac{(V_1 - V_2)^2}{2_g}$							

where V_1 & V_2 = entering and leaving velocities

and g = 32.17 ft/sec^2

Adapted from "Numbers", Bill Holladay and Cy Otterholm, 1985

Figure 3.44.0   Equivalent length of pipe for valves and fittings. (*By permission: Loren Cook, Springfield, MO.*)

Pounds of Dry Saturated Steam per Boiler Horsepower

Feed-Water Temp.	Gauge Pressure — psig																	
	0	2	10	15	20	40	50	60	80	100	120	140	150	160	180	200	220	240
30	29.0	29.0	28.8	28.7	28.6	28.4	28.3	28.2	28.2	28.1	28.0	28.0	27.9	27.9	27.9	27.9	27.9	27.8
40	29.3	29.2	29.1	29.0	28.9	28.7	28.6	28.5	28.4	28.3	28.2	28.2	28.2	28.2	28.2	28.1	28.1	28.1
50	29.6	29.5	29.3	29.2	29.1	28.9	28.8	28.8	28.7	28.6	28.5	28.5	28.4	28.4	28.4	28.3	28.3	28.3
60	29.8	29.8	29.6	29.5	29.4	29.2	29.1	29.0	28.9	28.8	28.8	28.7	28.7	28.6	28.6	28.6	28.6	28.5
70	30.1	30.0	29.9	29.8	29.7	29.5	29.4	29.3	29.2	29.1	29.0	29.0	28.9	28.9	28.9	28.8	28.8	28.8
80	30.4	30.3	30.1	30.0	3.0	29.8	29.6	29.6	29.5	29.3	29.2	29.2	29.2	29.2	29.1	29.1	29.1	29.0
90	30.6	30.6	30.4	30.3	30.2	30.0	29.9	29.8	29.7	29.6	29.5	29.5	29.4	29.4	29.4	29.3	29.3	29.3
100	30.9	30.8	30.6	30.6	30.5	30.3	30.2	30.1	30.0	29.8	29.8	29.8	29.7	29.7	29.7	29.6	29.6	29.6
110	31.2	31.2	30.9	30.8	30.8	30.6	30.4	30.3	30.2	30.0	30.0	30.0	30.0	30.0	29.9	29.9	29.8	29.8
120	31.5	31.4	31.2	31.2	31.1	30.8	30.7	30.6	30.5	30.4	30.3	30.3	30.2	30.2	30.2	30.1	30.1	30.1
130	31.8	31.7	31.5	31.4	31.4	31.1	31.0	30.9	30.8	30.7	30.6	30.6	30.5	30.5	30.4	30.4	30.4	30.4
140	32.1	32.0	31.8	31.7	31.6	31.4	31.3	31.2	31.1	31.0	30.9	30.8	30.8	30.8	30.8	30.7	30.7	30.6
150	32.4	32.4	32.1	32.0	31.9	31.7	31.6	31.5	31.4	31.2	31.2	31.2	31.1	31.1	31.0	31.0	30.9	30.9
160	32.7	32.7	32.4	32.4	32.3	32.0	31.9	31.8	31.7	31.5	31.4	31.4	31.4	31.4	31.3	31.3	31.2	31.2
170	33.0	33.0	32.7	32.6	32.6	32.3	32.2	32.1	32.0	31.8	31.7	31.7	31.7	31.6	31.6	31.6	31.5	31.5
180	33.4	33.3	33.0	33.0	32.9	32.6	32.5	32.4	32.3	32.2	32.1	32.0	32.0	32.0	31.9	31.9	31.8	31.8
190	33.8	33.7	33.4	33.3	33.2	32.9	32.8	32.7	32.6	32.5	32.4	32.4	32.3	32.3	32.2	32.2	32.1	32.1
200	34.1	34.0	33.7	33.6	33.5	33.2	33.1	33.0	32.9	32.8	32.7	32.6	32.6	32.6	32.6	32.5	32.4	32.4
212	34.5	34.4	34.2	34.1	33.0	33.6	33.5	33.4	33.3	33.2	33.1	33.0	33.0	33.0	32.9	32.9	32.8	32.8
220	34.8	34.7	34.4	34.3	34.2	33.9	33.8	33.7	33.5	33.4	33.3	33.3	33.2	33.2	33.1	33.1	33.1	33.0
227	35.0	34.9	34.7	34.5	34.4	34.1	34.0	33.9	33.8	33.7	33.6	33.5	33.5	33.4	33.4	33.3	33.3	33.3
230	35.2	35.0	34.8	34.7	34.5	34.2	34.1	34.0	33.9	33.8	33.7	33.6	33.6	33.5	33.5	33.4	33.4	33.4

Typical Units for Fuels

ITEM	GROSS HEATING VALUES
No. 2 Oil	140,000 Btu/gal.
No 5 Oil	148,000 Btu/gal.
No. 6 Oil	150,000 Btu/gal.
1 Therm	100,000 Btu
1 kW	3,413 Btu

Figure 3.45.0 Pounds of dry saturated steam per boiler horsepower.

Ultimate Analysis - Ultimate analysis is a statement of the quantities of the various elements of which a substance is composed. For fuel oils, this will likely state higher heating values and specific gravity in addition to the percentages by weight of each element.

Flash Point - The flash point of a fuel oil is an indication of the maximum temperature at which it can be stored and handled without serious fire hazard. The minimum permissible flash point is usually regulated by federal, state or municipal laws and is based on accepted practice in handling and use.

Pour Point - The pour point is an indication of the lowest temperature at which a fuel oil can be stored and still be capable of flowing under very low forces. The pour point is prescribed in accordance with the conditions of storage and use. Higher pour

Properties of saturated steam.

Gauge Pressure PSIG	Temper- ature °F	Heat in Btu/lb.			Specific Volume Cu. ft. per lb.	Gauge Pressure PSIG	Temper- ature °F	Heat in Btu/lb.			Specific Volume Cu. ft. per lb.
		Sensible	Latent	Total				Sensible	Latent	Total	
25	134	102	1017	1119	142	150	366	339	857	1196	2.74
20	162	129	1001	1130	73.9	155	368	341	885	1196	2.68
15	179	147	990	1137	51.3	160	371	344	853	1197	2.60
10	192	160	982	1142	39.4	165	373	346	851	1197	2.54
5	203	171	976	1147	31.8	170	375	348	849	1197	2.47
0	212	180	970	1150	26.8	175	377	351	847	1198	2.41
1	215	183	968	1151	25.2	180	380	353	845	1198	2.34
2	219	187	966	1153	23.5	185	382	355	843	1198	2.29
3	222	190	964	1154	22.3	190	384	358	841	1199	2.24
4	224	192	962	1154	21.4	195	386	360	839	1199	2.19
5	227	195	960	1155	20.1	200	388	362	837	1199	2.14
6	230	198	959	1157	19.4	205	390	364	836	1200	2.09
7	232	200	957	1157	18.7	210	392	366	834	1200	2.05
8	233	201	956	1157	18.4	215	394	368	832	1200	2.00
9	237	205	954	1159	17.1	220	396	370	830	1200	1.96
10	239	207	953	1160	16.5	225	397	372	828	1200	1.92
12	244	212	949	1161	15.3	230	399	374	827	1201	1.89
14	248	216	947	1163	14.3	235	401	376	825	1201	1.85
16	252	220	944	1164	13.4	240	403	378	823	1201	1.81
18	256	224	941	1165	12.6	245	404	380	822	1202	1.78
75	320	290	895	1185	4.91	350	435	414	790	1204	1.28
80	324	294	891	1185	4.67	355	437	416	789	1205	1.26
85	328	298	889	1187	4.44	360	438	417	788	1205	1.24
90	331	302	886	1188	4.24	365	440	419	786	1205	1.22
95	335	305	883	1188	4.05	370	441	420	785	1205	1.20
100	338	309	880	1189	3.89	375	442	421	784	1205	1.19
105	341	312	878	1190	3.74	380	443	422	783	1205	1.18
110	344	316	875	1191	3.59	385	445	424	781	1205	1.16
115	347	319	873	1192	3.46	390	446	425	780	1205	1.14
120	350	322	871	1193	3.34	395	447	427	778	1205	1.13
125	353	325	868	1193	3.23	400	448	428	777	1205	1.12
130	356	328	866	1194	3.12	450	460	439	766	1205	1.00
140	361	333	861	1194	2.92	500	470	453	751	1204	.89
145	363	336	859	1195	2.84	550	479	464	740	1204	.82
						600	489	475	728	1203	.74

In Vac. { brackets applied to rows 25, 20, 15, 10, 5 }

Figure 3.46.0 Properties of saturated steam. (*By permission: McGraw-Hill*, Plumber's and Pipefitter's Calculations Manual.)

Properties of saturated steam. (Cont'd.)

20	**259**	**227**	**939**	**1166**	**11.9**	**250**	**406**	**382**	**820**	**1202**	**1.75**
22	262	230	937	1167	11.3	255	408	383	819	1202	1.72
24	265	233	934	1167	10.8	260	409	385	817	1202	1.69
26	268	236	933	1169	10.3	265	411	387	815	1202	1.66
28	271	239	930	1169	9.85	270	413	389	814	1203	1.63
30	**274**	**243**	**929**	**1172**	**9.46**	**275**	**414**	**391**	**812**	**1203**	**1.60**
32	277	246	927	1173	9.10	280	416	392	811	1203	1.57
34	279	248	925	1173	8.75	285	417	394	809	1203	1.55
36	282	251	923	1174	8.42	290	418	395	808	1203	1.53
38	284	253	922	1175	8.08	295	420	397	806	1203	1.49
40	**286**	**256**	**920**	**1176**	**7.82**	**300**	**421**	**398**	**805**	**1203**	**1.47**
42	289	258	918	1176	7.57	305	423	400	803	1203	1.45
44	291	260	917	1177	7.31	310	425	402	802	1204	1.43
46	293	262	915	1177	7.14	315	426	404	800	1204	1.41
48	295	264	914	1178	6.94	320	427	405	799	1204	1.38
50	**298**	**267**	**912**	**1179**	**6.68**	**325**	**429**	**407**	**797**	**1204**	**1.36**
55	300	271	909	1180	6.27	330	430	408	796	1204	1.34
60	307	277	906	1183	5.84	335	432	410	794	1204	1.33
65	312	282	901	1183	5.49	340	433	411	793	1204	1.31
70	316	286	898	1184	5.18	345	434	413	791	1204	1.29

Figure 3.46.0   *(Continued)*

Heating & Refrigeration

Pump Bodies

Two basic types of pump bodies are:

Horizontal Split Case - split down centerline of pump horizontal axis. Disassembled by removing top half of pump body. Pump impeller mounted between bearings at center of shaft. Requires two seals. Usually double suction pump. Suction and discharge are in straight-line configuration.

Vertical Split Case - single-piece body casting attached to cover plate at the back of pump by capscrews. Pump shaft passes through seal and bearing in coverplate. Impeller is mounted on end of shaft. Suction is at right angle to discharge.

Pump Mounting Methods

The three basic types of pump mounting arrangements are:

Base Mount-Long Coupled - pump is coupled to base-mount motor. Motor can be removed without removing the pump from piping system. Typically standard motors are used.

Base Mount-Close Coupled - pump impeller is mounted on base mount motor shaft. No separate mounting is necessary for pump. Usually special motor necessary for replacement. More compact than long-coupled pump.

Line Mount - mounted to and supported by system piping. Usually resilient mount motor. Very compact. Usually for low flow requirements.

Pump Construction Types

The two general pump construction types are:

Bronze-fitted Pumps

- cast iron body
- brass impeller
- brass metal seal assembly components

Uses: Closed heating/chilled water systems, low-temp fresh water.

All-Bronze Pumps

- all wetted parts are bronze

Uses: Higher temp fresh water, domestic hot water, hot process water.

Pump Impeller Types

Single Suction - fluid enters impeller on one side only.
Double Suction - fluid enters both sides of impeller.
Closed Impeller - has a shroud which encloses the pump vanes, increasing efficiency. Used for fluid systems free of large particles which could clog impeller.
Semi-Open Impeller - has no inlet shroud. Used for systems where moderate sized particles are suspended in pumped fluid.
Open Impeller - has no shroud. Used for systems which have large particles suspended in pumped fluid, such as sewage or sludge systems.

Figure 3.47.0 Common pump bodies and types. (*By permission: Loren Cook, Springfield, MO.*)

Heating & Refrigeration

Common Pump Formulas

Formula for	I-P Units
Head	$H = psi \times 2.31/SG^* \text{ (ft)}$
Output power	$P_o = Q_v \times H \times SG^*/3960 \text{ (hp)}$
Shaft power	$P_s = \dfrac{Q_v \times H \times SG^*}{39.6 \times E_p} \text{ (hp)}$
Input power	$P_i = P_s \times 74.6/E_m \text{ (kw)}$
Utilization Q_D= design flow Q_A= actual flow H_D= design head H_A= actual head	$\eta \mu = 100 \dfrac{Q_D H_D}{Q_A H_A}$

*SG = specific gravity

Water Flow and Piping

Pressure drop in piping varies approx as the square of the flow:

$$\frac{h_2}{h_1} = \left(\frac{Q_2}{Q_1}\right)^2$$

The velocity of water flowing in a pipe is

$$v = \frac{gpm \times 0.41}{d^2}$$

Where V is in ft/sec and d is inside diameter, in.

Nom size	1/2	3/4	1	1-1/4	1-1/2	2	2-1/2	3	4
ID in.	0.622	0.824	1.049	1.380	1.610	2.067	2.469	3.068	4.02
d^2	0.387	0.679	1.100	1.904	2.59	4.27	6.10	9.41	16.21

Quiet Water Flows

Nom size	1/2	3/4	1	1-1/4	1-1/2	2	2-1/2	3	4
Gpm	1.5	4.	8.	14	22	44	75	120	240

Six fps is a reasonable upper limit for water velocity in pipes. The relationship between pressure drop and flow rate can also be expressed:

$$h_2 = h_1 \times \left(\frac{Q_2}{Q_1}\right)^2 \text{ or } Q_2 = Q_1 \times \sqrt{\frac{h_2}{h_1}}$$

Figure 3.47.1 Common pump formulas. (*By permission: Loren Cook, Springfield, MO.*)

Heating & Refrigeration

Estimated Seasonal Efficiencies of Heating Systems

Systems	Seasonal Efficiency
Gas Fired Gravity Vent Unit Heater	62%
Energy Efficient Unit Heater	80%
Electric Resistance Heating	100%
Steam Boiler with Steam Unit Heaters	65%-80%
Hot Water Boiler with HYD Unit Heaters	65%-80%
Oil Fired Unit Heaters	78%
Municipal Steam System	66%
INFRA Red (High Intensity)	85%
INFRA Red (Low Intensity)	87%
Direct Fired Gas Make Up Air	94%
Improvement with Power Ventilator Added to Gas Fired Gravity Vent Unit Heater	4%
Improvement with Spark Pilot Added to Gas Fired Gravity Vent Unit Heater	1/2%-3%
Improvement with Automatic Flue Damper and Spark Pilot Added to Gravity Vent Unit Heater	8%

Annual Fuel Use

Annual fuel use may be determined for a building by using one of the following formulas:

Electric Resistance Heating

$$H/(\Delta T \times 3413 \times E) \times D \times 24 \times C_D = KWH/YEAR$$

Natural Gas Heating

$$H/(\Delta T \times 100{,}000 \times E) \times D \times 24 \times C_D = THERMS/YEAR$$

Propane Gas Heating

$$H/(\Delta T \times 21739 \times E) \times D \times 24 \times C_D = POUNDS/YEAR$$

$$H/(\Delta T \times 91911 \times E) \times D \times 24 \times C_D = GALLONS/YEAR$$

Oil Heating

$$H/(\Delta T \times 140{,}000 \times E) \times D \times 24 \times C_D = GALLONS/YEAR$$

Where: ΔT = Indoor Design Minus Outdoor Design Temp.
H = Building Heat Loss
D = Annual Degree Days
E = Seasonal Efficiency (See Above)
C_D = Correlation Factor C_D vs. Degree-Days

Heating & Refrigeration

Annual Fuel Use (cont.)

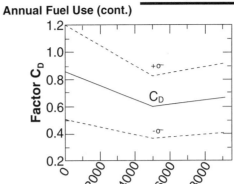

Figure 3.48.0 Estimated seasonal efficiencies of heating systems. (*By permission: Loren Cook, Springfield, MO.*)

Cooling with Ice

MODES OF OPERATION

The modular ICE CHILLER® Thermal Storage Unit can operate in any of five distinct operating modes. These modes of operation provide the flexibility required by building operators to meet their daily HVAC cooling requirements.

ICE BUILD In this operating mode, ice is built by circulating a 25% solution (by weight) of inhibited ethylene glycol through the coils contained in the ICE CHILLER® Thermal Storage Unit. Figure 2 illustrates typical chiller supply temperatures for 8, 10 and 12 hour build cycles. For a typical 10–hour build time, the supply glycol temperature is never lower than 22°F. As the graph illustrates, for build times exceeding 10 hours, the minimum glycol temperature is greater than 22°F. For build times less than 10 hours, the minimum glycol temperature will be lower than 22°F at the end of the build cycle. This performance is based on a chiller flow rate associated with a 5°F range. When a larger temperature range is the basis of the chiller selection, the chiller supply temperatures will be lower than shown in Figure 2.

ICE BUILD WITH COOLING When cooling loads exist during the ice build period, some of the cold ethylene glycol used to build ice is diverted to the cooling load to provide the required cooling. The amount of glycol diverted is determined by the building loop set point temperature. BAC recommends that this mode of operation be applied on systems using primary/secondary pumping. This reduces the possibility of damaging the cooling coil or heat exchanger by pumping cold glycol, lower than 32°F, to this equipment.

COOLING – ICE ONLY In this operating mode the chiller is off. The warm return ethylene glycol solution is cooled to the desired set point temperature by melting ice stored in the modular ICE CHILLER® Thermal Storage Unit.

COOLING – CHILLER ONLY In this operating mode the chiller supplies all the building cooling requirements. Glycol flow is diverted around the thermal storage equipment to allow the cold supply glycol to flow directly to the cooling load. Temperature set points are maintained by the chiller.

COOLING – ICE WITH CHILLER In this operating mode, cooling is provided by the combined operation of the chiller and thermal storage equipment. The glycol chiller precools the warm return glycol. The partially cooled glycol solution then passes through the ICE CHILLER® Thermal Storage Unit where it is cooled by the ice to the design temperature.

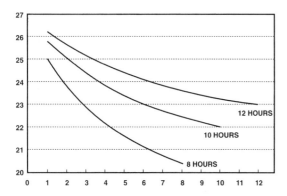

Figure 3.49.0 Cooling with ice—providing alternative means for air conditioning in five different operating modes. (*By permission: Baltimore Aircoil Company, Baltimore, MD.*)

LOAD PROFILE

A daily load profile is the hour–by–hour representation of cooling loads for a 24–hour period. Most HVAC applications use a daily load profile to determine the amount of storage required. Some HVAC systems apply a weekly load profile. For conventional air–conditioning systems, chillers are selected based on the peak cooling load. For ice storage systems, the chillers are selected based on the ton–hours of cooling required and a defined operating strategy. Thermal storage systems provide much flexibility for varying operating strategies as long as the total ton–hours selected are not exceeded. This is why an accurate load profile must be provided when designing an ice storage system.

Load profiles take many different shapes based on the application. Figure 1 illustrates a typical HVAC load profile for an office building with a 500–ton peak cooling load and a 12–hour cooling requirement. The shape of this curve is representative of most HVAC applications. For preliminary equipment selections, BAC's ICE CHILLER® Thermal Storage Unit Selection Program can generate a similar load profile. Information required is the estimated building peak cooling load and duration of the cooling load.

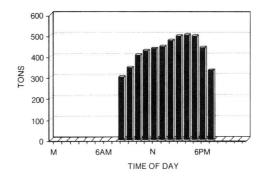

Typical HVAC Load Profile

The Air–Conditioning & Refrigeration Institute (ARI) has published Guideline T, "Specifying the Thermal Performance of Cool Storage Equipment." The purpose of Guideline T is to establish the minimum user specified data and supplier specified performance data. Design data provided by the engineer includes: System Loads, Flow Rates and Temperatures. Table 1 details the user specified data.

Figure 3.49.1 Typical load profile for conventional air conditioning systems during a normal workday. (*By permission: Baltimore Aircoil Company, Baltimore, MD.*)

SYSTEM SCHEMATICS

Two basic flow schematics are applied to select BAC's ICE CHILLER® Thermal Storage Units. Figure 3 illustrates a single piping loop with the chiller installed upstream of the thermal storage equipment. This

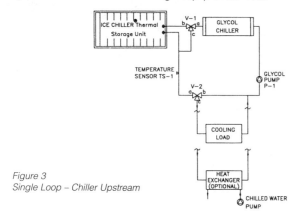

Figure 3
Single Loop – Chiller Upstream

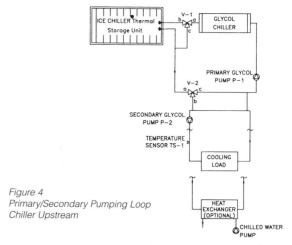

Figure 4
Primary/Secondary Pumping Loop
Chiller Upstream

design allows the thermal storage system to operate in four of the five possible operating modes. They are Ice Build, Cooling-Ice Only, Cooling-Chiller Only and Cooling–Ice With Chiller. For this schematic the following control logic is applied:

MODE	CHILLER	P–1	V–1	V–2
Ice Build	On	On	A–B	A–B
Cooling–Ice Only	Off	On	Modulate	A–C
Cooling–Chiller Only	On	On	A–C	A–C
Cooling–Ice With Chiller	On	On	Modulate	A–C

Valve V-1 modulates in response to temperature sensor, TS-1. Valve V-2 could be positioned to either maintain a constant flow, less than P-1, or modulate in response to the return glycol temperature from the cooling load.

When the building loop contains chilled water, a heat exchanger must be installed to separate the glycol loop from the building's chilled water loop. On applications where an existing water chiller is available, it can be installed in the chilled water loop to reduce the load on the thermal storage system.

This design should not be used when there is a requirement to build ice and provide cooling. This would require the cold return glycol from the thermal storage equipment be pumped to the cooling load or heat exchanger. Since the glycol temperature is below 32°F, the cooling coil or heat exchanger is subject to freezing.

The flow schematic illustrated in Figure 4 details a primary/secondary pumping loop with the chiller located upstream of the thermal storage equipment. This design allows the system to operate in all five operat-

ing modes. For this schematic, the following control logic is applied:

MODE	CHILLER	P–1	P–2	V–1	V–2
Ice Build	On	On	Off	A–B	A–C
Ice Build With Cooling	On	On	On	A–B	Modulate
Cooling–Chiller Only	On	On	On	A–C	A–B
Cooling–Ice Only	Off	On	On	Modulate	A–B
Cooling–Ice with Chiller	On	On	On	Modulate	A–B

Valve V-1 and Valve V-2 modulate, depending on the operating mode, in response to temperature sensor, TS-1. The benefit provided by the primary/secondary pumping loop is that the system can build ice and provide cooling without fear of freezing a cooling coil or heat exchanger. This system design also allows for different flow rates in each of the pumping loops. When the flow rates in the pumping loops are different, the glycol flow rate in the primary loop should be greater than or equal to the glycol flow rate in the secondary loop. As in the single loop schematic, a heat exchanger and a base water chiller can be added to the system schematic.

Variations to these schematics are possible but these are the most common for thermal storage systems. One common variation positions the chiller downstream of the thermal storage equipment. This design is used when the glycol temperatures off the ice cannot be maintained for the entire cooling period. By positioning the chiller downstream of the ice, the chiller is used to maintain the required supply temperature. In Figure 3 and Figure 4, the chiller is installed upstream of the ice. This offers two significant advantages compared to system designs that locate the chiller downstream of the ice. First, the chiller operates at higher glycol temperatures to precool the return glycol. This enables the chiller to operate at a higher capacity which reduces the amount of ice required. Second, since the chiller is operating at higher evaporator temperatures, the efficiency (kw/TR) of the chiller is improved.

Figure 3.49.2 Schematics of an ice storage system. (*By permission: Baltimore Aircoil Company, Baltimore, MD.*)

UNIT PIPING

Piping to the ICE CHILLER® Thermal Storage Unit should follow established piping guidelines. The coil connections on the unit are galvanized steel and are grooved for mechanical coupling.

For single tank applications, each pair of manifolded coil connections should include a shut off valve so the unit can be isolated from the system. Figure 8 illustrates the valve arrangement for a single unit. It is recommended that the piping include a bypass circuit to allow operation of the system without the ICE CHILLER® Thermal Storage Unit in the piping loop. This bypass can be incorporated into the piping design by installing a three way/modulating valve. This valve can also be used to control the leaving glycol temperature from the thermal storage unit. Temperature and pressure taps should be installed to allow for easier flow balancing and system troubleshooting. A relief valve, set at a maximum of 150-psi, must be installed between the shut off valves and the coil connections to protect the coils from excessive pressures due to hydraulic expansion. The relief valve should be vented to a portion of the system which can accommodate expansion.

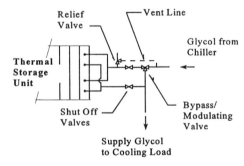

Figure 8
Single Unit Valve Arrangement

CAUTION: The system must include an expansion tank to accommodate changes in fluid volume. Adequately sized air vents must be installed at the high points in the piping loop to remove trapped air from the system.

Figure 9 illustrates reverse return piping for multiple units installed in parallel. The use of reverse return piping is recommended to ensure balanced flow to each unit. Shut off valves at each unit can be used as balancing valves.

When large quantities of ICE CHILLER® Thermal Storage Units are installed, the system should be divided into groups of units. Then, balancing of each unit can be eliminated and a common balancing valve for each group of units installed. Shut off valves for isolating individual units should be installed but not used for balancing glycol flow to the unit.

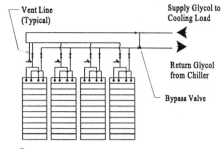

Figure 9
Reverse Return Piping

CONTROLS

To ensure efficient operation of the ICE CHILLER® Thermal Storage Units, each system is provided with factory installed Operating Controls. A brief description of the controls follow.

Once the ice build cycle has been initiated, the glycol chiller should run at full capacity without cycling or unloading until the ICE CHILLER® Thermal Storage Units are fully charged. When the units are fully charged, the chiller should be turned off and not allowed to re-start until cooling is required. The ice build cycle is terminated by the Operating Control Assembly. This assembly includes a low water cut-out, a shut-off switch and a safety switch. The low water cut-out prevents the ice build mode from starting if there is insufficient water in the tank. The shut-off switch will terminate the build cycle when the units are fully charged and will prevent the next ice build mode from starting until 15% of the ice is melted. The safety switch is provided to terminate the build cycle should the operating controls fail to function correctly.

Inventory controls that provide either a 4 - 20 mA or 1 - 5 Vdc are still available. These controls should be used for determining the amount of ice in inventory but not to terminate the ice build cycle. Complete operating control details are provided in the Installation, Operation and Maintenance Manual.

Figure 3.49.3 Piping and control considerations when installing an ice storage system. (*By permission: Baltimore Aircoil Company, Baltimore, MD.*)

INSTALLATION

ICE CHILLER® Thermal Storage Units must be installed on a continuous flat level surface. The pitch of the slab must not exceed 1/8" over a 10–foot span. Figure 5 details ICE CHILLER® Thermal Storage Unit layout guidelines. The units should be positioned so there is sufficient clearance between units and adjacent walls to allow easy access. When multiple units are installed, a minimum of 18" is recommended side-to-side and 3'-0" end-to-end for access to the operating controls.

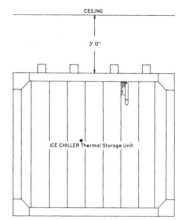

Figure 6
Recommended Overhead Clearance

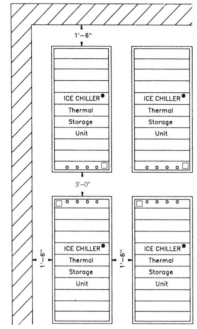

Figure 5
ICE CHILLER® Thermal Storage Unit
Layout Guidelines

There may be occasions when the thermal storage units must be installed outside and the equipment's visibility reduced. If a fenced enclosure or landscaping does not provide the desired screening when the ICE CHILLER® Thermal Storage Unit is installed on a concrete slab, the unit can be partially buried. Caution: When this equipment is buried, attention must be given to excavation, drainage, concrete pad design, placement of the unit and backfilling to prevent damage to the protective bitumastic coating on the unit. The concrete slab must be designed by a qualified engineer.

When installed indoors, the access and slab requirements described above also apply. The units should be placed close to a floor drain in the event they need to be drained. The minimum height requirement above the tank for proper pipe installation is 3 feet. Figure 6 illustrates the recommended overhead clearance for ICE CHILLER® Thermal Storage Units.

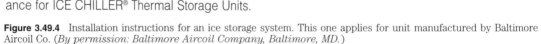

Figure 3.49.4 Installation instructions for an ice storage system. This one applies for unit manufactured by Baltimore Aircoil Co. (*By permission: Baltimore Aircoil Company, Baltimore, MD.*)

General Maintenance Information

The services required to maintain a cooling tower are primarily a function of the quality of the air and water in the locality of the installation:

AIR:

The most harmful atmospheric conditions are those with unusual quantities of industrial smoke, chemical fumes, salt or heavy dust. Such airborne impurities are carried into the cooling tower and absorbed by the recirculating water to form a corrosive solution.

WATER:

The most harmful conditions develop as water evaporates from the cooling tower, leaving behind the dissolved solids originally contained in the make-up water. These dissolved solids may be either alkaline or acidic and, as they are concentrated in the circulating water, can produce scaling or accelerated corrosion.

The extent of impurities in the air and water determines the frequency of most maintenance services and also governs the extent of water treatment which can vary from a simple continuous bleed and biological control to a sophisticated treatment system. (See "Water Treatment".)

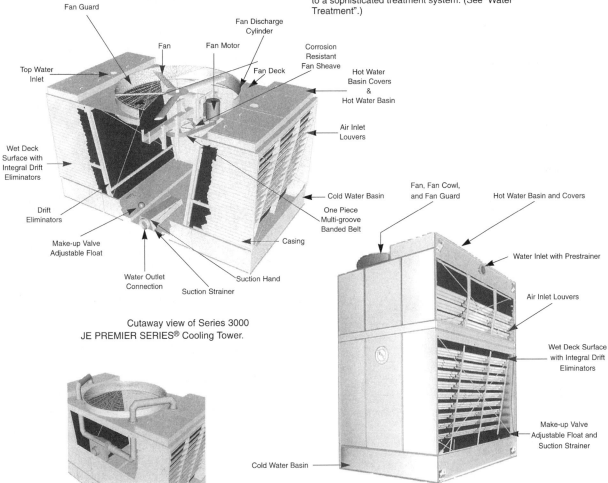

Cutaway view of Series 3000
JE PREMIER SERIES® Cooling Tower.

Series 3000 and JE PREMIER SERIES® Cooling Tower

Series 1500 Cooling Tower

Figure 3.50.0 General maintenance requirements for cooling tower operational efficiency. (*By permission: Baltimore Aircoil Company, Baltimore, MD.*)

Water Treatment

Corrosion and Scale Control

In cooling towers, cooling is accomplished by evaporation of a portion of the process water as it flows through the tower. As this water evaporates, the impurities originally present remain in the recirculating water. The concentration of the dissolved solids increases rapidly and can reach unacceptable levels. In addition, airborne impurities are often introduced into the recirculating water, intensifying the problem. If these impurities and contaminants are not effectively controlled. they can cause scaling, corrosion, and sludge accumulations which reduce heat transfer efficiency and increase system operating costs.

The degree to which dissolved solids and other impurities build up in the recirculating water may be defined as the cycles of concentration. Specifically, cycles of concentration is the ratio of dissolved solids (for example — TDS, chlorides, sulfates) in the recirculating water to dissolved solids in the make-up water. For optimal heat transfer efficiency and maximum equipment life, the cycles of concentration should be controlled such that the recirculating water is maintained within the guidelines listed below.

Recirculated Water Quality Guideline,

	Stainless Steel or BALTIBOND® Corrosion Protection System	G235 Galvanized Steel[1]
pH	6.5 to 9.0	7.30 to 9.0[2]
Hardness as CaCO₃	30 to 500 ppm	30 to 500 ppm
Alkalinity as CaCO₃	500 ppm max.	500 ppm max.
Total Dissolved Solids	1200 ppm max.	1000 ppm max.
Chlorides	250 ppm max.	125 ppm max.

[1]Units manufactured in Canada will be constructed of the metric equivalen, Z700 galvanized steel.

[2]Units having galvanized steel construction and a circulating water ph of 8.3 or higher will require periodic passivazation of the galvanized steel to prevent 'white rust', the accumulation of white, waxy, non-protective zinc corrosion products on galvanized steel surfaces.

In order to control the cycles of concentration such that the above guidelines are maintained, it will be necessary to "bleed" or blowdown" a small amount of recirculating water from the system. This "bleed" water is replenished with fresh make-up water, thereby limiting the buildup of impurities.

Typically the bleed is accomplished automatically through a solenoid valve controlled by a condictivity meter. The conductivity meter set point is the water conduc-

tivity at the desired cycles of concentration and should be determined by a competent water treatment expert. (Note: the solenoid valve and conductivity meter must be supplied by others.) Alternatively a bleed line with a valve can be used to continuously bleed from the system. (Note: the bleed line and valve must be supplied by others.) In this arrangement, the rate of bleed can be adjusted using the valve in the bleed line and measured by filling a container of known volume while noting the time period. The bleed rate and water quality should be periodically checked to ensure that adequate control of the water quality is being maintained.

The required continuous bleed rate may be calculated by the formula,

$$\text{Bleed Rate} = \frac{\text{Evaporation Rate}}{\text{Number of Cycles of Concentration-1}}$$

The evaporation rate can be determined by one of the following:

1. The evaporation rate is approximately 2 GPM per 1 million BTU/HR of heat rejection.

2. The evaporation rate is approximately 3 GPM per 100 tons of refrigeration.

3. Evaporation Rate = Water Flow Rate x Range x .001.

Example: At a flow rate of 900 GPM and a cooling range of 10°F, the evaporation rate is 9 GPM (900 GPM x 100°F x .001 = 9 GPM).

If the site conditions are such that constant bleed-off will not control scale or corrosion and maintain the water quality within the guidelines, chemical treatment may be necessary. If a chemical water treatment program is used, it must meet the following requirements:

1. The chemicals must be compatible with the unit construction (zinc coated) steel as well as all other materials used in the system (pipe, heat exchanger, etc.)

2. Chemicals to inhibit scale and corrosion should be added to the recirculating water by an automatic feed system on a continuously metered basis. This will prevent localized high concentrations of chemicals which may cause corrosion. It is recommended that the chemical be fed into the system at the discharge of the recirculating pump. They should *not* be batch fed directly into the cold water basin.

3. Acid water treatment is *not* recommended unless the unit(s) have been furnished with the BALTIBOND® Corrosion Protection System or is constructed of stainless steel — in which cases acid treatment can be used provided the requirements of paragraph 1 and 2 above are maintained.

Figure 3.50.1 Corrosion, scale, and biological controls required for extended life of a cooling tower. (*By permission: Baltimore Aircoil Company, Baltimore, MD.*)

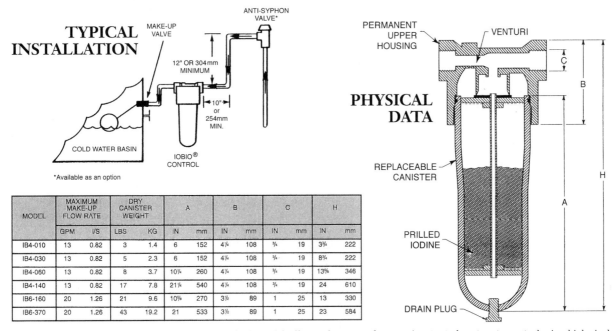

MODEL	MAXIMUM MAKE-UP FLOW RATE		DRY CANISTER WEIGHT		A		B		C		H	
	GPM	l/S	LBS	KG	IN	mm	IN	mm	IN	mm	IN	mm
IB4-010	13	0.82	3	1.4	6	152	4¼	108	¾	19	3¾	222
IB4-030	13	0.82	5	2.3	6	152	4¼	108	¾	19	8¾	222
IB4-060	13	0.82	8	3.7	10¼	260	4¼	108	¾	19	13⅝	346
IB4-140	13	0.82	17	7.8	21¼	540	4¼	108	¾	19	24	610
IB6-160	20	1.26	21	9.6	10⅝	270	3½	89	1	25	13	330
IB6-370	20	1.26	43	19.2	21	533	3½	89	1	25	23	584

Figure 3.50.2   Protecting the cooling tower from the harmful effects of untreated water. A patented system to control microbiological growth in the tower's make-up water supply.

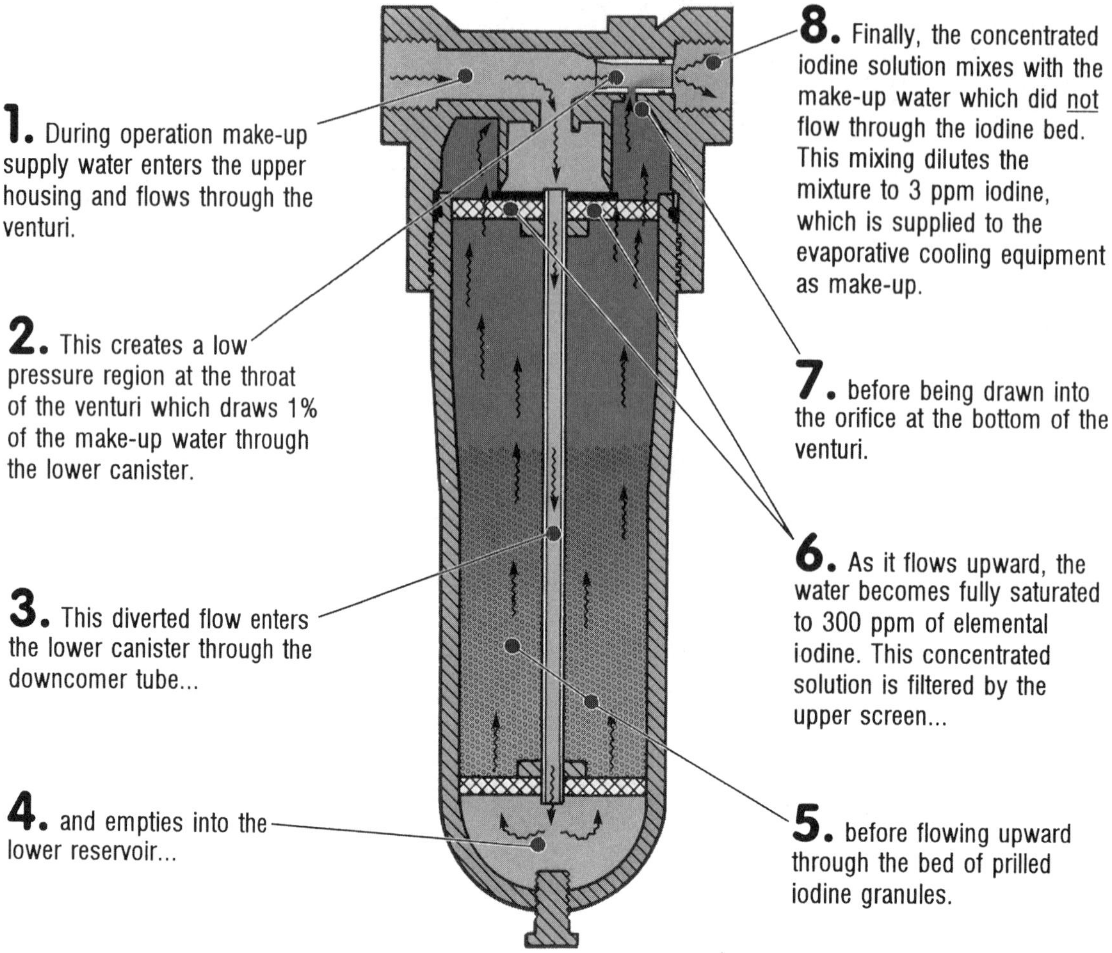

1. During operation make-up supply water enters the upper housing and flows through the venturi.

2. This creates a low pressure region at the throat of the venturi which draws 1% of the make-up water through the lower canister.

3. This diverted flow enters the lower canister through the downcomer tube...

4. and empties into the lower reservoir...

8. Finally, the concentrated iodine solution mixes with the make-up water which did <u>not</u> flow through the iodine bed. This mixing dilutes the mixture to 3 ppm iodine, which is supplied to the evaporative cooling equipment as make-up.

7. before being drawn into the orifice at the bottom of the venturi.

6. As it flows upward, the water becomes fully saturated to 300 ppm of elemental iodine. This concentrated solution is filtered by the upper screen...

5. before flowing upward through the bed of prilled iodine granules.

* Patent Pending

Figure 3.50.3   BAC's patent pending system of iodine injection to control bacteria, slime, and algae in the cooling tower's water supply.

Contents

4.0.0   Common Electrical Terminology

Amp (A)
A measurement of the rate of flow of electrons along a wire. If electricity can be likened to plumbing, amps would be the same as gallons-per-second. Watts ÷Volts = Amps.

American Wire Gauge (AWG)
AWG refers to common wire sizes and ratings.

CO/ALR
15 or 20 A devices which can be used with copper or aluminum wire. Higher-rated devices appropriate for direct connection to aluminum or copper wire are marked "AL-CU".

Circuit
The path electricity follows as it moves along a conductor. Branch circuits distribute power to the parts of the home where it's needed.

Circuit Breaker
A resettable safety device that automatically stops electrical flow in a circuit when an overload or short circuit occurs. Either circuit breakers or fuses are located in the home's load center.

Conductor
A material capable of carrying electricity's energy. Opposite of Insulator.

Current
The rate of flow of electrons through a conductor, measured in Amps.

Electron
An invisible particle of negatively-charged matter that moves at the speed of light through an electrical circuit.

Fed Spec
Devices which comply with Federal Specifications such as W-C-596 for connecting devices and W-S-896 for switches. Fed Spec Standards for switches and connecting devices include NEMA Performance Standards.

Fuse
A non-resettable safety device that automatically stops electrical flow in a circuit when an overload or short circuit occurs. Either fuses or circuit breakers are located in the home's load center.

Ground
Refers literally to <u>earth</u> which has an electrical potential (voltage) of zero.

Ground Fault Circuit Interrupter (GFCI or GFI)
A safety device that senses shock hazard to a far greater degree than fuses or circuit breakers. Automatically stops electrical flow in a circuit.

Grounding Wire
The conductor used to connect the electrical equipment to ground (or earth) at the service entrance point, minimizing the potential for electrical shock. Either clad in green insulation or unclad.

Hospital Grade
UL-established criteria for devices used in hospitals. To obtain that listing and carry the Hospital Grade green dot identification, devices must pass many of the same tests as those included in NEMA Performance Standards and must go beyond in ability to withstand impact, crushing and continuous torture without loss of grounding path continuity. The highest grade attainable is Hospital Grade.

Hot Wire
The ungrounded conductor that carries electricity from the utility to a load center, or from a branch circuit to a receptacle or switch. It is normally clad in red or black insulation.

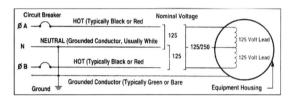

Insulation
A non-conductive covering that protects wires and other conductors of electricity.

Isolated Ground
In an isolated ground device, the grounding path is isolated from the device's mounting bracket. This "isolated ground" provides an electrical noise shield so that electromagnetic radiation waves will not turn into ground path noise which can disrupt sensitive electronics and can cause equipment malfunction.

Kilowatt (kw)
A thousand watts. (Watt is the measure of power that a electrical device consumes.) A kilowatt hour is the measurement most utilities use to measure electrical consumption. It indicates how many kilowatts are consumed for a full hour.

Knock-outs
Tabs that can be removed to make openings for wires and/or conduit in device and junction boxes or electrical panels.

Load Center
A home's fuse box or circuit breaker box. It divides the power into various branch circuits for distribution throughout the home.

4.1.0   Selecting Wire and Cable

Several criteria are considered when selecting wire and cable:

Voltage Rating

The voltage at which the cable is to operate determines the type and size of the conductor. Cable size is sometimes determined by voltage drop and the amount of heat that will be generated by the passage of electrical current. The ampacity of kVA loading is the primary consideration in determining final conductor size. The current load, kVA load, or the kilowatt load and power factor must all be known before the conductor size can be finalized.

Solid conductors are generally used for wire or cables in sizes No. 10 AWG and smaller. Stranded conductors are available in almost all cable construction.

Thickness and Type of Insulation

The thickness and type of insulation is probably the most important component of a wire or cable and its selection is governed by requirements for stability and long life, resistance to ionization and corona, resistance to high temperatures, resistance to moisture, mechanical strength, and flexibility.

Type XLP insulation is a general purpose insulation constructed of cross-linked polyethylene. It combines the best properties of rubber, is resistant to moisture and chemicals, rated at 90°F and exhibits excellent thermal, electrical, and physical properties.

Ethylene–propylene (EP) rubber is also a general purpose insulating material. It is flexible and has excellent heat, zone, and moisture resistant qualities. Cables insulated with EP are frequently used for power and control circuits installed in conduit, ducts, aerially, and direct burial cable.

Polyvinyl chloride (PVC) insulation is a thermoplastic and exhibits resistance to moisture, acids, alkalis, and oil. It can be used on most cables without any outer covering, and Underwriters listings such as type THHN and THWN use PVC insulation for wiring up to 600 volts.

Polyethylene is another thermoplastic insulation material with outstanding electrical properties and moisture resistance. It does, however, exhibit poor heat and flame resistance, which limits its use. Polyethylene insulated wire or cables are frequently used for high frequency communication cables, where capacitance and attenuation must be kept to a minimum.

Protective Coatings

Rubber-insulated cables require a covering such as neoprene or PVC for mechanical protection. Neoprene jacketing is unaffected by oils, acids, or alkalis and is tough, flexible, and moisture and flame resistant. EP, chlorosulfanated polyethylene, and nitrile–PVC compounds are also used as jacketing materials. Interlocked armor coverings provide excellent mechanical protection and can be constructed of galvanized steel, aluminum, or bronze. Where corrosive environments are expected, the armoring can be made of PVC.

Lead is available as a sheath material for many types of wire and cable insulation. This material is moisture proof, durable, and is often installed in underground ducts.

4.1.1 Conductor Properties (AWG Size 18 to 2000)

Size AWG/ kcmil	Area Cir. Mils	Stranding Quan-tity	Stranding Diam. In.	Overall Diam. In.	Overall Area In.²	Copper Uncoated ohm/MFT	Copper Coated ohm/MFT	Aluminum ohm/ MFT
18	1620	1	—	0.040	0.001	7.77	8.08	12.8
18	1620	7	0.015	0.046	0.002	7.95	8.45	13.1
16	2580	1	—	0.051	0.002	4.89	5.08	8.05
16	2580	7	0.019	0.058	0.003	4.99	5.29	8.21
14	4110	1	—	0.064	0.003	3.07	3.19	5.06
14	4110	7	0.024	0.073	0.004	3.14	3.26	5.17
12	6530	1	—	0.081	0.005	1.93	2.01	3.18
12	6530	7	0.030	0.092	0.006	1.98	2.05	3.25
10	10380	1	—	0.102	0.008	1.21	1.26	2.00
10	10380	7	0.038	0.116	0.011	1.24	1.29	2.04
8	16510	1	—	0.128	0.013	0.764	0.786	1.26
8	16510	7	0.049	0.146	0.017	0.778	0.809	1.28
6	26240	7	0.061	0.184	0.027	0.491	0.510	0.808
4	41740	7	0.077	0.232	0.042	0.308	0.321	0.508
3	52620	7	0.087	0.260	0.053	0.245	0.254	0.403
2	66360	7	0.097	0.292	0.067	0.194	0.201	0.319
1	83690	19	0.066	0.332	0.087	0.154	0.160	0.253
1/0	105600	19	0.074	0.373	0.109	0.122	0.127	0.201
2/0	133100	19	0.084	0.419	0.138	0.0967	0.101	0.159
3/0	167800	19	0.094	0.470	0.173	0.0766	0.0797	0.126
4/0	211600	19	0.106	0.528	0.219	0.0608	0.0626	0.100
250	—	37	0.082	0.575	0.260	0.0515	0.0535	0.0847
300	—	37	0.090	0.630	0.312	0.0429	0.0446	0.0707
350	—	37	0.097	0.681	0.364	0.0367	0.0382	0.0605
400	—	37	0.104	0.728	0.416	0.0321	0.0331	0.0529
500	—	37	0.116	0.813	0.519	0.0258	0.0265	0.0424
600	—	61	0.099	0.893	0.626	0.0214	0.0223	0.0353
700	—	61	0.107	0.964	0.730	0.0184	0.0189	0.0303
750	—	61	0.111	0.998	0.782	0.0171	0.0176	0.0282
800	—	61	0.114	1.03	0.834	0.0161	0.0166	0.0265
900	—	61	0.122	1.09	0.940	0.0143	0.0147	0.0235
1000	—	61	0.128	1.15	1.04	0.0129	0.0132	0.0212
1250	—	91	0.117	1.29	1.30	0.0103	0.0106	0.0169
1500	—	91	0.128	1.41	1.57	0.00858	0.00883	0.0141
1750	—	127	0.117	1.52	1.83	0.00735	0.00756	0.0121
2000	—	127	0.126	1.63	2.09	0.00643	0.00662	0.0106

These resistance values are valid ONLY for the parameters as given. Using conductors having coated strands, different stranding type, and especially, other temperatures, change the resistance.

Formula for temperature change: $R_2 = R_1 [1+\alpha(T_2-75)]$ where: $\alpha_{cu} = 0.00323$, $\alpha_{AL} = 0.00330$.

Conductors with compact and compressed stranding have about 9 percent and 3 percent, respectively, smaller bare conductor diameters than those shown. See Table 5A for actual compact cable dimensions.

The IACS conductivities used: bare copper = 100%, aluminum = 61%.

Class B stranding is listed as well as solid for some sizes. Its overall diameter and area is that of its circumscribing circle.

(FPN): The construction information is per NEMA WC8-1976 (Rev 5-1980). The resistance is calculated per National Bureau of Standards Handbook 100, dated 1966, and Handbook 109, dated 1972.

(Reprinted with permission from NFPA 70-1996, the National Electrical Code® National Fire Protection Association, Quincy, Massachusetts. National Electrical Code® and NEC® are registered trademarks of the National Fire Protection Association, Inc., Quincy, MA 02269)

4.2.0   Maximum Number of Conductors in Trade Sizes of Conduit or Tubing THWN, THHN Standard Conductors

Type Letters	Conductor Size AWG, kcmil	½	¾	1	1¼	1½	2	2½	3	3½	4	5	6
THWN,	14	13	24	39	69	94	154						
	12	10	18	29	51	70	114	164					
	10	6	11	18	32	44	73	104	160				
	8	3	5	9	16	22	36	51	79	106	136		
THHN, FEP (14 through 2), FEPB (14 through 8), PFA (14 through 4/0) PFAH (14 through 4/0) Z (14 through 4/0) XHHW (4 through 500 kcmil)	6	1	4	6	11	15	26	37	57	76	98	154	
	4	1	2	4	7	9	16	22	35	47	60	94	137
	3	1	1	3	6	8	13	19	29	39	51	80	116
	2	1	1	3	5	7	11	16	25	33	43	67	97
	1		1	1	3	5	8	12	18	25	32	50	72
	1/0		1	1	3	4	7	10	15	21	27	42	61
	2/0		1	1	2	3	6	8	13	17	22	35	51
	3/0		1	1	1	3	5	7	11	14	18	29	42
	4/0			1	1	2	4	6	9	12	15	24	35
	250			1	1	1	3	4	7	10	12	20	28
	300			1	1	1	3	4	6	8	11	17	24
	350			1	1	1	2	3	5	7	9	15	21
	400				1	1	1	3	5	6	8	13	19
	500				1	1	1	2	4	5	7	11	16
	600					1	1	1	3	4	5	9	13
	700						1	1	1	3	4	8	11
	750						1	1	1	2	3	7	11
XHHW	6	1	3	5	9	13	21	30	47	63	81	128	185
	600				1	1	1	1	3	4	5	9	13
	700					1	1	1	3	4	5	7	11
	750					1	1	1	2	3	4	7	10

Note: This table is for concentric stranded conductors only. For cables with compact conductors, the dimensions in Table 5A shall be used.

4.3.0   Percent of Cross-Section of Conduit and Tubing for Conductors

Number of Conductors	1	2	3	4	Over 4
All conductor types except lead-covered	53	31	40	40	40
Lead-covered conductors	55	30	40	38	35

Note 1. See Tables 3A, 3B, and 3C for number of conductors all of the same size in trade sizes of conduit or tubing ½ inch through 6 inch.

Note 2. For conductors larger than 750 kcmil or for combinations of conductors of different sizes, use Tables 4 through 8, Chapter 9, for dimensions of conductors, conduit and tubing.

Note 3. Where the calculated number of conductors, all of the same size, includes a decimal fraction, the next higher whole number shall be used where this decimal is 0.8 or larger.

Note 4. When bare conductors are permitted by other sections of this Code, the dimensions for bare conductors in Table 8 of Chapter 9 shall be permitted.

Note 5. A multiconductor cable of two or more conductors shall be treated as a single conductor cable for calculating percentage conduit fill area. For cables that have elliptical cross section, the cross-sectional area calculation shall be based on using the major diameter of the ellipse as a circle diameter.

4.4.0 Maximum Number of Concentric Stranded Conductors in Trade Sizes of Conduit or Tubing (RWH and RHH Conductors with Outer Covering)

Type Letters	Conductor Size AWG, kcmil	½	¾	1	1¼	1½	2	2½	3	3½	4	5	6
RHW,	14	3	6	10	18	25	41	58	90	121	155		
	12	3	5	9	15	21	35	50	77	103	132		
	10	2	4	7	13	18	29	41	64	86	110		
	8	1	2	4	7	9	16	22	35	47	60	94	137
RHH	6	1	1	2	5	6	11	15	24	32	41	64	93
(with	4	1	1	1	3	5	8	12	18	24	31	50	72
outer	3	1	1	1	3	4	7	10	16	22	28	44	63
covering)	2		1	1	3	4	6	9	14	19	24	38	56
	1		1	1	1	3	5	7	11	14	18	29	42
	1/0		1	1	1	2	4	6	9	12	16	25	37
	2/0			1	1	1	3	5	8	11	14	22	32
	3/0			1	1	1	3	4	7	9	12	19	28
	4/0			1	1	1	2	4	6	8	10	16	24
	250				1	1	1	3	5	6	8	13	19
	300				1	1	1	3	4	5	7	11	17
	350				1	1	1	2	4	5	6	10	15
	400				1	1	1	1	3	4	6	9	14
	500					1	1	1	3	4	5	8	11
	600						1	1	2	3	4	6	9
	700						1	1	1	3	3	6	8
	750						1	1	1	3	3	5	8

Note: This table is for concentric stranded conductors only.

(*Reprinted with permission from NFPA 70-1996, the National Electrical Code® National Fire Protection Association, Quincy, Massachusetts.* National Electrical Code® and NEC® are registered trademarks of the National Fire Protection Association, Inc., Quincy, MA 02269)

4.5.0 Dimensions of Rubber- and Thermoplastic-Covered Conductors

Size AWG kcmil	Types RFH-2, RH, RHH, RHW, SF-2 Approx. Diam. Inches	Approx. Area Sq. In.	Types TF, THW, TW Approx. Diam. Inches	Approx. Area Sq. In.	Types TFN, THHN, THWN Approx. Diam. Inches	Approx. Area Sq. In.	Types FEP, FEPB, FEPW, TFE, PF, PFA, PFAH, PGF, PTF, Z, ZF, ZFF Approx. Diam. Inches	Approx. Area Sq. Inches	Type XHHW, ZW Approx. Diam. Inches	Approx. Area Sq. In.	Types KF-1, KF-2, KFF-1, KFF-2 Approx. Diam. Inches	Approx. Area Sq. In.
Col. 1	Col. 2	Col. 3	Col. 4	Col. 5	Col. 6	Col. 7	Col. 8	Col. 9	Col. 10	Col. 11	Col. 12	Col. 13
18	.146	.0167	.106	.0088	.089	.0062	.081	.0052	…	….	.065	.0033
16	.158	.0196	.118	.0109	.100	.0079	.092	.0066	…	….	.070	.0038
14	30 mils .171	.0230	.131	.0135	.105	.0087	.105 .105	.0087 .0087	…	….	.083	.0054
14	45 mils .204*	.0327*	…	….	…	….	… …	…. ….	…	….	…	….
14	……	….	.162†	.0206†	…	….	… …	…. ….	.129	.0131	…	….
12	30 mils .188	.0278	.148	.0172	.122	.0117	.121 .121	.0115 .0115	…	….	.102	.0082
12	45 mils .221*	.0384*	…	….	…	….	… …	…. ….	…	….	…	….
12	……	….	.179†	.0252†	…	….	… …	…. ….	.146	.0167	…	….
10	…… .242	.0460	.168	.0222	.153	.0184	.142 .142	.0158 .0158	…	….	.124	.0121
10			.199†	.0311†	…	….			.166	.0216		
8	…… .328	.0845	.245	.0471	.218	.0373	.206 .186	.0333 .0272	…	….	…	….
8	…… ….	….	.276†	.0598†	…	….			.241	.0456	…	….
6	.397	.1238	.323	.0819	.257	.0519	.244 .302	.0468 .0716	.282	.0625	…	….
4	.452	.1605	.372	.1087	.328	.0845	.292 .350	.0670 .0962	.328	.0845	…	….
3	.481	.1817	.401	.1263	.356	.0995	.320 .378	.0804 .1122	.356	.0995	…	….
2	.513	.2067	.433	.1473	.388	.1182	.352 .410	.0973 .1320	.388	.1182	…	….
1	.588	.2715	.508	.2027	.450	.1590	.420 …	.1385 ….	.450	.1590	…	….
1/0	.629	.3107	.549	.2367	.491	.1893	.462 …	.1676 ….	.491	.1893	…	….
2/0	.675	.3578	.595	.2781	.537	.2265	.498 …	.1948 ….	.537	.2265	…	….
3/0	.727	.4151	.647	.3288	.588	.2715	.560 …	.2463 ….	.588	.2715	…	….
4/0	.785	.4840	.705	.3904	.646	.3278	.618 …	.3000 ….	.646	.3278	…	….

(*Reprinted with permission from NFPA 70-1996, the National Electrical Code® National Fire Protection Association, Quincy, Massachusetts.* National Electrical Code® and NEC® are registered trademarks of the National Fire Protection Association, Inc., Quincy, MA 02269)

4.6.0 Maximum Number of Conductors in Trade Sizes of Conduit or Tubing for TW, XHHW, RHW Conductors

Type Letters	Conductor Size AWG, kcmil	½	¾	1	1¼	1½	2	2½	3	3½	4	5	6
TW, XHHW (14 through 8)	14	9	15	25	44	60	99	142					
	12	7	12	19	35	47	78	111	171				
	10	5	9	15	26	36	60	85	131	176			
	8	2	4	7	12	17	28	40	62	84	108		
RHW and RHH (without outer covering), THW	14	6	10	16	29	40	65	93	143	192			
	12	4	8	13	24	32	53	76	117	157			
	10	4	6	11	19	26	43	61	95	127	163		
	8	1	3	5	10	13	22	32	49	66	85	133	
TW,	6	1	2	4	7	10	16	23	36	48	62	97	141
	4	1	1	3	5	7	12	17	27	36	47	73	106
THW,	3	1	1	2	4	6	10	15	23	31	40	63	91
	2	1	1	2	4	5	9	13	20	27	34	54	78
	1		1	1	3	4	6	9	14	19	25	39	57
FEPB (6 through 2), RHW and RHH (without outer covering)	1/0		1	1	2	3	5	8	12	16	21	33	49
	2/0		1	1	1	3	5	7	10	14	18	29	41
	3/0		1	1	1	2	4	6	9	12	15	24	35
	4/0			1	1	1	3	5	7	10	13	20	29
	250			1	1	1	2	4	6	8	10	16	23
	300			1	1	1	2	3	5	7	9	14	20
	350				1	1	1	3	4	6	8	12	18
	400				1	1	1	2	4	5	7	11	16
	500					1	1	1	3	4	6	9	14
	600					1	1	1	3	4	5	7	11
	700					1	1	1	2	3	4	7	10
	750					1	1	1	2	3	4	6	9

Note: This table is for concentric stranded conductors only.

4.7.0 Maximum Number of Fixture Wires in Trade Sizes of Conduit or Tubing

(40 Percent Fill Based on Individual Diameters)

Wire Types	½					¾					1					1¼					1½					2				
	18	16	14	12	10	18	16	14	12	10	18	16	14	12	10	18	16	14	12	10	18	16	14	12	10	18	16	14	12	10
PTF, PTFF, PGFF, PGF, PFF, PF, PAF, PAFF, ZF, ZFF	23	18	14			40	31	24			65	50	39			115	90	70			157	122	95			257	200	156		
TFFN, TFN	19	15				34	26				55	43				97	76				132	104				216	169			
SF-1	16					29					47					83					114					186				
SFF-1	15					26					43					76					104					169				
TF	11	10				20	18				32	30				57	53				79	72				129	118			
RFH-1	11					20					32					57					79					129				
TFF	11	10				20	17				32	27				56	49				77	66				126	109			
AF	11	9	7	4	3	19	16	12	7	5	31	26	20	11	8	55	46	36	19	15	75	63	49	27	20	123	104	81	44	34
SFF-2	9	7	6			16	12	10			27	20	17			47	36	30			65	49	42			106	81	68		
SF-2	9	8	6			16	14	11			27	23	18			47	40	32			65	55	43			106	90	71		
FFH-2	9	7				15	12				25	19				44	34				60	46				99	75			
RFH-2	7	5				12	10				20	16				36	28				49	38				80	62			
KF-1, KFF-1, KF-2, KFF-2	36	32	22	14	9	64	55	39	25	17	103	89	63	41	28	182	158	111	73	49	248	216	152	100	67	406	353	248	163	110

4.8.0   Conductor Size Increases from Copper to Aluminum

When substituting aluminum, increase the sizes of conductors in accordance with the following table:

Copper Size Conductor	Minimum Substitute Aluminum Size Conductor
#2	2/0
#1	2/0
1/0	4/0
2/0	4/0
3/0	300 MCM
4/0	350 MCM
250 MCM	400 MCM
300 MCM	500 MCM
350 MCM	600 MCM
400 MCM	600 MCM
500 MCM	750 MCM

Utilization	Acceptable Types
Conductors #1 AWG and smaller	THWN, THHN, XHHW
Conductors 1/0 and larger in "dry" locations	THHW, THHN, XHHW
Conductors 1/0 and larger in "wet" locations	THHW-2, XHHW-2, THWN-2

4.9.0   Minimum Radii Bends in Conduit

Bends in conduit shall have minimum radii:
1. For primary feeder - - - - - - -15'-0", except where specifically indicated otherwise or where turning up at termination point.
2. For primary feeder - - - - - - -4'-0" turning up at termination point
3. For secondary feeder   - - - -4'-0" all bends.
4. For communications   - - - - -4'-0" and/or signal wiring all bends

4.10.0 Aluminum Building Wire Nominal Dimensions

Bare Conductor**			Type THW		Type THHN		Type XHHW		
Size AWG or kcmil	Number of Strands	Diam. Inches	Approx. Diam. Inches	Approx. Area Sq. In.	Approx. Diam. Inches	Approx. Area Sq. In.	Approx. Diam. Inches	Approx. Area Sq. In.	Size AWG or kcmil
8	7	.134	.255	.0510	—	—	.224	.0394	8
6	7	.169	.290	.0660	.240	.0452	.260	.0530	6
4	7	.213	.335	.0881	.305	.0730	.305	.0730	4
2	7	.268	.390	.1194	.360	.1017	.360	.1017	2
1	19	.299	.465	.1698	.415	.1352	.415	.1352	1
1/0	19	.336	.500	.1963	.450	.1590	.450	.1590	1/0
2/0	19	.376	.545	.2332	.495	.1924	.490	.1885	2/0
3/0	19	.423	.590	.2733	.540	.2290	.540	.2290	3/0
4/0	19	.475	.645	.3267	.595	.2780	.590	.2733	4/0
250	37	.520	.725	.4128	.670	.3525	.660	.3421	250
300	37	.570	.775	.4717	.720	.4071	.715	.4015	300
350	37	.616	.820	.5281	.770	.4656	.760	.4536	350
400	37	.659	.865	.5876	.815	.5216	.800	.5026	400
500	37	.736	.940	.6939	.885	.6151	.880	.6082	500
600	61	.813	1.050	.8659	.985	.7620	.980	.7542	600
700	61	.877	1.110	.9676	1.050	.8659	1.050	.8659	700
750	61	.908	1.150	1.0386	1.075	.9076	1.090	.9331	750
1000	61	1.060	1.285	1.2968	1.255	1.2370	1.230	1.1882	1000

* Dimensions are from industry sources
** Compact conductor per ASTM B 400

(Reprinted with permission from NFPA 70-1996, the National Electrical Code® National Fire Protection Association, Quincy, Massachusetts. National Electrical Code® and NEC® are registered trademarks of the National Fire Protection Association, Inc., Quincy, MA 02269)

4.11.0 Expansion Characteristics of PVC Rigid Nonmetallic Conduit

Expansion Characteristics Of PVC Rigid Nonmetallic Conduit Coefficient Of Thermal Expansion = 3.38 × 10⁻⁵ In/In/°F

Temperature Change In Degrees F	Length Change In Inches per 100 ft. of PVC Conduit	Temperature Change in Degrees F	Length Change in inches per 100 ft. of PVC Conduit	Temperature Change in Degrees F	Length Change in inches per 100 ft. of PVC Conduit	Temperature Change in Degrees F	Length Change in inches per 100 ft. of PVC Conduit
5	0.2	55	2.2	105	4.2	155	6.3
10	0.4	60	2.4	110	4.5	160	6.5
15	0.6	65	2.6	115	4.7	165	6.7
20	0.8	70	2.8	120	4.9	170	6.9
25	1.0	75	3.0	125	5.1	175	7.1
30	1.2	80	3.2	130	5.3	180	7.3
35	1.4	85	3.4	135	5.5	185	7.5
40	1.6	90	3.6	140	5.7	190	7.7
45	1.8	95	3.8	145	5.9	195	7.9
50	2.0	100	4.1	150	6.1	200	8.1

(Reprinted with permission from NFPA 70–1996, the National Electrical Code® National Fire Protection Association, Quincy, Massachusetts. National Electrical Code® and NEC® are registered trademarks of the National Fire Protection Association, Inc., Quincy, MA 02269)

4.12.0 Maximum Number of Compact Conductors in Conduit or Tubing

Insulation Type / Conduit Trade Size

Conductor Size AWG or kcmil	1 in. THW	1 in. THHN	1 in. XHHW	1¼ in. THW	1¼ in. THHN	1¼ in. XHHW	1½ in. THW	1½ in. THHN	1½ in. XHHW	2 in. THW	2 in. THHN	2 in. XHHW	2½ in. THW	2½ in. THHN	2½ in. XHHW	3 in. THW	3 in. THHN	3 in. XHHW	3½ in. THW	3½ in. THHN	3½ in. XHHW	4 in. THW	4 in. THHN	4 in. XHHW
6	5	7	6	9	13	11	12	18	15	15	18	18												
4	4	4	4	7	8	8	9	11	11	11	13	13												
2	3	4	3	5	6	6	7	8	8	8	10	10												
1	3	3	3	3	4	4	5	6	6				11	14	14									
1/0				3	3	3	4	5	5	7	8	8	9	12	12	12	15	15						
2/0					3	3	3	4	4	5	7	7	8	10	10	10	13	13						
3/0							3	3	3	5	6	6	7	8	8	9	10	10						
4/0								3	3	4	5	5	6	7	7				12	14	14			
250										3	4	4	4	5	5	7	8	8	9	11	11	10	12	12
300										3	3	3	4	4	4	6	7	7	8	9	10	9	11	11
350											3	3	3	4	4	5	6	6	7	8	8	8	9	10
400													3	3	3	5	5	6	6	7	8			
500													3	3	3	4	4	4	5	6	6	7	8	8
600																4	4	4	4	5	5	6	6	7
700																3	3	3	4	5	5	5	6	6
750																3	3	3	4	5	4	5	5	5
1000																	3	3	3	4	3	4	4	4

(Reprinted with permission from NFPA 70-1996, the National Electrical Code® National Fire Protection Association, Quincy, Massachusetts. National Electrical Code® and NEC® are registered trademarks of the National Fire Protection Association, Inc., Quincy, MA 02269)

4.13.0 Maximum Rating of Motor-Branch Circuit, Short-Circuit, and Ground-Fault Protection Devices

Type of Motor	Percent of Full-Load Current			
	Nontime Delay Fuse	Dual Element (Time-Delay) Fuse	Instan-taneous Trip Breaker	* Inverse Time Breaker
Single-phase, all types				
No code letter	300	175	700	250
All ac single-phase and polyphase squirrel-cage and synchronous motors† with full-voltage, resistor or reactor starting:				
No code letter	300	175	700	250
Code letter F to V	300	175	700	250
Code letter B to E	250	175	700	200
Code letter A	150	150	700	150
All ac squirrel-cage and synchronous motors† with autotransformer starting:				
Not more than 30 amps				
No code letter	250	175	700	200
More than 30 amps				
No code letter	200	175	700	200
Code letter F to V	250	175	700	200
Code letter B to E	200	175	700	200
Code letter A	150	150	700	150
High-reactance squirrel-cage				
Not more than 30 amps				
No code letter	250	175	700	250
More than 30 amps				
No code letter	200	175	700	200
Wound-rotor —				
No code letter	150	150	700	150
Direct-current (constant voltage)				
No more than 50 hp				
No code letter	150	150	250	150
More than 50 hp				
No code letter	150	150	175	150

* The values given in the last column also cover the ratings of nonadjustable inverse time types of circuit breakers.

† Synchronous motors of the low-torque, low-speed type (usually 450 rpm or lower), such as are used to drive reciprocating compressors, pumps, etc. that start unloaded, do not require a fuse rating or circuit-breaker setting in excess of 200 percent of full-load current.

(Reprinted with permission from NFPA 70-1996, the National Electrical Code® National Fire Protection Association, Quincy, Massachusetts. National Electrical Code® and NEC® are registered trademarks of the National Fire Protection Association, Inc., Quincy, MA 02269)

4.14.0   Maximum Number of Conductors Allowed in Metal Boxes

The maximum number of conductors permitted shall be computed using the volume per conductor listed in the table, with the deductions provided for, and these volume deductions shall be based on the largest conductor entering the box. Boxes described in the table have a larger cubic inch capacity than is designated in the table shall be permitted to have their cubic inch capacity marked as required by this section and the maximum number of conductors permitted shall be computed using the volume per conductor listed.

Metal Boxes

Box Dimension, Inches Trade Size or Type	Min. Cu. In. Cap.	Maximum Number of Conductors						
		No. 18	No. 16	No. 14	No. 12	No. 10	No. 8	No. 6
4 x 1¼ Round or Octagonal	12.5	8	7	6	5	5	4	2
4 x 1½ Round or Octagonal	15.5	10	8	7	6	6	5	3
4 x 2⅛ Round or Octagonal	21.5	14	12	10	9	8	7	4
4 x 1¼ Square	18.0	12	10	9	8	7	6	3
4 x 1½ Square	21.0	14	12	10	9	8	7	4
4 x 2⅛ Square	30.3	20	17	15	13	12	10	6
4¹¹/₁₆ x 1¼ Square	25.5	17	14	12	11	10	8	5
4¹¹/₁₆ x 1½ Square	29.5	19	16	14	13	11	9	5
4¹¹/₁₆ x 2⅛ Square	42.0	28	24	21	18	16	14	8
3 x 2 x 1½ Device	7.5	5	4	3	3	3	2	1
3 x 2 x 2 Device	10.0	6	5	5	4	4	3	2
3 x 2 x 2¼ Device	10.5	7	6	5	4	4	3	2
3 x 2 x 2½ Device	12.5	8	7	6	5	5	4	2
3 x 2 x 2¾ Device	14.0	9	8	7	6	5	4	2
3 x 2 x 3½ Device	18.0	12	10	9	8	7	6	3
4 x 2⅛ x 1½ Device	10.3	6	5	5	4	4	3	2
4 x 2⅛ x 1⅞ Device	13.0	8	7	6	5	5	4	2
4 x 2⅛ x 2⅛ Device	14.5	9	8	7	6	5	4	2
3¾ x 2 x 2½ Masonry Box/Gang	14.0	9	8	7	6	5	4	2
3¾ x 2 x 3½ Masonry Box/Gang	21.0	14	12	10	9	8	7	4
FS—Minimum Internal Depth 1¾ Single Cover/Gang	13.5	9	7	6	6	5	4	2
FD—Minimum Internal Depth 2⅜ Single Cover/Gang	18.0	12	10	9	8	7	6	3
FS—Minimum Internal Depth 1¾ Multiple Cover/Gang	18.0	12	10	9	8	7	6	3
FD—Minimum Internal Depth 2⅜ Multiple Cover/Gang	24.0	16	13	12	10	9	8	4

Volume Required per Conductor

Size of Conductor	Free Space Within Box for Each Conductor
No. 18	1.5 cubic inches
No. 16	1.75 cubic inches
No. 14	2. cubic inches
No. 12	2.25 cubic inches
No. 10	2.5 cubic inches
No. 8	3. cubic inches
No. 6	5. cubic inches

4.15.0 Electrical Duct Bank Sizes for One to Nine Ducts

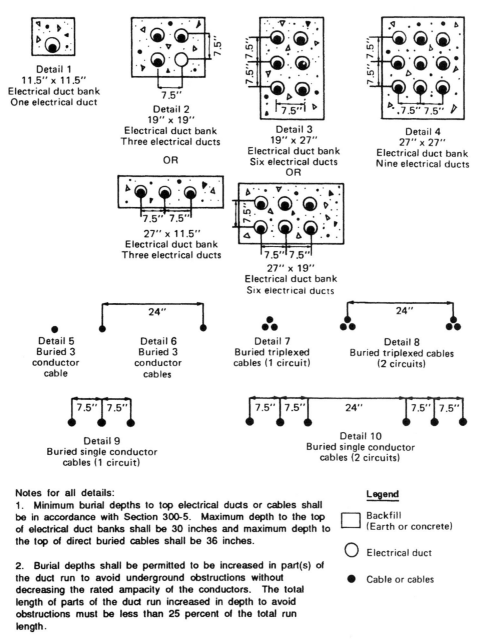

Detail 1
11.5" x 11.5"
Electrical duct bank
One electrical duct

Detail 2
19" x 19"
Electrical duct bank
Three electrical ducts

OR

Detail 3
19" x 27"
Electrical duct bank
Six electrical ducts
OR

Detail 4
27" x 27"
Electrical duct bank
Nine electrical ducts

27" x 11.5"
Electrical duct bank
Three electrical ducts

27" x 19"
Electrical duct bank
Six electrical ducts

Detail 5
Buried 3
conductor
cable

Detail 6
Buried 3
conductor
cables

Detail 7
Buried triplexed
cables (1 circuit)

Detail 8
Buried triplexed cables
(2 circuits)

Detail 9
Buried single conductor
cables (1 circuit)

Detail 10
Buried single conductor
cables (2 circuits)

Notes for all details:

1. Minimum burial depths to top electrical ducts or cables shall be in accordance with Section 300-5. Maximum depth to the top of electrical duct banks shall be 30 inches and maximum depth to the top of direct buried cables shall be 36 inches.

2. Burial depths shall be permitted to be increased in part(s) of the duct run to avoid underground obstructions without decreasing the rated ampacity of the conductors. The total length of parts of the duct run increased in depth to avoid obstructions must be less than 25 percent of the total run length.

3. For SI units: one inch = 25.4 millimeters; one foot = 305 millimeters.

Legend

☐ Backfill (Earth or concrete)

○ Electrical duct

● Cable or cables

(Reprinted with permission from NFPA 70-1996, the National Electrical Code® National Fire Protection Association, Quincy, Massachusetts. National Electrical Code® and NEC® are registered trademarks of the National Fire Protection Association, Inc., Quincy, MA 02269)

4.16.0 Minimum Cover Requirements for 0 to 600-Volt Conductors

(Cover is defined as the shortest distance measured between a point on the top surface of any direct
buried conductor, cable, conduit or other raceway and the top surface of finished grade, concrete, or similar cover.)

Location of Wiring Method or Circuit	Type of Wiring Method or Circuit				
	Direct Burial Cables or Conductors	Rigid Metal Conduit or Intermediate Metal Conduit	Rigid Nonmetallic Conduit Approved for Direct Burial Without Concrete Encasement or Other Approved Raceways	Residential Branch Circuits Rated 120 Volts or less with GFCI Protection and Maximum Overcurrent Protection of 20 Amperes	Circuits for Control of Irrigation and Landscape Lighting Limited to Not More than 30 Volts and Installed with Type UF or In Other Identified Cable or Raceway
All Locations Not Specified Below	24	6	18	12	6
In Trench Below 2-Inch Thick Concrete or Equivalent	18	6	12	6	6
Under a Building	0 (In Raceway Only)	0	0	0 (In Raceway Only)	0 (In Raceway Only)
Under Minimum of 4-Inch Thick Concrete Exterior Slab with no vehicular traffic and the slab extending not less than 6 inches beyond the underground installation	18	4	4	6 (Direct Burial) 4 (In Raceway)	6 (Direct Burial) 4 (In Raceway)
Under Streets, Highways, Roads, Alleys, Driveways, and Parking Lots	24	24	24	24	24
One- and Two-Family Dwelling Driveways and Parking areas, and Used for No Other Purpose	18	18	18	12	18
In or Under Airport Runways Including Adjacent Areas Where Trespassing Prohibited	18	18	18	18	18
In Solid Rock Where Covered by Minimum of 2 Inches Concrete Extending Down to Rock	2 (In Raceway Only)	2	2	2 (In Raceway Only)	2 (In Raceway Only)

Note 1. For SI Units: one inch = 25.4 millimeters
Note 2. Raceways approved for burial only where concrete encased shall require concrete envelope not less than 2 inches thick.
Note 3. Lesser depths shall be permitted where cables and conductors rise for terminations or splices or where access is otherwise required.
Note 4. Where one of the conduit types listed in columns 1-3 is combined with one of the circuit types in columns 4 and 5, the shallower depth of burial shall be permitted.

4.17.0 Demand Loads for Various Types of Residential Electrical Appliances

Demand Factors for Household Electric Clothes Dryers

Number of Dryers	Demand Factor Percent
1	100
2	100
3	100
4	100
5	80
6	70
7	65
8	60
9	55
10	50
11-13	45
14-19	40
20-24	35
25-29	32.5
30-34	30
35-39	27.5
40 & over	25

Where two or more single-phase ranges are supplied by a three-phase, 4-wire feeder, the total load shall be computed on the basis of twice the maximum number connected between any two phases. kVA shall be considered equivalent to kW for loads computed under this section.

Demand Loads for Household Electric Ranges, Wall-Mounted Ovens, Counter-Mounted Cooking Units, and Other Household Cooking Appliances over 1¾ kW Rating. Column A to be used in all cases except as otherwise permitted

NUMBER OF APPLIANCES	Maximum Demand (See Notes)	Demand Factors Percent (See Note 3)	
	COLUMN A (Not over 12 kW Rating)	COLUMN B (Less than 3½ kW Rating)	COLUMN C (3½ kW to 8¾ kW Rating)
1	8 kW	80%	80%
2	11 kW	75%	65%
3	14 kW	70%	55%
4	17 kW	66%	50%
5	20 kW	62%	45%
6	21 kW	59%	43%
7	22 kW	56%	40%
8	23 kW	53%	36%
9	24 kW	51%	35%

4.18.0 General Lighting Loads by Occupancy*

Type of Occupancy	Unit Load per Sq. Ft. (Volt-Amperes)
Armories and Auditoriums	1
Banks	3½**
Barber Shops and Beauty Parlors	3
Churches	1
Clubs	2
Court Rooms	2
*Dwelling Units	3
Garages — Commercial (storage)	½
Hospitals	2
*Hotels and Motels, including apartment houses without provisions for cooking by tenants	2
Industrial Commercial (Loft) Buildings	2
Lodge Rooms	1½
Office Buildings	3½**
Restaurants	2
Schools	3
Stores	3
Warehouses (storage)	¼
In any of the above occupancies except one-family dwellings and individual dwelling units of two-family and multifamily dwellings: Assembly Halls and Auditoriums Halls, Corridors, Closets, Stairways Storage Spaces	1 ½ ¼

4.19.0 Selection of Overcurrent Protection and Switching Devices

Category of Application		Acceptable Device Types (See Legend Below)
Individually mounted service disconnect unit	(8–800 amps) (above 800 amps)	SW-QMQB/CF SW-BP/CF
Service disconnect unit in main switchboard	(0–800 amps) (above 800 amps)	SW-QMQB/CF SW-BP/CF
Feeder unit in main switchboard	(0–800 amps) (above 800 amps)	SW-QMQB/CF SW-BP/CF
Main or branch unit in 265/460 (277/480) volt distribution panel or power panel		SW-QMQB/CF except CLCB-MC if needed in order to meet the specified series connected rating of downstream lighting or appliance panel.
Main unit in 265/460 (277/480) volt lighting or appliance panel		CB-SMC, except CLCB-MC if needed in order to meet the specified series connected rating of the panel.
Branch unit in 265/460 (277/480) volt lighting or appliance panel		CB-SMC
Branch unit in 120/208 volt lighting or appliance panel.		CB-SMC
Main or branch unit in 120/208 volt distribution panel or power panel		SW-QMQB/CF, except CLCB if needed in order to meet the specified series connected rating of downstream lighting or appliance panel.
Main unit in 120/208 volt lighting or appliance panel		CB-SMC, except CLCB of needed in order to meet specified series connected rating of the panel.
Branch unit in panelette		CB-CMC
Main unit in metering assembly		QMQB/CF
Tenant main unit in metering assembly		CB-SMC
Individually mounted unit	(0–1200 amps)	SW-QMQB/CF except CLCB-MC if needed in order to meet the specified series connected rating of downstream lighting or appliance panel.
Individually mounted unit without overcurrent protection	(0–1200 amps) (above 1200 amps)	SW-QMQB SW-BP
Motor starting fusing		CF

Explanation of abbreviations used above is as follows:

ABBREVIATION	DESCRIPTION
SW-BP	Distribution switch; bolted pressure type.
SW-QMQB	Distribution switch; quick-make, quick-break type.
/	Fusible—fused with.
CF	Cartridge fuses.
CB-SMC	Circuit breaker, standard molded case type.
CB-CMC	Circuit breaker, compact molded case type.

4.20.0   Size of Equipment and Raceway Grounding Conductors for 15- to 400-Amp Overcurrent Devices

SIZING OF EQUIPMENT AND RACEWAY GROUNDING CONDUCTORS AND LOAD SIDE OF SERVICE BONDING JUMPERS

OVERCURRENT DEVICE FUSE OR TRIP SIZE (AMPS)	GROUNDING CONDUCTOR OR BONDING JUMPER SIZE--CU **	(AL)
15,20	#12	-
25-60	#10	-
70-100	#8	-
110-200	#4	(#4)
225-400	#2	(1/0)
500,600	* 2 x #1	(2 x 2/0)
700,800	* 2 x 1/0	(2 x 3/0)
1000	* 3 x 2/0	(3 x 4/0)
1200	* 4 x 3/0	(4 x 250) MCM
1600	* 5 x 4/0	(5 x 350) MCM
2000	* 6 x 250 MCM	(6 x 400) MCM
2500	* 7 x 350 MCM	(7 x 600) MCM
3000	* 8 x 400 MCM	(8 x 600) MCM
4000	* 11 x 500 MCM	(11 x750) MCM

*   Adjust quantity (if needed) to match number of conduits in run.

**   Where phase leg conductor ampacity exceeds overcurrent device, increase grounding conductor as if the overcurrent device size matched the phase leg ampacity.

CC.   Grounding electrode conductors and conductors used for bonding on the supply side of the service device shall be sized in accordance with the following table:-

SIZING OF GROUNDING ELECTRODE CONDUCTORS AND MAIN (AND SUPPLY SIDE OF SERVICE) BONDING JUMPERS

SERVICE CONDUCTOR SIZE CABLE CU	(AL)	BUS	BROUNDING ELECTRODE CONDUCTOR SIZE--CU	(AL)	BONDING JUMPER SIZE---------- CU	(AL)
#2	(1/0)max.	100	#8	-	#8	-
3/0	(250)MCM max.	200	#4	#2	#4	(#2)
500	(700)MCM max.	400	1/0	3/0	1/0	(3/0)
2x350	(2x500)MCM max.	600	2/0	4/0	2/0	(4/0)
2x500	(2x700)MCM max.	800	2/0	4/0	2/0	(350)MCM
4x350	(4x500)MCM max.	1200	3/0	250MCM	250	(600)MCM
5x400	(5x600)MCM max.	1600	3/0	250MCM	400	(2x250)MCM
6x500	(6x750)MCM max.	2000	3/0	250MCM	500	(2x250)MCM
8x500	(8x750)MCM max.	3000	3/0	250MCM	2x500	(2x250)MCM
11x500	(11x750)MCM max.	4000	3/0	250MCM	2x500	(2x250)MCM

4.21.0 Enclosures for Nonhazardous Locations (NEMA Designations)

For a degree of protection against:	Designed to meet tests no. ❶	NEMA Type							
		For indoor use			Outdoor use		Indoor or outdoor		
		1	12	13	3R	3	4	4X	6P
Incidental contact with enclosed equipment	6.2	✓	✓	✓	✓	✓	✓	✓	✓
Falling dirt	6.2	✓	✓	✓	✓	✓	✓	✓	✓
Rust	6.8	✓	✓	✓	✓	✓	✓	✓	✓
Circulating dust, lint, fibers and flyings ❷	6.5.1.2 (2)		✓	✓			✓	✓	✓
Windblown dust	6.5.1.1 (2)					✓	✓	✓	✓
Falling liquids and light splashing	6.3.2.2		✓	✓			✓	✓	✓
Rain (test evaluated per 6.4.2.1)	6.4.2.1				✓	✓	✓	✓	✓
Rain (test evaluated per 6.4.2.2)	6.4.2.2					✓	✓	✓	✓
Snow and sleet	6.6.2.2				✓	✓	✓	✓	✓
Hosedown and splashing water	6.7						✓	✓	✓
Occasional prolonged submersion	6.11 (2)								✓
Oil and coolant seepage	6.3.2.2		✓	✓					
Oil or coolant spraying and splashing	6.12			✓					
Corrosive agents	6.9							✓	✓

❶ See below for abridged description of NEMA enclosure test requirements. Refer to NEMA Standards Publication No. 250 for complete test specifications.

❷ Non-hazardous materials, not Class III ignitable or combustible.

Rod Entry Test—A ⅛" diameter rod must not be able to enter enclosure except at locations where nearest live part is more than 4" from an opening—such opening shall not permit a ½" diameter rod to enter.

Drip Test—Water is dripped onto enclosure for 30 minutes from an overhead pan having uniformly spaced spouts, one every 20 square inches of pan area each spout having a drip rate of 20 drops per minute.
Evaluation 6.3.2.2: No water shall have entered enclosure.

Rain Test—Entire top and all exposed sides are sprayed with water at a pressure of 5 psi from nozzles for one hour at a rate to cause water to rise 18 inches in a straight-sided pan beneath the enclosure.
Evaluation 6.4.2.1: No water shall have reached live parts, insulation or mechanisms.
Evaluation 6.4.2.2: No water shall have entered enclosure.

Outdoor Dust Test (Alternate Method)—Enclosure and external mechanisms are subjected to a stream of water at 45 gallons per minute from a 1" diameter nozzle, directed at all joints from all angles from a distance of 10 to 12 feet. Test time is 48 seconds times the test length (height + width + depth of enclo-

sure in feet), or a minimum of 5 minutes. No water shall enter enclosure.

Indoor Dust Test (Alternate Method)—Atomized water at a pressure of 30 psi is sprayed on all seams, joints and external operating mechanisms from a distanc of 12 to 15 inches at a rate of three gallons per hour. No less than five ounces of water per linear foot of test length (height + length + depth of enclosure) is applied. No water shall enter enclosure.

External Icing Test—Water is sprayed on enclosure for one hour in a cold room (2° C): then room temperature is lowered to approximately –5° C and water spray is controlled so as to cause ice to buil up at a rate of ¼" per hour until ¾" thick ice has formed on top surface of a 1" diameter metal test bar, then temperature is maintained at –5° C for 3 hours.
Evaluation 6.6.2.2: Equipment shall be undamaged after ice has melted (external mechanisms not required to be operable while ice-laden).

Hosedown Test—Enclosure and external mechanisms are subjected to a stream of water at 65 gallons per minute from a 1" diameter nozzle, directed at all joints from all angles from a distance of 10 to 12 feet. Test time is 48 seconds times the test length (height + width + depth of enclosure in feet), or a minimum of 5 minutes. No water shall enter enclosure.

Rust Resistance Test (Applicable only to enclosures incorporating external ferrous parts)—Enclosure is subjected to a salt spray (fog) for 24 hours, using water with five parts by weight of salt (NaCl), at 35°C, then rinsed and dried. There shall be no rust except where protection is impractical (e.g., machined mating surfaces, sliding surfaces of hinges, shafts, etc.)

Corrosion Protection—Sheet steel enclosures are evaluated per UL 50, Part 13 (test for equivalent protection as G-90 commercial zinc coated sheet steel). Other materials per UL 508, 5.9 or 5.10.

(2) Air Pressure Test (Alternate Method)—Enclosure is submerged in water at a pressure equal to water depth of six feet, for 24 hours. No water shall enter enclosure.

Oil Exclusion Test—Enclosure is subjected to a stream of test liquid for 30 minutes from a ⅜" diameter nozzle at two gallons a minute. Water with 0.1% wetting agent is directed from all angles from a distance of 12 to 18 inches, while any externally operated device is operated at 30 operations per minute. No test liquid shall enter the enclosure.

4.22.0   Enclosures for Hazardous Locations (NEMA Designations)

For a degree of protection against atmospheres typically containing: [3]	Designed to meet tests: [2]	Class (National Electrical Code)	NEMA Type						
			7, Class I Group:				9, Class II Group:		
			A	B	C	D	E	F	G
Acetylene		I	✓						
Hydrogen, manufactured gas	Explosion test	I	✓	✓					
Diethyl ether, ethylene, hydrogen sulfide	Hydrostatic test	I			✓				
Acetone, butane, gasoline, propane, toluene	Temperature test	I			✓	✓			
Metal dusts and other combustible dusts with resistivity of less than 10^5 ohm-cm.		II					✓		
Carbon black, charcoal, coal or coke dusts with resistivity between 10^2–10^8 ohm-cm.	Dust penetration test	II						✓	
Combustible dusts with resistivity of 10^5 ohm-cm or greater	Temperature test with dust blanket	II							✓
Fibers, flyings	[4]	III							✓

[1] For indoor locations only unless cataloged with additional NEMA Type enclosure number(s) suitable for outdoor use as shown in table on Page 14. Some control devices (if so listed in the catalog) are suitable for Division 2 hazardous location use in enclosures for non-hazardous locations. For explanation of CLASSES, DIVISIONS and GROUPS, refer to the National Electrical Code.
Note: Classifications of hazardous locations are subject to the approval of the authority having jurisdiction. Refer to the National Electrical Code.

[2] See abridged description of test requirements below. For complete requirements, refer to UL Standard 698, compliance with which is required by NEMA enclosure standards.

[3] For listing of additional materials and information noting the properties of liquids, gases and solids, refer to NFPA 479M-1986, Classification of Gases, Vapors, and Dusts for Electrical Equipment in Hazardous (Classified) Locations.

[4] UL 698 does not include test requirements for Class III. Products that meet Class II, Group G requirements are acceptable for Class III.

4.23.0 Motor-Controller Enclosure Types (Indoor and Outdoor Use)

The table provides the basis for selecting enclosures for use in specific nonhazardous locations. The enclosures are not intended to protect against conditions such as condensation, icing, corrosion or contamination which may occur within the enclosure or enter via the conduit or unsealed openings. These internal conditions require special consideration by the installer and/or user.

Motor Controller Enclosure Selection Table

For Outdoor Use

Provides a Degree of Protection Against the Following Environmental Conditions	Enclosure Type Number†						
	3	3R	3S	4	4X	6	6P
Incidental contact with the enclosed equipment	X	X	X	X	X	X	X
Rain, snow and sleet	X	X	X	X	X	X	X
Sleet*	—	—	X	—	—	—	—
Windblown dust	X	—	X	X	X	X	X
Hosedown	—	—	—	X	X	X	X
Corrosive agents	—	—	—	—	X	—	X
Occasional temporary submersion	—	—	—	—	—	X	X
Occasional prolonged submersion	—	—	—	—	—	—	X

* Mechanism shall be operable when ice covered.

For Indoor Use

Provides a Degree of Protection Against the Following Environmental Conditions	Enclosure Type Number†										
	1	2	4	4X	5	6	6P	11	12	12K	13
Incidental contact with the enclosed equipment	X	X	X	X	X	X	X	X	X	X	X
Falling dirt	X	X	X	X	X	X	X	X	X	X	X
Falling liquids and light splashing	—	X	X	X	X	X	X	X	X	X	X
Circulating dust, lint, fibers and flyings	—	—	X	X	—	X	X	—	X	X	X
Settling airborne dust, lint, fibers and flyings	—	—	X	X	X	X	X	—	X	X	X
Hosedown and splashing water	—	—	X	X	—	X	X	—	—	—	—
Oil and coolant seepage	—	—	—	—	—	—	—	—	X	X	X
Oil or coolant spraying and splashing	—	—	—	—	—	—	—	—	—	—	X
Corrosive agents	—	—	—	X	—	—	X	X	—	—	—
Occasional temporary submersion	—	—	—	—	—	X	X	—	—	—	—
Occasional prolonged submersion	—	—	—	—	—	—	X	—	—	—	—

† Enclosure type number, except type number 1, shall be marked on the motor controller enclosure.

(Reprinted with permission from NFPA 70-1996, the National Electrical Code® National Fire Protection Association, Quincy, Massachusetts. National Electrical Code® and NEC® are registered trademarks of the National Fire Protection Association, Inc., Quincy, MA 02269)

4.24.0   Voltage-Drop Tables for 6- and 12-Volt Equipment

The National Electrical Code limits voltage drop to a maximum of 5% of nominal. Thus, circuit runs must be of sufficient size to maintain operating voltage when remote fixtures and/or exit signs are connected to the emergency lighting equipment. The table below shows the length of wire run based on system voltage, wire gauge and total wattage on the run. To determine loads or lengths of wire runs not listed, divide the *known* value into the *constant* value at the bottom of the appropriate row.

Total Watts on Wire Run	6 Volt System Wire Gauge				Total Watts on Wire Run	12 Volt System Wire Gauge			
	12	10	8	6		12	10	8	6
	Length of Wire Run (Feet)					Length of Wire Run (Feet)			
6	94	150	238	379	6	378	600	955	1518
7	81	129	204	325	7	324	515	818	1301
8	70	112	179	284	8	283	450	716	1138
10	56	90	143	227	10	226	360	570	910
12	44	70	112	178	12	178	283	450	715
14	40	64	102	162	14	162	257	409	650
16	33	53	84	134	16	133	212	338	538
18	30	47	75	119	18	119	189	300	477
20	28	45	71	114	20	113	180	286	455
21	27	43	68	108	21	108	171	273	434
24	24	38	60	95	24	89	141	225	357
25	21	34	54	86	25	86	136	216	344
30	19	30	48	76	30	75	120	190	303
35	15	25	39	63	35	65	103	164	260
40	13	21	33	53	40	53	85	135	214
48	11	17	28	44	48	44	70	112	178
50	11	17	27	43	50	43	68	108	172
75	7	11	18	29	75	28	45	72	115
100	5	8	14	21	100	21	34	54	86
125	4	7	11	17	125	17	27	43	69
150	3	5	9	14	150	14	23	36	57
175	3	5	8	12	175	12	19	31	49
200	2	4	6	10	200	10	16	27	42
225	2	4	6	10	225	10	16	25	40
250	2	3	5	9	250	9	14	22	36
CONSTANT	567	901	1432	2277	CONSTANT	2267	3604	5730	9109

Example 1— A 12V system uses 8-gauge wire and will operate three 7W exit signs. Total watts on wire run is 21, length of run from table is 273'.

Longer Wire Runs

If loads are uniformly spaced along circuit path (equal watts, equal distances), the lengths in the table can be increased by certain values.

Number of fixtures	2	3	4	5
Multiplier	1.33	1.5	1.6	1.67

Example 2— Exit signs from example 1 will be uniformly spaced. Multiplier is 1.5 for three fixtures. Maximum permissible length of wire run is 273' X 1.5, or 409".

4.25.0   Seismic Restraints and Bracing

All seismic restraint and isolation devices, braces, and supports shall be capable of accepting without failure forces produced by seismic acceleration (expressed in multiples of the acceleration of gravity "G") based on the level above grade of the attachment of the equipment support system. For design purposes, the following acceleration levels shall be used:

DESIGN LEVEL OF ACCELERATION AT EQUIPMENT CENTER OF GRAVITY			
ELEVATION ABOVE GRADE	RIGIDLY FLOOR OR WALL MOUNTED EQUIPMENT	RESILIENTLY MOUNTED AND/OR SUPPORTED FROM CEILING OR STRUCTURE ABOVE	LIFE SAFETY EQUIPMENT (FIRE ALARM, HOSPITAL COMMUNICATIONS, EMERGENCY
BELOW GRADE UP TO 20 FEET ABOVE GRADE	0.125 "G"	0.500 "G"	1.000 "G"
21 FEET AND UP	0.500 "G"	0.750 "G"	

SEISMIC BRACING TABLE			
EQUIPMENT	ON CENTER SPACING		WITHIN EACH CHANGE OF DIRECTION
	TRANSVERSE	LONGITUDINAL	
CONDUIT	40 FEET	80 FEET	10 FEET OR 15 DIAMETERS

For all seismically supported trapeze supported <u>conduit, the individual conduits shall be</u> <u>transversely and vertically restrained to the trapeze support</u> at the designated restraint locations. Restrain at least every third trapeze hanger transversely and every fifth one longitudinally as well as the trapeze on both sides of every change of direction.

For overhead supported equipment, overstress of the building structure must not occur. Bracing may occur from:

1)   Flanges of structural steel beams.

2)   Upper truss chords in bar joists.

4.26.0 Full Load Current (in Amperes) for Single-Phase, Two-Phase, Three-Phase, and Direct-Current Motors

Full-Load Currents (in Amperes) for Single-Phase Alternating-Current Motors

The following values of full-load currents are for motors running at usual speeds and motors with normal torque characteristics. Motors built for especially low speeds or high torques may have higher full-load currents, and multispeed motors will have full-load current varying with speed, in which case the nameplate current ratings shall be used.

HP	115V	200V	208V	230V
⅙	4.4	2.5	2.4	2.2
¼	5.8	3.3	3.2	2.9
⅓	7.2	4.1	4.0	3.6
½	9.8	5.6	5.4	4.9
¾	13.8	7.9	7.6	6.9
1	16	9.2	8.8	8
1½	20	11.5	11	10
2	24	13.8	13.2	12
3	34	19.6	18.7	17
5	56	32.2	30.8	28
7½	80	46	44	40
10	100	57.5	55	50

*(Reprinted with permission from NFPA 70-1996, the National Electrical Code®
National Fire Protection Association, Quincy, Massachusetts. National Electrical
Code® and NEC® are registered trademarks of the National Fire Protection Association, Inc., Quincy, MA 02269)*

4.26.0 Full Load Current (in Amperes) for Single-Phase, Two-Phase, Three-Phase, and Direct-Current Motors

Full-Load Current* for
Three-Phase Alternating-Current Motors

HP	Induction Type Squirrel-Cage and Wound-Rotor Amperes							Synchronous Type †Unity Power Factor Amperes			
	115V	200V	208V	230V	460V	575V	2300V	230V	460V	575V	2300V
½	4	2.3	2.2	2	1	.8					
¾	5.6	3.2	3.1	2.8	1.4	1.1					
1	7.2	4.1	4.0	3.6	1.8	1.4					
1½	10.4	6.0	5.7	5.2	2.6	2.1					
2	13.6	7.8	7.5	6.8	3.4	2.7					
3		11.0	10.6	9.6	4.8	3.9					
5		17.5	16.7	15.2	7.6	6.1					
7½		25.3	24.2	22	11	9					
10		32.2	30.8	28	14	11					
15		48.3	46.2	42	21	17					
20		62.1	59.4	54	27	22					
25		78.2	74.8	68	34	27		53	26	21	
30		92	88	80	40	32		63	32	26	
40		119.6	114.4	104	52	41		83	41	33	
50		149.5	143.0	130	65	52		104	52	42	
60		177.1	169.4	154	77	62	16	123	61	49	12
75		220.8	211.2	192	96	77	20	155	78	62	15
100		285.2	272.8	248	124	99	26	202	101	81	20
125		358.8	343.2	312	156	125	31	253	126	101	25
150		414	396.0	360	180	144	37	302	151	121	30
200		552	528.0	480	240	192	49	400	201	161	40

*These values of full-load current are for motors running at speeds usual for belted motors with normal torque characteristics. Motors built for especially low speeds or high torques may require more running current, and multispeed motors will have full-load current varying with speed, in which case the nameplate current rating shall be used.

†For 90 and 80 percent power factor the above figures shall be multiplied by 1.1 and 1.25 respectively.

The voltages listed are rated motor voltages. The currents listed shall be permitted for system voltage ranges of 110 to 120, 220 to 240, 440 to 480, and 550 to 600 volts.

(Reprinted with permission from NFPA 70-1996, the National Electrical Code® National Fire Protection Association, Quincy, Massachusetts. National Electrical Code® and NEC® are registered trademarks of the National Fire Protection Association, Inc., Quincy, MA 02269)

4.26.0 Full Load Current (in Amperes) for Single-Phase, Two-Phase, Three-Phase, and Direct-Current Motors (Continued)

Full-Load Currents (in Amperes) for Direct-Current Motors

The following values of full-load currents* are for motors running at base speed.

HP	Armature Voltage Rating*					
	90V	120V	180V	240V	500V	550V
¼	4.0	3.1	2.0	1.6		
⅓	5.2	4.1	2.6	2.0		
½	6.8	5.4	3.4	2.7		
¾	9.6	7.6	4.8	3.8		
1	12.2	9.5	6.1	4.7		
1½		13.2	8.3	6.6		
2		17	10.8	8.5		
3		25	16	12.2		
5		40	27	20		
7½		58		29	13.6	12.2
10		76		38	18	16
15				55	27	24
20				72	34	31
25				89	43	38
30				106	51	46
40				140	67	61
50				173	83	75
60				206	99	90
75				255	123	111
100				341	164	148
125				425	205	185
150				506	246	222
200				675	330	294

*These are average direct-current quantities.

(Reprinted with permission from NFPA 70-1996, the National Electrical Code® National Fire Protection Association, Quincy, Massachusetts. National Electrical Code® and NEC® are registered trademarks of the National Fire Protection Association, Inc., Quincy, MA 02269)

4.26.0 Full Load Current (in Amperes) for Single-Phase, Two-Phase, Three-Phase, and Direct-Current Motors (Continued)

Full-Load Current for Two-Phase Alternating-Current Motors (4-Wire)

The following values of full-load current are for motors running at speeds usual for belted motors and motors with normal torque characteristics. Motors built for especially low speeds or high torques may require more running current, and multispeed motors will have full-load current varying with speed, in which case the nameplate current rating shall be used. Current in the common conductor of a 2-phase, 3-wire system will be 1.41 times the value given.

The voltages listed are rated motor voltages. The currents listed shall be permitted for system voltages ranges of 110 to 120, 220 to 240, 440 to 480, and 550 to 600 volts.

HP	Induction Type Squirrel-Cage and Wound-Rotor Amperes				
	115V	230V	460V	575V	2300V
½	4	2	1	.8	
¾	4.8	2.4	1.2	1.0	
1	6.4	3.2	1.6	1.3	
1½	9	4.5	2.3	1.8	
2	11.8	5.9	3	2.4	
3		8.3	4.2	3.3	
5		13.2	6.6	5.3	
7½		19	9	8	
10		24	12	10	
15		36	18	14	
20		47	23	19	
25		59	29	24	
30		69	35	28	
40		90	45	36	
50		113	56	45	
60		133	67	53	14
75		166	83	66	18
100		218	109	87	23
125		270	135	108	28
150		312	156	125	32
200		416	208	167	43

(Reprinted with permission from NFPA 70-1996, the National Electrical Code® National Fire Protection Association, Quincy, Massachusetts. National Electrical Code® and NEC® are registered trademarks of the National Fire Protection Association, Inc., Quincy, MA 02269)

Current Carrying Capacities of Copper and Aluminum and Copper-Clad Aluminum Conductors From National Electrical Code (NEC), 1996 Edition (NFPA70-1996)

Table 310-16: Allowable Ampacities of Insulated Conductors Rated 0-2000 Volts, 60° to 90°C (140° to 194°F)
Not More Than Three Conductors in Raceway or Cable or Earth (Directly Buried), Based on Ambient Temperature of 30°C (86°F)

Size AWG kcmil	Temperature Rating of Conductor. See Table 310-13.						Size AWG kcmil
	60°C (140°F)	75°C (167°F)	90°C (194°F)	60°C (140°F)	75°C (167°F)	90°C (194°F)	
	Types TW†, UF†	Types FEPW†, RH†, RHW†, THHW†, THW†, THWN†, XHHW†, USE†, ZW†	Types TBS, SA, SIS, FEP†, FEPB†, MI, RHH†, RHW-2, THHN†, THHW†, THW-2, THWN-2, USE-2, XHH, XHHW†, XHHW-2, ZW-2	Types TW†, UF†	Types RH†, RHW†, THHW†, THW†, THWN†, XHHW†, USE†	Types TBS, SA, SIS, THHN†, THHW†, THW-2, THWN-2, RHH†, RHW-2, USE-2, XHH, XHHW, XHHW-2, ZW-2	
	Copper			Aluminum or Copper-Clad Aluminum			
18	....	....	14	....	....	....	....
16	....	....	18	....	....	....	....
14	20†	20†	25†	....	....	....	....
12	25†	25†	30†	20†	20†	25†	12
10	30	35†	40†	25	30†	35†	10
8	40	50	55	30	40	45	8
6	55	65	75	40	50	60	6
4	70	85	95	55	65	75	4
3	85	100	110	65	75	85	3
2	95	115	130	75	90	100	2
1	110	130	150	85	100	115	1
1/0	125	150	170	100	120	135	1/0
2/0	145	175	195	115	135	150	2/0
3/0	165	200	225	130	155	175	3/0
4/0	195	230	260	150	180	205	4/0
250	215	255	290	170	205	230	250
300	240	285	320	190	230	255	300
350	260	310	350	210	250	280	350
400	280	335	380	225	270	305	400
500	320	380	430	260	310	350	500
600	355	420	475	285	340	385	600
700	385	460	520	310	375	420	700
750	400	475	535	320	385	435	750
800	410	490	555	330	395	450	800
900	435	520	585	355	425	480	900
1000	455	545	615	375	445	500	1000
1250	495	590	665	405	485	545	1250
1500	520	625	705	435	520	585	1500
1750	545	650	735	455	545	615	1750
2000	560	665	750	470	560	630	2000

Correction Factors

Ambient Temp. °C	For ambient temperatures other than 30°C (86°F), multiply the allowable ampacities shown above by the appropriate factor shown below.						Ambient Temp. °F
21-25	1.08	1.05	1.04	1.08	1.05	1.04	70-77
26-30	1.00	1.00	1.00	1.00	1.00	1.00	78-86
31-35	.91	.94	.96	.91	.94	.96	87-95
36-40	.82	.88	.91	.82	.88	.91	96-104
41-45	.71	.82	.87	.71	.82	.87	105-113
46-50	.58	.75	.82	.58	.75	.82	114-122
51-55	.41	.67	.76	.41	.67	.76	123-131
56-60	....	.58	.71	....	.58	.71	132-140
61-70	....	.33	.58	....	.33	.58	141-158
71-80	....	....	.41	....	....	.41	159-176

†Unless otherwise specifically permitted elsewhere in this Code, the overcurrent protection for conductor types marked with an obelisk (†) shall not exceed 15 amperes for No. 14, 20 amperes for No. 12, and 30 amperes for No. 10 copper; or 15 amperes for No. 12 and 25 amperes for No. 10 aluminum and copper-clad aluminum after any correction factors for ambient temperature and number of conductors have been applied.

Note: For applications 2000 volts and below under conditions of use other than covered by the above table, and for applications over 2000 volts, see Article 310 and additional tables in NEC.

See NEC for complete notes to Table 310-16. Some of the most important are summarized in part below.

8. Adjustment Factors

a. More Than Three Current-Carrying Conductors in a Raceway or Cable.
Where the number of current-carrying conductors in a raceway or cable exceeds three, the allowable ampacities shall be reduced as shown in the following table:

Number of Current-Carrying Conductors	Percent of Values in Tables as Adjusted for Ambien Temperature if Necessary
4 through 6	80
7 through 9	70
10 through 20	50
21 through 30	45
31 through 40	40
41 and above	35

Where single conductors or multiconductor cables are stacked or bundled longer than 24 inches (610 mm) without maintaining spacing and are not installed in raceways, the allowable ampacity of each conductor shall be reduced as shown in the above table.

Exception No. 1: Where conductors of different systems, as provided in Section 300-3, are installed in a common raceway or cable, the derating factors shown above shall apply to the number of power and lighting (Articles 210, 215, 220, and 230) conductors only.
Exception No. 2: For conductors installed in cable trays, the provisions of Section 318-11 shall apply.
Exception No. 3: Derating factors shall not apply to conductors in nipples having a length not exceeding 24 inches (610 mm).
Exception No. 4: Derating factors shall not apply to underground conductors entering or leaving an outdoor trench if those conductors have physical protection in the form of rigid metal conduit, intermediate metal conduit, or rigid nonmetallic conduit having a length not exceeding 10 feet (3.05 m) and the number of conductors does not exceed four.
Exception No. 5: For other loading conditions, adjustment factors and ampacities shall be permitted to be calculated under Section 310-15(b).
(FPN): See Appendix B, Table B-310-11 for adjustment factors for more than three current-carrying conductors in a raceway or cable with load diversity.

b. More Than One Conduit, Tube or Raceway. Spacing between conduits, tubing or raceways shall be maintained.

9. Overcurrent Protection.
Where the standard ratings and settings of overcurrent devices do not correspond with the ratings and settings allowed for conductors, the next higher standard rating and setting shall be permitted.
Exception: As limited in Section 240-3.

Note 9: Overcurrent protection.
Where the standard ratings and settings of overcurrent devices do not correspond with the ratings and settings allowed for conductors, the next higher standard rating and setting shall be permitted, except as limited in Section 240-3 (not above a rating of 800A).

Note 10: Neutral Conductor
a. A neutral conductor which carries only the unbalanced current from other conductors, as in the case of normally balanced circuits of three or more conductors, shall not be counted when applying the provisions of Note 8.
b. In a 3-wire circuit consisting of 2-phase wires and the neutral of a 4-wire, 3-phase wye-connected system, a common conductor carries approximately the same current as the line to neutral load currents of the other conductors and shall be counted when applying the provisions of Note 8.
c. On a 4-wire, 3-phase wye circuit where the major portion of the load consists of non linear loads, there are harmonic currents present in the neutral conductor and the neutral shall be considered to be a current-carrying conductor.

Figure 4.27.0 Current carrying capacities of copper and aluminum conductors. (*By permission: Cutler Hammer, Pittsburgh, PA.—a part of Eaton Corporation.*)

Note 11: Grounding or Bonding Conductor
A grounding or bonding conductor shall not be counted when applying the provisions of Note 8.

Note: UL listed circuit breakers rated 125A or less shall be marked as being suitable for 60°C (140°F), 75°C (167°F) only or 60/75°C (140/167°F) wire. All Westinghouse listed breakers rated 125A or less are marked 60/75°C. All UL listed circuit breakers rated over 125A are suitable for 75°C conductors. Conductors rated for higher temperatures may be used, but must not be loaded to carry more current than the 75°C ampacity of that size conductor for equipment marked or rated 75°C or the 65°C ampacity of that size conductor for equipment marked or rated 65°C. However, the full 90°C ampacity may be used when applying derated factors, so long as the actual load does not exceed the lower of the derated ampacity or the 75°C or 60°C ampacity that applies.

Conduit Fill

Reproduced From 1993 NEC. For estimate only – see 1996 NEC, Chapter 9, Tables 1-10 for exact code requirements.

Table 3A: Maximum Number of Conductors in Trade Sizes of Conduit or Tubing (Based on Table 1, Chapter 9)

Type Letters	Conductor Size AWG/kcmil	½	¾	1	1¼	1½	2	2½	3	3½	4	5	6
TW, XHHW (14 through 8) RH (14 + 12)	14	9	15	25	44	60	99	142					
	12	7	12	19	35	47	78	111	171				
	10	5	9	15	26	36	60	85	131	176			
	8	2	4	7	12	17	28	40	62	84	108		
RHW and RHH (without outer covering), RH (10 + 8) THW, THHW	14	6	10	16	29	40	65	93	143	192			
	12	4	8	13	24	32	53	76	117	157			
	10	4	6	11	19	26	43	61	95	127	163		
	8	1	3	5	10	13	22	32	49	66	85	133	
TW,	6	1	2	4	7	10	16	23	36	48	62	97	141
	4	1	1	3	5	7	12	17	27	36	47	73	106
THW,	3	1	1	2	4	6	10	15	23	31	40	63	91
	2	1	1	2	4	5	9	13	20	27	34	54	78
	1		1	1	3	4	6	9	14	19	25	39	57
FEPB (6 through 2), RHW and RHH (without outer covering)	1/0		1	1	2	3	5	8	12	16	21	33	49
	2/0		1	1	1	3	5	7	10	14	18	29	41
	3/0		1	1	1	2	4	6	9	12	15	24	35
	4/0			1	1	1	3	5	7	10	13	20	29
	250			1	1	1	2	4	6	8	10	16	23
	300			1	1	1	2	3	5	7	9	14	20
	350				1	1	1	3	4	6	8	12	18
	400				1	1	1	2	4	5	7	11	16
	500				1	1	1	1	3	4	6	9	14
RH, THHW	600					1	1	1	3	4	5	7	11
	700					1	1	1	2	3	4	7	10
	750					1	1	1	2	3	4	6	9

Note 1. This table is for concentric stranded conductors only. For cables with compact conductors, the dimensions in Table 5A shall be used.
Note 2. Conduit fill for conductors with a -2 suffix is the same as for those types without the suffix.

Reproduced From 1993 NEC

Table 3B: Maximum Number of Conductors in Trade Sizes of Conduit or Tubing (Based on Table 1, Chapter 9)

Type Letters	Conductor Size AWG/kcmil	½	¾	1	1¼	1½	2	2½	3	3½	4	5	6
THWN,	14	13	24	39	69	94	154						
	12	10	18	29	51	70	114	164					
	10	6	11	18	32	44	73	104	160				
	8	3	5	9	16	22	36	51	79	106	136		
THHN, FEP (14 through 2), FEPB (14 through 8), PFA (14 through 4/0) PFAH (14 through 4/0) Z (14 through 4/0) XHHW (4 through 500 kcmil)	6	1	4	6	11	15	26	37	57	76	98	154	
	4	1	2	4	7	9	16	22	35	47	60	94	137
	3	1	1	3	6	8	13	19	29	39	51	80	116
	2	1	1	3	5	7	11	16	25	33	43	67	97
	1		1	1	3	5	8	12	18	25	32	50	72
	1/0		1	1	3	4	7	10	15	21	27	42	61
	2/0		1	1	2	3	6	8	13	17	22	35	51
	3/0		1	1	1	3	5	7	11	14	18	29	42
	4/0		1	1	1	2	4	6	9	12	16	24	35
	250			1	1	1	3	4	7	10	12	20	28
	300			1	1	1	3	4	6	8	11	17	24
	350			1	1	1	2	3	5	7	9	15	21
	400				1	1	1	3	5	6	8	13	19
	500				1	1	1	2	4	5	7	11	16
	600				1	1	1	1	3	4	5	9	13
	700					1	1	1	3	4	5	8	11
	750					1	1	1	2	3	4	7	11
XHHW	6	1	3	5	9	13	21	30	47	63	81	128	185
	600				1	1	1	1	3	4	5	9	13
	700					1	1	1	3	4	5	7	11
	750					1	1	1	2	3	4	7	10

Note 1. This table is for concentric stranded conductors only. For cables with compact conductors, the dimensions in Table 5A shall be used.
Note 2. Conduit fill for conductors with a -2 suffix is the same as for those types without the suffix.

Figure 4.27.0 *(Continued)*

Section View of Typical Structure

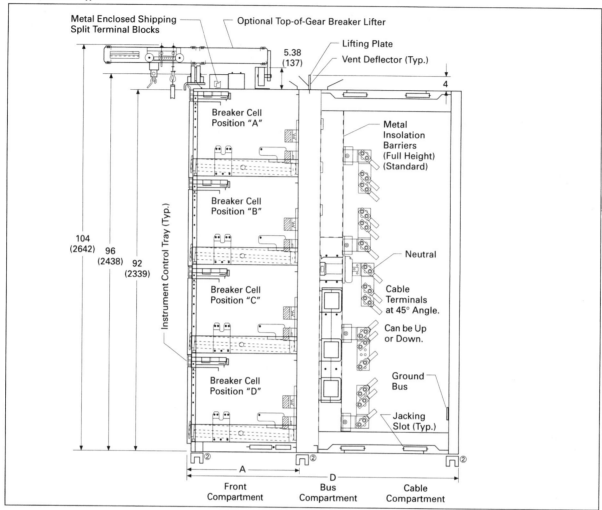

Figure 4.28.0 Low voltage switchgear—typical components. (*By permission: Cutler Hammer, Pittsburgh, PA.—a part of Eaton Corporation.*)

Distribution Section — Front View

Distribution Section — Rear View

① Glass polyester circuit breaker compartment
② Insulated copper load side runbacks
③ Full length barrier isolating the cable compartment
④ Horizontal cross bus

⑤ Tandem mounted circuit breakers through 400 amperes
⑥ Isolating bus compartment
⑦ Available zero sequence ground fault
⑧ Angled neutral connections

⑨ A, B, C phase connections
⑩ Anti-turn lugs
⑪ Movable cable support
⑫ Generous conduit space

Figure 4.28.1 Low voltage distribution switchboard—typical components. (*By permission: Cutler Hammer, Pittsburgh, PA.—a part of Eaton Corporation.*)

Standard Features

① Cover – Welded to tank

② Cooling tubes

③ Bolted handhole on cover

④ Automatic resealing mechanical relief device

⑤ HV bushing, 3 total, located in ANSI Segment 2①

⑥ LV bushing, 4 total (Wye connected), located in ANSI Segment 4①

⑦ Z-Bar flange

⑧ Lifting loops – 2 for lifting cover only

⑨ Lifting hooks – 4 for lifting complete unit

⑩ Jacking provisions on tank or base

⑪ Ground pad – 2 total

⑫ Drain valve – for combination lower filter press connection and complete drain with ⅜-inch sampler

⑬ Base (may be flat or formed)

⑭ Control cabinet for alarm lead termination

⑮ Diagram instruction nameplate with warning nameplate

⑯ De-energized tap changer with padlock provisions

⑰ Liquid temperature indicator with maximum indicating hand

⑱ 1-inch valve for upper filter press connection

⑲ Magnetic liquid level gauge

⑳ Vacuum pressure gauge with air test and Sealedaire® valve

Liquid Filled Primary Unit Substation Transformer With Wall-Mounted HV and LV Bushings

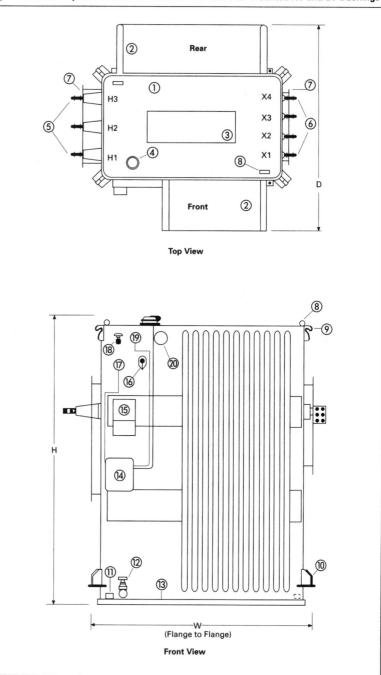

Figure 4.29.0 Primary unit substation—typical components. (*By permission: Cutler Hammer, Pittsburgh, PA.—a part of Eaton Corporation.*)

Liquid-Filled Substation Transformers

Application

Liquid-filled substation transformers are custom designed power transformers suitable for both indoor and outdoor applications.

The transformers are of the sealed tank design and suitable for use in coordinated unit substation in most any type of application and environment. Typical applications of liquid-filled transformers are:

- Utility substations
- Pulp and paper mills
- Steel mills
- Chemical plants/refineries
- General industry
- Commercial buildings

Benefits

- Custom design flexibility to meet special customer needs and applications
- Computerized loss-evaluated designs for specific customer load and evaluation criteria

- Available in higher kVA ratings
- High short circuit strength
- ANSI short time overload capability
- Aluminum or copper windings
- Impervious to the environment through sealed design
- Lowest first cost and comparable cost of ownership to cast/dry designs
- Available as mineral oil-filled or with less-flammable liquids, such as silicone or R-Temp fluids

Design and Technology

Liquid-filled transformers are custom designed and manufactured. Coils are of the rectangular design with Insuldur layer insulation. Primary windings are comprised of strap conductors, either aluminum or copper. Secondary windings are either full height sheet conductors or strap conductor dependent on the voltage and kVA rating. The turn-to-turn insulation is coated with a diamond pattern of B-stage epoxy adhesive which cures during processing to form a high strength bond. This bond restrains the windings during operation and under short circuit stresses.

Liquid-filled transformers are suitable for use up to 65°C average rise over a maximum ambient temperature of 40°C, not to exceed 30°C average for any 24-hour period. The transformer may be specified as 55°C rise, in which case the transformer has a self-cooled (OA) overload capability of 112% without loss of life.

Material used for cores is non-aging, cold rolled, high permeability, grain oriented silicone steel. Cores are rigidly braced to reduce sound levels and losses in the finished product.

The core and coil assembly is immersed in either mineral oil, silicone or R-Temp® fluid and is contained in a sealed tank.

Flat, tubular or panel radiators are mounted on the front and back of the tank. The liquid circulates though the tank and radiators by means of natural convection and effectively cools the core and coil assembly.

Standard Features and Accessories

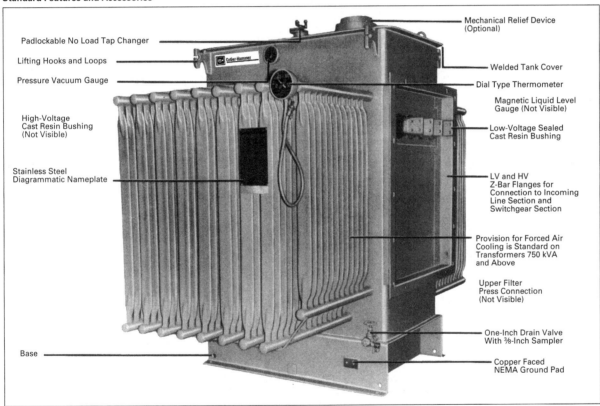

Liquid-Filled Transformer

Figure 4.29.1 Secondary unit substation—typical components. (*By permission: Cutler Hammer, Pittsburgh, PA.—a part of Eaton Corporation.*)

a. Panels used as service entrance equipment must be located near the point where the supply conductors enter the building.

b. A panelboard having main lugs only shall have a maximum of six service disconnects to de-energize the entire panelboard from the supply conductors. Where more than six disconnects are required, a main service disconnect must be provided.

c. Must include connector for bonding and grounding neutral conductor.

d. A service-entrance-type UL label must be factory installed.

e. Ground Fault Protection of Equipment shall be provided for solidly grounded Wye electrical services of more than 150 volts to ground, but not exceeding 600 volts phase-to-phase for each service disconnecting means rated 1000 amperes or more.

Service entrance panels must be identified as such on the order entry to the manufacturing location.

Panelboard Classifications
Panelboards are classified by NEC and UL in two general categories: Lighting and Appliance Panels and Power Distribution Panels.

Lighting and Appliance Panelboards
Lighting and appliance branch circuit panelboards are defined in NEC (Article 384) as "One having more than 10 percent of its overcurrent devices rated 30 amperes or less for which neutral connections are provided". Article 384 also limits the number of overcurrent devices (branch circuit poles) to a maximum of 42 in any one cabinet. This rule applies to all panel types including PRL4 when the panel circuitry comes within the scope of this definition. When the 42 poles are exceeded, two or more separate panels are required.

Power Distribution Panelboards
Distribution panelboards (all others not defined as a lighting and appliance branch circuit panelboards) are restricted only to practical physical limitations such as standard box heights and widths.

It is important to note that the NEC requires that the operating handle of the topmost mounted device to be no more than 6 feet 6 inches above the finished floor.

Additional boxes and fronts are required when the components required for one panelboard exceed the standard box dimensions.

Column Type Panelboards
The same general code restrictions apply as for standard width panels except where trough extensions are used.

Multi-Section Panelboards
When more than 42 overcurrent protective devices are required (see Lighting and Appliance Panelboards), two or more separate enclosures may be required. Separate fronts for each box are standard. A common front can be furnished at additional charge.

Interconnecting Multi-Section Panelboards
When a panelboard, for connection to one feeder, must be furnished in more than one section (Box), each section must be furnished with main bus and terminals of the same rating, unless a main overcurrent device is provided in each section.

There are three commonly used methods for interconnecting multi-section panels: gutter tapping, sub-feeding, and through-feeding.

Gutter Taps (Figure 1)
The use of gutter taps is one method of connecting each section of a multi-section panelboard to the feeder. The gutter tap devices are not furnished as part of the panelboard but are available from several manufacturers. In some instances, increased gutter space may be required. Gutter taps are available on main lug only panels.

Sub-feed Lugs (Figure 2)
Sub-feed lugs are another means of interconnecting multi-section panels. The sub-feed (or second set) of lugs are mounted directly beside the main lugs. These are required in each section except the last panel in the lineup. The feeder cables are brought into the wiring gutter of the first section and connected to the main lugs. Another set of the same size cables are connected to the sub-feed lugs (Section 1) and are carried over to the main lugs of the adjacent panel. Cross connection cables are not furnished by Cutler-Hammer. Sub-feed lugs are available on main lug only panels.

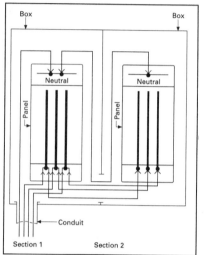

Figure 2. Sub-feed Lugs

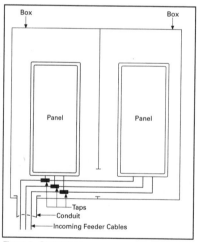

Figure 1. Gutter Taps

Figure 4.30.0   Service entrance equipment—NEC required components. (*By permission: Cutler Hammer, Pittsburgh, PA.—a part of Eaton Corporation.*)

Through-Feed Lugs (See Figure 3)

A third method involves the use of through-feed lugs. The incoming feeder cables are connected to the main lugs or main breaker at the bottom of panel Section 1. Another set of lugs (through-feed) are located at the opposite end of the main bus. The interconnecting cables are connected to the through-feed lugs in Section 1 and are carried over to the main lugs in Section 2. The connection arrangement could be reversed, i.e., main lugs at top; through-feed lugs at bottom end of panel. Cross cables are not furnished by Cutler-Hammer.

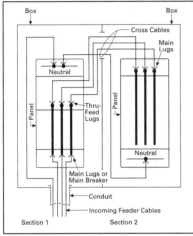

Figure 3. Through-Feed Lugs

Multiple Section Panelboard

Shown below is standard method for flush mounting multiple section lighting and distribution panelboards using standard flush trims.

Branch Circuit Loading for Lighting and Appliance Panels

The size of mains and branches should be selected based on the following:

a. Lighting circuits: NEC Article 210, 215, 220, and 240.
b. Distribution circuits, actual or continuous loads: NEC Article 384-16 (c).
c. Motor circuits: NEC Article 430.
d. Diversity factor.
e. Provision for future loading.

Overcurrent Protection

The following requirements will be found in Article 384-16 (a) of the NEC:

Each lighting and appliance branch circuit panelboard shall be individually protected on the supply side by not more than two main circuit breakers or two sets of fuses having a combined rating not greater than that on the panelboard.

> Exception Number 1 – Individual protection for a lighting and appliance panelboard is not required when the panelboard feeder has overcurrent protection not greater than that of the panelboard.

> Exception Number 2 – For existing installations, individual protection for lighting and appliance branch circuit panelboards is not required where such panelboards are used as service equipment in supplying an individual residential occupancy.

Ambient Temperatures

The primary function of an overcurrent device is to protect the conductor and its insulation against overheating. In selecting the size of the devices and conductors, consideration should be given to the ambient temperature surrounding the conductors within and external to the panelboard. Cumulative heating within the panelboard may cause premature operation of the overcurrent protective devices.

Underwriters Laboratories, Inc. test procedures are based, in part, on 80% loading of panelboard branch circuit devices. Article 384-16 (c) of the NEC limits the loading of overcurrent devices in panelboards to 80% of rating where in normal operation the load will continue for three hours or more. Further derating may be required, depending on such factors as ambient temperature, duty cycle, frequency or altitude.

Special Conditions

Standard panelboards, assembled with standard components, are adequate for most applications. However, special consideration should be given to those required for application under special conditions such as:

a. Subject to excessive vibration or shock
b. Frequencies above 60 cycles
c. Altitudes above 6600 feet
d. Possible fungus growth
e. Compliance with federal, state, and municipal electrical codes and standards
f. Seismic considerations

Harmonic Currents

Article 210-4 (a) (FPN), of the NEC, states:

> "A 3-phase, 4-wire power system used to supply power to computer systems or other similar electronic loads may necessitate that the power system design allow for the possibility of high harmonic neutral currents."

Standard panelboard neutrals are rated for 100% of the panelboard current. However, since harmonic currents can cause overheated neutrals, an option is provided for neutrals to be rated at 200% (1200-Amp maximum neutral for 600-Amp main bus) of the panelboard phase current. Panelboards with the 200% rated neutral are UL listed as suitable for use with nonlinear loads. When specifying 200% rated neutrals in panelboards, be sure to specify 200% rated neutral conductors feeding that panelboard.

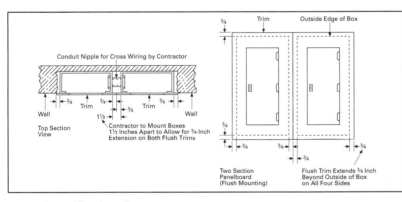

Figure 4.30.0 *(Continued)*

Impact of Arc Fault Breakers

The AFCI represents a major step forward in electrical safety: a step which is expected to have a major impact on the quoted fire statistics. In particular, the Consumer Product Safety Commission (CPSC) has reported that more than 35 percent of all electrical wiring fires are associated with the fixed wiring in Zone 1, and the present AFCI will respond to both parallel and series arcs in this zone (see Figure 1). The Task Force of the NEMA Molded Case Circuit Breaker Section also analyzed fire statistics provided by a major insurance company. Figure 2 shows the percentage of electrical fires associated with the various zones. This figure again indicates that many fires are associated with Zone 1, with statistics similar to those reported by CPSC.

In addition to detecting and interrupting potentially dangerous arcs in Zone 1, the present AFCI would also detect parallel arcs in Zones 2 and 3, and respond to all arcs to ground (see Figure 1). As such, the circuit breaker and AFCI represent a major safety improvement over the present day circuit breakers.

It must be noted that AFCIs will mitigate the effect of arcing-faults, but will not eliminate them completely.

Even under optimum conditions, there will always be at least one arcing half-cycle, and this could cause ignition at high currents.

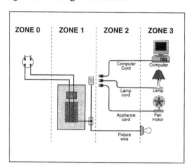

Figure 1. Division of residential wiring into four zones. Zone 0 is associated with the meter, meter socket and service cable, Zone 1 with the loadcenter and the fixed premise wiring, Zone 2 with the wiring between the receptacles and the loads, and Zone 3 with the appliances and other loads.

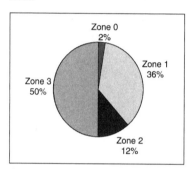

Figure 2. Percentage of Electrical Fires Associated with the Zones Defined in Figure 1.

Typical Causes of Arcing Faults

Arcing faults can occur in homes, apartments, or any other residential dwellings where there is deterioration in wire insulation caused by one or more of these hazardous situations.

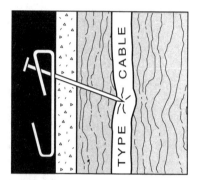

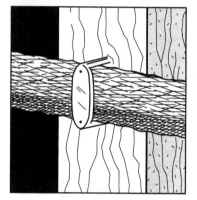

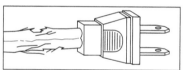

Figure 4.31.0 Causes of arcing and the impact of arc fault breakers. (*By permission: Cutler Hammer, Pittsburgh, PA.—a part of Eaton Corporation.*)

Arc Chute

There are three basic means of extinguishing an arc: lengthening the arc path; cooling by gas blast or contraction; deionizing or physically removing the conduction particles from the arc path. It was the discovery by Westinghouse of this last method which made the first large power air circuit breaker possible.

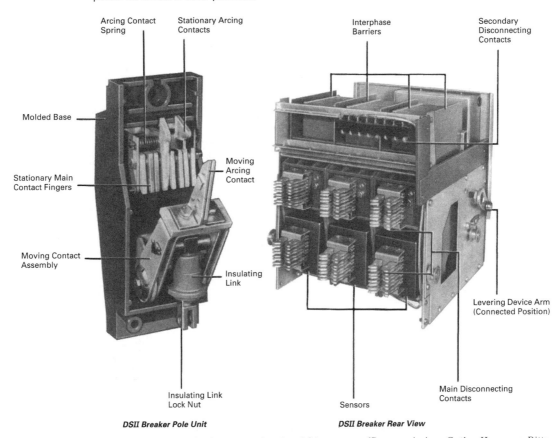

Arcing Contact Spring

Stationary Arcing Contacts

Molded Base

Stationary Main Contact Fingers

Moving Arcing Contact

Moving Contact Assembly

Insulating Link

Insulating Link Lock Nut

DSII Breaker Pole Unit

Interphase Barriers

Secondary Disconnecting Contacts

Levering Device Arm (Connected Position)

Main Disconnecting Contacts

Sensors

DSII Breaker Rear View

Figure 4.31.1   Arc chute—one of three basic means of extinguishing an arc. (*By permission: Cutler Hammer, Pittsburgh, PA.—a part of Eaton Corporation.*)

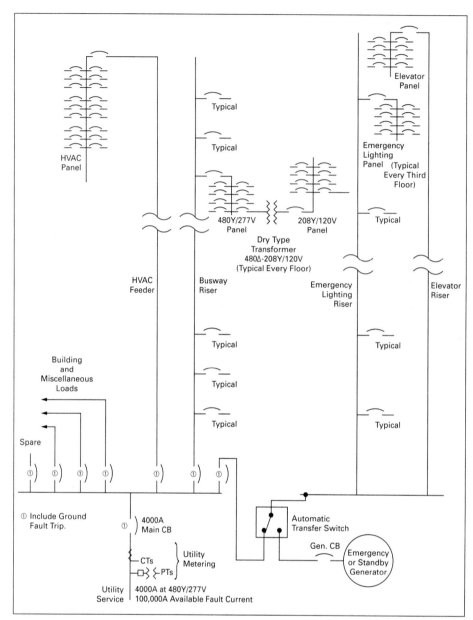

Figure 4.32.0 Typical power distribution and riser diagram for commercial buildings. (*By permission: Cutler Hammer, Pittsburgh, PA.—a part of Eaton Corporation.*)

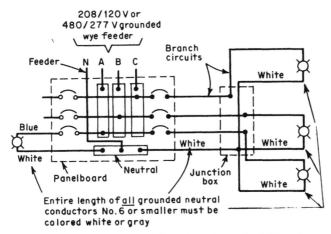

Generally any grounded circuit conductor that is No. 6 size or smaller must have a continuous white or natural gray outer finish. [Sec. 200-6(a).]

Figure 4.33.0 Identifying ground conductor for No. 6 or smaller. (*By permission: The McGraw-Hill Company*, National Electrical Code Handbook, *23rd Ed.*)

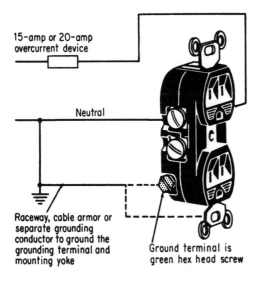

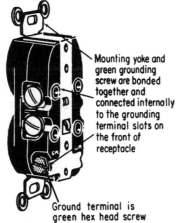

Figure 4.33.1 Grounding of 15/20 amp receptacle. (*By permission: The McGraw-Hill Company*, National Electrical Code Handbook, *23rd Ed.*)

Any receptacle on a 15- or 20-A branch circuit must be a grounding type. (Sec. 210-7.)

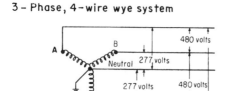

Single-phase, 3-wire system

Transformer secondary

240 volts — Neutral — 120 volts / 120 volts

Voltage-to-ground is 120 volts from either ungrounded leg to the neutral conductor or to any grounded metal

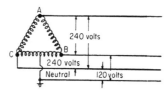

3 - Phase, 4-wire wye system

A — B — 480 volts — Neutral 277 volts — 277 volts — 480 volts — C

Voltage-to-ground is the voltage from any phase leg to the grounded neutral— 277 volts, in this case

3-Phase, 4-wire delta system

A — 240 volts — B — C — 240 volts — Neutral — 120 volts

Voltage-to-ground is 208 volts from phase A conductor and 120 volts from phase B or C to the grounded neutral

Figure 4.33.2 Grounding of single- and three-phase systems. (*By permission: The McGraw-Hill Company,* National Electrical Code Handbook, *23rd Ed.*)

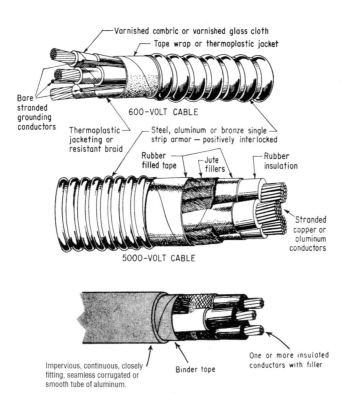

Varnished cambric or varnished glass cloth
Tape wrap or thermoplastic jacket
Bare stranded grounding conductors
600-VOLT CABLE

Thermoplastic jacketing or resistant braid
Steel, aluminum or bronze single strip armor — positively interlocked
Rubber filled tape
Jute fillers
Rubber insulation
Stranded copper or aluminum conductors
5000-VOLT CABLE

Impervious, continuous, closely fitting, seamless corrugated or smooth tube of aluminum.
Binder tape
One or more insulated conductors with filler.
ALS cable

These are some of the constructions in which Type MC cable is available. (Sec. 334-1.)

Figure 4.34.0 MC Cable cut-away. (*By permission: The McGraw-Hill Company,* National Electrical Code Handbook, *23rd Ed.*)

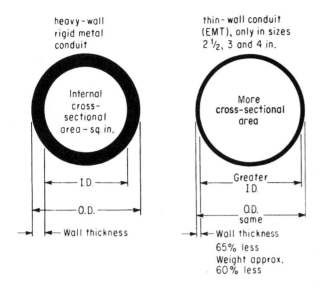

Trade size rigid and EMT	Inches outer dia. (O.D.) EMT and rigid	Wall thickness in.		Inside cross-sectional area sq. in.		More C.S.A. % for EMT
		Rigid	EMT	Rigid	EMT	
2½	2.875	0.203	0.072	4.79	5.85	22%
3	3.500	0.216	0.072	7.38	8.84	19%
4	4.500	0.237	0.083	12.72	14.75	16%

Larger sizes of EMT have same outside diameters as rigid and IMC. (Sec. 348-7.)

Figure 4.34.1 Rigid and EMT ID and OD dimensions. (*By permission: The McGraw-Hill Company*, National Electrical Code Handbook, *23rd Ed.*)

WHEN BUILDING CONTAINS ONLY ONE SYSTEM VOLTAGE FOR CIRCUITS:

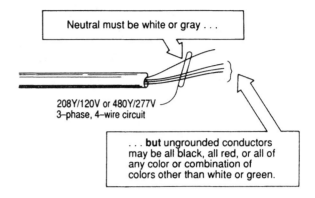

IF THERE ARE TWO SYSTEM VOLTAGES:

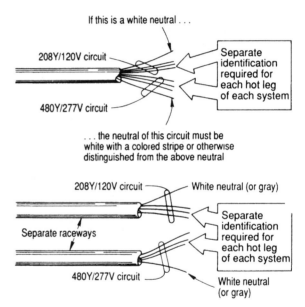

Separate identification of ungrounded conductors is required only if a building utilizes more than one nominal voltage system. Neutrals must be color-distinguished if circuits of two voltage systems are used in the same raceway (Sec. 210-5), but *not* if different voltage systems are run in separate raceways. [Sec. 210-4(d).]

Figure 4.35.0 Color of neutral wire in one and two voltage systems. (*By permission: The McGraw-Hill Company*, National Electrical Code Handbook, *23rd Ed.*)

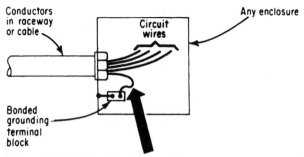

Equipment grounding conductor may be bare, or covered to show a green color or green with one or more yellow stripes.

Sec. 250-134(c)

But, for a grounding conductor larger than No. 6, an insulated conductor of other than green color or green with yellow stripes may be used provided *one* of the following steps is taken:

1. Stripping the insulation from an insulated conductor of another color (say, black) for the entire length that is exposed in the box or other enclosure— so that the conductor appears as a bare conductor.

2. Painting the exposed insulation green for its entire length within the enclosure.

3. Wrapping the entire length of exposed insulation with green-colored tape or green-colored adhesive labels.

VIOLATION?

Equipment grounding conductor for branch circuit. Although previously permitted, such application seems to be prohibited. [Sec. 210-5(b).]

Figure 4.36.0 Grounding conductors—proper usage. (*By permission: The McGraw-Hill Company*, National Electrical Code Handbook, *23rd Ed.*)

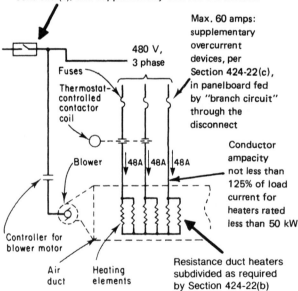

Rules on disconnects for heating equipment demand careful study for HVAC systems with duct heaters and supplementary overcurrent protective devices. (Sec. 424-19.)

Figure 4.37.0 Disconnects required for HVAC equipment. (*By permission: The McGraw-Hill Company*, National Electrical Code Handbook, *23rd Ed.*)

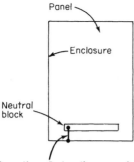

BONDING is the insertion of a bonding screw into the panel neutral block to connect the block to the panel enclosure, or it is use of a bonding jumper from the neutral block to an equipment grounding block that is connected to the enclosure.

NOTE: Bonding – the connection of the neutral terminal to the enclosure or to the ground terminal that is, itself, connected to the enclosure – might also be done in an individual switch or CB enclosure.

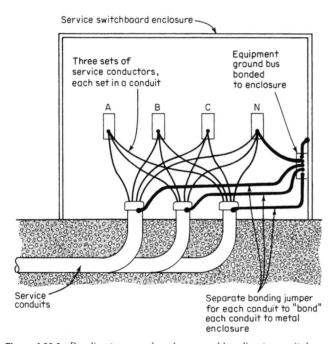

Figure 4.38.0 Bonding to a panel enclosure and bonding to a switchgear enclosure. (*By permission: The McGraw-Hill Company,* National Electrical Code Handbook, *23rd Ed.*)

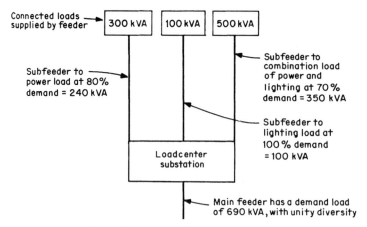

How demand factors are applied to connected loads. (Sec. 220-11.)

Figure 4.39.0   How demand factors are applied to connected loads. (*By permission: The McGraw-Hill Company,* National Electrical Code Handbook, *23rd Ed.*)

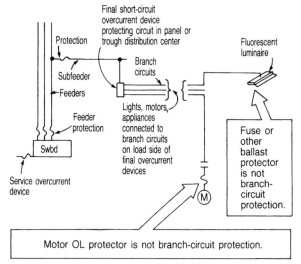

Figure 4.40.0   Motor overload protectors are not branch circuit protection. (*By permission: The McGraw-Hill Company,* National Electrical Code Handbook, *23rd Ed.*)

All major components of AMPGARD starters — mechanical isolating switch, vacuum contactor, current transformers and control transformer — were designed specifically to function together as an integrated starter unit.

One of the most important design features, however, is the component-to-component circuit concept employed to eliminate 50% of the current carrying junctions.

The flow of power through a vacuum-break controller can be traced by referring to the lower portion of this figure where the starter is shown in the energized position. The line stab assembly mounted at the back of the enclosure also serves as the starter line terminals (1). The stabs themselves are engaged by the fuse jaws (2) of the isolating switch which is mounted on rails at the top of the enclosure. The line ferrules (3) of the current-limiting motor-starting power fuses (4), clip into the fuse jaws, and the load ferrules (5) fit into the fuse holders (6) which are part of the contactor line terminals. Power flow through the contactor is from the load ferrules of the power fuse, through the shunts (7), and the vacuum interrupters (bottles) of the contactor (8), to the contactor load terminals (9).

Spring loaded contact jaws mounted on the contactor load terminals plug into the lower stab assembly (10), providing a convenient connection through the current transformers to the motor load terminals mounted on the left hand side wall of the enclosure.

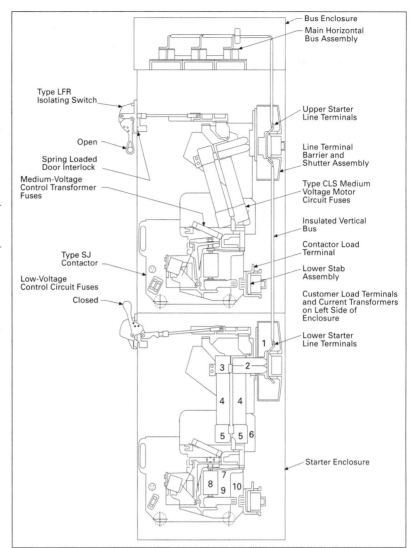

Section View of 400 Ampere Two-High Starters

Figure 4.41.0 Medium voltage starter components. (*By permission: Cutler Hammer, Pittsburgh, PA.—a part of Eaton Corporation.*)

IEC

The IEC (International Electro-technical Commission) is a commission that publishes recommendations for certain product design parameters and laboratory test procedures. Manufacturers may test per and publish technical information based on the IEC parameters. This information can be used by customers to provide a basis for product comparison at a given rating.

NEMA

NEMA (National Electrical Manufacturers' Association) has developed product design standards and test specifications for device qualification, many of which have been adopted by UL. Neither NEMA nor IEC perform any device testing.

NEC (National Electrical Code)

The NEC is the guideline used by electrical inspectors in North America to ensure equipment safety. Electrical equipment is deemed suitable for use when the "the equipment has been listed by a qualified electrical testing laboratory" (NEC 1995 article 90-7). Equipment that does not bear a listing could face rejection by the inspector.

Choosing the Proper Product for the Needs

The table below shows issues that should be considered when making a choice between IEC type and NEMA type starters. For more information on comparison of IEC to NEMA, refer to DI184A.

Product Comparison – IEC Type and NEMA Type

	Issue	IEC Type	NEMA Type
1	Starter Size	Smaller/horsepower (rating)	Larger/horsepower (rating)
2	Contactor Performance	Electrical life – 1 million AC-3 operations 30,000 AC-4 operations when tested per IEC recommendations	Electrical life typically 2.5 to 4 times higher than equivalently rated IEC device on the same test
3	Contactor Application	Application sensitive – greater knowledge and care necessary	Application easier – fewer parameters to consider
4	Overload Relay Trip Characteristics	Class 10 typical – designed for use with motors per IEC recommendations	Class 20 typical – designed for use with motors per NEMA standards
5	Overload Relay Adjustability	Fixed heaters. Adjustable to suit different motors at the same horsepower (heaters not field interchangeable)	Field changeable heaters allow adjustments to motors of different horsepowers
6	Overload Relay Reset Mechanism Characteristics	Some use RESET/STOP dual function operating mechanism	RESET ONLY mechanism typical
		Hand/Auto Reset typical	Hand Reset only typical
7	Fault Withstandability	Typically designed for use with fast acting, current limiting fuses	Designed for use with common domestic fuses and circuit breakers

Figure 4.42.0 Choosing the proper starters. (*By permission: Cutler Hammer, Pittsburgh, PA.—a part of Eaton Corporation.*)

Enclosure Types

Enclosures provide mechanical and electrical protection for operator and equipment. Brief descriptions of the various types of enclosures offered by Cutler-Hammer are given below. See NEMA Standards Publication No. 250 for more comprehensive descriptions, definitions and/or test criteria.

NEMA 1

NEMA Type 1 (conforms to IP-40) – For Indoor Use

Suitable for most applications where unusual service conditions do not exist and where a measure of protection from accidental contact with enclosed equipment is required. Designed to meet tests for Rod Entry and Rust Resistance. Enclosure is sheet steel, treated to resist corrosion. Depending on the size, knockouts are provided on the top, bottom and sometimes on the side.

NEMA 3R

NEMA Type 3R (conforms to IP-52) – For Outdoor Use

Primarily intended for applications where falling rain, sleet or external ice formations are present. Gasketed cover. Designed to meet tests for Rain, Rod Entry, External Icing and Rust Resistance. Enclosure is sheet steel, treated to resist corrosion. Depending on the size, a blank cover plate is attached to the top (for a conduit hub) and knockouts are provided on the bottom.

Cover-mounted pilot device holes are provided and covered with hole plugs.

NEMA Type 4 (conforms to IP-65) – For Indoor or Outdoor Use

Provide measure of protection from splashing water, hose-directed water, and wind blown dust or rain. Constructed of sheet steel with gasketed cover.

Figure 4.42.0 *(Continued)*

Designed to meet tests for Hosedown, External Icing and Corrosion Protection. Enclosure has two watertight hubs (power) installed top and bottom and one control hub installed in bottom – depending on size.

Cover-mounted pilot device holes are provided and covered with hole plugs.

NEMA 4X

NEMA Type 4X (conforms to IP-65) – For Indoor or Outdoor Use

Provide measure of protection from splashing water, hose-directed water, wind blown dust, rain and corrosion. Constructed of stainless steel with gasketed cover. Designed to meet same tests as Type 4 except enclosure must pass a 200-hour salt spray corrosion resistance test.

NEMA Type 12 (conforms to IP-62) – For Indoor Use

Provide a degree of protection from dripping liquids (non-corrosive), falling dirt and dust. Designed to meet tests for Drip, Dust and Rust Resistance. Constructed of sheet steel. Hole plugs cover pilot device holes. There are no knockouts, hub cover plates or hubs installed.

NEMA 12

Many Cutler-Hammer NEMA Type 12 enclosures are suitable for use in Class II, Division 2, Group G and Class III, Divisions 1 and 2 locations as defined in the National Electrical Code.

NEMA Type 12 – Modified for Outdoor Use

NEMA Type 12 enclosures can be modified for NEMA Type 3R outdoor use as follows: NEMA Type 3R – Use watertight conduit hub or equivalent provision for watertight connection at conduit entrance. A provision for drainage is made. (1/8-inch diameter)

The Type 12 enclosure can be ordered with a safety interlock on the door that can be padlocked off. A vault-type door latch system is used. A tapered plate holds the gasketed door tight against the case edge to provide a positive seal. The special door interlock consists of the door handle and a screwdriver operated cover defeater.

The cover defeater and the disconnect interlock defeater are both recessed screwdriver operated devices which cannot be manipulated with other types of tools.

IEC IP Index of Protection Ratings

1st Number	Description	2nd Number	Description
0	No protection	0	No protection
1	Protection against solid objects greater than 50 mm	1	Protection against vertically falling drops of water
2	Protection against solid objects greater than 12 mm	2	Protection against dripping water when tilted up to 15 degrees
3	Protection against solid objects greater than 2.5 mm	3	Protection against spraying water
4	Protection against solid objects greater than 1 mm	4	Protection against splashing water
5	Total protection against dust – limited ingress (dust protected)	5	Protection against water jets
6	Total protection against dust (dusttight)	6	Protection against heavy seas
		7	Protection against the effects of immersion
		8	Protection against submersion

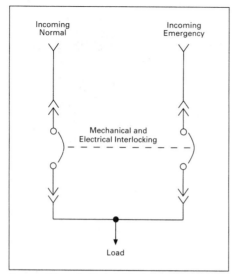

Draw-out Transfer Switch Single Line Diagram

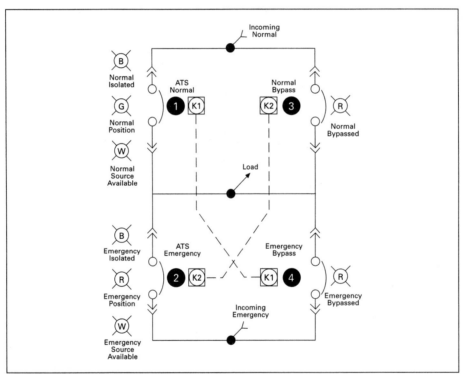

Single Line Diagram of SPB Draw-out Bypass Isolation Automatic Transfer Switch

Figure 4.43.0 Transfer switch operations—line diagrams. (*By permission: Cutler Hammer, Pittsburgh, PA.—a part of Eaton Corporation.*)

Sound Levels

Sound Levels of Electrical Equipment for Offices, Hospitals, Schools and Similar Buildings

Insurance underwriters and building owners desire and require that the electrical apparatus be installed for maximum safety and the least interference with the normal use of the property. Architects should take particular care with the designs for hospitals, schools and similar buildings to keep the sound perception of such equipment as motors, blowers and transformers to a minimum.

Even though transformers are relatively quiet, resonant conditions may exist near the equipment which will amplify their normal 120 Hertz hum. Therefore, it is important that consideration be given to the reduction of amplitude and to the absorption of energy at this frequency. This problem begins in the designing stages of the equipment and the building. There are two points worthy of consideration: 1) What sound levels are desired in the normally occupied rooms of this building? 2) To effect this, what sound level in the equipment room and what type of associated acoustical treatment will give the most economical installation overall?

A relatively high sound level in the equipment room does not indicate an abnormal condition within the apparatus. However, absorption may be necessary if sound originating in an unoccupied equipment room is objectionable outside the room. Furthermore, added absorption material usually is desirable if there is a "build-up" of sound due to reflections.

Some reduction or attenuation takes place through building walls, the remainder may be reflected in various directions, resulting in a build-up or apparent higher levels, especially if resonance occurs because of room dimensions or material characteristics.

Area Consideration

In determining permissible sound levels within a building, it is necessary to consider how the rooms are to be used and what levels may be objectionable to occupants of the building. The ambient sound level values given in Table A22 are representative average values and may be used as a guide in determining suitable building levels.

Table A22: Typical Sound Levels

Radio, Recording and TV Studios	25-30 db
Theatres and Music Rooms	30-35
Hospitals, Auditoriums and Churches	35-40
Classrooms and Lecture Rooms	35-40
Apartments and Hotels	35-45
Private Offices and Conference Rooms	40-45
Stores	45-55
Residence (Radio, TV Off) and Small Offices	53
Medium Office (3 to 10 Desks)	58
Residence (Radio, TV On)	60
Large Store (5 or More Clerks)	61
Factory Office	61
Large Office	64
Average Factory	70
Average Street	80

Decrease in sound level varies at an approximate rate of 6 decibels for each doubling of the distance from the source of sound to the listener. For example, if the level six feet from a transformer is 50 db, the level at a distance of twelve feet would be 44 db and at 24 feet the level decreases to 38 db, etc. However, this rule applies only to equipment in large areas equivalent to an out-of-door installation, with no nearby reflecting surfaces.

Transformer Sound Levels

Transformers emit a continuous 120 Hertz hum with harmonics when connected to 60 Hertz circuits. The fundamental frequency is the "hum" which annoys people primarily because of its continuous nature. For purposes of reference, sound measuring instruments convert the different frequencies to 1000 Hertz and a 40 db level. Transformer sound levels based on NEMA publication TR-1 are listed in Table A23.

Table A23: Maximum Average Sound Levels - Decibels

kVA	Liquid-Filled Transformers		Dry-Type Transformers	
	Self-Cooled Rating (OA)	Forced-Air Cooled Rating (FA)	Self-Cooled Rating (AA)	Forced-Air Cooled Rating (FA)
300	55	. .	58	67
500	56	67	60	67
750	58	67	64	67
1000	58	67	64	67
1500	60	67	65	68
2000	61	67	66	69
2500	62	67	68	71
3000	63	67	68	71
3750	64	67	70	73
5000	65	67	71	73
6000	66	68	72	74
7500	67	69	73	75
10000	68	70	. .	76

Since values given in Table A23 are in general higher than those given in Table A22, the difference must be attenuated by distance and by proper use of materials in the design of the building. An observer may believe that a transformer is noisy because the level in the room where it is located is high. Two transformers of the same sound output in the same room increase the sound level in the room approximately 3 db, and 3 transformers by about 5 db, etc.

Sounds due to structure-transmitted vibrations originating from the transformer are lowered by mounting the transformers on vibration dampeners or isolators. There are a number of different sound vibration isolating materials which may be used with good results. Dry-type power transformers are often built with an isolator mounted between the transformer support and case members. The natural period of the core and coil structure when mounted on vibration dampeners is about 10% of the fundamental frequency. The reduction in the transmitted vibration is approximately 98%. If the floor or beams beneath the transformer are light and flexible, the isolator must be softer or have improved characteristics in order to keep the transmitted vibrations to a minimum. (Enclosure covers and ventilating louvers are often improperly tightened or gasketed and produce unnecessary noise.) The building structure will assist the dampeners if the transformer is mounted above heavy floor members or if mounted on a heavy floor slab. Positioning of the transformer in relation to walls and other reflecting surfaces has a great effect on reflected noise and resonances. Often, placing the transformer at an angle to the wall, rather than parallel to it, will reduce noise. Electrical connections to a substation transformer should be made with flexible braid or conductors; connections to an individually-mounted transformer should be in flexible conduit.

Figure 4.44.0 Sound levels of various electrical components. (*By permission: Cutler Hammer, Pittsburgh, PA.—a part of Eaton Corporation.*)

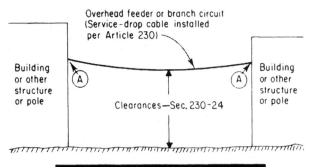

Overhead feeder or branch circuit
(Service-drop cable installed
per Article 230)

Building
or other
structure
or pole

A

Clearances—Sec. 230-24

A

Building
or other
structure
or pole

AT THE POINTS "A" ABOVE, CON-
NECTION MUST BE MADE IN AN
APPROVED MANNER, SUCH AS THE
FOLLOWING.

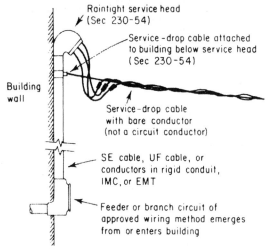

Raintight service head
(Sec 230-54)

Service-drop cable attached
to building below service head
(Sec 230-54)

Building
wall

Service-drop cable
with bare conductor
(not a circuit conductor)

SE cable, UF cable, or
conductors in rigid conduit,
IMC, or EMT

Feeder or branch circuit of
approved wiring method emerges
from or enters building

Figure 4.45.0 Aerial cable clearances per NEC. (*By permission:
The McGraw-Hill Company,* National Electrical Code Handbook,
23rd Ed.)

Clearance Needed in Direction of Access to Live Parts in Enclosures for Switchboards, Panelboards, Switches, CBs, or Other Electrical Equipment—Plan Views [Sec. 110-26(a)]

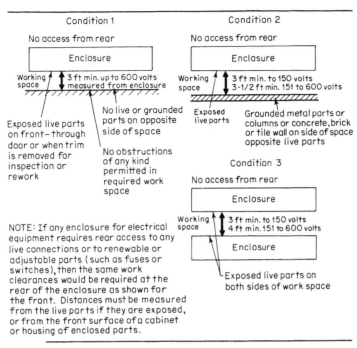

Figure 4.46.0 Clearance levels for enclosures per NEC. (*By permission: The McGraw-Hill Company,* National Electrical Code Handbook, *23rd Ed.*)

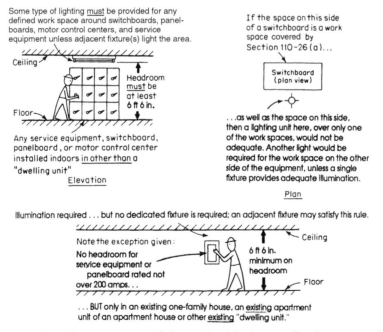

Electrical equipment requires lighting and 6½-ft headroom at *all* work spaces around certain equipment. [Secs. 110-26(d) and (e).]

Figure 4.46.1 Headroom and lighting requirements around switchboards per NEC. (*By permission: The McGraw-Hill Company,* National Electrical Code Handbook, *23rd Ed.*)

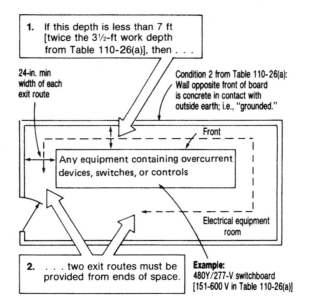

1. If this depth is less than 7 ft [twice the 3½-ft work depth from Table 110-26(a)], then . . .

24-in. min width of each exit route

Condition 2 from Table 110-26(a): Wall opposite front of board is concrete in contact with outside earth; i.e., "grounded."

Front

Any equipment containing overcurrent devices, switches, or controls

Electrical equipment room

2. . . . two exit routes must be provided from ends of space.

Example: 480Y/277-V switchboard [151-600 V in Table 110-26(a)]

(A) Complies

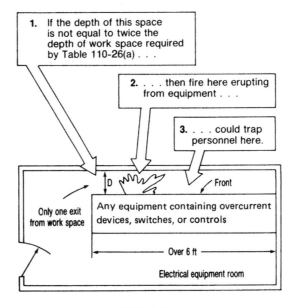

1. If the depth of this space is not equal to twice the depth of work space required by Table 110-26(a) . . .

2. . . . then fire here erupting from equipment . . .

3. . . . could trap personnel here.

Only one exit from work space

D

Front

Any equipment containing overcurrent devices, switches, or controls

Over 6 ft

Electrical equipment room

(B) Violation

There must be two paths out of the workspace required in front of any equipment containing fuses, circuit breakers, motor starters, switches, and/or any other control or protective devices, where the equipment is rated 1,200 A or more and is over 6 ft (1.83 m) wide. [Sec. 110-26(c).]

Figure 4.46.2   Workspace required in front of electrical equipment per NEC. (*By permission: The McGraw-Hill Company*, National Electrical Code Handbook, *23rd Ed.*)

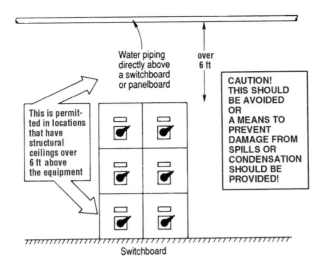

Water pipes and other "foreign" piping must not be located less than 6 ft above switchboard. (Sec. 110-26(f).)

Figure 4.46.3 Location of water pipes above switchboard. (*By permission: The McGraw-Hill Company,* National Electrical Code Handbook, *23rd Ed.*)

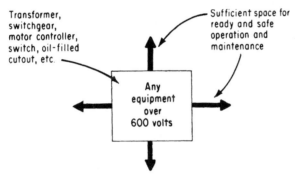

This is the general rule for work space around any high-voltage equipment.

Sufficient headroom and adequate lighting are essential to safe operation, maintenance, and repair of high-voltage equipment.

Figure 4.46.4 Space requirements around high-voltage equipment. (*By permission: The McGraw-Hill Company,* National Electrical Code Handbook, *23rd Ed.*)

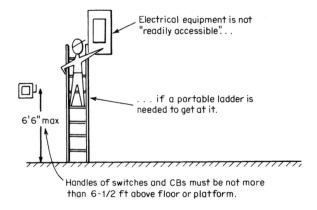

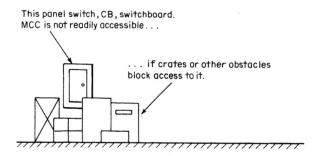

In general, aside from the cases noted above, fuses and CBs must **not** be used above a suspended ceiling because they would be not readily accessible. But equipment that is not required by a **Code** rule to be "accessible" or to be "readily accessible" may be mounted above a suspended ceiling.

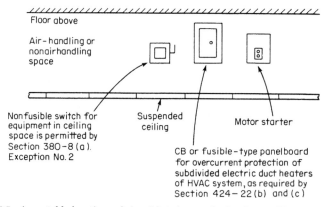

Figure 4.46.5 Acceptable locations of circuit breakers and other items. (*By permission: The McGraw-Hill Company,* National Electrical Code Handbook, *23rd Ed.*)

TRANSFORMERS RATED 112½ KVA OR LESS

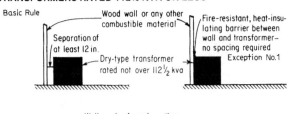

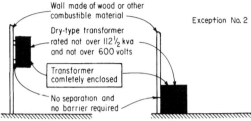

TRANSFORMERS RATED OVER 112½ KVA

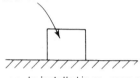

... may be installed in any room or area (need not be fire-resistant)

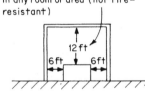

Dry-type transformer with 80°C rise or higher insulation but not enclosed and ventilated

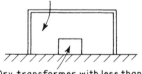

Dry transformer with less than 80°C rise insulation

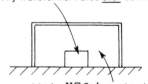

...must be in *NE Code* constructed transformer vault (Part C, Art.450)

Construction of dry-type transformer affects indoor installation rules. (Sec. 450-21.)

Figure 4.46.6 Clearances required around transformers. (*By permission: The McGraw-Hill Company*, National Electrical Code Handbook, *23rd Ed.*)

Condensed General Criteria for Preliminary Consideration

General Need	Specific Need	Maximum Tolerance Duration of Power Failure	Recommended Minimum Auxiliary Supply Time	Type of Auxiliary Power System		System Justification
				Emergency	Standby	
Lighting	Evacuation of personnel	Up to 10 s, preferably not more than 3 s	2 h	×		Prevention of panic, injury, loss of life Compliance with building codes and local, state, and federal laws Lower insurance rates Prevention of property damage Lessening of losses due to legal suits
	Perimeter and security	10 s	10–12 h during all dark hours	×	×	Lower losses from theft and property damage Lower insurance rates Prevention of injury
	Warning	From 10 s up to 2 or 3 min	To return to prime power source	×		Prevention or reduction of property loss Compliance with building codes and local, state, and federal laws Prevention of injury and loss of life
	Restoration of normal power system	1 s to indefinite depending on available light	Until repairs completed and power restored	×	×	Risk of extended power and light outage due to a longer repair time
	General lighting	Indefinite; depends on analysis and evaluation	Indefinite; depends on analysis and evaluation		×	Prevention of loss of sales Reduction of production losses Lower risk of theft Lower insurance rates

General Need	Specific Need	Maximum Tolerance Duration of Power Failure	Recommended Minimum Auxiliary Supply Time	Type of Auxiliary Power System		System Justification
				Emergency	Standby	
	Hospitals and medical areas	Uninterruptible to 10 s ANSI/NFPA 99-1984 [10], 101-1985 [11] allow 10 s for alternate power source to start and transfer	To return of prime power	×	×	Facilitate continuous patient care by surgeons, medical doctors, nurses, and aids Compliance with all codes, standards, and laws Prevention of injury or loss of life Lessening of losses due to legal suits
	Orderly shutdown time	0.1 s to 1 h	10 min to several hours	×		Prevention of injury or loss of life Prevention of property loss by a more orderly and rapid shutdown of critical systems Lower risk of theft Lower insurance rates
Startup power	Boilers	3 s	To return of prime power	×	×	Return to production Prevention of property damage due to freezing Provision of required electric power
	Air compressors	1 min	To return of prime power		×	Return to production Provision for instrument control
Transportation	Elevators	15 s to 1 min	1 h to return of prime power		×	Personnel safety Building evacuation Continuation of normal activity
	Material handling	15 s to 1 min	1 h to return of prime power		×	Completion of production run Orderly shutdown Continuation of normal activity
	Escalators	15 s to no requirement for power	Zero to return of prime power		×	Orderly evacuation Continuation of normal activity

Figure 4.47.0   Emergency power criteria. (*By permission: McGraw-Hill*, Electrical Engineer's Portable Handbook, *Bob Hickey.*)

Condensed General Criteria for Preliminary Consideration (*Continued*)

General Need	Specific Need	Maximum Tolerance Duration of Power Failure	Recommended Minimum Auxiliary Supply Time	Type of Auxiliary Power System		System Justification
				Emergency	Standby	
	Conveyors	15 s to 1 min	As analyzed and economically justified		×	Completion of production run Completion of customer order Orderly shutdown Continuation of normal activity
Mechanical utility systems	Water (cooling and general use)	15 s	½ h to return of prime power		×	Continuation of production Prevention of damage to equipment Supply of fire protection
	Water (drinking and sanitary)	1 min to no requirement	Indefinite until evaluated		×	Providing of customer service Maintaining personnel performance
	Boiler power	0.1 s	1 h to return of prime power	×	×	Prevention of loss of electric generation and steam Maintaining production Prevention of damage to equipment
	Pumps for water, sanitation, and production fluids	10 s to no requirement	Indefinite until evaluated		×	Prevention of flooding Maintaining cooling facilities Providing sanitary needs Continuation of production Maintaining boiler operation
	Fans and blowers for ventilation and heating	0.1 s to return of normal power	Indefinite until evaluated	×	×	Maintaining boiler operation Providing for gas-fired unit venting and purging Maintaining cooling and heating functions for buildings and production
Heating	Food preparation	5 min	To return of prime power		×	Prevention of loss of sales and profit Prevention of spoilage of in-process preparation

General Need	Specific Need	Maximum Tolerance Duration of Power Failure	Recommended Minimum Auxiliary Supply Time	Type of Auxiliary Power System		System Justification
				Emergency	Standby	
	Process	5 min	Indefinite until evaluated; normally for time for orderly shutdown, or to return of prime power		×	Prevention of in-process product damage Prevention of property damage Continued production Prevention of payment to workers during no production Lower insurance rates
Refrigeration	Special equipment or devices which have critical warmup (cryogenics)	5 min	To return of prime power		×	Prevention of equipment or product damage
	Depositories of critical nature (blood banks, etc)	5 min (10 s per ANSI/NFPA 99-1984 [10]	To return of prime power		×	Prevention of loss of material stored
	Depositories of noncritical nature (meat, produce, etc)	2 h	Indefinite until evaluated		×	Prevention of loss of material stored Lower insurance rates
Production	Critical process power (sugar factory, steel mills, chemical processes, glass products, etc)	1 min	To return of prime power or until orderly shutdown		×	Prevention of product and equipment damage Continued normal production Reduction of payment to workers on guaranteed wages during nonproductive period Lower insurance rates Prevention of prolonged shutdown due to nonorderly shutdown

Figure 4.47.0 (*Continued*)

Condensed General Criteria for Preliminary Consideration (*Continued*)

General Need	Specific Need	Maximum Tolerance Duration of Power Failure	Recommended Minimum Auxiliary Supply Time	Type of Auxiliary Power System		System Justification
				Emergency	Standby	
	Process control power	Uninterruptible (UPS) to 1 min	To return of prime power	×	×	Prevention of loss of machine and process computer control program Maintaining production Prevention of safety hazards from developing Prevention of out-of-tolerance products
Space conditioning	Temperature (critical application)	10 s	1 min to return of prime power	×	×	Prevention of personnel hazards Prevention of product or property damage Lower insurance rates Continuation of normal activities Prevention of loss of computer function
	Pressure (critical) pos/neg atmosphere	1 min	1 min to return of prime power	×	×	Prevention of personnel hazards Continuation of normal activities Prevention of product or property damage Lower insurance rates Compliance with local, state, and federal codes, standards, and laws
	Humidity (critical)	1 min	To return of prime power		×	Prevention of loss of computer functions Maintenance of normal operations and tests Prevention of explosions or other hazards

General Need	Specified Need	Maximum Tolerance Duration of Power Failure	Recommended Minimum Auxiliary Supply Time	Type of Auxiliary Power System		System Justification
				Emergency	Standby	
	Static charge	10 s or less	To return of prime power	×	×	Prevention of static electric charge and associated hazards Continuation of normal production (printing press operation, painting spray operations)
	Building heating and cooling	30 min	To return of prime power		×	Prevention of loss due to freezing Maintenance of personnel efficiency Continuation of normal activities
	Ventilation (toxic fumes)	15 s	To return of prime power or orderly shutdown	×	×	Reduction of health hazards Compliance with local, state, and federal codes, standards, and laws Reduction of pollution
	Ventilation (explosive atmosphere)	10 s	To return of prime power or orderly shutdown	×	×	Reduction of explosion hazard Prevention of property damage Lower insurance rates Compliance with local, state, and federal codes, standards, and laws Lower hazard of fire Reduce hazards to personnel
	Ventilation (building general)	1 min	To return of prime power		×	Maintaining of personnel efficiency Providing make-up air in building

Figure 4.47.0   (*Continued*)

Condensed General Criteria for Preliminary Consideration (*Continued*)

General Need	Specified Need	Maximum Tolerance Duration of Power Failure	Recommended Minimum Auxiliary Supply Time	Type of Auxiliary Power System		System Justification
				Emergency	Standby	
	Ventilation (special equipment)	15 s	To return of prime power or orderly shutdown	×	×	Purging operation to provide safe shutdown or startup Lowering of hazards to personnel and property Meeting requirements of insurance company Compliance with local, state, and federal codes, standards, and laws Continuation of normal operation
	Ventilation (all categories noncritical)	1 min	Optional		×	Maintaining comfort Preventing loss of tests
	Air pollution control	1 min	Indefinite until evaluated; compliance or shutdowns are options	×	×	Continuation of normal operation Compliance with local, state, and federal codes, standards, and laws
Fire protection	Annunciator alarms	1 s	To return of prime power	×		Compliance with local, state, and federal codes, standards, and laws Lower insurance rates Minimizing life and property damage
	Fire pumps	10 s	To return of prime power		×	Compliance with local, state, and federal codes, standards, and laws Lower insurance rates Minimizing life and property damage

General Need	Specified Need	Maximum Tolerance Duration of Power Failure	Recommended Minimum Auxiliary Supply Time	Type of Auxiliary Power System		System Justification
				Emergency	Standby	
	Auxiliary lighting	10 s	5 min to return of prime power	×	×	Servicing of fire pump engine should it fail to start Providing visual guidance for fire-fighting personnel
Data processing	CPU memory tape/disk storage, peripherals	½ cycle	To return of prime power or orderly shutdown	×	×	Prevention of program loss Maintaining normal operations for payroll, process control, machine control, warehousing, reservations, etc
	Humidity and temperature control	5 to 15 min (1 min for water-cooled equipment)	To return of prime power or orderly shutdown		×	Maintenance of conditions to prevent malfunctions in data processing system Prevention of damage to equipment Continuation of normal activity
Life support and life safety systems (medical field, hospitals, clinics, etc)	X-ray	Milliseconds to several hours	From no requirement to return of prime power, as evaluated	×	×	Maintenance of exposure quality Availability for emergencies
	Light	Milliseconds to several hours	To return of prime power	×	×	Compliance with local, state, and federal codes, standards, and laws Preventing interruption to operation and operating needs
	Critical to life, machines, and services	½ cycle to 10 s	To return of prime power	×	×	Maintenance of life Prevention of interruption of treatment or surgery Continuation of normal activity Compliance with local, state, and federal codes, standards, and laws

Figure 4.47.0 (*Continued*)

Condensed General Criteria for Preliminary Consideration (*Continued*)

General Need	Specified Need	Maximum Tolerance Duration of Power Failure	Recommended Minimum Auxiliary Supply Time	Type of Auxiliary Power System		System Justification
				Emergency	Standby	
	Refrigeration	5 min	To return of prime power		×	Maintaining blood, plasma, and related stored material at recommended temperature and in prime condition
Communication systems	Teletypewriter	5 min	To return of prime power		×	Maintenance of customer services Maintenance of production control and warehousing Continuation of normal communication to prevent economic loss
	Inner building telephone	10 s	To return of prime power	×		Continuation of normal activity and control
	Television (closed circuit and commercial)	10 s	To return of prime power		×	Continuation of sales Meeting of contracts Maintenance of security Continuation of production
	Radio systems	10 s	To return of prime power	×	×	Maintenance of security and fire alarms Providing evacuation instructions Continuation of service to customers Prevention of economic loss Directing vehicles normally
	Intercommunication systems	10 s	To return of prime power	×	×	Providing evacuation instructions Directing activities during emergency Providing for continuation of normal activities Maintaining security

General Need	Specific Need	Maximum Tolerance Duration of Power Failure	Recommended Minimum Auxiliary Supply Time	Type of Auxiliary Power System		System Justification
				Emergency	Standby	
	Paging systems	10 s	½ h	×	×	Locating of responsible persons concerned with power outage Providing evacuation instructions Prevention of panic
Signal circuits	Alarms and annunciation	1 to 10 s	To return of prime power	×	×	Prevention of loss from theft, arson, or riot Maintaining security systems Compliance with codes, standards, and laws Lower insurance rates Alarm for critical out-of-tolerance temperature, pressure, water level, and other hazardous or dangerous conditions Prevention of economic loss
	Land-based aircraft, railroad, and ship warning systems	1 s to 1 min	To return of prime power	×	×	Compliance with local, state, and federal codes, standards, and laws Prevention of personnel injury Prevention of property and economic loss

Figure 4.47.0 (*Continued*)

Typical Emergency and Standby Lighting Recommendations

Standby*	Immediate, Short-Term†	Immediate, Long-Term‡
Security lighting	Evacuation lighting	Hazardous areas
Outdoor perimeters	Exit signs	Laboratories
Closed circuit TV	Exit lights	Warning lights
Night lights	Stairwells	Storage areas
Guard stations	Open areas	Process areas
Entrance gates	Tunnels	
	Halls	Warning lights
Production lighting		Beacons
Machine areas	Miscellaneous	Hazardous areas
Raw materials storage	Standby generator areas	Traffic signals
Packaging	Hazardous machines	
Inspection		Health care facilities
Warehousing		Operating rooms
Offices		Delivery rooms
		Intensive care areas
Commercial lighting		Emergency treatment areas
Displays		
Product shelves		Miscellaneous
Sales counters		Switchgear rooms
Offices		Elevators
		Boiler rooms
Miscellaneous		Control rooms
Switchgear rooms		
Landscape lighting		
Boiler rooms		
Computer rooms		

* An example of a standby lighting system is an engine-driven generator.
† An example of an immediate short-term lighting system is the common unit battery equipment.
‡ An example of an immediate long-term lighting system is a central battery bank rated to handle the required lighting load only until a standby engine-driven generator is placed on-line.

Figure 4.47.1 Typical emergency and standby lighting recommendations. (*By permission: McGraw-Hill, Electrical Engineer's Portable Handbook, Bob Hickey.*)

Emergency and Standby Power Systems

Code Letters on AC Motors

NEMA Code Letter	SKVA per hp	Mid-Value
A	0.00- 3.14	1.57
B	3.15- 3.54	3.34
C	3.55- 3.99	3.77
D	4.00- 4.49	4.24
E	4.50- 4.99	4.74
F	5.00- 5.59	5.30
G	5.60- 6.29	5.94
H	6.30- 7.09	6.70
J	7.10- 7.99	7.54
K	8.00- 8.99	8.50
L	9.00- 9.99	9.50
M	10.00-11.19	10.60
N	11.20-12.49	11.84
P	12.50-13.99	13.24
R	14.00-15.99	15.00
S	16.00-17.99	17.00
T	18.00-19.99	19.00
U	20.00-22.39	21.20
V	22.40-	

Use 6.0 if code letter unknown

Wound Rotor Motor has no code letter

Figure 4.48.0 Code letter on emergency standby AC motors. (*By permission: McGraw-Hill, Electrical Engineer's Portable Handbook, Bob Hickey.*)

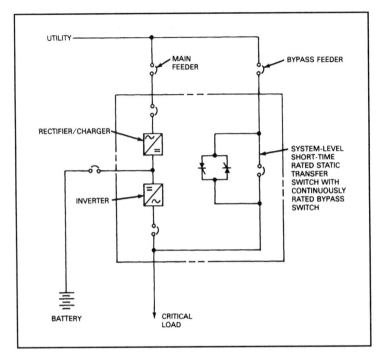

Figure 4.49.0 Uninterrupted power systems (UPS)—single module type. (*By permission: McGraw-Hill*, Electrical Engineer's Portable Handbook, *Bob Hickey.*)

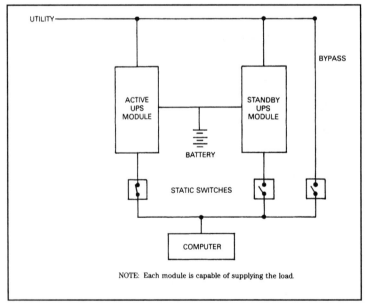

Figure 4.49.1 Dual redundant UPS system electrical layout. (*By permission: McGraw-Hill*, Electrical Engineer's Portable Handbook, *Bob Hickey.*)

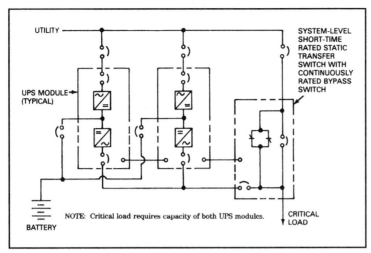

Figure 4.49.2 Parallel capacity UPS system. (*By permission: McGraw-Hill,* Electrical Engineer's Portable Handbook, *Bob Hickey.*)

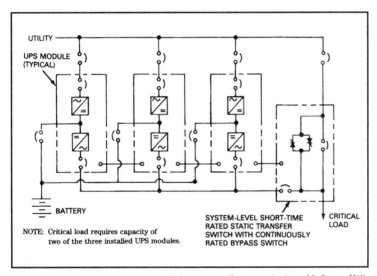

Figure 4.49.3 Parallel redundant UPS system. (*By permission: McGraw-Hill,* Electrical Engineer's Portable Handbook, *Bob Hickey.*)

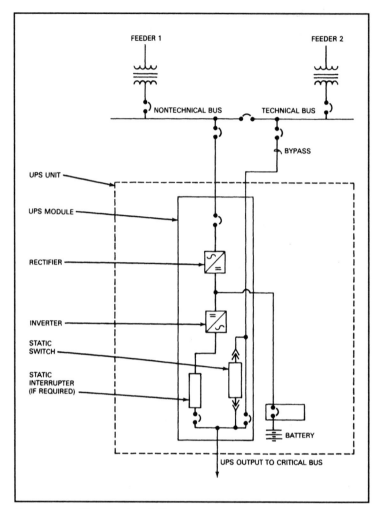

Figure 4.49.4   Nonredundant UPS system configuration. (*By permission: Mc-Graw-Hill,* Electrical Engineer's Portable Handbook, *Bob Hickey.*)

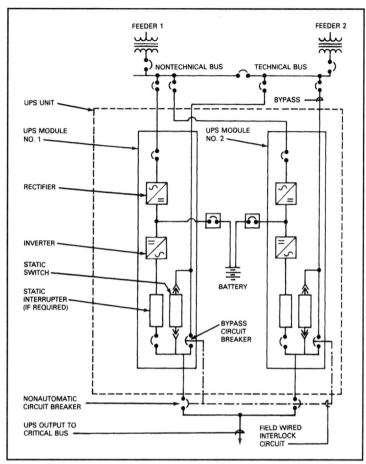

FEEDER 1

FEEDER 2

NONTECHNICAL BUS   TECHNICAL BUS

BYPASS

UPS UNIT

UPS MODULE
NO. 1

UPS MODULE
NO. 2

RECTIFIER

INVERTER

STATIC
SWITCH

STATIC
INTERRUPTER
(IF REQUIRED)

BATTERY

BYPASS
CIRCUIT
BREAKER

NONAUTOMATIC
CIRCUIT BREAKER

UPS OUTPUT TO
CRITICAL BUS

FIELD WIRED
INTERLOCK
CIRCUIT

Figure 4.49.5 "Cold" standby redundant UPS system. (*By permission: Mc-Graw-Hill*, Electrical Engineer's Portable Handbook, *Bob Hickey.*)

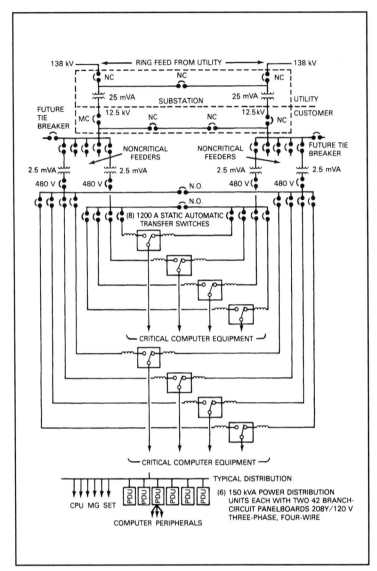

Figure 4.49.6 UPS with dual utility sources and static transfer switches. (*By permission: McGraw-Hill,* Electrical Engineer's Portable Handbook, *Bob Hickey.*)

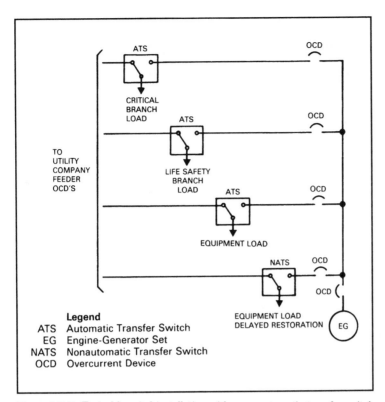

Legend

ATS   Automatic Transfer Switch
EG   Engine-Generator Set
NATS Nonautomatic Transfer Switch
OCD   Overcurrent Device

Figure 4.49.7 Typical hospital installation with a nonautomatic transfer switch and several automatic transfer switches. (*By permission: McGraw-Hill*, Electrical Engineer's Portable Handbook, *Bob Hickey.*)

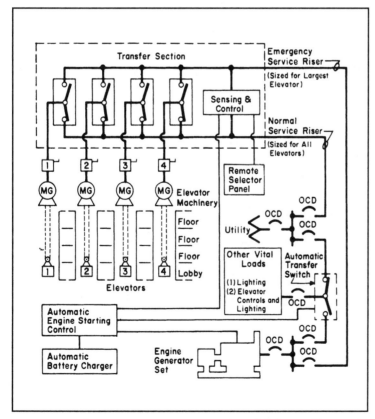

Figure 4.49.8 Elevator emergency power transfer switch layout. (*By permission: McGraw-Hill*, Electrical Engineer's Portable Handbook, *Bob Hickey.*)

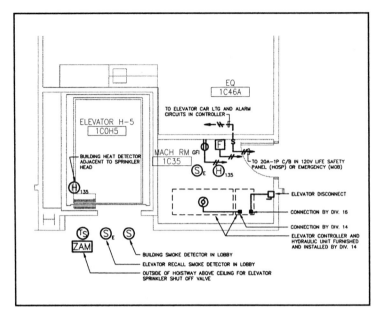

Figure 4.49.9   Typical elevator hoistway/machine room device installation. (*By permission: McGraw-Hill,* Electrical Engineer's Portable Handbook, *Bob Hickey.*)

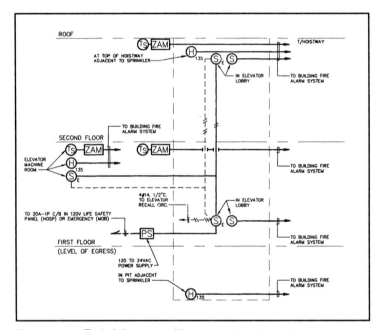

Figure 4.49.10   Typical elevator recall/emergency shutdown schematic. (*By permission: McGraw-Hill,* Electrical Engineer's Portable Handbook, *Bob Hickey.*)

Motor and Drive Basics

Motor Service Factors

Some motors can be specified with service factors other than 1.0. This means the motor can handle loads above the rated horsepower. A motor with a 1.15 service factor can handle a 15% overload, so a 10 horsepower motor can handle 11.5 HP of load. In general for good motor reliability, service factor should not be used for basic load calculations. By not loading the motor into the service factor under normal use the motor can better withstand adverse conditions that may occur such as higher than normal ambient temperatures or voltage fluctuations as well as the occasional overload.

Locked Rotor KVA/HP

Locked rotor kva per horsepower is a rating commonly specified on motor nameplates. The rating is shown as a code letter on the nameplate which represents various kva/hp ratings.

Code Letter	kva/hp	Code Letter	kva/hp
A	0 - 3.15	L	9.0 - 10.0
B	3.15 - 3.55	M	10.0 - 11.2
C	3.55 - 4.0	N	11.2 - 12.5
D	4.0 - 4.5	P	12.5 - 14.0
E	4.5 - 5.0	R	14.0 - 16.0
F	5.0 - 5.6	S	16.0 - 18.0
G	5.6 - 6.3	T	18.0 - 20.0
H	6.3 - 7.1	U	20.0 - 22.4
J	7.1 - 8.0	V	22.4 and up
K	8.0 - 9.0		

The nameplate code rating is a good indication of the starting current the motor will draw. A code letter at the beginning of the alphabet indicates a low starting current and a letter at the end of the alphabet indicates a high starting current. Starting current can be calculated using the following formula:

Starting current = (1000 x hp x kva/hp) / (1.73 x Volts)

Motor and Drive Basics

Definitions and Formulas

Alternating Current: electric current that alternates or reverses at a defined frequency, typically 60 cycles per second (Hertz) in the U.S.

Breakdown Torque: the maximum torque a motor will develop with rated voltage and frequency applied without an abrupt drop in speed.

Efficiency: a rating of how much input power an electric motor converts to actual work at the rotating shaft expressed in percent.

% efficiency = (power out / power in) x 100

Horsepower: a rate of doing work expressed in foot-pounds per minute.

HP = (RPM x torque) / 5252 lb-ft.

Locked Rotor Torque: the minimum torque that a motor will develop at rest for all angular positions of the rotor with rated voltage and frequency applied.

Rated Load Torque: the torque necessary to produce rated horsepower at rated-load speed.

Single Phase AC: typical household type electric power consisting of a single alternating current at 110-115 volts.

Slip: the difference between synchronous speed and actual motor speed. Usually expressed in percent slip.

$$\% \ slip = \frac{(synchronous \ speed - actual \ speed)}{synchronous \ speed} \times 100$$

Synchronous speed: the speed of the rotating magnetic field in an electric motor.

Synchronous Speed = (60 x 2f) / p
Where: f = frequency of the power supply
p = number of poles in the motor

Three Phase AC: typical industrial electric power consisting of 3 alternating currents of equal frequency differing in phase of 120 degrees from each other. Available in voltages ranging from 200 to 575 volts for typical industrial applications.

Torque: a measure of rotational force defined in foot-pounds or Newton-meters.

Torque = (HP x 5252 lb-ft.) / RPM

Figure 4.50.0 Motor and drive basics—definitions and formulas and motor service factors. (*By permission: Loren Cook Company, Springfield, MO.*)

Motor and Drive Basics

Allowable Ampacities of Not More Than Three Insulated Conductors

AWG kcmil	Temperature Rating of Aluminum or Copper-Clad Conductor		
	60°C (140°F) Types TW†, UF†	75°C (167°F) Types RH†, RHW†, THHW†, THW†, THWN†, XHHW†, USE†	90°C (194°F) Types TA, TBS, SA, SIS, THHN†, THHW†, THW-2, THWN-2, RHH†, RHW-S, USE-2, XHH, XHHW, XHHW-2, ZW-2
12	20†	20†	25†
10	25	30†	35†
8	30	40	45
6	40	50	60
4	55	65	75
3	65	75	85
2	75	90	100
1	85	100	115
1/0	100	120	135
2/0	115	135	150
3/0	130	155	175
4/0	150	180	205
250	170	205	230
300	190	230	255
350	210	250	280
400	225	270	305
500	260	310	350
600	285	340	385
700	310	375	420
750	320	385	435
800	330	395	450
900	355	425	480
1000	375	445	500
1250	405	485	545
1500	435	520	585
1750	455	545	615
2000	470	560	630

†Unless otherwise specifically permitted elsewhere in this Code, the overcurrent protection for conductor types marked with an obelisk (†) shall not exceed 15 amperes for No. 14, 20 amperes for No. 12, and 30 amperes for No. 10 copper, or 15 amperes for No. 12 and 25 amperes for No. 10 aluminum and copper-clad aluminum after any correction factors for ambient temperature and number of conductors have been applied.

Figure 4.50.1 Allowable ampacities with copper aluminum conductors. (*By permission: Loren Cook Company, Springfield, MO.*)

Motor and Drive Basics

Allowable Ampacities of Not More Than Three Insulated Conductors

Rated 0-2000 Volts, 60° to 90°C (140° to 194°F), in Raceway or Cable or Earth (directly buried). Based on ambient air temperature of 30°C (86°F).

AWG kcmil	Temperature Rating of Copper Conductor		
	60°C (140°F) Types TW†, UF†	75°C (167°F) Types FEPW†, RH†, RHW†, THHW†, THW†, THWN†, XHHW†, USE†, ZW†	90°C (194°F) Types TA, TBS, SA, SIS, FEP†, FEPB†, MI, RHH†, RHW-2, THHN†, THHW†, THW-2, USE-2, XHH, XHHW†, XHHW-2, ZW-2
18	—	—	14
16	—	—	18
14	20†	20†	25†
12	25†	25†	30†
10	30	35†	40†
8	40	50	55
6	55	65	75
4	70	85	95
3	85	100	110
2	95	115	130
1	110	130	150
1/0	125	150	170
2/0	145	175	195
3/0	165	200	225
4/0	195	230	260
250	215	255	290
300	240	285	320
350	260	310	350
400	280	335	380
500	320	380	430
600	355	420	475
700	385	460	520
750	400	475	535
800	410	490	555
900	435	520	585
1000	455	545	615
1250	495	590	665
1500	520	625	705
1750	545	650	735
2000	560	665	750

Figure 4.50.2 Allowable ampacities copper conductors only. (*By permission: Loren Cook Company, Springfield, MO.*)

Motor and Drive Basics

General Effect of Voltage and Frequency Variations on Induction Motor Characteristics

Characteristic	Voltage	
	110%	90%
Starting Torque	Up 21%	Down 19%
Maximum Torque	Up 21%	Down 19%
Percent Slip	Down 15-20%	Up 20-30%
Efficiency - Full Load	Down 0-3%	Down 0-2%
3/4 Load	0 - Down Slightly	Little Change
1/2 Load	Down 0-5%	Up 0-1%
Power Factor - Full Load	Down 5-15%	Up 1-7%
3/4 Load	Down 5-15%	Up 2-7%
1/2 Load	Down 10-20%	Up 3-10%
Full Load Current	Down Slightly to Up 5%	Up 5-10%
Starting Current	Up 10%	Down 10%
Full Load - Temperature Rise	Up 10%	Down 10-15%
Maximum Overload Capacity	Up 21%	Down 19%
Magnetic Noise	Up Slightly	Down Slightly

Characteristic	Frequency	
	105%	95%
Starting Torque	Down 10%	Up 11%
Maximum Torque	Down 10%	Up 11%
Percent Slip	Up 10-15%	Down 5-10%
Efficiency - Full Load	Up Slightly	Down Slightly
3/4 Load	Up Slightly	Down Slightly
1/2 Load	Up Slightly	Down Slightly
Power Factor - Full Load	Up Slightly	Down Slightly
3/4 Load	Up Slightly	Down Slightly
1/2 Load	Up Slightly	Down Slightly
Full Load Current	Down Slightly	Up Slightly
Starting Current	Down 5%	Up 5%
Full Load - Temperature Rise	Down Slightly	Up Slightly
Maximum Overload Capacity	Down Slightly	Up Slightly
Magnetic Noise	Down Slightly	Up Slightly

Figure 4.50.3 General effect of voltage and frequency variations on induction motors. (*By permission: Loren Cook Company, Springfield, MO.*)

Motor and Drive Basics

Motor Efficiency and EPAct

As previously defined, motor efficiency is a measure of how much input power a motor converts to torque and horsepower at the shaft. Efficiency is important to the operating cost of a motor and to overall energy use in our economy. It is estimated that over 60% of the electric power generated in the United States is used to power electric motors. On October 24, 1992, the U.S. Congress signed into law the Energy Policy Act (EPAct) that established mandated efficiency standards for general purpose, three-phase AC industrial motors from 1 to 200 horsepower. EPAct became effective on October 24, 1997.

Department of Energy
General Purpose Motors
Required Full-Load Nominal Efficiency
Under EPACT-92

Motor HP	Nominal Full-Load Efficiency					
	Open Motors			Enclosed Motors		
	6 Pole	4 Pole	2 Pole	6 Pole	4 Pole	2 Pole
1	80.0	82.5		80.0	82.5	75.5
1.5	84.0	84.0	82.5	85.5	84.0	82.5
2	85.5	84.0	84.0	86.5	84.0	84.0
3	86.5	86.5	84.0	87.5	87.5	85.5
5	87.5	87.5	85.5	87.5	87.5	87.5
7.5	88.5	88.5	87.5	89.5	89.5	88.5
10	90.2	89.5	88.5	89.5	89.5	89.5
15	90.2	91.0	89.5	90.2	91.0	90.2
20	91.0	91.0	90.2	90.2	91.0	90.2
25	91.7	91.7	91.0	91.7	92.4	91.0
30	92.4	92.4	91.0	91.7	92.4	91.0
40	93.0	93.0	91.7	93.0	93.0	91.7
50	93.0	93.0	92.4	93.0	93.0	92.4
60	93.6	93.6	93.0	93.6	93.6	93.0
75	93.6	94.1	93.0	93.6	94.1	93.0
100	94.1	94.1	93.0	94.1	94.5	93.6
125	94.1	94.5	93.6	94.1	94.5	94.5
150	94.5	95.0	93.6	95.0	95.0	94.5
200	94.5	95.0	94.5	95.0	95.0	95.0

Figure 4.50.4 Motor efficiency and EPA Act of 1992. (*By permission: Loren Cook Company, Springfield, MO.*)

Motor and Drive Basics

Types of Alternating Current Motors

Single Phase AC Motors

This type of motor is used in fan applications requiring less than one horsepower. There are four types of motors suitable for driving fans as shown in the chart below. All are single speed motors that can be made to operate at two or more speeds with internal or external modifications.

Single Phase AC Motors (60hz)

Motor Type	HP Range	Efficiency	Slip	Poles/RPM	Use
Shaded Pole	1/6 to 1/4 hp	low (30%)	high (14%)	4/1550 6/1050	small direct drive fans (low start torque)
Perm-split Cap.	Up to 1/3 hp	medium (50%)	medium (10%)	4/1625 6/1075	small direct drive fans (low start torque)
Split-phase	Up to 1/2 hp	medium-high (65%)	low (4%)	2/3450 4/1725 6/1140 8/850	small belt drive fans (good start torque)
Capacitor-start	1/2 to 34 hp	medium-high (65%)	low (4%)	2/3450 4/1725 6/1140 8/850	small belt drive fans (good start torque)

Three-phase AC Motors

The most common motor for fan applications is the three-phase squirrel cage induction motor. The squirrel-cage motor is a constant speed motor of simple construction that produces relatively high starting torque. The operation of a three-phase motor is simple: the three phase current produces a rotating magnetic field in the stator. This rotating magnetic field causes a magnetic field to be set up in the rotor. The attraction and repulsion of these two magnetic fields causes the rotor to turn.

Squirrel cage induction motors are wound for the following speeds:

Number of Poles	60 Hz Synchronous Speed	50 Hz Synchronous Speed
2	3600	3000
4	1800	1500
6	1200	1000
8	900	750

Types of Alternating Current Motors

Actual motor speed is somewhat less than synchronous speed due to slip. A motor with a slip of 5% or less is called a "normal slip" motor. A normal slip motor may be referred to as a constant speed motor because the speed changes very little with load variations. In specifying the speed of the motor on the nameplate most motor manufacturers will use the actual speed of the motor which will be less than the synchronous speed due to slip.

NEMA has established several different torque designs to cover various three-phase motor applications as shown in the chart.

NEMA Design	Starting Current	Locked Rotor	Breakdown Torque	% Slip
B	Medium	Medium Torque	High	Max. 5%
C	Medium	High Torque	Medium	Max. 5%
D	Medium	Extra-High Torque	Low	5% or more

NEMA Design	Applications
B	Normal starting torque for fans, blowers, rotary pumps, compressors, conveyors, machine tools. Constant load speed.
C	High inertia starts - large centrifugal blowers, fly wheels, and crusher drums. Loaded starts such as piston pumps, compressors, and conveyers. Constant load speed.
D	Very high inertia and loaded starts. Also considerable variation in load speed. Punch presses, shears and forming machine tools. Cranes, hoists, elevators, and oil well pumping jacks.

Motor Insulation Classes

Electric motor insulation classes are rated by their resistance to thermal degradation. The four basic insulation systems normally encountered are Class A, B, F, and H. Class A has a temperature rating of 105°C (221°F) and each step from A to B, B to F, and F to H involves a 25° C (77° F) jump. The insulation class in any motor must be able to withstand at least the maximum ambient temperature plus the temperature rise that occurs as a result of continuous full load operation.

Figure 4.50.5 Types of single-phase and three-phase AC motors. (*By permission: Loren Cook Company, Springfield, MO.*)

Motor and Drive Basics

Full Load Current†
Three Phase Motors

A-C Induction Type-Squirrel Cage and Wound Rotor Motors*

HP	115V	200V	230V	460V	575V	2300V	4000V
1/2	4	2.3	2	1	0.8		
3/4	5.6	3.2	2.8	1.4	1.1		
1	7.2	4.15	3.6	1.8	1.4		
1-1/2	10.4	6	5.2	2.6	2.1		
2	13.6	7.8	6.8	3.4	2.7		
3		11	9.6	4.8	3.9		
5		17.5	15.2	7.6	6.1		
7-1/2		25	22	11	9		
10		32	28	14	11		
15		48	42	21	17		
20		62	54	27	22		
25		78	68	34	27		
30		92	80	40	32		
40		120	104	52	41		
50		150	130	65	52		
60		177	154	77	62	15.4	8.8
75		221	192	96	77	19.2	11
100		285	248	124	99	24.8	14.3
125		358	312	156	125	31.2	18
150		415	360	180	144	36	20.7
200		550	480	240	192	48	27.6
Over 200 hp Approx. Amps/hp		2.75	2.4	1.2	0.96	.24	.14

† Branch-circuit conductors supplying a single motor shall have an ampacity not less than 125 percent of the motor full-load current rating.

*Based on Table 430-150 of the **National Electrical Code®**, 1993. For motors running at speeds usual for belted motors and with normal torque characteristics.*

* For conductor sizing only

Figure 4.50.6 Full load current on three-phase motors. *(By permission: Loren Cook Company, Springfield, MO.)*

Motor and Drive Basics

Full Load Current†
Single Phase Motors

HP	115V	200V	230V
1/6	4.4	2.5	2.2
1/4	5.8	3.3	2.9
1/3	7.2	4.1	3.6
1/2	9.8	5.6	4.9
3/4	13.8	7.9	6.9
1	16	9.2	8
1-1/2	20	11.5	10
2	24	13.8	12
3	34	19.6	17
5	56	32.2	28
7-1/2	80	46	40
10	100	57.5	50

† Based on Table 430-148 of the National Electric Code®, 1993. For motors running at usual speeds and motors with normal torque characteristics.

Figure 4.50.7 Full load current on single-phase motors. *(By permission: Loren Cook Company, Springfield, MO.)*

Motor and Drive Basics

Bearing Life

Bearing life is determined in accordance with methods prescribed in ISO 281/1-1989 or the Anti Friction Bearing Manufacturers Association (AFBMA) Standards 9 and 11, modified to follow the ISO standard. The life of a rolling element bearing is defined as the number of operating hours at a given load and speed the bearing is capable of enduring before the first signs of failure start to occur. Since seemingly identical bearings under identical operating conditions will fail at different times, life is specified in both hours and the statistical probability that a certain percentage of bearings can be expected to fail within that time period.

Example:

A manufacturer specifies that the bearings supplied in a particular fan have a minimum life of L-10 in excess of 40,000 hours at maximum cataloged operating speed. We can interpret this specification to mean that a minimum of 90% of the bearings in this application can be expected to have a life of at least 40,000 hours or longer. *To say it another way, we should expect less than 10% of the bearings in this application to fail within 40,000 hours.*

L-50 is the term given to Average Life and is simply equal to 5 times the Minimum Life. For example, the bearing specified above has a life of L-50 in excess of 200,000 hours. *At least 50% of the bearings in this application would be expected to have a life of 200,000 hours or longer.*

Figure 4.50.8 Bearing life criteria. *(By permission: Loren Cook Company, Springfield, MO.)*

Motor and Drive Basics

Belt Drives

Most fan drive systems are based on the standard "V" drive belt which is relatively efficient and readily available. The use of a belt drive allows fan RPM to be easily selected through a combination of AC motor RPM and drive pulley ratios.

In general select a sheave combination that will result in the correct drive ratio with the smallest sheave pitch diameters. Depending upon belt cross section, there may be some minimum pitch diameter considerations. Multiple belts and sheave grooves may be required to meet horsepower requirements.

$$Drive\ Ratio = \frac{Motor\ RPM}{desired\ fan\ RPM}$$

V-belt Length Formula

Once a sheave combination is selected we can calculate approximate belt length. Calculate the approximate V-belt length using the following formula:

$$L = 2C + 1.57\ (D+d) + \frac{(D-d)^2}{4C}$$

L = Pitch Length of Belt
C = Center Distance of Sheaves
D = Pitch Diameter of Large Sheave
d = Pitch Diameter of Small Sheave

Belt Drive Guidelines

1. Drives should always be installed with provision for center distance adjustment.
2. If possible centers should not exceed 3 times the sum of the sheave diameters nor be less than the diameter of the large sheave.
3. If possible the arc of contact of the belt on the smaller sheave should not be less than 120°.
4. Be sure that shafts are parallel and sheaves are in proper alignment. Check after first eight hours of operation.
5. Do not drive sheaves on or off shafts. Be sure shaft and keyway are smooth and that bore and key are of correct size.
6. Belts should never be forced or rolled over sheaves. More belts are broken from this cause than from actual failure in service.
7. In general, ideal belt tension is the lowest tension at which the belt will not slip under peak load conditions. Check belt tension frequently during the first 24-48 hours of operation.

Motor and Drive Basics

Estimated Belt Drive Loss†

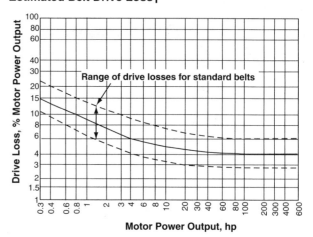

Higher belt speeds tend to have higher losses than lower belt speeds at the same horsepower.

Drive losses are based on the conventional V-belt which has been the "work horse" of the drive industry for several decades.

Example:
*   Motor power output is determined to be 13.3 hp.
*   The belts are the standard type and just warm to the touch immediately after shutdown.
*   From the chart above, the drive loss = 5.1%
*   Drive loss   = 0.051 x 13.3 = 0.7 hp
*   Fan power input   = 13.3 - 0.7 hp = 12.6 hp

Figure 4.50.9 Belt drives and estimated drive losses. (*By permission: Loren Cook Company, Springfield, MO.*)

4.51.0   Fiber Optics—The Latest Means of Transmitting Data Over Land Lines

Fiber cables are available in two designs—loose tube and tight tube. Loose tube fiber is used for long distances where low attenuation and wide variations in temperatures are to be expected. Several fibers are incorporated into a tube that is reinforced by Kevlar to reduce the cable's elongation and stress. Tight fiber is used in interior building applications where temperature variations are minimal. Bundles of fiber optic cables are helically wrapped with a special glass fiber yarn to insure its integrity against elongation as it is pulled through the conduit during installation.

Fiber Optic Types

This form of cabling is classified as either single mode or multimode. A singlemode fiber optic cable has a core as small as 9 microns surrounded by an outer casing or cladding of 125 micron diameter. The extremely small core directs the light to travel in a straight line. Multimode fiber cabling generally has a core of 62.5 microns, with the outer cladding ranging in thickness from 50 to 125 microns. This larger core permits light to travel in multiple paths and therefore serve multiple modes.

Important Characteristics of Fiber Optic Cables

The term *attenuation,* one of the cable characteristics, can be compared to the *resistance* found in conventional copper cabling and refers to the degree of signal loss or noise that occurs at a specific wavelength over a given distance. The attenuation loss of a single mode fiber has been calculated at 0.5 decibel/kilometer at a wavelength of 1,310 nanometers (nm). This attenuation loss is expressed as "db/km."

Fiber Optic Networking Protocols

At this time, there are seven networking protocols used with fiber optic cabling systems:

Ethernet

Ethernet nodes constantly send and receive information causing signal congestion that is counteracted by a collision detection system. Ethernet is generally used on light- to medium-traffic networks and performs best when a network transmits data in short bursts. Ethernet data transmisison operates at speeds of 10Mbs or 100 Mbs.

Token Ring

Developed by IBM, this system passes data from node to node in a circular fashion by using a "token" that pulls the data from one station or node to the other. Token ring data transmission rates are either 4 Mbs or 16 Mbs.

Escon

Also developed by IBM, this system is the successor to token ring, where there is direct communication between hubs and peripherals and transmission speeds approach 150 Mbs.

Fiber Data Distribution System

Fiber data distribution information (FDDI) is the first system designed strictly for fiber optic communications and it uses a ring topology with a back-up system that allows continued transmission even if a break in the main ring occurs. This system operates at 100 Mbs.

Asynchronous Transfer Method

Asynchronous transfer method (ATTM) of transmission can support voice and data services or digital video and can receive gigabits of data per second.

Synchronous Optical Network

Synchronous optical network (SONET) is a single-mode fiber network capable of tremendous variations in speed from 51 Mbs to 2.48 gigabits, and can also handle voice and data transmission.

Time Division Multiplexing

Time division multiplexing (TDM) combines information from various types into a single signal then sends it down the wire, where the signal is demultiplexed into its original components.

Common Fiber Optic Cable Products

Category 1 cable—This cable can support up to 1 Mbps and in its unshielded twisted pair (UTP) form it is suitable for voice and low speed applications.
Category 3 cable—This cable is offered as unshielded twisted pair (UTP) or shielded twisted pair (STP) and supports data rates up to 16 Mpbs. This cable is frequently used-in 10Base-T Ethernet networks.
Category 5 cable

UTP vs STP

Although it would appear that STP would offer better immunity against interference because it relies on the integrity of its shielding and its earthin, UTP has offered the industry greater reliability than STP wiring, and therefore is usually recommended by designers.

Schematic of an Optical Cable

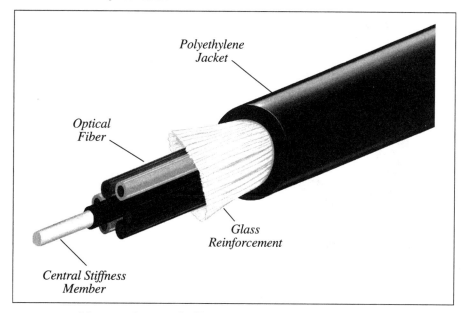

Figure 4.51.1   Schematic of an optical cable.

Schematic of an Optical Cable Jacketing Line

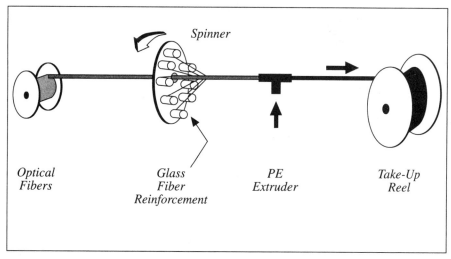

Figure 4.51.2   Schematic of jacketing optical cables.

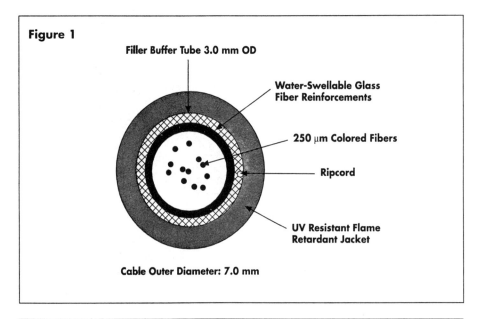

Figure 1

Filler Buffer Tube 3.0 mm OD

Water-Swellable Glass
Fiber Reinforcements

250 μm Colored Fibers

Ripcord

UV Resistant Flame
Retardant Jacket

Cable Outer Diameter: 7.0 mm

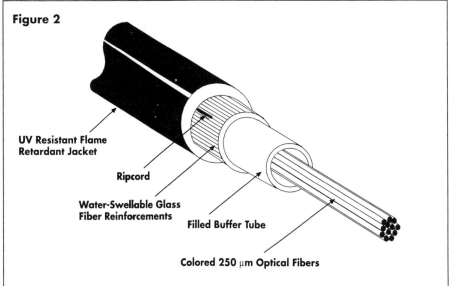

Figure 2

UV Resistant Flame
Retardant Jacket

Ripcord

Water-Swellable Glass
Fiber Reinforcements

Filled Buffer Tube

Colored 250 μm Optical Fibers

Figure 4.51.3 Sections through optical cable with jacket and fiber reinforcement.

Schematic of a Glass Impregnation Line

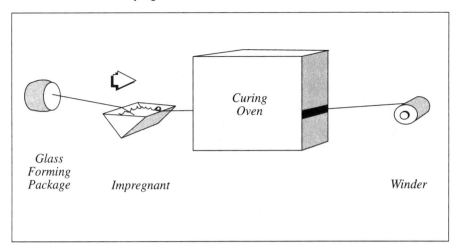

Curing Oven

Glass Forming Package

Impregnant

Winder

Figure 4.51.4   Schematic of a glass impregnation production line.

Cable Cross Section of a Conventional Single-Pair Telephone Drop Wire

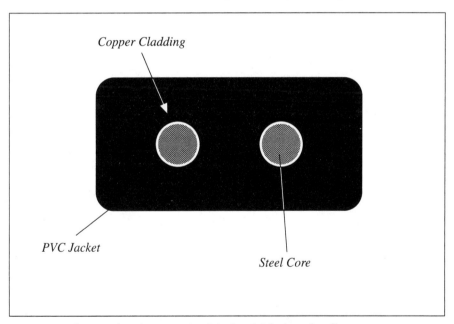

Copper Cladding

PVC Jacket

Steel Core

Figure 4.51.5   Cross section of a conventional single pair telephone drop line.

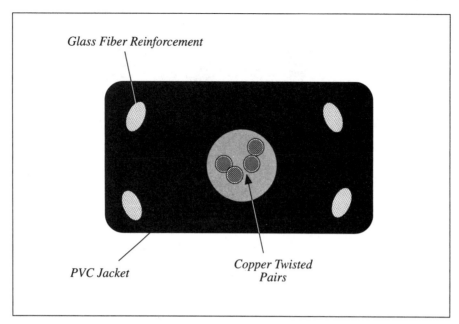

Figure 4.51.6 Cross section of a fiberglass reinforced telephone drop line.

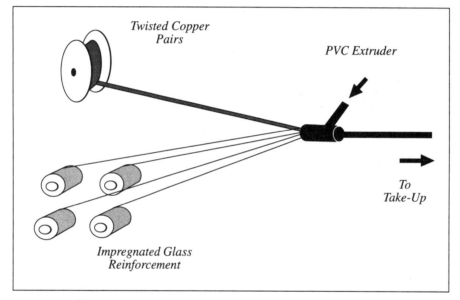

Figure 4.51.7 Schematic of drop wire extrusion production line.

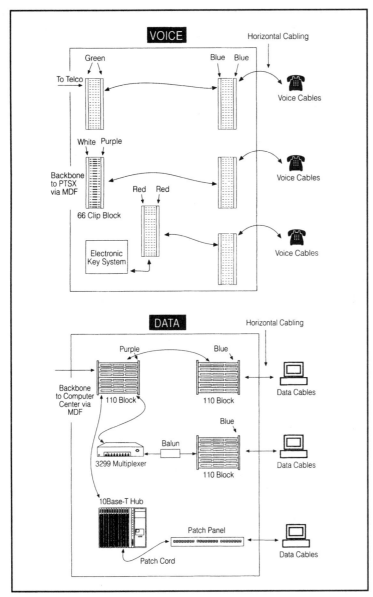

Figure 4.52.0 Voice and data backboard installation in a typical telecommunications closet. (*By permission: McGraw-Hill,* Electrical Engineer's Portable Handbook, *Bob Hickey.*)

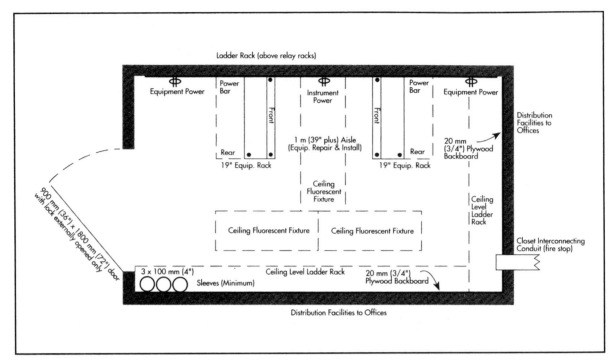

Figure 4.52.1 Another typical telecommunications closet layout. (*By permission: McGraw-Hill,* Electrical Engineer's Portable Handbook, *Bob Hickey.*)

Separation of Telecommunications Pathways from 480-Volt or Less Power Lines

Condition	Minimum Separation Distance		
	< 2 kVA	2-5 kVA	> 5 kVA
Unshielded power lines or electrical equipment in proximity to open or nonmetal pathways.	127 mm (5 in)	305 mm (12 in)	610 mm (24 in)
Unshielded power lines or electrical equipment in proximity to a grounded metal conduit pathway	64 mm (2.5 in)	152 mm (6 in)	305 mm (12 in)
Power lines enclosed in a grounded metal conduit (or equivalent shielding) in proximity to a grounded metal conduit pathway.	- -	76 mm (3 in)	152 mm (6 in)

Figure 4.53.0 Separation of telecommunications from 480 volt power lines. (*By permission: McGraw-Hill,* Electrical Engineer's Portable Handbook, *Bob Hickey.*)

Controls for HVAC Systems

Contents

5.0.0   HVAC Controls and Their Components

We are all familiar with HVAC control systems. Just look at the homes we live in—from the simple 220-volt electric baseboard heat or hot air furnace wired directly to a wall thermostat, or the "two-wire" millivolt system controlling a gas-fired wall or floor heater or the heat pump running a compressor forward to create cooling and running backward to create heat, connected to a thermostat via a five-conductor cable to control the fan, heat, and cool cycles.

Although industrial and commercial HVAC installations are more complex than these, control systems all operate on the same basic principle, requiring a control loop with a sensor, a controller, and a controlled device, the object of all three components being to maintain a heating or cooling system at some preset value. This preset value being known as a *set point*.

5.0.1   Control Loops Have Four Components

1. A controller (a thermostat)

   Controllers receive information form the sensors, analyze this information, and compare the information with the set point and send instructions on to the control device—a damper or a valve.

   The control device's function is to respond to the controller and position or reposition a heating or cooling system component device according to the requirement of the set point. These control devices are mechanical in nature and regulate the flow of steam, air, or liquids in an HVAC system, allowing that system to respond to the needs of the set point. Control devices, valves, and dampers include an actuator, usually an integral part of the controller, or at least a bolt-on accessory, that provides an energy source to change the position of the control device.

2. A controlled medium (steam or hot or cold air or water)

3. A sensing device, such as a pressure or temperature transmitter, or even a paddle inserted in a liquid stream.

These sensors measure temperature, humidity, pressure, or flow and can take many forms, from sophisticated, solid-state assemblies to rather straightforward mechanical or electromechanical devices (Fig. 5.0.3).

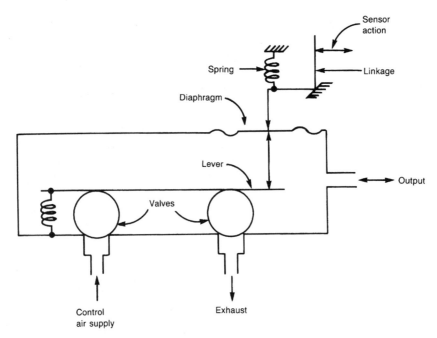

Figure 5.0.3.1   A pneumatic device whereby downward movement on the lever allows the air supply valve to open, increasing air pressure in the chamber. The flexible diaphram pushes up agaisnt the sensor until the air supply valve closes at some balance point. (*By permission: McGraw-Hill,* HVAC Systems Design Handbook, *R. W. Haines and C. L. Wilson.*)

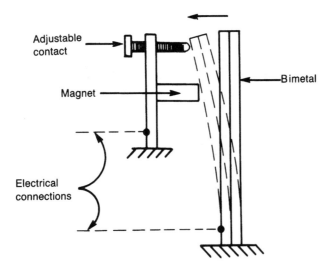

Figure 5.0.3.2   A bimetallic sensor used as a simple controller. The set point is created by adjusting the threaded contact, which regulates the travel of the bimetal sensor. When contact is made with the magnet, a make-or-break action is obtained. (*By permission: McGraw-Hill,* HVAC Systems Design Handbook, *R. W. Haines and C. L. Wilson.*)

Movement when heated

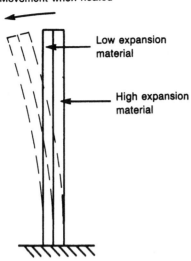

Low expansion
material

High expansion
material

Figure 5.0.3.3 The bimetal temperature sensor is constructed of two strips of metal, each with a different coefficient of expansion. When temperatures change, one metal expands or contracts more than the other, creating a bending action and generating a signal. (*By permission: McGraw-Hill,* HVAC Systems Design Handbook, *R. W. Haines and C. L. Wilson.*)

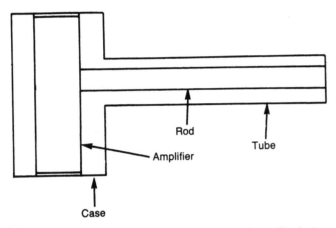

Rod

Tube

Amplifier

Case

Figure 5.0.3.4 The rod and tube sensor is another bimetallic device. The rod is made of one type metal and the tube is made of another with a different coefficient of expansion. (*By permission: McGraw-Hill,* HVAC Systems Design Handbook, *R. W. Haines and C. L. Wilson.*)

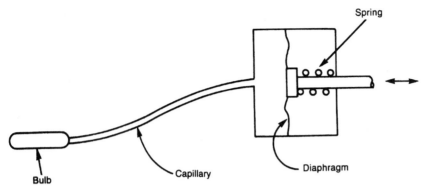

Figure 5.0.3.5 The bulb and capillary temperature sensor contains a fluid inside the bulb and capillary. When a portion of the bulb is exposed to freezing temperatures, the refrigerant in the bulb condenses, creating a drop in sensor pressure that, in turn, can open a two-way switch to stop a fan, for example. (*By permission: McGraw-Hill,* HVAC Systems Design Handbook, *R. W. Haines and C. L. Wilson.*)

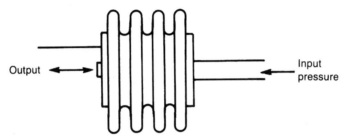

Figure 5.0.3.6 Another pressure sensitive sensor, the bellows pressure sensor, is vapor filled and reacts to changes in temperature by expanding or contracting, signalling a switch to open or close. (*By permission: McGraw-Hill,* HVAC Systems Design Handbook, *R. W. Haines and C. L. Wilson.*)

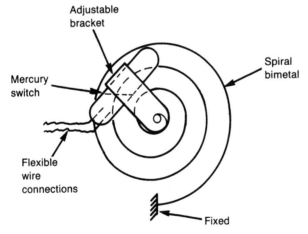

Figure 5.0.3.7 A bimetallic sensor that on expanding or contracting, triggers a mercury switch. (*By permission: McGraw-Hill,* HVAC Systems Design Handbook, *R. W. Haines and C. L. Wilson.*)

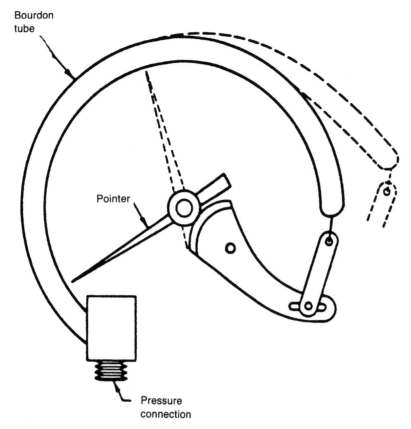

Bourdon
tube

Pointer

Pressure
connection

Figure 5.0.3.8 A Bourdon tube sensor is a sealed tube that attempts to straighten itself as pressure increases. Used most frequently in pressure gauges rather than control systems. (*By permission: McGraw-Hill*, HVAC Systems Design Handbook, *R. W. Haines and C. L. Wilson.*)

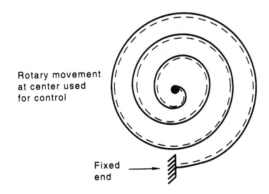

Rotary movement
at center used
for control

Fixed
end

Figure 5.0.3.9 The spiral bimetal temperature sensor is another bimetallic sensor, but this time in the form of a spiral. (*By permission: McGraw-Hill*, HVAC Systems Design Handbook, *R. W. Haines and C. L. Wilson.*)

5.0.2   Types of Control Systems

Self-Contained

A pressure relief valve is an example of a self-contained control system. When fitted to a steam boiler, for example, to control pressure, once the preset pressure set point has been reached, any further build up of steam within the boiler is controlled by venting excess pressure to the atmosphere.

Pneumatic Control System

Air pressure, usually provided by an air compressor, delivers constant-pressure air to a controller that regulates the amount of air pressure to be transmitted to the controlled device. This system usually provides for modulation of a diaphragm- or piston-operated control device.

Electromechanical

Electrical current, either low voltage or line current, is used to power the controller in a two-position mode—on or off, or, in the case of a variable speed motor, regulate the speed of the motor in order to comply with the demand of the controller.

Electronic

Solid-state components in electronic circuits generate signals in response to sensor input and direct the action of the controlled device.

Direct Digital Control Systems

A system in which all control logic is performed by computers that detect, amplify, and analyze sensor information based on their software program and then send signals to the HVAC system. The digital signals, generated by the direct digital control (DDC) system via a transducer, are converted into either pneumatic or electrical impulses that operate the controlled device.

5.0.4   Manual, Semiautomated, and Automated Systems

Manual Control Systems

These systems rely on an operator to collect data, evaluate it, and make adjustments as required. For example, a wall-mounted thermostat in a private office that allows the occupant to determine whether they wish to have more or less heat, or more or less cooling, and by adjusting the thermostat achieve their own comfort level, is a rather simple example of a manual control. In years past, when coal-fired steam boilers provided heat throughout a large building, the plant "engineer" would monitor the boiler gauges and control the flow and pressure of steam from the central plant to the terminal devices, adding more coal to the furnace as required to maintain desired pressure and temperature levels.

Semiautomatic Controls

An example of a semiautomatic control system, a combination of manual methods and automated controls, would be a system in which a facility manager monitoring a chiller's efficiency via a computer could manually adjust the flow through the chiller by altering valve openings or pump flow or pressure to alter operating conditions.

Automated Controls

Introduced in the construction industry in the mid-1980s, the fully automated control systems encompassing DDC has revolutionized the industry. These fully automated systems use preprogrammed procedures to collect, analyze, and respond to set point requirements without being prompted by a human observer. Components within these systems are often referred to as *automated diagnosticians*. The DDC system has the ability to accurately and precisely control the temperature of air and water flows that previously had been controlled by pneumatic and electric control systems.

5.0.5   Direct Digital Control (DDC) Systems

DDC systems provide savings to owners in a number of ways:

- The accuracy and precision of the DDC system eliminates the old problems of overshooting or hunting that prevail in pneumatic systems.
- The ability to respond to an almost unlimited range of sensor signals and system demands.
- By proving the inherent stability of solid-state components is superior to mechanical components.
- By creating relatively simple or even automatic recalibration to ensure accuracy across the entire system.
- Because the DDC system is software based, changes or modifications are often possible without major hardware changes or modifications.
- Finally, because the monitoring of DDC systems can be performed remotely within the structure where it has been installed or even in another part of the country, it is ideal for centralized management purposes.

Type	Type of Control Systems	
	Power Source	Signal Output
Self-contained	Vapor	Expansion as a result of pressure
	Liquid filled	
Pneumatic	15 psi (104 kPa)	0–15 psi (0–104 kPa)
	20 psi (136 kPa)	0–20 psi (0–136 kPa)
Electric	24 VAC	0–24 VAC
	120 VAC	0–120 VAC
	220 VAC	0–220 VAC
Electronic	5, 12, or 15 VDC	0–5 VDC
	12 VAC	0–12 VAC
	24 VAC	0–24 VAC
Digital	24 VAC	0–20 milliamps
	120 VAC	0–10 VDC
	220 VAC	Direct digital

5.0.6   Hydronic Systems Controls

A hydronic HVAC system is composed of a primary heating/cooling source, such as a boiler (heat) and chiller (cold), pumps to transfer the heated/cooled liquid through a piping and valve system, and terminal devices to disperse the heat or cooling throughout the structure as required.

In these types of systems, the pump(s) takes on a critical role as it is called on to regulate both flow and pressure. Pump curves are used to graphically display the relationship between flow and pressure. Pump curves are used to graphically display the relationship between flow and pressure and control of both functions is necessary to maintain an efficient system (Fig. 5.0.7).

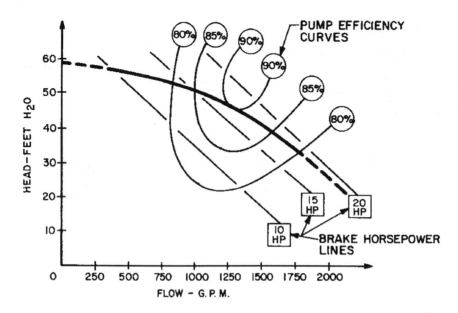

CENTRIFUGAL PUMP CURVE

Figure 5.0.7   Pump efficiency curves showing the relationship between flow and pressure. (*By permission: Johnson Controls, Milwaukee, WI.*)

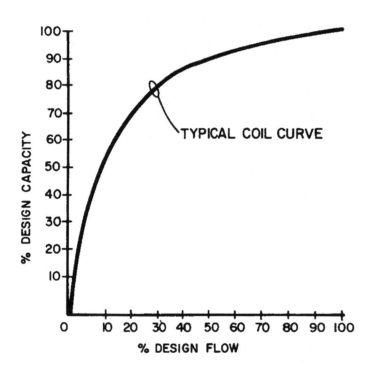

TYPICAL COIL PERFORMANCE

Figure 5.0.10   A typical coil performance curve reveals how significantly coil capacity increases as flow increases. (*By permission: Johnson Controls, Milwaukee, WI.*)

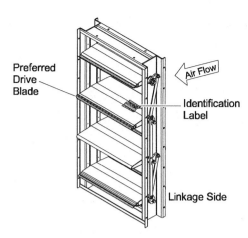

Preferred Drive Blade

Air Flow

Identification Label

Linkage Side

Drive Blade Identification

Figure 5.0.11 Dampers that regulate the volume of air allowed to pass through. (*By permission: Johnson Controls, Milwaukee, WI.*)

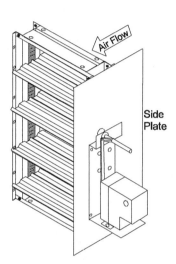

Air Flow

Side Plate

Figure 5.0.12 A smoke damper with an electric actuator attachment. (*By permission: Johnson Controls, Milwaukee, WI.*)

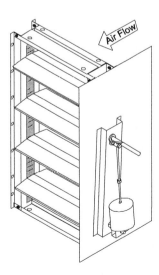

Air Flow

Figure 5.0.13 A smoke damper with a pneumatic actuator attached. (*By permission: Johnson Controls, Milwaukee, WI.*)

The wide use of variable-frequency drive motors permits increased flexibility relating to regulating pump(s) flow and pressure.

Pumps play an important role in the operation of chillers because most chiller manufacturers establish minimum and maximum water velocity limits for their evaporator tube bundles. Accepted good practice dictates that chilled water be maintained at a constant flow rate to reduce problems arising out of control instability, due primarily to flow switches.

It is recommended that chilled water velocity should not drop below 4 fps; otherwise, the potential for freeze-up increases dramatically; nor should chilled water flow exceed 10 fps, because premature tube failure may occur due to water side erosion.

Resistance to flow created by pipe and fittings, referred to as the *pipe–friction factor*, is an important factor considered by systems designers when calculating flow through coils as the design of a hydronics control system is being proposed. Because water flow rate affect

- Dampers that are fire-rated and whose purpose it is to restrict the passage of flame through a building when installed in a fire-rated partition or floor or ceiling. This type of damper closes completely in order to maintain the integrity of the fire separation within the sector of the building in which it has been installed (Fig. 5.0.16).

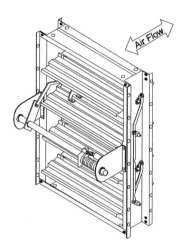

Figure 5.0.16 Fire damper. (*By permission: Johnson Controls, Milwaukee, WI.*)

5.0.14 Four Damper Functions

- Barometric dampers open and close to maintain a predetermined differential pressure across their face. A typical example of this form of damper is the type used to insure proper air flow in boiler flue exhaust.

- Fire/smoke dampers as discussed previously. Fire dampers are generally activated when a fusible link in its frame melts, allowing a high-torque spring to close the damper blades. Smoke dampers can be directed to close on receiving a signal from a smoke detection device, which may also signal the building's air handling unit to shut down.

- Round dampers are used to balance the air flow in the duct system (Fig. 5.0.17).

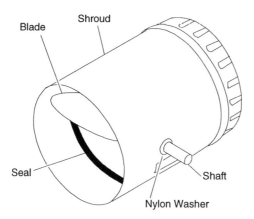

The round damper consists of five major components:

- Shroud--provides structural support

- Blade--provides the proper resistance to airflow

- Shaft--drives the blade to position

- Washer--provides support for blade movement on diameters larger than 8 inches

- Seals--prevents airflow when blade is closed

After rolling and forming, the shroud is mechanically joined for strength.

Figure 5.0.17 A round damper and its components. (*By permission: Johnson Controls, Milwaukee, WI.*)

- Rectangular dampers are most often used for proportional control applications and consist of parallel blade dampers (Fig. 5.0.18) or opposed blade dampers (Fig. 5.0.19). Characteristics of opposed blade dampers differ from parallel blade linkage as depicted in the typical inherent flow curves of opposed and parallel blades displayed in Fig. 5.0.20 prepared by Johnson Controls.

Figure 5.0.18 Schematic of operation of parallel-blade dampers. (*By permission: McGraw-Hill,* HVAC Systems Design Handbook, *R. W. Haines and C. L. Wilson.*)

Figure 5.0.19 Schematic of operation of opposed-blade dampers. (*By permission: McGraw-Hill—* HVAC Systems Design Handbook, *R. W. Haines and C. L. Wilson.*)

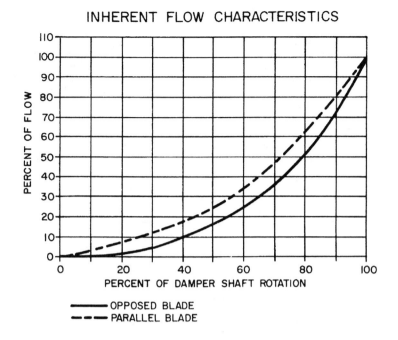

The inherent flow characteristic curves shown in Figure 1 are valid only when the damper is operating with a constant differential pressure difference across it. In a real fan/duct system; the ductwork, balancing dampers, filters, coils, and fittings will also have a pressure drop across them. As the damper throttles, the flow rate and pressure drop through the other components of duct system will be reduced. The differential pressure across the damper will increase by the same amount as the differential pressure across the other components of duct system decreased. Thus a trade-off occurs at the damper.

Figure 5.0.20 Characteristics of opposed-blade and parallel-blade dampers. (*By permission: Johnson Controls, Milwaukee, WI.*)

5.0.15 Five Basic Damper Applications

- Two position: This type damper is either fully open or fully closed. Dampers of this type are generally designed to provide the lowest wide-open pressure drop possible.

- Static pressure control: A static control damper is one that is modulated to maintain a static pressure set point at some downstream location in the ductwork system. With the widespread use of variable speed drives, the use of static pressure control dampers has been greatly diminished. The pressure drop across the damper compensates for the increase in pressure developed by the fan and the reduction in pressure drop across the ductwork as the air flow rate is decreased. Figure 5.0.21 is a graphic representation of the change in pressure drop as the system flow rate is decreased.

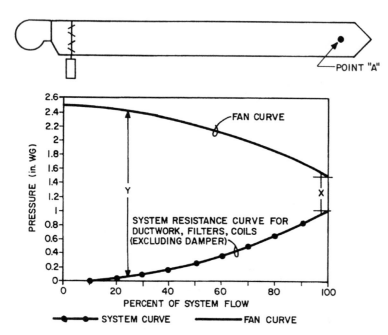

Figure 5.0.21 Schematic of pressure drop across the damper. (*By permission: Johnson Controls, Milwaukee, WI.*)

- Temperature control: The most common application of a temperature control damper is when it is used to regulate the amount of conditioned supply air into an area. Temperature control dampers often operate in pairs working together to mix air streams and maintain a temperature set point (Fig. 5.0.22).

- Face and bypass: Face and bypass dampers use a parallel-blade configuration (Fig. 5.0.18) to provide better downstream air mixing. These dampers are sized so that the combined flow rate through them is relatively constant. To achieve this relatively constant flow rate, the same full flow rate through the coil face and bypass dampers is necessary. A typical installation of a face and bypass damper is shown in Fig. 5.0.23.

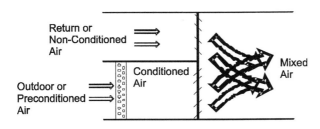

Temperature Control Application

Temperature control dampers operate as pairs, working together to maintain a predetermined temperature set point. The dampers simultaneously regulate two different airstreams, each having a different temperature.

Figure 5.0.22 Schematic of operation of temperature control dampers. (*By permission: Johnson Controls, Milwaukee, WI.*)

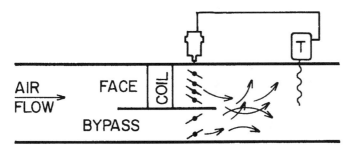

Figure 5.0.23 Schematic of typical face and bypass damper. (*By permission: Johnson Controls, Milwaukee, WI.*)

- Mixed air control: Oftentimes, dampers are used to regulate the flow rate of two airstreams so that the temperature of this mixed airstream can be controlled. An economizer cycle is one example of the use of mixed air control dampers Mixed air dampers must be matched so that the combined flow is relatively constant, regardless of their position. Figure 5.0.24 shows the combined flow characteristics of a well-matched set of dampers in which percentage of flow remains nearly constant. Figure 5.0.25 illustrates the combined flow characteristics of a poorly matched set of dampers and the wide fluctuations in the percentage of air flow.

In order for a damper to be controlled properly, it must first be constructed properly. The linkage and the linkage operator must be built to withstand the numerous cycles and the rigors for which it has been designed.

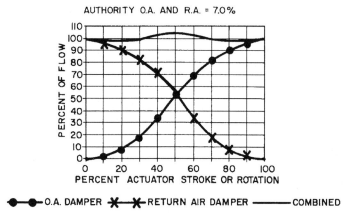

Figure 5.0.24 Combined flow characteristics of a well-matched set of dampers. (*By permission: Johnson Controls, Milwaukee, WI.*)

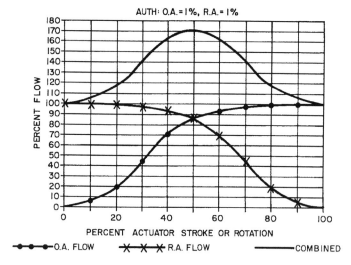

Figure 5.0.25 Combined flow characteristics of a poorly matched set of dampers. (*By permission: Johnson Controls, Milwaukee, WI.*)

5.0.26 Damper Construction

A strong damper frame provides the rigid framework for the efficient and multicycled operation of the damper blades. Rectangular frame components are shown in Fig. 5.0.27. The Johnson Control damper configuration as reflected in Fig. 5.0.28 is available in various sizes, damper heights, blade and seal configurations, and materials of construction, similar to other makers of quality dampers.

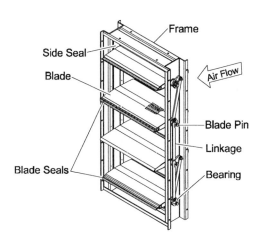

Figure 5.0.27 Components of a rectangular damper. (*By permission: Johnson Controls, Milwaukee, WI.*)

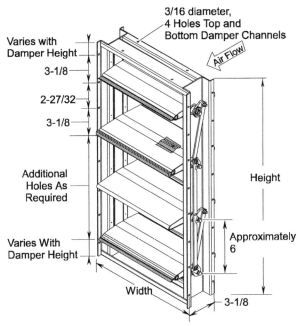

Figure 5.0.28 A Johnson Controls damper configuration with variable sized component availability. (*By permission: Johnson Controls, Milwaukee, WI.*)

Damper frame construction should meet the following criteria:

- Minimize vibration and noise when exposed to high pressures and airflows
- Present less potential for damage during shipment or rough handling during storage and installation.
- Ability to remain square, assuring smooth operation of the blades and linkage.
- Allow for "twinning" of multiple sections in the field when these configurations are required, ensuring that the multisection unit remains true to individual component shape and overall shape of the assembly (Fig. 5.0.29).
- Blade stops at top and bottom that provide proper sealing and retain long life.

Damper blades are another key component of the damper assembly. They must be capable of withstanding a wide range of static pressures and air velocities. They must ensure low leakage over an extended period of time and perform with minimal noise generation.

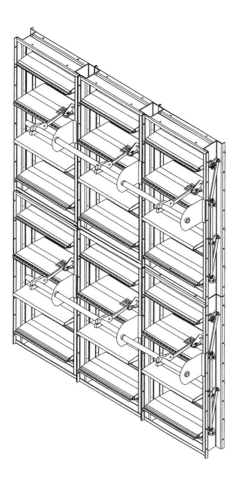

Figure 5.0.29 Assembly of individual dampers to create a multiple section configuration. (*By permission: Johnson Controls, Milwaukee, WI.*)

Various types of blade profiles are available: 16-gauge, double-piece, and airfoil blade configurations are shown in Fig. 5.0.30. The leakage resistance, Class I, II, and III blades, measured by static pressure, are set forth in Fig. 5.0.31.

Blades

Damper blades must withstand different levels of static pressure and velocities; some must provide low leakage rates, withstand high static pressures, or perform with minimal noise generation. The Johnson Controls family of dampers allows you to achieve these results by selecting 16-gauge, double-piece, or airfoil blades, as shown in Figures 4, 5, and 6.

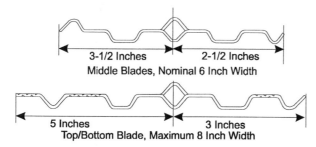

3-1/2 Inches   2-1/2 Inches
Middle Blades, Nominal 6 Inch Width

5 Inches   3 Inches
Top/Bottom Blade, Maximum 8 Inch Width

16-Gauge Blade Profile

- 16-gauge blades are made of 16-gauge rolled sheet metal.

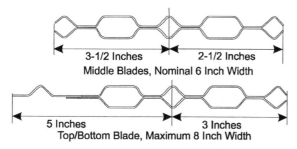

3-1/2 Inches   2-1/2 Inches
Middle Blades, Nominal 6 Inch Width

5 Inches   3 Inches
Top/Bottom Blade, Maximum 8 Inch Width

Double-Piece Blade Profile

- Double-piece blades are made from two layers of 22-gauge rolled sheet metal, mechanically joined.

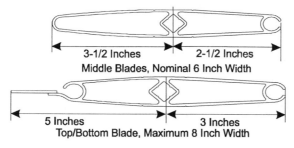

3-1/2 Inches   2-1/2 Inches
Middle Blades, Nominal 6 Inch Width

5 Inches   3 Inches
Top/Bottom Blade, Maximum 8 Inch Width

Airfoil Blade Profile

Figure 5.0.30 Profiles of 16-gauge, double-piece, and airfoil blades for various operations. (*By permission: Johnson Controls, Milwaukee, WI.*)

Leakage Resistance Classes

Class	Static Pressure (inches water)			
	1	4	8	12
I	4	8	11	14
II	10	20	28	35
III	40	80	112	140

Leakage in cfm/sq ft

Figure 5.0.31 Leakage resistance classifications of damper blades. (*By permission: Johnson Controls, Milwaukee, WI.*)

Damper blades can operate as either opposed or parallel (Fig. 5.0.32). In parallel operation, all blades rotate in the same direction; in an opposed operation, one blade will operate in a clockwise position while the other operates in a counterclockwise direction. Linkage is another important component when dampers are to be fitted with actuators and used in automatic operations. Linkage is meant to connect the drive blade to the other blades assuring simultaneous rotation of all blades.

Linkage must be rugged and self-lubricating, with either acetal or bronze bearings tested for at least 100,000 cycles. The linkage located in a side channel of the damper frame reduces noise, friction, and pressure drop. Damper manufacturers should be able to provide data substantiating a proven record of low linkage maintenance.

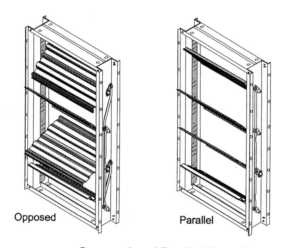

Opposed   Parallel

Opposed and Parallel Operation

The blades can operate parallel or opposed. In parallel operation, all the blades rotate in the same direction as shown in Figure 9 (right). In opposed operation, one blade rotates clockwise while the adjacent blade operates counterclockwise as shown in Figure 9 (left).

The flow characteristics for the parallel and opposed operation differ significantly, but the pressure drop across the parallel and opposed dampers is identical.

Figure 5.0.32 Damper assemblies containing opposed- and parallel-blade components. (*By permission: Johnson Controls, Milwaukee, WI.*)

5.0.33 Selection of the Proper Control Damper

Because the type, size, and function of control dampers can vary with each design application, selection must consider the following criteria:

- What is the anticipated maximum velocity in the duct?
- What is the anticipated maximum static pressure?
- What leakage rates will be acceptable?
- What are the anticipated minimum and maximum temperature ranges in the duct?
- What control application is going to be used?

5.0.34 Control Valves

Control valves are an integral part of any hydronic system and play an equally important role in air distribution heating and cooling systems. They regulate flow, temperature, and pressure from condensing units, chillers, hot water, and steam heating sources. Without properly sized valves and the proper selection of valve types and compatible components, a system may have difficulty in maintaining an efficient operating level.

5.0.35 Valve Types and Components Used in HVAC Control Systems

Plug valves have various plug configurations, and the size and configuration of the plug body must also take into consideration the type of plug required to fit its specific application. Plug valve configurations can vary depending on the proposed use of the valve. Plug shapes for quick opening, linear, and equal performance flow are shown in Fig. 5.0.36. The percentage of flow rate versus percentage of valve stroke for quick opening, linear, and equal performance plugs is displayed in Figure 5.0.37.

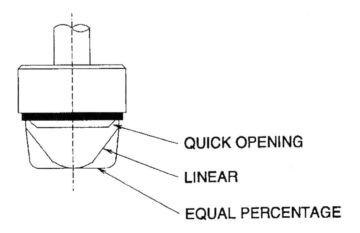

The most common types of plugs are the equal percentage, linear, and quick opening plug. Typically JCI offers the equal percentage characteristic.

Figure 5.0.36 Variation in plug types for quick opening, linear and equal percentage plug valves. (*By permission: Johnson Controls, Milwaukee, WI.*)

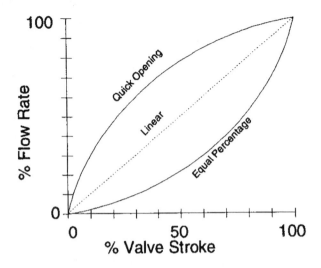

Typical Valve Inherent Flow Characteristics

Figure 5.0.37 Percentage of flow rate versus percentage of valve stroke for quick opening, linear, and equal percentage valves. (*By permission: Johnson Controls, Milwaukee, WI.*)

The plug in this type of valve is configured to provide one of three types of flow:

* Quick opening: These types of plug valves are generally used when a "full open" or "full closed" position is required. They produce a large increase in flow for a small change in stem rotation.

* Linear: Flow through the linear type plug varies directly in proportion to the position of the valve stem. This type of plug valve is often used in process control systems in which it is necessary to control the flow of liquids or gases into or out of a process.

* Equal percentage: Flow through a plug valve with an equal percentage plug allows flow to change in equal increments with respect to valve stem rotation travel, thereby maintaining a constant percentage of the total flow as stem changes take place. Figure 5.0.39 shows the relationship between

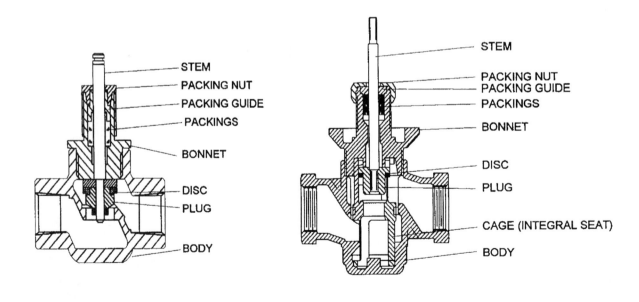

N.O. VT Valve Body   **N.O. Cage Trim Valve Body**

Figure 5.0.38   Cut-away sections of two types of Globe valves. (*By permission: Johnson Controls, Milwaukee, WI.*)

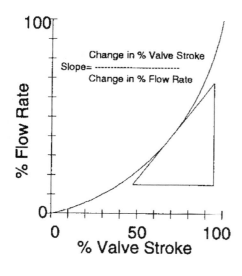

Figure 5.0.39 Relationship between flow rate and valve stroke for equal percentage valves. (*By permission: Johnson Controls, Milwaukee, WI.*)

Equal Percentage Valve Gain

flow rate and valve stroke in the equal percentage valve gain. These valve plugs are often used to control coil flow and hydronic system heat exchangers. Equal percentage flow characteristic curves apply only when the control valve is operating at a constant pressure drop. The amount of actual deviation from the equal percentage curve is determined by a property known as *valve authority,* which is defined as the ratio of pressure drop through the control valve to the total system pressure drop at design flow.

Valve authority is to be taken into consideration by the systems designer in the sizing of the valves and whether single or multiple valves are to be used. According to Johnson Controls, the pressure drop across each branch and valve increases at the same ratio, therefore the valve authority is dependent on the initial sizing of the control valve and does not change when pressures shifts occur in the system.

5.0.35.1 Valve Components

The plug valve can be fitted to bodies of varying configurations (Fig. 5.0.36) and a stem to change the position of the plug. The shape of the plug determines the flow characteristics of the valve. Stem packing prevents leakage from the valve through its valve guide and packing nut.

Globe valves (Fig. 5.0.38) have four basic components, as follows:

1. The body containing the orifice and that main housing that direct flow through the valve

2. The trim—comprising the "guts" of the valve—the valve seat, plug, disc, disc holder and stem.

3. The bonnet—a guide through which the stem will protrude containing a centerpiece, packing, packing guide and a packing nut (See Figure 5.0.40). The packing provides the seal to prevent leakage from the bonnet and is generally made of EPDM (Ethylene Propylene Diene Monomer) for long life. Teflon has also been used as stem packing and older valves often used graphite impregnated packing material.

4. The actuator—A wheel attached to the protruding stem provides a manual actuating device—turning it by hand opens or closes the globe valve in varying degrees. However in automated control systems, the actuator becomes a key part of the valve and will provide the method by which the valve responds to the demands of the system.

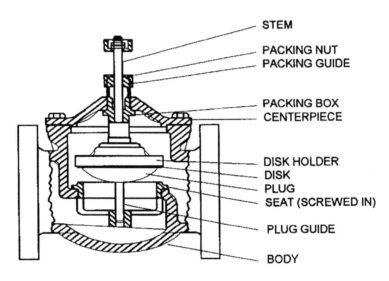

N.O. Cast Iron Flanged Style Valve Body

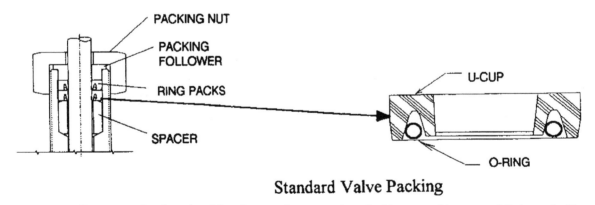

Standard Valve Packing

Figure 5.0.40 Cut-away section through a globe valve; an enlargement of standard bonnet packing nuts and O-ring seals. (*By permission: Johnson Controls, Milwaukee, WI.*)

5.0.41 Butterfly Valves

The two predominant types of valves installed in today's heating, ventilating and air conditioning systems are the globe valve, described above and the butterfly valve. The butterfly valve (Figure 5.0.41a) consists of a circular casting designated "A" in this figure. The valve neck "B" supports the valve stem "C" and the stem bushing "D", stem seal "E" and stem retaining ring "F" insure stem alignment. The stem seals in a butterfly valve, contrary to those in a globe valve which prevent leakage from inside the valve, are called upon to prevent contamination from entering the valve bore. The inner surface of the circular butterfly valve casting contains the valve seat "G." The valve disc "H" fits within this seat and provides the elastic surface to ensure a bubble-tight closure. These butterfly valves have many advantages:

- They are compact and require much less room in a piping installation
- They are more cost-effective than other types of valves for control applications
- They have few parts and are easier to maintain
- They are available in very large sizes (up to 20 inches)
- They are bubble-tight in the closed position
- They can provide accurate, stable, and modulating flow control
- Their actuator sizing is not dependent on the differential pressure across the system, therefore problems with the valve being unable to close when exposed to differential pressures in the system are not a concern
- The elastic seat of a butterfly valve provides the seal between the flanges that connects it to the piping system
- Valve packing is no longer a maintenance item because this is a dry stem design and the stem is not exposed to the media flowing through the valve body

Butterfly valves do have some disadvantages when installed in a control system: They have lower valve recovery coefficients than globe valves and may be more easily subjected to cavitation problems; they also have a greater water hammer potential. Because butterfly valves are relative "newcomers" to the HVAC industry, there is a certain amount of reticence in replacing globe valves with butterfly valves.

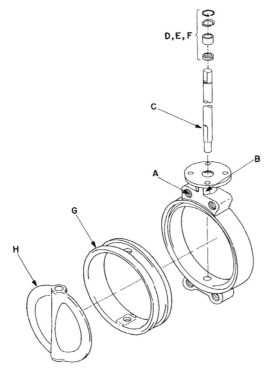

Figure 5.0.41.1 Exploded view of typical butterfly valve. (*By permission: Johnson Controls, Milwaukee, WI.*)

5.0.41.2 Physical Components of the Butterfly Valve

The valve bodies of most butterfly valves are made of cast iron to meet ASTMA A-126 Class B standards.

Valve Stem Options

Phosphate-coated carbon steel (ASTM A-108)
Type 416 Stainless steel (ASTM A-582 Type 416)
Type 304 Stainless steel (ASTM A-276 Type 304)
Type 316 Stainless steel (ASTMA A-276 Type 316)

Valve Seat Materials

Ethylene propylene diene monomer (EDPM) is the most common valve seat material because it is resistant to alcohol, acids, alkalies, sodium hypochlorite, inorganic acids, neutral salts, and saltwater. EPDM is not resistant, however, to hydrocarbons and petroleum-based oils and is therefore not suitable for those applications.

Valve Discs

The most common material of construction for butterfly valve discs is ductile iron coated with Nylon 11, a thermoplastic resistant to most acids, alkalies, and solvents. Further investigation is required if the Nylon 11 disc is to be exposed to some inorganic acids, phenols, and chlorinated solvents because its resistance to these materials may be limited.

5.0.42 Normally Open, Normally Closed, and Three-Way Mixing Valves

Normally open control valves have an internal spring force in the range of 2 to 5 psig to maintain its normally open position, and therefore requires an actuating pressure of at least 5 psig to close. The normally closed control valve generally has an internal spring force of 9 to 13 psig to maintain its closed position. The spring of the normally closed valve must be able to overcome the force of the differential pressure across the valve. Figure 5.0.41.1 reveals cut-away sections of normally open and normally closed valves.

Three way mixing valves are frequently used to mix liquids flowing through the lines that connect to the valve or the third port of a three way mixing valve can function as a bypass port (Figure 5.0.42.2).

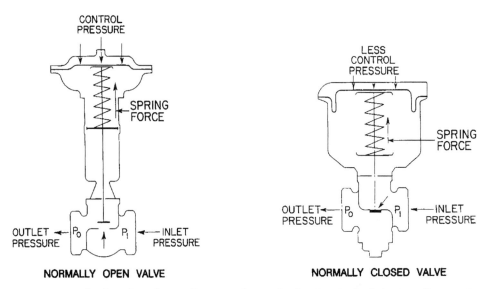

Figure 5.0.42.1 Sections through normally open and normally closed spring-loaded valves. (*By permission: Johnson Controls, Milwaukee, WI.*)

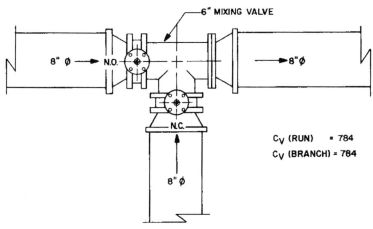

Figure 5.0.42.2 Schematic of three-way valve installation—operation of a three-way mixing valve. (*By permission: Johnson Controls, Milwaukee, WI.*)

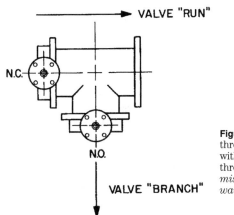

Figure 5.0.42.2 Schematic of three-way valve installation—A tee with two butterfly valves creating a three-way mixing valve. (*By permission: Johnson Controls, Milwaukee, WI.*)

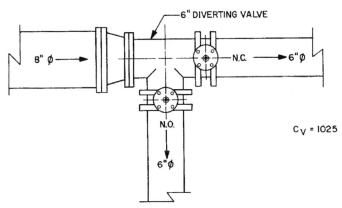

Figure 5.0.42.2 Schematics of three-way valve installations—three-way diverting valve. (*By permission: Johnson Controls, Milwaukee, WI.*)

5.0.43   Valve Actuators

Valves installed in an HVAC system must be actuated in response to the control system demands and the actuator's function is to provide the force necessary to move the plug or disc or the diaphram in a valve to accomplish the desired flow or pressure dictated by the system's set points. The type and size of actuator is determined primarily by the dynamic pressure that must be overcome in order to modulate the valve; the force required of an actuator being directly proportional to the effective area of the actuator diaphragm. Actuators are activated by either pneumatic pressure or electric current.

5.0.43a   Pneumatic Actuators

Pneumatic actuators contain various types of diaphrams with encapsulated springs, which, when subjected to air pressure, open and when the pressure diminishes or is released entirely, the encapsulated spring functions in a manner similar to a fully closed valve. Examples of pneumatic actuator construction are shown in Figure 5.0.43.

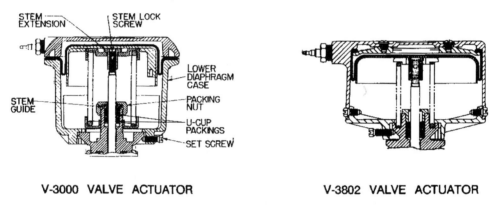

V-3000 VALVE ACTUATOR   **V-3802 VALVE ACTUATOR**

Figure 5.0.43.2 Sections through a pneumatic valve actuator. (*By permission: Johnson Controls, Milwaukee, WI.*)

5.0.44   Electrical Actuators

Electric actuators depend upon an electric motor to activate the valve. Depending upon the signal passed on by the control system, the motor would move a valve stem up or down or clockwise or counterclockwise. For proportional control, the actuator will return to zero when the controller stops sending a signal. Shut-off is affected by a force sensor that stops the motor when the valve stem limit has been reached. Figure 5.0.44.1 depicts cut-away sections an electric actuator and one method of adjusting a lower spring rate.

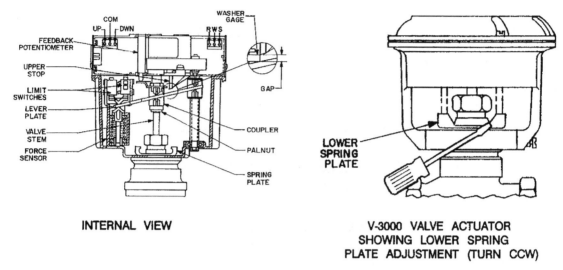

INTERNAL VIEW

V-3000 VALVE ACTUATOR
SHOWING LOWER SPRING
PLATE ADJUSTMENT (TURN CCW)

Figure 5.0.44.1   Sections through electrically operated actuators. (*By permission: Johnson Controls, Milwaukee, WI.*)

5.0.45   Valve Cavitation

Cavitation occurs in a valve when the local velocity of the liquid in the valve is high enough to allow the liquid to begin to vaporize (Fig. 5.0.46). As this liquid continues to flow through the valve, velocity drops and the bubbles in the vapor begin to collapse, creating substantial pressure changes within the walls of the valve. This is the first stage of cavitation and, if allowed to persist, the second stage of cavitation will ultimately damage the valve. During the second stage of cavitation, very high pressures are generated by these tiny implosions, often reaching 100,000 psi. The tiny shock waves created will travel to the solid parts of the valve acting as tiny hammer blows. If this condition is allowed to continue, metal fatigue will occur on the surface of the body and begin to chip away portions of the metal. Although low cavitation has a much less detrimental effect on the physical portion of the valve, unless corrected, it can increase in intensity and damage both valve trim and valve body.

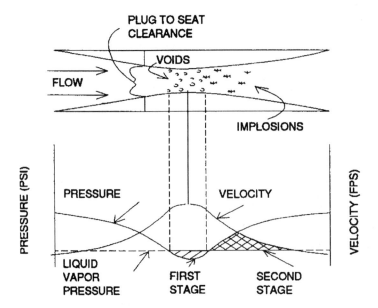

Variations of Pressure and Velocity for Points
Along a Flow Stream Through a Restriction.

Cavitation is a two-stage phenomenon which can greatly shorten the life of the valve trim in a control valve. Whenever a given quantity of liquid passes through a restricted area such as an orifice or a valve port, the velocity of the fluid increases. As the velocity increases, the static pressure decreases. If this velocity continues to increase, the pressure at the orifice will decrease below the vapor pressure of the liquid and vapor bubbles will form in the liquid. This is the first stage of cavitation.

As the liquid moves downstream, the velocity decreases with a resultant increase in pressure. If the downstream pressure is maintained above the vapor pressure of the liquid, the voids or cavities will collapse or implode. This is the second stage of cavitation. (See Figure 9)

The second stage of cavitation is detrimental to valve. Because of the tremendous pressures created by these implosions (sometimes as high as 100,000 psi), tiny shock waves are generated in the liquid. If these shock waves strike the solid portions of the valve they act as hammer blows on these surfaces. Repeated implosions on a minute surface will eventually cause fatigue of the metal surface and chip a portion of this surface off. Tests show that only those implosions close to the solid surfaces of a valve act on the valve in this manner.

Low degrees of cavitation are tolerable in a control valve. Minimum damage to the valve trim and little variation in flow occur at these levels. However, there is a point where the increasing cavitation becomes very detrimental to the valve trim and possibly even the valve body. It is at this point that the cavitation is beginning to choke the flow through the valve resulting in the flow rate staying the same regardless of increases in pressure drop.

Figure 5.0.46 Valve cavitation, illustrated and explained. (*By permission: Johnson Controls, Milwaukee, WI.*)

5.1.0 A Typical Specification for a DDC System

When a DDC system is being designed, the interaction and communication between designer, technician, and end-user, more than likely, will become intense, particularly if this is the end-user's first experience with the design and installation of a DDC system. The degree of complexity of design and function of the DDC system often produces detailed specifications similar to the following, which could also serve as the format for a request for proposal (RFP) for a DDC system.

Submittals

A. Provide complete operating data, systems drawings, wiring, schematic diagrams, and engineering data on each control system component, to include:
  1. Riser diagrams, field panel layouts, location of all interconnections
  2. Field panel specification sheets
  3. The proposed panel loading and spare capacity of the panel
  4. Field sensors and instrumentation specification sheets
  5. Valve, damper, and actuator schedules and specifications
  6. Instrument locations marked on a set of mechanical drawings
  7. Complete I/O summary included input and output devices and their connection point(s)
B. Provide schematic control diagrams for each building system to be controlled, to include:
  1. Sequence of operation
  2. I/O points
  3. Set points for each control loop
  4. Descriptions of all alarm algorithms, including alarm limits (where applicable)
  5. Device parts numbers
  6. Cable and tubing requirements
C. Schematic control diagrams furnished on mylar reproducibles
D. Manufacturer's installation instructions to be submitted for all equipment
E. Source listings for all site-specific DDC software. Software shall be provided in machine-readable format, and once communications are established with the owner's EMCS, new listings shall be provided whenever software changes are made.
F. The owner is to be notified of all software and hardware modifications made to the system after communications are established with the EMCS.

Guarantees

A. The Contractor(s) shall guarantee the control system to be free from defects in material and/or workmanship for a period of one (1) year from the date of completion and acceptance by the engineer.
B. The Contractor(s) shall guarantee the system to maintain any controlled temperature with one (1) degree of its set point at all times.
C. The Contractor(s) shall provide a one (1) year free maintenance service for the control components to commence concurrently with the acceptance of the system. The service shall include inspection and adjustment, on a bimonthly basis, of all operating controls and the replacement of parts or instruments found deficient or defective. At the end of the first six (6) months of operation and during the remaining six (6) months, the contractor(s) shall update the equipment with the latest modifications and improvements in software, firmware, and hardware that the manufacturer may have incorporated in the furnished equipment.

Products

A. All components of the system shall be electric or electronic except for actuators, which may be pneumatic.
B. Local room control, where applicable, shall have day/night setback thermostats. The E/P switch on the main control shall be controlled by the direct digital control system. The terminal controls may be pneumatic.

Control Devices

A. Sensors and transmitters
  1. Sensors appropriate to the type of physical quantity being monitored shall be provided. The Contractor(s) shall be responsible for the electrical, pneumatic, and hydronic connections. If transmitters are employed, the cable length shall not exceed one thousand (1,000) feet. Splices are not permitted in conduit. The transmitter shall produce a 4–20 ma signal that

complies with the Instrument Society of America Standard S50.1. All such transmitters shall be two (2) wire devices, powered by and presenting their signals on a single pair of wires. The Contractor(s) shall provide a regulated, protected 24-volt DC power supply with the ability to produce at least 33% more current than required by the installed transmitters. Output regulation shall be .005% for a 10-volt change in line voltage or .05% for a change from no load to full load. Ripple shall be less than .5 mv. There shall be no overshoot on turn on or turn off. Operating temperature range shall be −20 to +70 degrees Centigrade. The Contractor(s) will install the power supply in a NEMA 12 enclosure adjacent to the DDC panel.

2. In the case of differential pressure switches for air systems, the switch shall be (Specify supplier and model number). The switches shall be installed in accordance with the installation instructions contained in (Specify supplier and catalog/bulletin number).

   For liquids (Specify supplier and model number), differential pressure switches shall be provided. All switches shall be mounted in accessible and, to the extent practical, in vibration-free locations and not on ductwork. Do not use differential pressure switches to run status on VSD pumps. Use the VSD status to indicate a pump run condition.

   (Author's note: When adding onto a DDC system or expanding its scope, designers usually specify single-source suppliers to ensure compatibility between the existing installation and the new proposed installation.)

3. All temperature sensors shall be one of the following types of wire-wound RTDs:

Sensors	Leads	Nom Resistance	TCR	Lead length w/o transmitter
Platinum	3	100 ohms @ 0 degree C	.00385	Unlimited
Platinum	2	1000 ohms @ 0 degree C	.00385	100 feet
Nickel	2	1000 ohms @ 70 degree F	.00660	100 feet

4. Temperature transmitters with an accuracy of 1% of range shall be provided for:
   a. Outside air temperature
   b. Entering preheat coils
   c. After each preheat coil
   d. After cooling coil
   e. Return air
   f. Main water system supply and return water lines
   g. Supply fan discharge

5. All sensors shall be mounted in appropriate enclosures.

6. Electric to pneumatic transducers shall be compatible with the analog output signal produced by the controller.

7. Differential pressure transducers shall have a minimum accuracy of +/− 1% of full range, a repeatability of 0.5%, and shall be installed with a valve manifold and pressure/temperature test ports in lieu of pressure gauges. Water differential transmitters shall be 24 VDC powered MAMAC PR-282 rated for 200 psig static pressure and 50 psi differential, 4–20 mA output.

B. Thermostats

1. Thermostats shall be of the single-point controller electronic type, outputting a pneumatic signal to the appropriate final control element.

2. Thermostats shall be capable of being remotely reset, via the DDC system, from the installation's head end. The reset schedule for each thermostat shall be stored in a table in the DDC for display at the head end.

3. Each thermostat shall be capable of reporting the present space temperature and the present set point, via the DDC to the head end.

4. Communications with the DDC shall be via twisted pair.

5. Each thermostat shall have timed four (4) hour comfort override, accessible to the user, which will automatically return to the scheduled temperature set point at the conclusion of the override.

6. Each thermostat shall be capable of being rapidly and remotely reset by 10 degrees F to facilitate energy load shedding via a flag test algorithm in the DDC.

C. Direct Digital Controller

1. Control of the mechanical systems shall be performed by a field-programmable microprocessor-based DDC that incorporates closed loop control algorithms, all necessary energy management functions, and provides for digital display and convenient local adjustments of desired variables at the controller cabinet. Systems that require the existing user-defined

database to be reentered through the operator's terminal after a failure or power interruption shall not be acceptable.

2. Each DDC shall be capable of performing all specified control functions in a completely independent manner. Accordingly, DDCs shall be capable of being networked for single-point programming and for the sharing of information between panels, including, but not limited to, sensor values, calculated point values, control set points, tuning parameters, and control instructions.

3. Each DDC shall include its own microcomputer controller, power supply, input/output modules, termination modules, battery, and spare AC outlet. The battery shall be continuously charged and be capable of supporting all memory for a minimum of 24 hours or, if the battery is not of the rechargeable type, shall be continuously monitored and shall produce a system alarm whenever it is incapable of providing at least the required duration of standby power.

4. The unit shall be listed by Underwriters Laboratories (UL) against fire and shock hazard as a signal system appliance unit.

5. The DDC shall be enclosed in a metal cabinet. The cabinet shall be mounted and electrical terminations made during the construction phase of the project. The DDC electronics shall be separately packaged to be installed at a later date.

6. The DDC cabinet shall be provided with a key lock. All cabinets shall utilize one master key through the project.

7. Each DDC unit shall be capable of being directly interfaced to a variety of industry standard sensors and actuators. The Contractor shall provide sufficient I/O multiplexers to assure 15% spare capacity for each type of input and output points.

8. It shall be possible for each DDC to monitor the following types of inputs:
   a. Analog inputs
      4–20 MA
      0–10 vDC
      3–15 psig
      RTDs
   b. Binary inputs
      Dry contact closure
      Pulse accumulator

9. The DDC shall directly control pneumatic and electronic actuators and control devices. Each control unit shall be capable of providing the following outputs:
   a. Analog outputs
      4–20 MA
      0–10 vDC
      3–15 psig
   b. Binary outputs
      Low voltage contact closure
      Bipolar voltage pulse

   All start/stop control relays shall be deenergized to start the controlled equipment.

10. The DDC shall provide a local status indication for each binary input and ouput for constant, up-to-date verification of point condition separate and apart from the indication provided by a portable operator's terminal. The indicator shall be an LED or similar device acceptable to the owner. In addition, an operator shall have the ability to manually override all binary and analog outputs at the DDC via built-in local, point discrete hand/off/auto switches for binary points and gradual switches for analog points. These switches shall be operable even if the panel is not being supplied with AC line power.

11. Each unit shall be capable of performing the following energy management routines as a minimum:
    a. Time-of-day scheduling
    b. Start/stop time optimization
    c. Duty cycling (temperature compensated)
    d. Dry-bulb economizer control
    e. Supply air reset
    f. Chilled water reset
    g. Outdoor air reset

      h. Event-initiated programs

      i. Simultaneous heating and cooling monitoring

12. All control functions shall be executed within the control unit. The Owner shall be able to customize control strategies and sequences of control and shall be able to define appropriate control loop algorithms and choose the optimum loop parameters for loop control. Control loops shall support any of the following control modules:

      a. Two position

      b. Proportional

      c. Proportional plus integral

      d. Proportional plus integral plus derivative

13. In addition, the Owner shall be able to create customized control strategies based on arithmetic, Boolean, or time-delay logic. The arithmetic functions shall permit simple relationships between variables (+, −, ×, :) as well as more complex relationships, i.e., square root and exponential

5.2.0   Glossary of Frequently Used DDC Terminology

Authority, Valve   The ratio of valve pressure drop to total branch pressure drop at design flow. The total branch pressure drop includes the valve, piping, coil, fittings, etc.

Cavitation   The forming and imploding of vapor bubbles in a liquid due to decreased, then increased, pressure as the liquid flows through a restriction.

Control Loop   Chain of components which makes up a control system. If feedback is incorporated it is a closed loop; if there is no feedback, it is an open loop system.

Controlled Fluid   When applied to a valve, this term refers to whatever fluid is being regulated. For heating-cooling systems, this fluid can be hot or chilled water, steam or refrigerant.

Critical Pressure Drop   This applies to gases and vapors. It is the pressure drop which causes the maximum possible velocity through the valve. Higher pressure drops will not increase the flow velocity.

Dynamic Pressure   The pressure of a fluid resulting from its motion. Total Pressure - Static Pressure = Dynamic Pressure (Pump Head)

Flow Characteristic   Relation between flow through the valve and percent rated travel as the latter is varied from zero to 100 percent.

Flow Characteristic, Inherent	Flow characteristic when constant pressure drop is maintained across the valve.
Flow Characteristic, Installed	Flow characteristic when pressure drop across valve varies as dictated by flow and related conditions in system in which valve is installed.
Flow Coefficient (C_v)	Number of US. gallons per minute of 60F water that will flow through a fully open valve with a 1 psi drop across it.
Rangeability	Ratio of maximum to minimum controllable flow at which specified flow characteristic prevails.
Rated Flow	For a coil this is the flow through the coil which will produce full rated heat output of the coil.
Spring Range	Control pressure range through which the signal applied must change to produce total movement of the controlled device from one extreme position to the other.
Actual Spring Range	Control pressure range that causes total movement under actual conditions to overcome forces due to spring force, fluid flow, friction etc.
Nominal Spring Range	Control pressure range that causes total movement when there is no external force opposing the actuator.
Static Pressure	The pressure with respect to a surface at rest in relation to the surrounding field.
Stroke	This is synonymous to lift, travel, and percent open. These are terms used when referring to the amount a valve has moved from either extreme of fully open or fully closed.
Total Pressure	The sum of the Static Pressure and the Dynamic Pressure.
Three Way Valve	Valve with three connections, one of which is a common and two flow paths.
Bypass or Diverting Valve	Common connection is the only inlet: Fluid entering this connection is diverted to either outlet.

Mixing Valve   Two connections are inlets and the common is the outlet. Fluid from either or both inlets is selected to go out the common connection.

Two Way Valve   Valve with two connections and a single flow path.

Uncontrollable Flow   The flow rate at low load conditions that causes the valve to hunt or cycle. Typically occurs within the first 10% of valve stroke.

Valve Pressure Drop   Portion of the system pressure drop which appears across the valve. For valve sizing this drop is across a fully open valve.

(By permission: Johnson Controls, Milwuakee, WI.)

5.3.0   How Things Work

This section contains a series of schematic drawings of various HVAC control systems representative of those generally encountered in today's commercial, industrial, and institutional construction projects. The increased demand for DDC systems in the marketplace now permits building owners and facility managers to employ the concepts of these basic control systems where stand-alone or centrally located head end units use microprocessors to receive and analyze signals and direct motors, pumps, valves, and dampers to respond to predesignated set points.

Acknowledgment

The writer expresses his appreciation to Mr. John L. Levenhagen, author of *HVAC Control System Design Diagrams,* and Mrs. R. W. Haines and C. L. Wilson, coauthors of *HVAC Systems Design Handbook,* for their contribution to a better understanding of how things work; and to McGraw-Hill, the publisher of both volumes, for granting permission to reprint the following schematic drawings.

Heating Only—Electric Controls

This diagram shows a simple fan coil system that is supplied with hot water in the heating season. There is no cooling with this unit. The room thermostat T-1 cycles the unit on and off on demand and the strap-on T-2 makes sure there is hot water available before the unit cycles on.

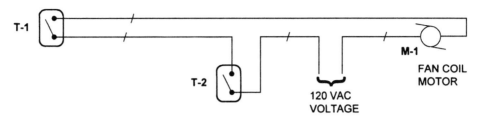

LEGEND:

T-1 = LINE VOLTAGE ROOM THERMOSTAT
T-2 = STRAP-ON THERMOSTAT
M-1 = LINE VOLTAGE FAN COIL MOTOR

Figure 5.3.1 Simple fan coil where thermostat controls heat on/off via line voltage. (*By permission: McGraw-Hill,* HVAC Control Systems Design Diagrams, *John I. Levenhagen.*)

Heating Only—Pneumatic Controls

This diagram shows a simple fan coil system that is supplied with hot water in the heating season. There is no cooling with this unit. The room thermostat T-1 cycles the unit on and off on demand through the PE switch PE-1 and the strap-on T-2 makes sure there is hot water available before the unit cycles on. The controls are pneumatic/electric in this case.

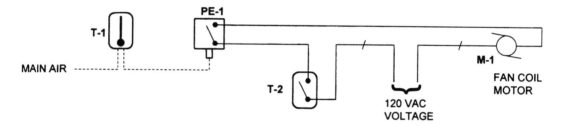

LEGEND:

T-1 = LINE VOLTAGE ROOM THERMOSTAT
T-2 = STRAP-ON THERMOSTAT
M-1 = LINE VOLTAGE FAN COIL MOTOR
PE-1 = PRESSURE/ELECTRIC SWITCH

Figure 5.3.2 Fan coil unit, supplied by hot water, but whose thermostat is pneumatically operated. (*By permission: McGraw-Hill,* HVAC Control Systems Design Diagrams, *John I. Levenhagen.*)

This diagram shows a simple system where room thermostats modulate valves V-1 on the radiation in the space. The thermostats and valves are electric.

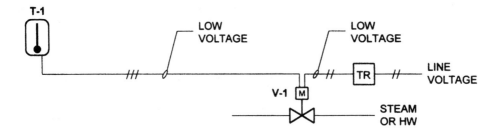

LEGEND:

T-1   = PROPORTIONAL MODULATING ROOM THERMOSTAT
V-1   = MODULATING 3-WIRE STEAM OR HW VALVE
TR   = LINE/LOW VOLTAGE TRANSFORMER

Figure 5.3.3   Steam or hot water radiation controlled by simple thermostatically actuated modulating valves. (*By permission: McGraw-Hill,* HVAC Control Systems Design Diagrams, *John I. Levenhagen.*)

This diagram shows a simple system where room thermostats modulate valves V-1 on the radiation in the space. The thermostats and valves are pneumatic.

LEGEND:

T-1   = MODULATING PNEUMATIC ROOM THERMOSTAT
V-1   = PNEUMATIC STEAM OR HW VALVE ON RADIATION

Figure 5.3.4   Similar system as Fig. 5.3.3, except that thermostats and valves are operated pneumatically rather than electrically. (*By permission: McGraw-Hill,* HVAC Control Systems Design Diagrams, *John I. Levenhagen.*)

This diagram shows typical controls on a low-pressure hot water boiler being used with room controls. The room thermostat T-1 controls the boiler through its aquastat directly. The circulator runs as long as there is hot water sensed by a strapon on the boiler discharge line.

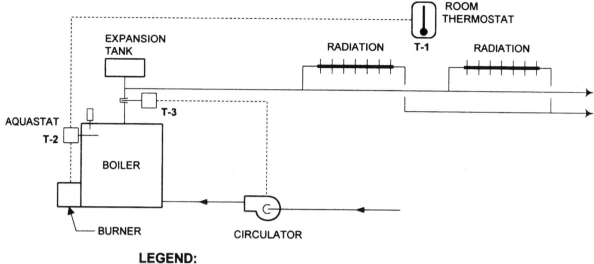

LEGEND:

T-1 = TWO POSITION ROOM THERMOSTAT
T-2 = BOILER AQUASTAT
T-3 = STRAP-ON THERMOSTAT

Figure 5.3.5 Thermostatic control of a boiler's aquastat. (*By permission: McGraw-Hill*, HVAC Control Systems Design Diagrams, *John I. Levenhagen.*)

This diagram shows typical controls on a low-pressure steam boiler with feed pump and condensate tank. An outdoor thermostat, T-2, resets the control point of a room thermostat, T-1, which controls a zone valve, V-1, to feed the radiators. The boiler is operated under a constant pressure by the pressure stat on the boiler.

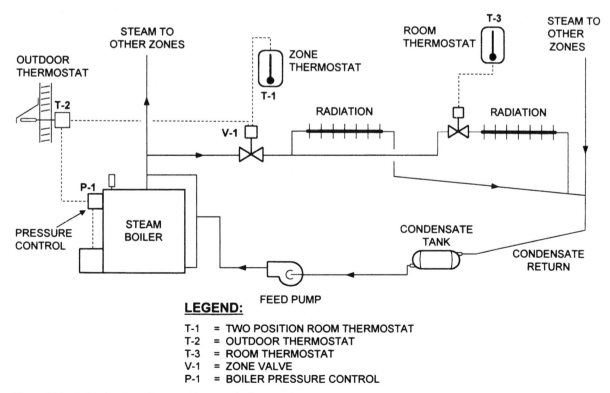

LEGEND:

T-1 = TWO POSITION ROOM THERMOSTAT
T-2 = OUTDOOR THERMOSTAT
T-3 = ROOM THERMOSTAT
V-1 = ZONE VALVE
P-1 = BOILER PRESSURE CONTROL

Figure 5.3.6 Indoor/outdoor thermostatic control of zone valve feeding radiators. (*By permission: McGraw-Hill*, HVAC Control Systems Design Diagrams, *John I. Levenhagen.*)

This diagram shows typical controls on a low-pressure hot water boiler. There are two pressure controls—an operating one and a high-limit safety control. There is an "altitude gage" and a low-water cutoff. The boiler, unlike steam boilers, must be filled with water at all times.

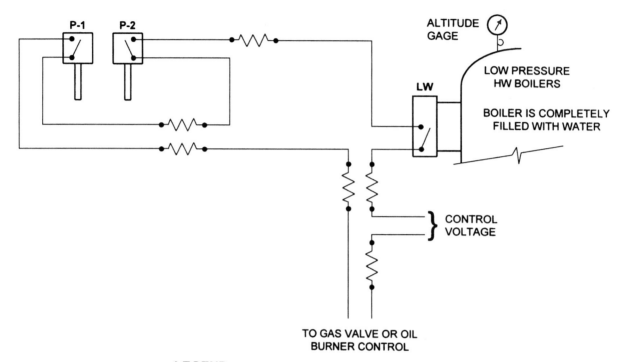

LEGEND:

P-1 = OPERATING TEMPERATURE CONTROL
P-2 = HI LIMIT TEMPERATURE CONTROL
LW = LOW WATER CUTOFF

Figure 5.3.7   Low-pressure hot water boiler with low water cut-off control, pressure, and high-limit safety controls. (*By permission: McGraw-Hill*, HVAC Control Systems Design Diagrams, *John I. Levenhagen.*)

This diagram shows typical controls on a low-pressure hot water boiler being used with room zone controls. The room thermostats T-1 and T-2 operate zone valves V-1 and V-2 with the circulator controlled so that if any one valve opens, the circulator runs. The boiler is fired at a constant temperature by its aquastat.

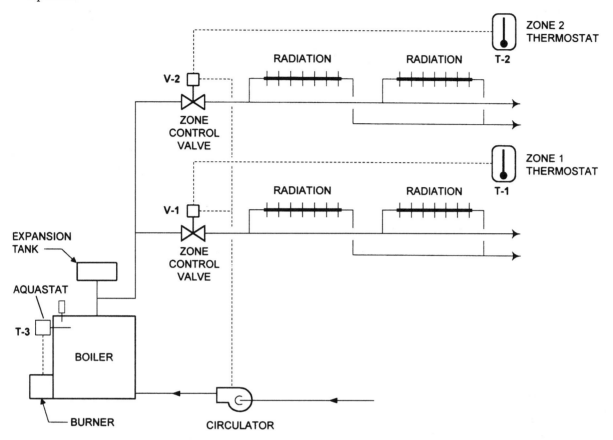

LEGEND:

T-1	= MODULATING ROOM THERMOSTAT
T-2	= MODULATING ROOM THERMOSTAT
T-3	= BOILER THERMOSTAT
V-1	= ZONE VALVE
V-2	= ZONE VALVE

Figure 5.3.8 Low-pressure hot water boiler with room zone controls. (*By permission: McGraw-Hill*, HVAC Control Systems Design Diagrams, *John I. Levenhagen.*)

This diagram shows typical controls on a low-pressure hot water boiler being used with the zone room thermostats T-1 and T-2 controlling the two circulators directly. The aquastat T-3 fires the boiler at a constant temperature.

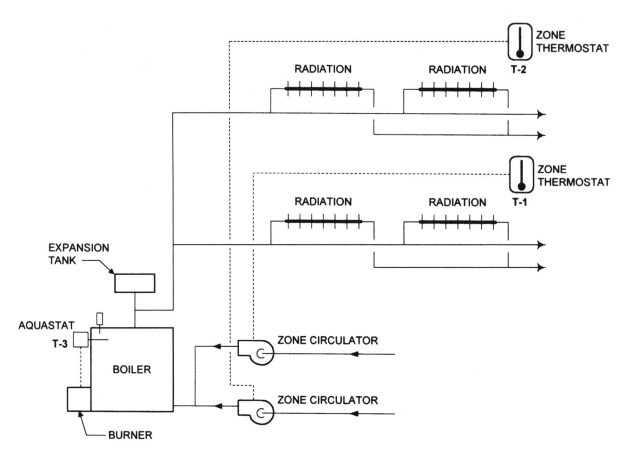

LEGEND:

T-1   = TWO POSITION ZONE ROOM THERMOSTAT
T-2   = TWO POSITION ZONE ROOM THERMOSTAT
T-3   = BOILER AQUASTAT

Figure 5.3.9 Zone thermostats control circulating pumps as aquastat maintains constant boiler temperature. (*By permission: McGraw-Hill,* HVAC Control Systems Design Diagrams, *John I. Levenhagen.*)

This diagram shows typical controls on a low-pressure hot water boiler being used with room zone controls. The room thermostat T-1 controls a 3-way valve V-1 and the circulator runs all the time. The boiler is fired at a constant temperature by its aquastat.

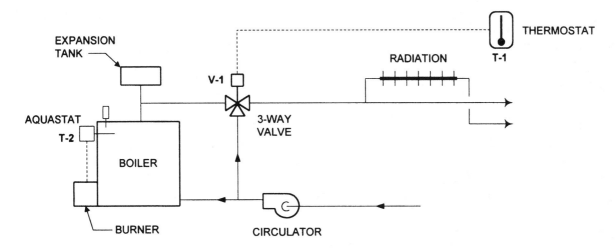

LEGEND:

T-1   = MODULATING ROOM THERMOSTAT
T-2   = BOILER AQUASTAT
V-1   = 3-WAY VALVE

Figure 5.3.10   Three-way control valves can act as mixing valves or bypass valves. In this case, it acts as a bypass valve. (*By permission: McGraw-Hill,* HVAC Control Systems Design Diagrams, *John I. Levenhagen.*)

This diagram shows a fan coil system that is supplied with both hot water and chilled water all the time. The room thermostat T-1 controls a 3-way valve V-1 on the supply to the coil of the unit. The valve has a special "hesitation spring" in it so that at no time does the valve mix the hot water and chilled water.

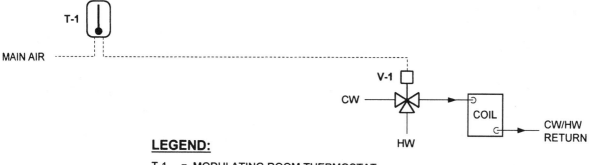

LEGEND:

T-1   = MODULATING ROOM THERMOSTAT
V-1   = MODULATING "SPECIAL" HW/CW VALVE

Figure 5.3.11   Fan coil producing heating/cooling by means of three-way valve. (*By permission: McGraw-Hill,* HVAC Control Systems Design Diagrams, *John I. Levenhagen.*)

This diagram shows a fan coil system that is supplied with both hot water and chilled water all the time. The R.A. thermostat T-1 controls a 3-way valve V-1 on the supply to the coil of the unit. The valve has a special "hesitation spring" in it so that at no time does the valve mix the hot water and chilled water.

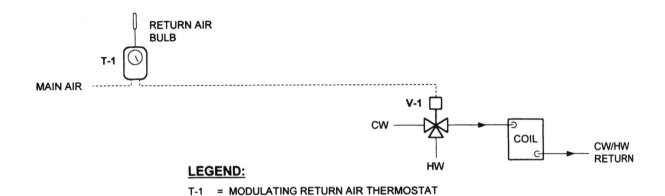

LEGEND:

T-1 = MODULATING RETURN AIR THERMOSTAT
V-1 = MODULATING "SPECIAL" HW/CW VALVE

Figure 5.3.12 Fan coil producing heating/cooling means of three-way valve controlled by return air thermostat. (*By permission: McGraw-Hill*, HVAC Control Systems Design Diagrams, *John I. Levenhagen.*)

This diagram shows typical controls of a hot water, chilled water four pipe system with 4 valves on each coil. The valves have hesitation springs in them and the supply line valves and the return line valves operate as a team. There is a dead-band in the operation of both the supply and return line valves so that hot water and chilled water are never mixed. The hot water and the chilled water are modulated to the unit coil by the room controllers.

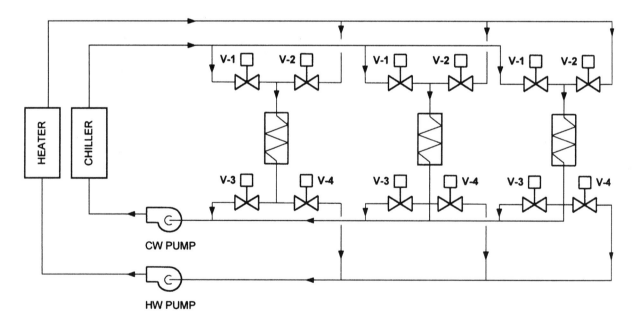

LEGEND:

V-1 = CW SUPPLY TWO WAY VALVE
V-2 = HW SUPPLY TWO WAY VALVE
V-3 = CW RETURN TWO WAY VALVE
V-4 = HW RETURN TWO WAY VALVE

Figure 5.3.13 Four-pipe heating and cooling system with valves on each coil modulated by room controllers. (*By permission: McGraw-Hill*, HVAC Control Systems Design Diagrams, *John I. Levenhagen.*)

Whenever the fan runs, the EP-1 is energized and O.A., R.A., and REL dampers are placed under automatic controls. When the fan stops, all dampers return to their normal positions.

Room thermostats T-1 control the multizone dampers to maintain space conditions. Submaster receiver controller RC-1, with sensors T-2 and T-3, maintain a variable hot deck temperature in accordance with O.A. temperature. Receiver controller RC-2 with sensor T-4 maintains a fixed cold deck temperature by controlling valve V-2 on the cooling coil. Duct thermostat T-5 controls the O.A., R.A., and REL dampers to maintain a fixed temperature in the mix chamber. Minimum position switch S-1 allows for a minimum position of the O.A. damper as long as the fan is running.

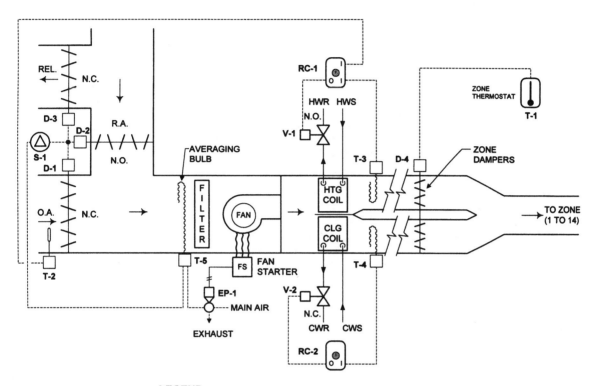

LEGEND:

T-1 = MODULATING ROOM ZONE THERMOSTAT
T-2 = OUTSIDE AIR MASTER TEMPERATURE SENSOR
T-3 = HOT DECK TEMPERATURE SENSOR
T-4 = COLD DECK TEMPERATURE SENSOR
T-5 = MIXED AIR CONTROLLER
V-1 = N.O. MODULATING 2-WAY HEATING COIL VALVE
V-2 = N.C. MODULATING 2-WAY COOLING COIL VALVE
D-1 = OUTSIDE AIR DAMPER MOTOR
D-2 = RETURN AIR DAMPER MOTOR
D-3 = RELIEF AIR DAMPER MOTOR
D-4 = ZONE DAMPER MOTOR
EP-1 = SOLENOID AIR VALVE
RC-1 = RECEIVER CONTROLLER
RC-2 = RECEIVER CONTROLLER
S-1 = MINIMUM POSITION SWITCH

Figure 5.3.14   A multizone air handling unit using hot water and chilled water, mixed air controls, and a series of zone dampers. (*By permission: McGraw-Hill,* HVAC Control Systems Design Diagrams, *John I. Levenhagen.*)

This diagram shows a fan coil system that is supplied with both hot water and chilled water all the time. The room thermostat T-1 controls a 3-way valve V-1 on the supply line and a 3-way valve V-2 on the return line. The valves are sequenced so that as the unit is fed hot water through the supply valve the return line valve insures that the hot water is sent back to the hot water return line. The valve on the return line is a 3-way bypass.

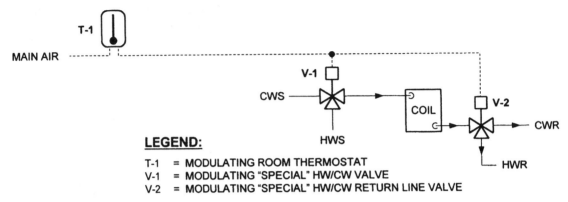

LEGEND:

T-1 = MODULATING ROOM THERMOSTAT
V-1 = MODULATING "SPECIAL" HW/CW VALVE
V-2 = MODULATING "SPECIAL" HW/CW RETURN LINE VALVE

Figure 5.3.15 Fan coil unit in a four-pipe system controlled by thermostat operating a three-way valve. (*By permission: McGraw-Hill*, HVAC Control Systems Design Diagrams, *John I. Levenhagen.*)

This diagram shows a fan coil system that is supplied with both hot water and chilled water all the time. The R.A. thermostat T-1 controls a 3-way valve V-1 on the supply line and a 3-way valve V-2 on the return line. The valves are sequenced so that as the unit is fed hot water through the supply valve the return line valve insures that the hot water is sent back to the hot water return line. The valve on the return line is a 3-way bypass.

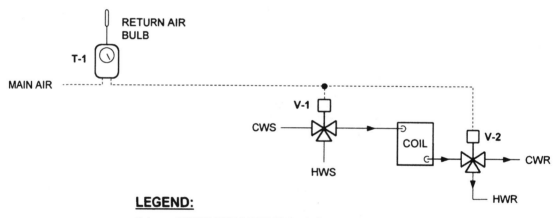

LEGEND:

T-1 = MODULATING RETURN AIR THERMOSTAT
V-1 = MODULATING "SPECIAL" HW/CW VALVE
V-2 = MODULATING "SPECIAL" HW/CW RETURN LINE VALVE

Figure 5.3.16 Fan coil unit in a four-pipe system in which a return-air thermostat controls a three-way valve. (*By permission: McGraw-Hill*, HVAC Control Systems Design Diagrams, *John I. Levenhagen.*)

This diagram shows a typical classroom unit ventilator controls where the unit is cycled at unoccupied periods (for night). The system requires a "dual" air supply pressure (15 psi for day and 20 psi for night). The cycle is a typical ASHRAE cycle II and the PE switch PE-1 with two circuits wired to the fan motor stops the fan when the supply air temperature reached 20 psi. The T-1 thermostat can close the circuit and restart the fan when the nighttime temperature falls below a certain point.

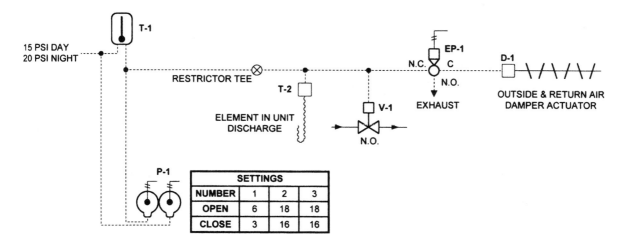

	SETTINGS		
NUMBER	1	2	3
OPEN	6	18	18
CLOSE	3	16	16

LEGEND:

T-1 = MODULATING DAY/NIGHT ROOM THERMOSTAT
T-2 = LOW LIMIT THERMOSTAT
V-1 = N.O. HW HEATING VALVE
D-1 = OUTSIDE & RETURN AIR DAMPER MOTOR
EP-1 = SOLENOID AIR VALVE
PE-1 = PRESSURE/ELECTRIC SWITCH

Figure 5.3.17 Typical classroom unit ventilator for heat only with night setback controls. (*By permission: McGraw-Hill,* HVAC Control Systems Design Diagrams, *John I. Levenhagen.*)

This diagram shows typical controls of a chiller with multiple loads by monitoring the valve V-1 positions on the chilled water coils. The thermostat T-2 acts as a low limit to protect the chiller.

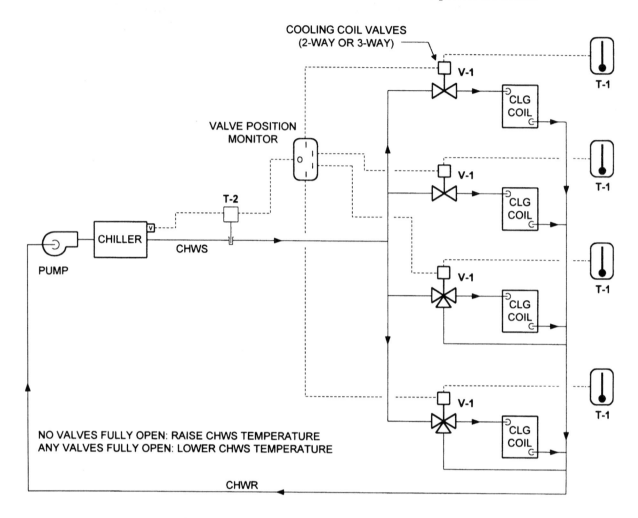

LEGEND:

T-1 = UNIT ROOM THERMOSTAT
T-2 = MODULATING LOW LIMIT CWS THERMOSTAT
V-1 = UNIT COIL VALVES(TWO WAY OR 3-WAY)

Figure 5.3.18 Chiller system using one chiller with multiple loads monitored by a series of valves and thermostats. (*By permission: McGraw-Hill,* HVAC Control Systems Design Diagrams, *John I. Levenhagen.*)

This diagram shows typical controls of multiple chillers with a constant flow system. The thermostats T-2 on the return lines to the chillers operate the chillers with low limit thermostats T-1 limiting the temperature of the chilled water. This system requires 3-way valves on the loads, as there is no variable flow system.

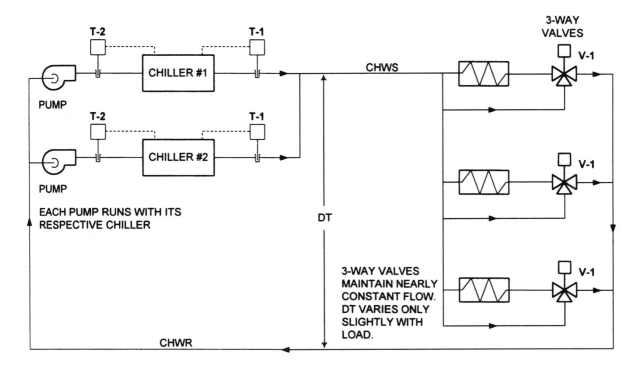

LEGEND:

T-1 = CHWS MODULATING USED WITH CENTRIFUGAL CHILLERS
T-2 = CHWR MODULATING USED WITH RECIPROCATING CHILLERS
V-1 = UNIT COIL 3-WAY VALVES

Figure 5.3.19 Chilled water system using multiple chillers and three-way valves to control loads. (*By permission: McGraw-Hill, HVAC Control Systems Design Diagrams, John I. Levenhagen.*)

Whenever the fan runs, the EP-1 is energized and the O.A. damper opens. If the unit is large, arrangements need to be made to be sure the damper is open before the fan runs.

Thermostat T-1 opens valve V-1 on the steam preheat coil, when freezing conditions exist. Room thermostat T-2 modulates valves V-2 and V-3 on heating and cooling coils in sequence to maintain space temperatures.

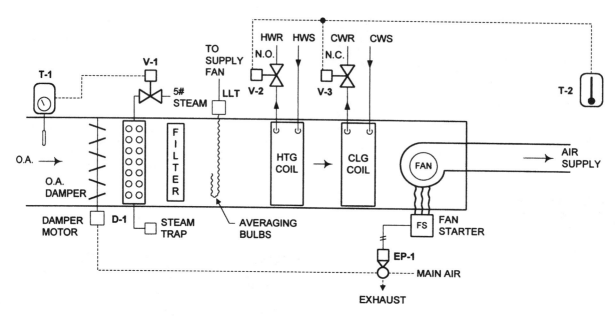

LEGEND:

T-1 = TWO POSITION CAPILLARY THERMOSTAT
T-2 = MODULATING ROOM THERMOSTAT
V-1 = TWO POSITION PREHEAT STEAM COIL VALVE
V-2 = N.O. MODULATING 2-WAY HEATING COIL VALVE
V-3 = N.C. MODULATING 2-WAY COOLING COIL VALVE
D-1 = TWO POSITION OA DAMPER MOTOR
EP-1 = SOLENOID AIR VALVE
LLT = LOW LIMIT ELECTRIC FREEZE PROTECTION THERMOSTAT

Figure 5.3.20 Heating and cooling with 100% outside air and preheat coil to prevent freeze-ups. (*By permission: McGraw-Hill, HVAC Control Systems Design Diagrams, John I. Levenhagen.*)

Whenever the fan runs, the EP-1 is energized and the O.A. damper opens. If the unit is large, arrangements need to be made to be sure the damper is open before the fan runs.

Thermostat T-1 controls valve V-1 on the hot water preheat coil. Low-limit thermostat LLT stops the fan on freezing conditions at that location. Room thermostat T-2 modulates valve V-2 along with PE switch PE-1. PE-1 stops the refrigeration when room conditions are satisfied.

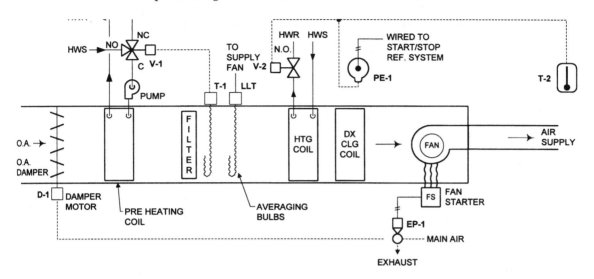

LEGEND:

T-1   = MODULATING CAPILLARY THERMOSTAT
T-2   = MODULATING ROOM THERMOSTAT
V-1   = MODULATING 3-WAY HOT WATER VALVE
V-2   = N.O. MODULATING 2-WAY HEATING COIL VALVE
D-1   = TWO POSITION OA DAMPER MOTOR
EP-1 = SOLENOID AIR VALVE
PE-1 = PRESSURE/ELECTRIC SWITCH
LLT   = LOW LIMIT ELECTRIC FREEZE PROTECTION THERMOSTAT

Figure 5.3.21 Heating and cooling using DX cooling coil, 100% outside air, and low-limit stat to stop fan when freezing conditions are encountered. (*By permission: McGraw-Hill,* HVAC Control Systems Design Diagrams, *John I. Levenhagen.*)

Whenever the fan runs, the EP-1 is energized and O.A., R.A., and REL dampers are placed under automatic controls. When the fan stops, all dampers return to their normal positions.

Room thermostat T-3 controls valve V-1 through the low-limit thermostat T-2 on the heating coil and DX cooling coil through PE switch PE-1. DX cooling coil and hot water coil must be manually operated summer and winter, so that they are not on at the same time. The O.A., R.A., and REL dampers are controlled by thermostat T-1 through O.A. high-limit thermostat T-4. T-1 and T-4 act as economizer thermostats to allow for up to 100% O.A. when conditions are correct. Minimum position relay S-1 can be set to maintain a minimum amount of O.A. through high signal selector R-1 as long as the fan is running. Room humidistat H-1 controls through high-limit duct humidistat H-2 valve V-2.

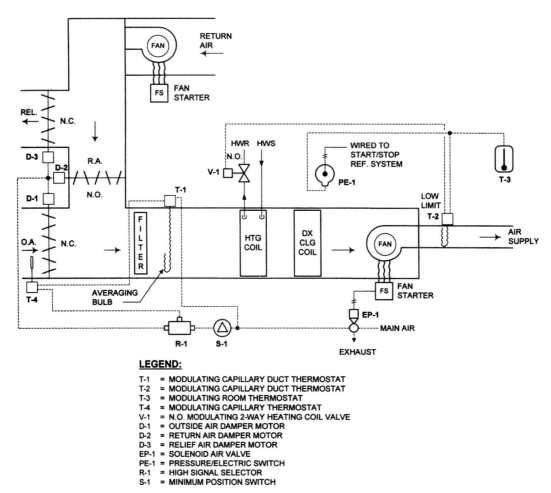

LEGEND:

T-1	=	MODULATING CAPILLARY DUCT THERMOSTAT
T-2	=	MODULATING CAPILLARY DUCT THERMOSTAT
T-3	=	MODULATING ROOM THERMOSTAT
T-4	=	MODULATING CAPILLARY THERMOSTAT
V-1	=	N.O. MODULATING 2-WAY HEATING COIL VALVE
D-1	=	OUTSIDE AIR DAMPER MOTOR
D-2	=	RETURN AIR DAMPER MOTOR
D-3	=	RELIEF AIR DAMPER MOTOR
EP-1	=	SOLENOID AIR VALVE
PE-1	=	PRESSURE/ELECTRIC SWITCH
R-1	=	HIGH SIGNAL SELECTOR
S-1	=	MINIMUM POSITION SWITCH

Figure 5.3.22 An air-handling unit with an economizer control cycle. (*By permission: McGraw-Hill,* HVAC Control Systems Design Diagrams, *John I. Levenhagen.*)

This diagram shows a typical cutoff VAV box that is pressure independent. The room thermostat T-1 controls the box damper to reduce the volume of air going into the space. This box is cooling only. The flow controller F-1 at the box maintains a minimum flow to the space.

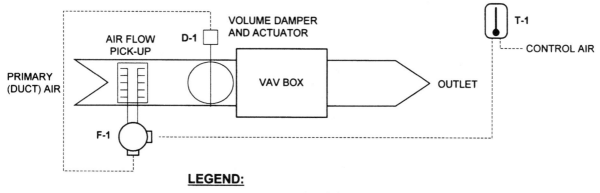

Figure 5.3.23 Variable air volume (VAV) box controls, pressure independent. (*By permission: McGraw-Hill*, HVAC Control Systems Design Diagrams, *John I. Levenhagen.*)

This diagram shows a typical cutoff VAV box that is pressure dependent. The room thermostat T-1 controls the box damper and valve V-1 (reheat valve) in sequence. This box is cooling with reheat.

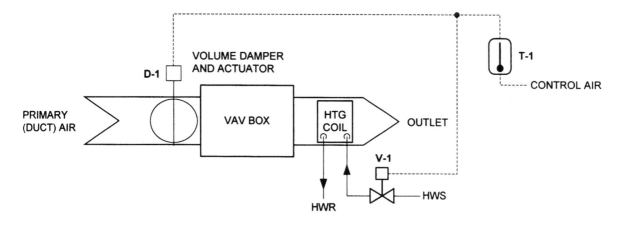

Figure 5.3.24 VAV system controls for heating only. (*By permission: McGraw-Hill*, HVAC Control Systems Design Diagrams, *John I. Levenhagen.*)

This diagram shows a typical cutoff VAV box that is pressure dependent (no pressure controls at the box). The room thermostat T-1 controls the box damper to reduce the volume of air going into the space. This box is cooling only.

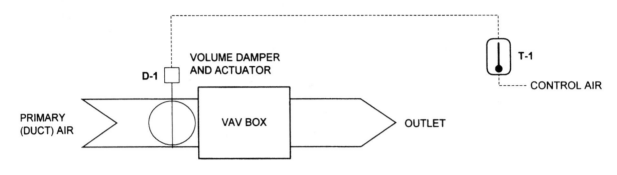

Figure 5.3.25   VAV pressure-dependent cooling system. (*By permission: McGraw-Hill,* HVAC Control Systems Design Diagrams, *John I. Levenhagen.*)

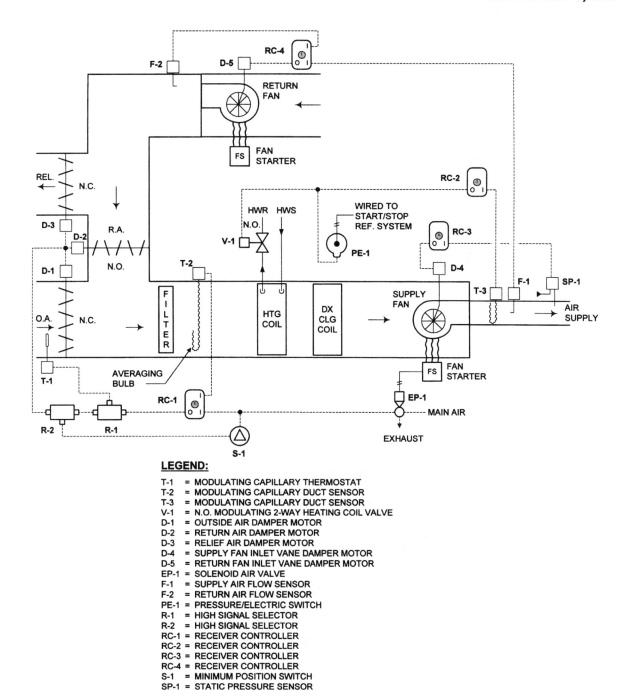

LEGEND:

T-1 = MODULATING CAPILLARY THERMOSTAT
T-2 = MODULATING CAPILLARY DUCT SENSOR
T-3 = MODULATING CAPILLARY DUCT SENSOR
V-1 = N.O. MODULATING 2-WAY HEATING COIL VALVE
D-1 = OUTSIDE AIR DAMPER MOTOR
D-2 = RETURN AIR DAMPER MOTOR
D-3 = RELIEF AIR DAMPER MOTOR
D-4 = SUPPLY FAN INLET VANE DAMPER MOTOR
D-5 = RETURN FAN INLET VANE DAMPER MOTOR
EP-1 = SOLENOID AIR VALVE
F-1 = SUPPLY AIR FLOW SENSOR
F-2 = RETURN AIR FLOW SENSOR
PE-1 = PRESSURE/ELECTRIC SWITCH
R-1 = HIGH SIGNAL SELECTOR
R-2 = HIGH SIGNAL SELECTOR
RC-1 = RECEIVER CONTROLLER
RC-2 = RECEIVER CONTROLLER
RC-3 = RECEIVER CONTROLLER
RC-4 = RECEIVER CONTROLLER
S-1 = MINIMUM POSITION SWITCH
SP-1 = STATIC PRESSURE SENSOR

Whenever the fan runs, the EP-1 is energized and O.A., R.A., and REL dampers are placed under automatic controls. When the fan stops, the dampers return to their normal positions.

Receiver controller RC-1 with sensors T-1 in the O.A. and sensor T-2 in the mix chamber controls the O.A., R.A., and REL dampers. The system acts as an economizer control cycle. R-1 relay allows the O.A. sensor T-1 to override the actions of T-2. Minimum pressure selector switch S-1 through relay R-2 controls the dampers to allow a minimum amount of ventilation air, regardless of the actions of the economizer thermostats. Receiver controller RC-3 with static pressure sensor SP-1 controls the supply fan inlet vanes. Receiver controller RC-2 with sensor T-3 in the fan discharge controls PE switch PE-1 on the DX cooling coil and heating coil valve V-1 in sequence. Flow controller F-1 resets the control point of receiver controller RC-4, which through flow controller F-2 controls the inlet vanes on the return fan.

This above system is designed to "match" the supply and return fans so that they act as a team and maintain space static pressure while at the same time modulating down from full capacity to reduced capacity as the VAV boxes in the system become satisfied.

Figure 5.3.26 VAV heating and cooling with economizer cycle and variable volume fan capacity control. (*By permission: McGraw-Hill,* HVAC Control Systems Design Diagrams, *John I. Levenhagen.*)

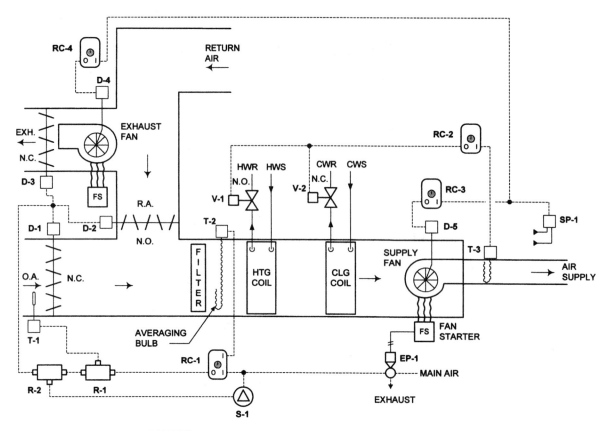

LEGEND:

T-1 = MODULATING CAPILLARY THERMOSTAT
T-2 = MODULATING CAPILLARY DUCT SESNOR
T-3 = MODULATING CAPILLARY DUCT SENSOR
V-1 = N.O. MODULATING 2-WAY HEATING COIL VALVE
V-2 = N.C. MODULATING 2-WAY COOLING COIL VALVE
D-1 = OUTSIDE AIR DAMPER MOTOR
D-2 = RETURN AIR DAMPER MOTOR
D-3 = EXHAUST AIR DAMPER MOTOR
D-4 = EXHAUST FAN INLET VANE DAMPER MOTOR
D-5 = SUPPLY FAN INLET VANE DAMPER MOTOR
EP-1 = SOLENOID AIR VALVE
R-1 = HIGH SIGNAL SELECTOR
R-2 = HIGH SIGNAL SELECTOR
RC-1 = RECEIVER CONTROLLER
RC-2 = RECEIVER CONTROLLER
RC-3 = RECEIVER CONTROLLER
RC-4 = RECEIVER CONTROLLER
S-1 = MINIMUM POSITION SWITCH
SP-1 = STATIC PRESSURE SENSOR

Whenever the fan runs, the EP-1 is energized and O.A., R.A., and REL dampers are placed under automatic controls. When the fan stops, the dampers return to their normal positions.

Receiver controller RC-1 with sensor T-2 in the mix chamber and sensor T-1 in the O.A. control the O.A., R.A., and REL dampers. The system acts as an economizer control cycle. Minimum position switch S-1 allows through relays R-1 and R-2 a minimum setting for the O.A. damper regardless of the actions of the economizer thermostats. The system does not have a return air fan but has an exhaust fan. Receiver controller RC-3 with static pressure sensor SP-1 controls the supply fan inlet vanes. Sensor SP-1 resets the control point of receiver controller RC-4 controlling the exhaust air fan. Receiver controller RC-2 with sensor T-3 controls valves V-1 and V-2 on the heating and cooling coils. The damper for the exhaust fan is operated in conjunction with the other dampers.

The control of the supply and exhaust fans is done through building static pressure controls with SP-1 sensing the building pressure. Receiver controller RC-2 through sensor T-3 in the fan discharge controls valve V-1 and valve V-2 on the heating and cooling coils.

Figure 5.3.27 VAV heating and cooling, economizer cycle and variable volume control by direct building pressure control. (*By permission: McGraw-Hill*, HVAC Control Systems Design Diagrams, *John I. Levenhagen.*)

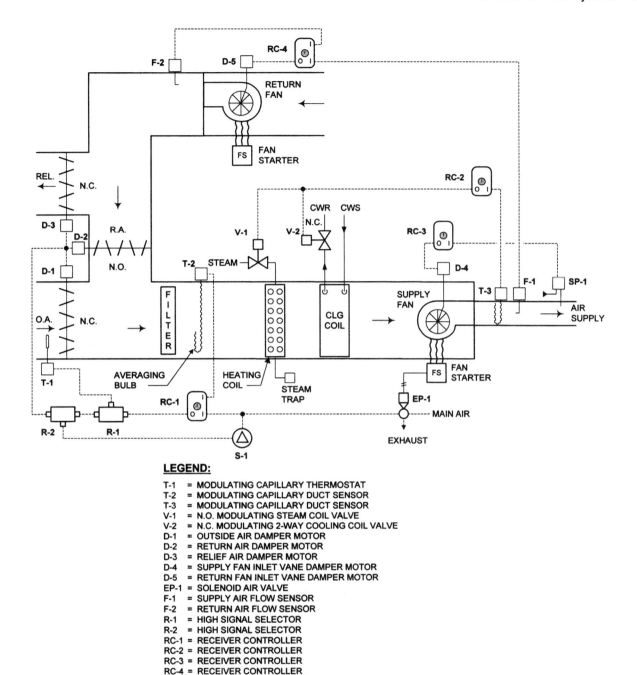

LEGEND:

T-1 = MODULATING CAPILLARY THERMOSTAT
T-2 = MODULATING CAPILLARY DUCT SENSOR
T-3 = MODULATING CAPILLARY DUCT SENSOR
V-1 = N.O. MODULATING STEAM COIL VALVE
V-2 = N.C. MODULATING 2-WAY COOLING COIL VALVE
D-1 = OUTSIDE AIR DAMPER MOTOR
D-2 = RETURN AIR DAMPER MOTOR
D-3 = RELIEF AIR DAMPER MOTOR
D-4 = SUPPLY FAN INLET VANE DAMPER MOTOR
D-5 = RETURN FAN INLET VANE DAMPER MOTOR
EP-1 = SOLENOID AIR VALVE
F-1 = SUPPLY AIR FLOW SENSOR
F-2 = RETURN AIR FLOW SENSOR
R-1 = HIGH SIGNAL SELECTOR
R-2 = HIGH SIGNAL SELECTOR
RC-1 = RECEIVER CONTROLLER
RC-2 = RECEIVER CONTROLLER
RC-3 = RECEIVER CONTROLLER
RC-4 = RECEIVER CONTROLLER
S-1 = MINIMUM POSITION SWITCH
SP-1 = STATIC PRESSURE SENSOR

Whenever the fan runs, the EP-1 is energized and O.A., R.A., and REL dampers are placed under automatic controls. When the fan stops, the dampers return to their normal positions.

Receiver controller RC-1 with sensors T-1 in the O.A. and sensor T-2 in the mix chamber controls the O.A., R.A., and REL dampers. The system acts as an economizer control cycle. R-1 relay allows the O.A. sensor T-6 to override the actions of T-2. Minimum pressure selector switch S-1 through relay R-2 controls the dampers to allow a minimum amount of ventilation air, regardless of the actions of the economizer thermostats. Receiver controller RC-3 with static pressure sensor SP-1 controls the supply fan inlet vanes. Receiver controller RC-2 with sensor T-3 in the fan discharge controls valves V-1 and V-2 on the heating and cooling coils in sequence.

Flow controller F-1 resets the control point of receiver controller RC-4, which through flow controller F-2 controls the inlet vanes on the return fan.

This above system is designed to "match" the supply and return fans so that they act as a team and maintain space static pressure while at the same time modulating down from full capacity to reduced capacity as the VAV boxes in the system become satisfied.

Figure 5.3.28 VAV heating and cooling, economizer cycle, variable volume fan control through volumetric matching of fans. (*By permission: McGraw-Hill*, HVAC Control Systems Design Diagrams, *John I. Levenhagen.*)

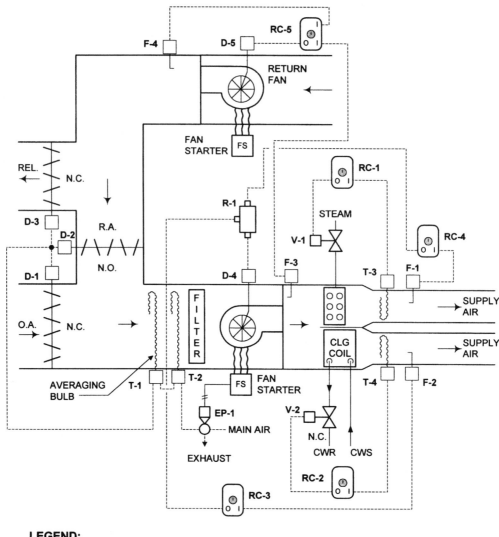

LEGEND:

T-1	= MODULATING CAPILLARY DUCT THERMOSTAT		EP-1	= SOLENOID AIR VALVE
T-2	= MODULATING CAPILLARY DUCT THERMOSTAT		F-1	= HOT DECK FLOW SENSOR
T-3	= MODULATING CAPILLARY DUCT THERMOSTAT		F-2	= COLD DECK FLOW SENSOR
T-4	= MODULATING CAPILLARY DUCT THERMOSTAT		F-3	= SUPPLY AIR FLOW SENSOR
V-1	= N.O. MODULATING STEAM COIL VALVE		F-4	= RETURN AIR FLOW SENSOR
V-2	= N.C. MODULATING 2-WAY COOLING COIL VALVE		RC-1	= RECEIVER CONTROLLER
D-1	= OUTSIDE AIR DAMPER MOTOR		RC-2	= RECEIVER CONTROLLER
D-2	= RETURN AIR DAMPER MOTOR		RC-3	= RECEIVER CONTROLLER
D-3	= RELIEF AIR DAMPER MOTOR		RC-4	= RECEIVER CONTROLLER
D-4	= SUPPLY FAN INLET VANE DAMPER MOTOR		RC-5	= RECEIVER CONTROLLER
D-5	= RETURN FAN INLET VANE DAMPER MOTOR		R-1	= HIGH SIGNAL SELECTOR

Dual-Duct VAV Heating and Cooling Air-Handling Unit, Steam Heating Coil, Chilled Water Cooling Coil, Inlet Vanes on Supply and Return Fans, O.A., R.A., and REL Fans Using Economizer Control Cycle, Static Pressure Control on Both Fans

Whenever the fan runs, the EP-1 is energized and O.A., R.A., and REL dampers are placed under automatic controls. When the fan stops, the dampers return to their normal positions.

Duct thermostats T-1 and T-2 in the mix chamber control the O.A., R.A., and REL dampers and act as economizer thermostats. They can allow for up to 100% O.A. when conditions are favorable. Receiver controller RC-1 with sensor T-3 in the hot air duct controls the valve V-1 on the heating coil. Receiver controller RC-2 with sensor in the cold duct controls valve V-2 on the cooling coil. Receiver controllers RC-3 and RC-4 with flow sensors F-1 and F-2, through averaging relay R-1, control the inlet vanes on the supply fan. Flow sensor F-3 in the fan discharge resets the control point of receiver controller RC-5, which is also sensing the return fan sensor F-4 in the return air fan discharge to control the return air fan inlet vane dampers.

Figure 5.3.29 Dual duct VAV heating/cooling system with static pressure control on both fans. (*By permission: McGraw-Hill, HVAC Control Systems Design Diagrams, John I. Levenhagen.*)

Whenever the fan runs, the EP-1 is energized and O.A., R.A., and REL dampers are placed under automatic controls. When the fan stops, the dampers return to their normal positions.

Duct thermostat T-2 controls the O.A., R.A., and REL dampers through mixed air thermostat T-4. T-2 and T-4 act as economizer thermostats. They act together to allow for up to 100% O.A. when conditions allow. T-2 and T-4 act through minimum position switch S-1, which can be set to maintain a minimum amount of O.A. as long as the fan is running, regardless of the actions of T-2 and T-4.

The ventilation comes through the deck where normally the warm air is passed. Thermostat T-3 controls chilled water coil, through valve V-2.

The zone dampers D-4 are controlled by room thermostat(s) T-1 in sequence with reheat coil valve(s) V-1. There can be up to 14 damper zones with 14 reheat or more coils.

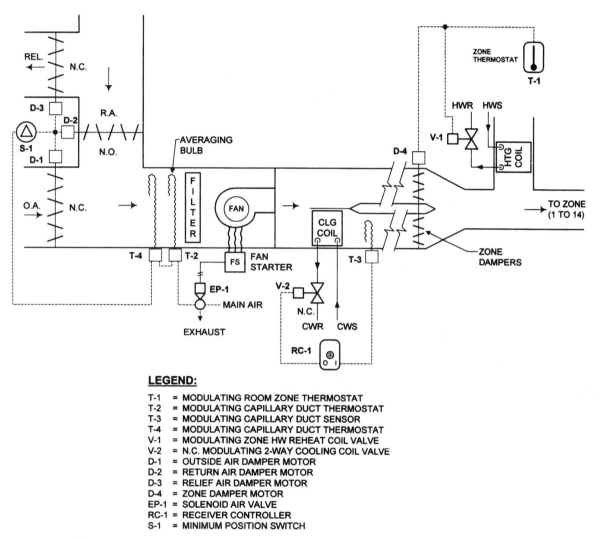

LEGEND:

T-1	=	MODULATING ROOM ZONE THERMOSTAT
T-2	=	MODULATING CAPILLARY DUCT THERMOSTAT
T-3	=	MODULATING CAPILLARY DUCT SENSOR
T-4	=	MODULATING CAPILLARY DUCT THERMOSTAT
V-1	=	MODULATING ZONE HW REHEAT COIL VALVE
V-2	=	N.C. MODULATING 2-WAY COOLING COIL VALVE
D-1	=	OUTSIDE AIR DAMPER MOTOR
D-2	=	RETURN AIR DAMPER MOTOR
D-3	=	RELIEF AIR DAMPER MOTOR
D-4	=	ZONE DAMPER MOTOR
EP-1	=	SOLENOID AIR VALVE
RC-1	=	RECEIVER CONTROLLER
S-1	=	MINIMUM POSITION SWITCH

Figure 5.3.30 Multizone heating/cooling system with economizer cycle zone dampers controlled by room thermostats. (*By permission: McGraw-Hill*, HVAC Control Systems Design Diagrams, *John I. Levenhagen.*)

Whenever the fan runs, the EP-1 is energized and the O.A. damper opens. If the unit is large, arrangements need to be made to be sure the damper is open before the fan runs.

Thermostat T-1 opens valve V-1 on the steam preheat coil, when freezing conditions exist. Thermostat T-2 modulates the face and bypass dampers on the coil to control downstream temperature. Low-limit thermostat LLT stops the fan on freezing conditions at that location. Room thermostat T-3 modulates valves V-2 and V-3 on the cooling and heating coils in sequence to maintain space temperatures.

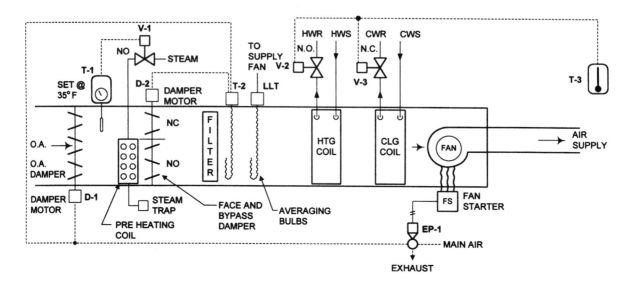

LEGEND:

T-1	=	TWO POSITION CAPILLARY THERMOSTAT
T-2	=	MODULATING CAPILLARY THERMOSTAT
T-3	=	MODULATING ROOM THERMOSTAT
V-1	=	TWO POSITION PREHEAT STEAM COIL VALVE
V-2	=	N.O. MODULATING 2-WAY HEATING COIL VALVE
V-3	=	N.C. MODULATING 2-WAY COOLING COIL VALVE
D-1	=	TWO POSITION OA DAMPER MOTOR
D-2	=	MODULATING FACE & BYPASS DAMPER MOTOR
EP-1	=	SOLENOID AIR VALVE
LLT	=	LOW LIMIT ELECTRIC FREEZE PROTECTION THERMOSTAT

Figure 5.3.31   Heating/cooling with 100% outside air, face, and bypass dampers to control downstream temperatures. (*By permission: McGraw-Hill*, HVAC Control Systems Design Diagrams, *John I. Levenhagen.*)

Whenever the fan runs, the EP-1 is energized and O.A., R.A., and REL dampers are placed under automatic controls. When the fan stops, the dampers return to their normal positions.

Room thermostat T-1 controls the DX cooling coil through PE switch PE-1 and the reheat coil valve V-2 through relay R-1. Humidistat H-1 also controls the cooling coil through relay R-1, so that either the thermostat or the humidistat can call for cooling. The thermostat can call for reheat through reheat coil valve V-2 if the humidistat calls for too much cooling trying to dehumidify. The humidistat also controls the humidifier valve V-2 through high-limit humidistat H-2 to add humidity when required. Thermostat T-1 also controls in sequence the O.A., R.A., and REL dampers to maintain a fixed mix chamber temperature. Minimum position switch S-1 sets the minimum percentage of O.A., regardless of the actions of other controllers as long as the fan is running. The heating coil is a "reheat" coil and must be after the cooling coil.

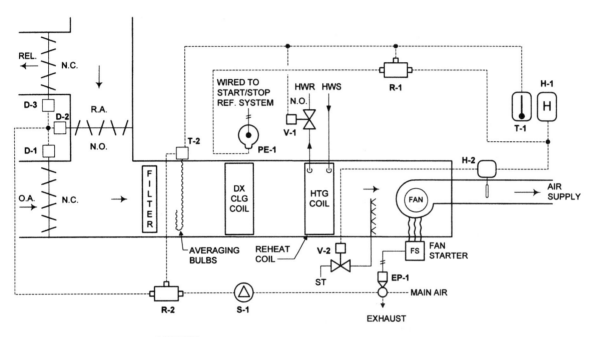

LEGEND:

T-1	=	MODULATING ROOM THERMOSTAT
T-2	=	MODULATING CAPILLARY DUCT THERMOSTAT
H-1	=	MODULATING ROOM HUMIDISTAT
H-2	=	HIGH LIMIT DUCT HUMIDISTAT
V-1	=	N.O. REHEAT COIL HOT WATER VALVE
V-2	=	N.C. MODULATING HUMIDIFIER STEAM VALVE
D-1	=	OUTSIDE AIR DAMPER MOTOR
D-2	=	RETURN AIR DAMPER MOTOR
D-3	=	RELIEF AIR DAMPER MOTOR
EP-1	=	SOLENOID AIR VALVE
PE-1	=	PRESSURE/ELECTRIC SWITCH
R-1	=	HIGH SIGNAL SELECTOR
R-2	=	HIGH SIGNAL SELECTOR
S-1	=	MINIMUM POSITION SWITCH

Figure 5.3.32 Constant temperature, constant humidity system using mixed air control cycles from room thermostats. (*By permission: McGraw-Hill,* HVAC Control Systems Design Diagrams, *John I. Levenhagen.*)

Condenser water supply temperature from the cooling tower is sensed by T-1, which controls bypass valve V-1 to send water over the tower or bypass the tower and send water back to the condenser. T-1 also can control the fans through a VFD, as an example.

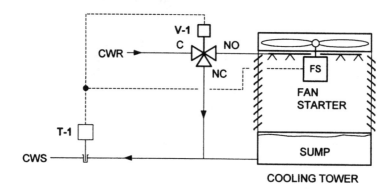

LEGEND:

T-1 = MODULATING CONDENSER WATER SUPPLY THERMOSTAT
V-1 = 3-WAY DIVERTING VALVE

Figure 5.3.33 Cooling tower controls using three-way bypass valves. (*By permission: McGraw-Hill*, HVAC Control Systems Design Diagrams, *John I. Levenhagen.*)

Condenser water supply temperature from the cooling tower is sensed by T-1, which controls valves V-1 and V-2 to send water over the tower or bypass the tower and send water to the sump. T-1 also can control the fans through a VFD, as an example.

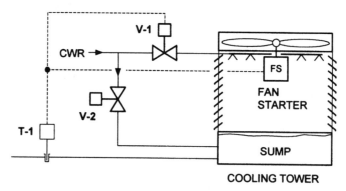

LEGEND:

T-1 = MODULATING CONDENSER WATER SUPPLY THERMOSTAT
V-1 = CONTROL VALVE TO COOLING TOWER
V-2 = CONTROL VALVE TO COOLING TOWER SUMP

Figure 5.3.34 Cooling tower controls using condenser water temperature. (*By permission: McGraw-Hill*, HVAC Control Systems Design Diagrams, *John I. Levenhagen.*)

Whenever the fan runs, the EP-1 is energized and the O.A. damper opens. If the unit is large, arrangements need to be made to be sure the damper is open before the fan runs.

Thermostat T-1 modulates the three-way valve on the hot water preheat coil. Low-limit thermostat LLT stops the fan on a freezing condition at that location.

The pump must run whenever the fan runs, and an alarm system is recommended should the pump stop when O.A. freezing conditions exist.

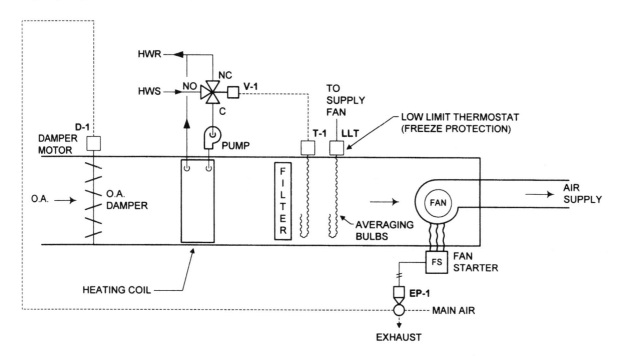

LEGEND:

T-1   = MODULATING CAPILLARY THERMOSTAT
V-1   = MODULATING 3-WAY HOT WATER VALVE
D-1   = TWO POSITION OA DAMPER MOTOR
EP-1 = SOLENOID AIR VALVE
LLT   = LOW LIMIT ELECTRIC FREEZE PROTECTION THERMOSTAT

Figure 5.3.35   Hot water preheat coil controlled by modulating valve and outside air damper controller to prevent freeze-ups. (*By permission: McGraw-Hill,* HVAC Control Systems Design Diagrams, *John I. Levenhagen.*)

Whenever the fan runs, the EP-1 is energized and the O.A. damper opens. If the unit is large, arrangements need to be made to be sure the damper is open before the fan runs.

Thermostat T-1 opens the steam coil valve on the preheat coil whenever the O.A. temperature is below the setting of the thermostat. Thermostat T-2 modulates the face and bypass dampers on the preheat coil to prevent overheating downstream. Low-limit thermostat LLT stops the fan on a freezing condition at that location.

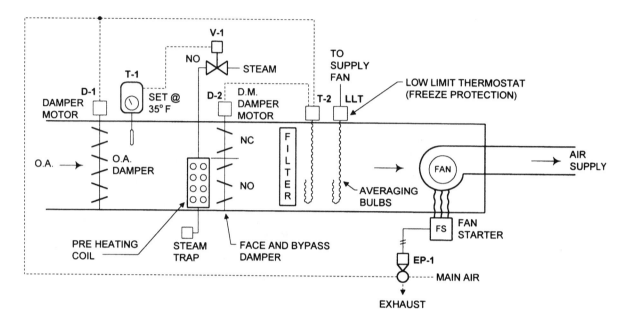

LEGEND:

T-1 = TWO POSITION CAPILLARY THERMOSTAT
T-2 = MODULATING CAPILLARY THERMOSTAT
D-1 = TWO POSITION OA DAMPER MOTOR
D-2 = MODULATING FACE & BYPASS DAMPER MOTOR
V-1 = TWO POSITION STEAM VALVE
EP-1 = SOLENOID AIR VALVE
LLT = LOW LIMIT ELECTRIC FREEZE PROTECTION THERMOSTAT

Figure 5.3.36 Steam preheat coils with thermostat controlling operation of face and bypass dampers to prevent downstream overheating. (*By permission: McGraw-Hill,* HVAC Control Systems Design Diagrams, *John I. Levenhagen.*)

This diagram shows a typical chilled water ceiling radiant panel system. The room thermostat T-1 modulates the 3-way valve V-1 feeding the ceiling coils. The thermostat T-2 is a special "dew point" thermostat that changes the signal to the 3-way valve should "sweating" on the pipes develop.

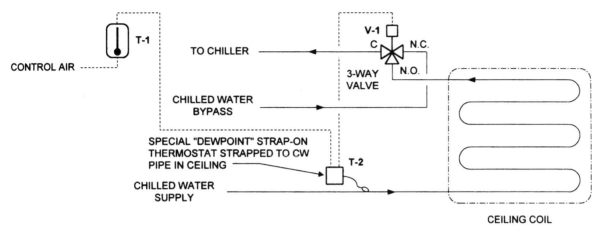

LEGEND:

T-1 = MODULATING ROOM THERMOSTAT
T-2 = SPECIAL DEWPOINT THERMOSTAT
V-1 = MODULATING 3-WAY VALVE ON CW

Figure 5.3.37 Chilled water ceiling radiant panel control system with special thermostat to reduce "sweating" of pipes. (*By per-mission: McGraw-Hill*, HVAC Control Systems Design Diagrams, *John I. Levenhagen.*)

This diagram shows a typical hot water floor radiant panel system. The room thermostat T-1 modulates the 3-way valve V-1 feeding the floor coils.

Typically the systems are zoned in schools where the children are on the floor a lot.

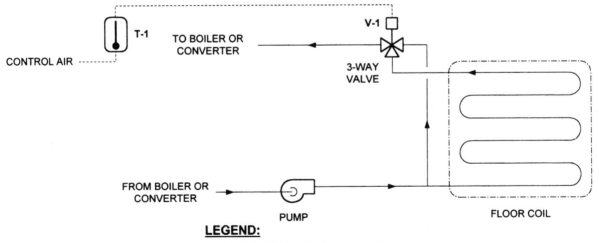

LEGEND:

T-1 = MODULATING ROOM THERMOSTAT
V-1 = MODULATING 3-WAY HW VALVE FOR ZONE

NOTE:
SEE SECTION I FOR PRIMARY CONTROL OF HW SUPPLIED
TO ZONE CONTROL VALVES

Figure 5.3.38 Radiant heating in the floor with three-way valve control. (*By permission: McGraw-Hill*, HVAC Control Systems Design Diagrams, *John I. Levenhagen.*)

Glossary of Terms Relating to Computer Technology, Local Area Networks, Fiber Optics, and the Internet

The world of fiber optics, local area networks, broad band technology, and Internet access has spawned an entirely new language. The following glossary compiled by Mr. Mark Hodgkinson will enlighten contractors and design consultants as they encounter these terms during their ongoing design-construct activities.

Glossary of Terms

Not just a glossary, but also hopefully useful facts and troubleshooting about how these subjects link together in the real world.

a b c d e f g h i j k l m n o p q r s t u v w x y z

Notes on use: Look through the list for abbreviations first, because they take precedence over the full name (i.e., look for FTP instead of file transfer protocol). Because most of us are addicted to abbreviations, I thought that would be most useful. The first instance of a topic covered elsewhere is listed in bold, which lets you know there is more you can read should the explanation not be enough, as well as provides you with a list of interconnected topics. Named anchors were too tedious to implement. To get around the page faster, letter links have been included with each heading.

Miscellaneous

a b c d e f g h i j k l m n o p q r s t u v w x y z

10Base-FL—A specification for Ethernet over fiber-optic cables. The 10Base-FL specification calls for a 10 Mbps data rate using baseband. The advantages of 10Base-FL are long cabling runs (10Base-FL supports a maximum cabling distance of about 2,000 M) and the elimination of any potential electrical complications.

10Base-T—The trend in wiring Ethernet Network is to use unshielded twisted pair (UTP). 10Base-T, which uses UTP, is one of the most popular implementations for Ethernet. It is based on the IEEE 802.3 standard. 10Base-T supports a data rate of 10 Mbps using baseband transmission. It is wired in a star topology. The nodes are wired to a central hub, which serves as a multiport repeater while it functions logically as a linear bus. The hub repeats the signal to all nodes as if connected along a linear bus. The cable uses 8 pin RJ45 connectors and the Network cards normally have RJ45 Jacks built into the back of them. 10BaseT segments can be connected by using Coaxial or Fiber-optic backbone segments. Some hubs will provide connectors for thicknet and thinnet cables (in addition to 10Base-T UTP type connectors). By attaching a 10Base-T transceiver to the AUI port of a network adapter, you can use a PC setup for Thicknet on a 10 Base-T network. **This network is reliable and easy to manage.** They use a concentrator (a centralized wiring hub). These hubs are intelligent in that they can detect defective cable segments and shut them down. **You can design and build your LAN one segment at a time.** This capability makes 10Base-T more flexible than many other LAN cabling options. **10Base-T is also relatively inexpensive to use.** In some cases, existing data-grade cable can be used for the LAN. **Star-based networks are significantly easier to troubleshoot and repair than bus wired networks.** With a star network, a problem node can be easily isolated from the rest of the network. The maximum number of computers on a single collision domain is 1,024. (Practical considerations such as traffic volume usually keeps it to a much smaller number.) The cabling should be UTP categories 3, 4, or 5 and the maximum unshielded cable segment length (hub to transceiver) is 100 M (328 feet).

100Base-X—Uses a star bus topology similar to 10Base-T. 100Base-X provides a data transmission speed of 100Mbps using baseband. It is sometimes referred to as Fast Ethernet. Like 100VG-AnyLAN, 100Base-X provides compatibility with existing 10Base-T systems (that were properly cabled), and this enables plug and play upgrades from 10Base-T.

100VG-AnyLAN—Defined in the IEEE 802.12 standard. IEEE802.12 is a standard for transmitting Ethernet and token ring packets (IEEE 802.3 and 802.5) at 100 Mbps. 100VG-AnyLAN is sometimes called 100Base-VG (VG stands for *voice grade*). 100VG-AnyLAN's demand priority access provides for two priority levels when resolving media access conflicts. 100VG-AnyLAN uses a cascaded star topology, which calls for a hierarchy of hubs. Computers are attached to child hubs, and the child hubs are connected to higher-level hubs called parent hubs. The maximum length for the two longest cables attached to 100VG-AnyLAN hub is 250M (820 ft). The specified cabling is Category 3, 4, or 5 twisted pair or fiber optic. 100VG-AnyLAN is compatible with 10Base-T cabling. Both 100VG-AnyLAN and 100Base-X can be installed as a plug-and-play upgrade to a 10Base-T system.

.386—Indicative of a Protected Mode Windows Driver.

5-4-3 Network Rule—States that the following can appear between any two nodes on an Ethernet Network: Up to 5 cable segments in series; up to 4 concentrators or repeaters; up to 3 segments of cables that contain nodes. (This really only applies to coaxial cables, however, because UTP and fiber-optic networks are always implemented in a point to point fashion.)

8228 MSAU—The original wiring hub developed by IBM for Token Ring networks. Each 8228 has 10 connectors, 8 of which accept cables to clients or servers. Ring in (RI) and Ring out (RO) connectors are used to connected multiple 8228's to form larger networks. It has mechanical devices that consist of relays and connectors. Their purpose is to switch clients in and out of the network. Each port is controlled by a relay powered by a voltage sent to the unit by the client. When an 8228 is first set up, each of these relays must be initialized with the setup tool shipped with the unit.

a b c d e f g h i j k l m n o p q r s t u v w x y z

Access Point—Wireless LAN terminology—a stationary transceiver connected to a cable based LAN that enables the cordless PC to communicate with the network. The access point acts as a conduit for the wireless PC.

Account Domain—*See* Trusted.

ACL—Access Control List—Each resource has an ACL that contains a list of SIDs who have access to the resource in a domain networking model.

ACPI—Advanced Configuration and Power Interface—Part of the OnNow design initiative and is how the OS and BIOS communicate to handle power issues, with particular relevance to Windows 98 and Windows 2000. It is a hardware specification that allows for the powering off of devices such as hard disks, CD-ROMs, and Monitors. Features include immediate power on when a button is pressed, but some applications that assume the power is always on may have a hard time with such standards.

ADMIN$—Default share created under Remote Administration on '98 that gives Administrators access to the hard disk.

Administrators—A local group in Windows NT. The Global Group Domain Admins typically is placed in this local group on each server in the domain.

AFP—Appletalk File Protocol—Provides remote file services. Responsible for enforcing file system security. Verifies and encrypts logon names and passwords during connection setup.

AGP—Advanced Graphics Port—A dedicated Graphics bus in a PC.

Alt + F5—Boots a Windows '95/'98 machine to the safe mode Command Prompt.

Allocation Unit—*See* Cluster.

Analog—Changes continuously and can take on many different values. Music is a good example. Analog signals vary constantly in one or more values and these changes in values can be used to represent data. Analog waveforms frequently take the form of sine waves that have the characteristics of frequency, amplitude, and phase. These characteristics can be used either individually, or together to encode data. *See also* Digital.

API—Application Programming Interface—A set of functions that give application programmers access to common system functions.

Apple Share—A client/server system for Macs. Provides file and print sharing services.

Apple Talk—The network computing architecture developed by Apple computer for the Mac family of personal computers.

Applet—A small program that is started for instance when you click an icon in the Windows Control Panel.

Application—Terms that encompass the programs and packages we run on PCs. 16-bit applications can send the processor only one thread of execution at any one time, and 32-bit applications can send multiple threads. Under '98, should one 16-bit application crash, it will hang up all other applications. When that application is closed however, the remaining applications will continue to run. Under NT, all 16-bit applications will fail. 32-bit applications run separately to each other in all cases and 16-bit applications under NT can be made to run in their own memory spaces, allowing other 16-bit applications to continue to run, unaffected by any other's failure.

Application Layer—*See* **OSI.**

Application Server—An application server runs all or part of an application on behalf of the client, then transmits the result to the client for further processing. Care must be taken when discussing application servers to make it clear about what type of system you are discussing. The term is sometimes used to refer to a server on which end-user applications are stored. This is generally considered a misuse of the term, because this is simply a file server that happens to contain executable files. An application server is a network resource, normally thought of as a server end of a client/server application. The application server is responsible for a share, often a large share, of application processing. It works with a client application running at the end-user workstation that provides the user interface and is normally responsible for preparing server requests and formatting returned results. Server system requirements can sometimes be significant, with the client requirements somewhat less. This makes for an efficient model, letting you organize machines to their role. In many cases it means that pieces are somewhat interchangeable. You will often have a choice of client applications as long as your selections are compatible with the server application. In many cases, upgrades to the server application do not necessarily mean you need to upgrade your client applications. This reduces the long-term costs of supporting the application. One of the most common applications of this model is through database servers.

ARP—Address Resolution Protocol—A protocol that TCP/IP hosts use to discover the hardware address of a destination node when only the IP address is known. A hardware address refers to a unique address on the network for the network adapter card. Windows '98 and NT include a command line tool for examining the systems local ARP cache. For more information, type ARP /? at the command prompt.

ARPA—Advanced Research Projects Institute—A government agency responsible for developing robust internetworking technologies that led to the ARP Anet, a predecessor of today's Internet.

ASCII—American Standard Code for Information Interchange.

AsyBEUI—Asynchronous NetBEUI—Created for early version of NT RAS.

ASF—Advanced Streaming Format—For MPEG4 audio/video playback.

ASP—Active Server Pages—If using MS Internet Information Server, you can test your pages' look and work right before deploying them to your server.

Asynchronous Transfer Mode—With USB, the general data transfer method of 1.5 Mbps.

Athlon—AMD processor to rival Pentium II and III. Until this point, PII and PIII had no competition from AMD because their offerings of the the K6-2 and K6-III just were not up to snuff in terms of floating-point performance, and they also lagged behind the Pentium II and III in clock speed. Its Athlon chip (also known as K7) offers superior performance to the Pentium III at the same clock speed allegedly. Although AMD and Intel have taken different approaches to chip design, when all is said and done, the Pentium III and the Athlon are quite similar in terms of chip features. For one thing, currently both chips are built on a .25-micron manufacturing process and use aluminum interconnects (IBM is pushing to have its Power PC chips built with a somewhat more advanced copper process). Intel's so-called Coppermine Pentium IIIs, set to debut in Autumn of 2000, will shift to a more advanced .18-micron process that should have lower power requirements and allow for increases in clock speed. Although AMD is sure to move Athlon to a .18-micron process, it has not announced that part of its roadmap. The Athlon's 128K of internal primary (L1) cache is larger than the Pentium III's 32K. That's certainly an advantage, but probably not as great of an advantage as the numbers might suggest; performance is also affected by how the chip's cache controller is implemented. Do not use it. More important, perhaps, is the L2 cache. Right now, both chips come with 512K of external L2 cache running on a backside bus at half the CPU's clock speed. (Note: The L2 cache of Intel's high-priced Xeon processors runs at full clock speed, and AMD is expected to come out with similarly configured Athlons.) With the upcoming Coppermine chip, however, Pentium III is set to gain a little advantage by incorporating 256K of internal L2 cache running at full clock speed. AMD's K6-III processor has a similar configuration, although AMD has not yet announced plans to incorporate cache into the Athlon. In addition to old-school MMX support, both CPUs include pro-prietary SIMD (single instruction multiple data) instruction sets. AMD introduced its 3DNow instruction in 1999 with its K6-2 processor, and Intel caught up in the year 2000 with its SSE instruction set in the Pentium III. Architecture-wise, these two are pretty similar, with each being able to perform four floating-point opera-tions in a single clock cycle. This can speed up the geometry stage of the graphics pipeline and could also be applied to more sophisticated physics modeling in games. Intel includes in its SSE some special integer in-structions useful for speeding video encoding and voice recognition. Through its MMX license, AMD has been able to match these instructions exactly in the Athlon. The two companies' SIMD floating-point instruction sets, however, are proprietary, and each must be supported separately by games and display drivers. It is im-portant to make sure that the graphics card you pair the processor with is optimized for the instruction set. This is especially true for accelerating OpenGL performance in games such as Quake III. A chip by itself does not a PC make; it is only one part of a system architecture. The design and features of that architecture de-termine a PC's performance. With the Athlon, AMD is introducing a 200-MHz system bus based on the Alpha EV6 bus protocol. This compares to the 100-MHz system bus of current Pentium III systems. This design should be especially helpful for multiprocessor Athlon systems when they are released. As fast as this EV6 bus may be in theory, Athlon systems are not yet ready to take full advantage of it. AMD's first chipset sup-ports only the 200-MHz speed between the CPU and the so-called northbridge of the chipset; system memory speed is limited to 100 MHz. This 200-MHz pipe may relieve some system bottlenecks, but a bigger boost in performance should come in 2001, when AMD adds support for faster memory speeds. Unlike the K6-2, which put the L2 cache on the relatively slow system bus, the Athlon follows the lead of the Pentium II/III in sup-porting a backside L2 cache that interfaces directly with the CPU, independent of the system bus. The cur-rent Pentium III runs with a 100-MHz system bus, but Intel is expected to release its "Camino" chipset in Sep-tember. Adding support for Rambus memory and raising the system bus speed to 133 MHz, Camino will also support the AGP 4X standard for graphics cards, offering twice the speed of the current AGP 2X. Athlon sys-tems supporting AGP 4X, however, are not expected until the beginning of the year 2001.

ATM—Asynchronous Transfer Mode—Network Protocol designed to run at very high speed for the transfer of not only data but also voice and real-time video. It is a connection orientated protocol and must establish a connection before it can transfer data. Once connected, the data is transferred using dedicated network bandwidth using three components. ATM Call Manager, which establishes the connection with the switch. ATM Lan Emulation Client is client software that provides connectivity to the ATM network. It enables com-munication between the ATM driver and **NDIS.** ATM Emulated LAN, which is used to implement a Virtual LAN (VLAN), which lets you portion segments of the LAN into smaller segments. It is considered the best choice for mixing voice, video, and data. It is a high-bandwidth switching technology developed by the ITU Telecommunications Standards Sector (ITU-TSS). An organization called the *ATM forum* is responsible for defining ATM implementation characteristics. ATM can be layered on other physical layer technologies such as SONET or SDH. It is based on fixed length 53-byte cells (with 5 bytes for the header and 48 bytes of data), whereas other technologies employ frames that can vary in length to accomodate different amounts of data. Because ATM cells are uniform in length, switching mechanisms can be easily implemented in hardware to operate at a high level of efficiency. This high level of efficiency results in very high data transfer rates. Some ATM systems can operate at several gigabits per second, but the most common two speeds for ATM are 155 Mbs and 622 Mbps. Asynchronous delivery refers to the characteristic in ATM in which transmission time slots do not occur periodically (as in the traditional telecommunications TDM environment), but are granted at irregular intervals as needed. Traffic that is time-critical, such as voice or video, can be given priority over data traffic that can be delayed slightly with no ill effect. A high-priority transmission need not be held until its next time-slot allocation. Instead, it might be required to wait only until the current 53-byte cell has been transmitted. As in other switched WAN services, devices communicate on ATM networks by establishing a vir-tual circuit. It is a relatively new technology and only a few suppliers provide the equipment necessary to sup-port it (it must use ATM-compatible switches, routers, etc.). An interesting advantage of ATM is that it makes it possible to use the same technology for both LANs and WANs. Some disadvantages, however, include the cost, limited availability of equipment, and present lack of expertise due to its recent arrival. The most com-pelling reason for its use is where you require WAN speeds of more than 100 Mbps.

ATM Forum—*See* ATM.

ATP—AppleTalk Transaction Protocol—*See* NBP—AppleTalk Transport Layer protocols.

Attenuation—A measure of how much a transmission medium weakens a signal. Attenuation measurements al-ways specify the frequency used to make the measurement because attentuation always varies with fre-quency. The higher the frequency, the greater the attenuation, generally. Too much attenuation weakens a sig-nal to the point where the signal is indistinguishable from the background noise.

AUI—Attachment Unit Interface—Connects a workstation to the Thicknet medium.

AUI cable—(also called a *tranceiver cable* or *drop cable*)—Connects a workstation to the thicknet medium.

AUI Connector—Used to connect to category 5 or category 3 twisted-pair cables through a media filter, which converts the 9-pin AUI interface to the RJ-45 interface.

Auto Dial—Automatically establishes a connection when accessing a remote computer, file, or printer.

Automatic Private IP Addressing—If a machine cannot access a DHCP server, the machine allocates itself an address of 169.254.x.x until the DHCP server comes on-line at any time, at which point the system will gain an IP address as normal. The only two ways a computer with this system can access the Internet is either through a proxy server or NAT gateway.

AWG—American Wire Gauge—A specification for wire gauges. The higher the gauge, the thinner the wire.

B

a b c d e f g h i j k l m n o p q r s t u v w x y z

B-ISDN—Broadband Integrated Services Digital Network—*See* Frame Relay, ISDN.

Backup—A backup policy fall's under one of the following categories. *See* Single Backup Server, Individual Tape Units, Independent, redundant backup network. *See also* Full, Differential, and Incremental Backups.

Backup Browser—To stop a Browse Master from being overwhelmed, any computer with at least file and print sharing enabled can become a Backup Browser. If there is only one computer on the network, then there are no Backup Browsers. With 2 to 15 machines, there is 1 Backup Browser, and after that there is another 1 created for every further 15 computers. **Note:** Windows 95 resource kit says 1 extra per 32 computers?!!? *See* **Browse Master**

Bad Sectors—Caused by normal wear-and-tear of the disk and also by mishandling. Under 95/98, you can use Scandisk to try and recover data that occur on bad sectors and to mark the sector so it is no longer used. Under NTFS on SCSI drives, hot fixing can move data on the fly from the bad sector so as not to cause any user problems at all, although a new hard drive should be considered as soon as possible.

Band—A group of contiguous frequencies in the electromagnetic spectrum, used for a common purpose or treated as a single frequency.

Bandwidth—Refers to the measure of the capacity of a medium to carry data.

Baseband—Devotes the entire capacity of the medium to one communication channel. Most LANs function in baseband mode.

Batch 98 (Installation)—The automated installation of Windows 98 using a batch file to input needed information that would normally be supplied by the user.

Baud—The term *baud* is sometimes incorrectly used to refer to bits per second (bps). Phone lines are limited to 2,400 baud, but modems operate at 28,800 bps or better. (*Baud* is the abbreviation for Baudot, which gets its name from J. M. Emile Baudot, 1845–1903, who invented the code.) The Baudot code is a special set of binary characters, which use 5 bits per character forming 32 combinations, which was increased to 62 through the use of two special shift characters. The Baudot code was mainly used to handle telex messages by common communication carriers such as Western Union. The main disadvantage of the Baudot code is its lack of an error-checking bit.

BDC—Backup Domain Controller—A fault tolerance initiative for Microsoft Domain networks. A BDC receives automatically updates from its PDC's SAM database while it is part of that domain. A BDC cannot be moved from domain to domain; the only way to move a BDC to another domain is to reinstall NT Server on the machine.

Bidirectional—Means two ways (of course), but the terminology can become confused with reference to printers. A bidirectional printer normally refers to some sort of raster or line printer, and after printing one line of data, the head does not waste time by returning to the carriage rest, but instead prints the next line of data from right to left instead of left to right. Bidirectional communications, to be more exact, means that the printer can talk back to the computer and advise of low toner, no paper, etc. Novell Print queues suffer from unidirectional communication, whereas NDPS (Novell Distributed Print Services) found in Novell 5 now features this ability. You need an IEEE1284 cable for bidirectional printing (two-way communication).

Bindery—The Novell name for a database of users or groups on a Netware 2.x or 3.x server. It stores all account, security, and other related information. The disadvantage of this method was that the Bindery was server-centric, which meant this information could not be shared between servers. To access resources on two servers, you would need two accounts. It was replaced with NDS in NetWare 4.x. *See* NDS.

Bindery Emulation—The process of a NetWare 4.x Server emulating the way that NetWare 2.x and 3.x servers provide and store security information. Available for backward compatibility.

Bit Pipes—*See* ISDN.

Black Box—Device used to link a printer for instance to the network when you cannot use a network card.

BOOTP Routing—It is possible to configure a DHCP server to provide support for multiple subnetworks. There is, however, a requirement. All routers between the client system and the DHCP server must be BOOTP compliant. Initially, and to some extent now, BOOTP (Bootstrap protocol) is used to boot and configure diskless workstations. DHCP was created to be an extension to BOOTP. Originally, BOOTP was defined in RFC 951. Currently, the BOOTP RFC is RFC 1542 that includes DHCP support. A workstation needs a variety of configuration parameters, including IP address and default gateway. The use of messaging to configure workstation(s) paved the way for DHCP as an extension. BOOTP and DHCP message formats are nearly identical, some fields are used by DHCP differently and DHCP has new fields that are not included in BOOTP. The major advantage of DHCP using the same messaging format as BOOTP is that DHCP messages can be relayed between two subnets by existing routers that act as BOOTP (RFC 1542) relay agents. This makes it possible to have IP addresses and configuration information for systems on both subnets configured on a single DHCP Server.

Bounded Communications—Wire-and-fiber-based communications. This is the type of media with which you will most often be working. Closely related subjects are network topologies, access protocols, and network adapter cards.

Bridge—Limits the traffic between two computers or networks. They work at the MAC sublayer of the data link layer of the OSI model and can be used to extend the maximum size of a network. They pass on only those signals targetted for a computer on the other side of the network, and it can make this determination because each device on the network is designated a unique address. For example: The bridge receives every packet on LANs A and B. It learns from the packets which device addresses are located on A and which are on B, which allows it to compile a table of information. Packets on A and addressed to A are discarded, as are packets on B addressed to B—they can be delivered without the bridge. Packets on A addressed to devices on B are transmitted to LAN B, and the same in reverse for packets on B addressed to A. Most bridges are now learning bridges, which means the table is built automatically—otherwise the table had to be programmed into the device. However, there are restrictions when using bridges. A network with bridges generally cannot make use of redundant paths. They cannot analyze the network to decide the fastest route over which to forward a packet, which is a very desirable thing to have in a WAN environment. They also do not filter out broadcast packets so they are of no use in broadcast storms, and bridges cannot join dissimilar types of LAN because they depend on the physical address of devices. The physical address is a function of the data-link layer, and different data-link layer protocols are used for each type of network, so a bridge cannot be used to join an Ethernet and Token Ring network. However, before accepting the statement that different links cannot be connected, *see* Heterogenous or Translating Bridges.

BIOS—Basic Input Output System—A program stored in the firmware of a computer that contains the basic information and drivers needed to get the computer up and running. Firmware is another way of referring to software stored on a ROM chip. This information is nonvolatile, meaning that it remains there when the power is switched off.

BIOS Bootstrap—Performs the **POST** operation, identifies and configures PnP devices and locates the bootable partition.

BNC Barrel—Connects two thinnet connectors.

BNC Cable Connector—Attaches cable segments to the T connectors.

BNC Connector—Bayonet–Neill–Concelman or British Naval Connector. Generally used on Thinnet cabling.

BNC T Connector—Must be used to connect the network card in the PC to the network. The T connector attaches directly to the network board.

BNC Terminator—A special kind of connector that includes a resistor carefully matched to the characteristic of the cabling system. Necessary to stop cable reflections.

Boot Sector—The sector on the hard disk that contains the basic information about the OS. This sector is the first sector on the hard disk that is read during the boot process.

Broadband—Enables two or more communication channels to share the bandwidth of the communication medium.

Broadband Optical Telepoint—This method uses broadband technology. Data transfer rates in this high end option are competitive with those of cable-based networks. *See also* Infrared Transmission.

Brouter—A router that can also act as a bridge. A brouter attempts to deliver packets based on network protocol information, but if a particular network layer protocol is not supported, the brouter bridges using device addresses. If you need to connect networks that are using a routable protocol such as TCP/IP and a non-routable protocol, such as NetBEUI, you must use a Brouter. This gives you the subnetting advantages of a router while maintaining the full connectivity of a bridge. Almost all dedicated hardware routers (including those from 3COM, Cisco, and Bay networks, for example) include an option to support bridging and can be considered brouters. Brouter is becoming an obsolete term in the networking industry.

Browse List—MS Networks—Essentially a phone book of the machines on line on the network at a given time, because having all machines advertising their presence all the time would quickly flood the network with broadcast traffic.

Browse Master—This machines keeps a centralized copy of the Browse List. There is always one Browse Master per workgroup or domain. By default all computers with File and Print sharing enabled have the capability to become Browse masters. *See also* **Backup Browsers, Browser Elections.** This is also a setting on '98 machines under Advanced—File and Print Sharing.

Browser—A NetBIOS client that attempts to contact other NetBIOS resources on a network, using the Browse list maintained by one or more browse masters. The Browse List contains a list of active systems identified by their NetBIOS network names. *See also* Browse Master.

Browser Elections—In the first instance, the election of Master Browser is decided by OS run by the machine (i.e., NT Server installed as PDC, NT Server, NT Workstation, '98, '95, Windows for Workgroups). If two computers are competing for the role of Browse Master and both are running the same OS, the one with the newest version is elected. If both machines are the same, then the one that is on-line the longest is elected. PDCs are normally the Browse Masters because they are the Preferred Master Browser.

Bus—An architecture for communication between multiple devices, whether on a system board or across a network. A bus architecture is a shared medium, where all devices on the bus must contest for communication time, and all devices can hear all transmissions.

Bus Topology—(or linear bus)—The simplest form of networking, making it the least expensive to implement. This topology consists of a single cable that connects all computers in a single line without any active components to amplify or modify the signal. This bus must be terminated at each end. Without termination, the signals on the bus reflect back on reaching each end of the bus. This causes serious network errors. The most common form of this topology is 10Base2. This is also referred to as *Thinnet* and uses RG58 cable that has a 50-ohm impedance and so must be terminated by a 50-ohm terminator at each end. The maximum segment distance is 185M or 607 feet. Another bus topology is 10Base5 or Thicknet. 10Base5 networks use RG6 cable, which is much thicker and harder to work with than RG58. The bus network is a passive technology because each computer only monitors the signals on the bus—they do not pass through the network card in the PC. To increase the distance of the bus, repeaters are used. These are active devices that regenerate incoming signals as they pass through the repeater.

C

a b c d e f g h i j k l m n o p q r s t u v w x y z

C2 Security—A measure of how secure a network server operating system is according to the National Security Agency in America. NT was awarded this status as a stand-alone machine and is currently being evaluated as a network OS.

Cable Testers—Cable testers can be used to check the condition of the cable. They can also be used to monitor network traffic. Traffic to or from a specific computer can be monitored as well. A cable tester can be used to check for excessive collisions or beaconing, or to locate bad cable segments or bad network adapter cards.

Cache—Commonly used data from one data storage medium stored on another faster medium. *See also* Caching.

Caching—The process of storing commonly used data from one data storage medium in another faster storage medium. This greatly increases the speed at which the computer can access information. By caching recently used information that may be used in a faster medium (such as RAM) rather than accessing it directly from the slower medium (such as the hard drive), access times can be greatly reduced.

Capture—Some MS-DOS programs do not understand UNC names. Therefore, printer ports must sometimes be captured, fooling the local application that the printer is local.

Cascaded Star—*See* 100VG-AnyLAN.

CBT—Computer Based Training—Instead of having an instructor, your PC teaches you what you need to know.

CDF—Channel Definition Format—The CDF allows a user to personalize information that is received and the method of delivery from a **channel.**

CDFS—Compact Disk File System—A file format used exclusively for CD ROMs.

Cell Relay—Another term for ATM. *See* ATM.

Cellular Networking—The mobile device sends and receives cellular digital packet data CDPD using cellular technology and the cellular phone network. Cellular networking provides very fast communications. *See* Mobile Computing.

Change Access Control Permission—Users can change security settings on MS networks. Equates to the NetWare Access Control Right.

Change File Attributes/Permission—On MS Networks, users can change the attributes of files and folders, read only, hidden, and so on. Equates to the Modify NetWare Right.

Channel—A web site that delivers content to your computer on a schedule. When you subscribe to a channel, information is automatically downloaded from the channel's web site to your desktop. A channel is a web site that includes a **Channel Definition Format** or **CDF.**

Chemical Epoxy—Used to splice fiber-optic cables. *See also* Laser Tape.

CIDR - Classless Interdomain Routing.   IP addresses are divided into classes. In the past, you would be assigned a certain class of IP addresses based on how many machines would need one. Now, the InterNIC has devised this new way of assigning addresses to avoid wasting large numbers of addresses.

CIR—Committed Information Rate—*See* Frame Relay.

Class A Addresses—The first octet of a Class A address ranges from 001 to 126, 127 being reserved for diagnostic reasons. When InterNIC assigns these addresses, only the first octet is assigned to you and the rest is left for you to assign. This could be as many as 16,777,254 computers—but for only 126 lucky class A address owners.

Class B Addresses—The InterNIC here would assign the first two octets to you, leaving the latter two octets for you to assign. The first octets range is from 128 to 191, and the second ranges from 0 to 255. This gives 65,534 possible assignments. Again, as with class A, all 16,384 class B assignments have been made.

Class C Addresses—InterNIC decides on the first three octets, leaving the final octet for you to modify. The first octet ranges from 192 to 223, the second and third from 0 to 255. This leaves you with 254 addresses to assign. You cannot use address 255 because it is reserved for broadcast use. There are still class C networks available from the pool of 2,097,152 possible combinations.

CLIK Drive—A 40Mb disk format introduced by IOMEGA for easy storage of digital camera data, PDAs, and even for their own Clik!Boy to play games with and the Clik!Phone with all the contact details held on the disk.

Client—Any computer that accesses a server for services or data.

Client 32—Because Novell does not make desktop OSs, this is their connectivity software, the main advantages of which are that it fully supports NDS and because it supports any NETX or VLM functions, which the Microsoft client will not, even with service for netware directory services. Also, if you need to use Novell utilities such as Novell IP Gateway, Novell Application Launcher (NAL), and so on. If you will be using NetWARE IP and you wish to use NetWARE NCP packet signature security. The NetWare Client 32 for NT fully supports both NDIS and ODI standards gives full NDS support and gives full access to NetWare services and Network security. Gives automatic reconnection and automatic client updates with Multiprotocol support. It requires 10Mb of free disk space, and if installing the additional Administration utilities will require an additional 16Mb of disk space.

Cluster—An OS allocates disk space for files in units of one or more sectors, these units are deemed clusters or allocation units. Clusters may vary in size depending on the size of the disk. On a disk, a 1 byte file is allocated one cluster of disk space, wasting the unused area of the cluster. A file that is 3.2 clusters large is given 4 clusters. Overall, a smaller cluster size means less waste. The cluster size of the FAT 16 file partition is significantly larger than that of the FAT 32 file system.

CODECS (Coder/Decoder)—In many respects CODECs function as mirror images. Whereas a modem converts digital data so that it can be transmitted on an analog communication medium, a COder/DECoder converts analog data for transmission on a digital transmission medium.

Cold Boot—The process of restarting a computer by completely cutting off the computer's power and then turning it back on.

Cold Docking—Connecting a laptop to a docking station when the laptop must be switched off before it can be docked or undocked.

Command Line Switch—An addition to the normal command line of a program. When running the program at a command prompt or using the Start/Run utility, you can specify that the program run in a different way, usually by adding a (/) and a letter or phrase. An example would be FORMAT /S. This command will format the disk and then add the system files to make it bootable. Without the /S, the command would simply format the disk specified.

Complete Trust Domain Model—This is the most difficult trust model to install and maintain on an NT domain. Here, each domain trusts every other domain with a two-way trust relationship. In order to calculate the number of trusts required for this model, you can use the formula $n(n - 1)$, where n is the number of domains. By looking at the formula, you can tell that the number of trust relationships can get enormous over a period of time; as such, you should only use the complete trust model where the number of domains will be relatively small. It is though a scalable model that permits the use of of centralized and decentralized administration for user accounts and resources. This can be desirable in large companies that prefer to have IS departments at each location manage their own resources. There is also a network security issue because you have no control over the membership of global groups that are outside the local domain administrator's control. In addition to the trust relationship between domains, this type of model requires complete trust between administrators.

CONFIG.POL—The file that makes up a Windows 95/98 policy. This file is usually stored in the Windows directory of the local computer. If the policy will be server based, it will be stored in the NETLOGON share of a Windows NT server or the Public folder of a NetWare server.

Connection Orientated—A communications system that assumes there will be communication errors between computers. With this in mind, these protocols are designed to make sure data is delivered in sequential order, error-free, to its destination. TCP/IP is an example of a connection orientated protocol. (*See* Connectionless Orientated.) The path from PC to PC is predetermined and can contain several links, forming a logical pathway called a *connection*. The nodes forwarding the data packet can track which packet is part of which connection. Internal nodes can therefore provide flow control as the data moves along the connection. Should a node detect a transmission error, it requests the preceding node to retransmit the bad packet. The nodes keep track of which packets belong to what connection, allowing several concurrent connections through a single node.

Connectionless Orientated—These systems assume data will reach its destination with no errors. Thus, there is no protocol overhead associated with these systems. Without this overhead, these systems are typically very fast. User datagram protocol (UDP) does not incorporate the level of internal control mechanisms found in connection orientated mode. Error recovery is delegated to the source and destination nodes that acknowledge the packet and retransmission of lost or bad packets. These internal nodes only forward packets without tracking connections and handling errors. However, in an error-prone environment with many links, this mode can become slower because the packet must be retransmitted from the source node rather than an internal node.

Contention—Where computers are contending for use of the transmission medium. Any computer in the network can transmit at any time (first-come, first-served access, however). *See* Token Passing.

Control Panel—A software utility that allows you to modify the settings and behavior of Windows products after the '95 release. These settings are stored in the registry.

Conventional Memory—The first 640K of memory.

Coaxial Cable—the first type of cable used on LANs. Gets its name because two conductors share a common axis—more frequently referred to as *coax*. There are two types of coaxial cable—Thinnet and Thicknet.

Cooperative Multitasking—Programs must take turns sending threads of code to the processor. The programs are in control here, cooperatively sending their threads of code. Unfortunately, some programs will not release the processor, which can lead to a very poor situation. Normally found in 16-bit applications. '98 supports cooperative multitasking for 16-bit applications, but should one crash, when that application is terminated, the others will restart due to their being a single message queue. Windows NT can allow 16-bit applications to run in their own protected memory spaces, and should one crash, the others will continue to function. If they all run in the same memory space, however, all the 16-bit applications are generally lost. *See* **Preemptive Multitasking.**

CRC—Cyclical Redundancy Check—For error-checking data transmissions.

Create Permission—On MS Networks, users with this permission can create files and folders on a resource. Equates to the NetWare Create right.

Cross-linked files—A FAT error! Where files attempt to use the same cluster(s) on a hard disk. The data in the cross-linked cluster is usually correct for only one of the cross-linked files or it will be incorrect for both.

Crosstalk—A special kind of EMI caused by having wires next to each other carrying data and "leaking" some of their signal as EMI. Crosstalk is of particular concern in high-speed networks using copper cables due to their typically being so many next to each other.

CSMA/CA—Carrier Sense Multiple Access/Collision Avoidance—A slower system than CSMA/CD; thus it is not as popular. A computer wanting to transmit waits a random amount of time after the last transmission on the cable. When this timer has expired, the computer transmits the data. The transmitting machine does not detect collisions; instead it waits for a response from the destination computer. If a response is not received in a certain amount of time, the computer waits for the cable to be idle and tries again. Local talk, which is the transmission medium for AppleTalk, uses CSMA/CA.

CSMA/CD—Carrier Sense Multiple Access/Collision Detection—CSMA/CD is a bus contention communication method used by networking technologies such as Ethernet. In an CSMA/CD network, workstations first determine whether a transmission is taking place (carrier sense) before trying to transmit. Once the wire is free, any workstation can attempt to transmit packets (multiple access). However, if multiple workstations attempt to transmit at the same time, a collision occurs (collision detection), and the workstation waits a random period of time to retry communication. CSMA/CD networks are not suitable for real-time data transmission because no workstation is guaranteed the capability to transmit its packets. However, CSMA/CD networks have proven simple enough and reliable enough for most data transmission needs.

CTS—Clear to Send.

CVF—Compressed Volume File—A single data file that comprises the entire contents of the compressed portion (or volume) of the hard disk. The data file is created when using the '98 or other MS equivalent disk compression utility. It would be named something like DRVSPACE.000 or DBLSPACE.000. *See* **Host Drive.**

D

a b c d e f g h i j k l m n o p q r s t u v w x y z

.DRV—Indicative of Real Mode Windows Drivers, usually found in the SYSTEM.INI file on a '95 or '98 machine.

Daily Copy—Type of backup in which only the files modified that day are stored. The files backed up are not marked as being backed up.

Database servers—Machines that store and retrieve data. Database servers differ from file servers in that certain search functions and some processing of the data can be performed on the server. A special term is used to describe the interaction between the clients and database servers—*client–server database management.* Typically, the client software formulates the request and process the responses, whereas the server database software evaluates the requests and return any required data. It is this division of labor that is the mark of a client–server database system. Network database services provide the following: Optimization of the computer running the database application for storing, searching, and retrieving database records; controlling where in the network the database is physically stored; local organization of the data; directory security; and fast access time. Network database servers must also offer distributed data coordination. Most companies distribute the responsibility for controlling and updating information. When using a common database, ownership and control of the data becomes an issue. To address this, we can use distributed data coordination services. Using this tactic, the different portions of the data are assigned to individual departmental computers for storage. This offers seamless data retrieval by allowing the database server to act as the database manager, retrieving and updating the data when necessary. *See* Replication of a Database.

Data-Link Layer—*See* OSI

DCD—Data Carrier Detect.

DCE—Data Communications Equipment—Provides access to the Telephone Network—*See* X.25

DDP—Datagram Delivery Protocol—The AppleTalk Network Layer protocol.

DDS—Digital Data Service—Usually implies a relatively low-speed digital service used for SNA connectivity. DDS circuits usually transmit data point-to-point at 2.4, 4.8, 9.6, or 56Kbps.

Defrag—*See* Disk Defragmentor.

Demand Paging—The process in which an OS moves files to and from memory and to and from the hard disk. This process frees up memory so it can be used by other applications. *See* Paging.

Demand Priority—An access method used with 100 Mbps 100VG-AnyLan. Although demand priority officially considered a contention-based access method, demand priority is considerably different from basic CSMA/CD. Network nodes are connected to hubs, and those hubs are duly connected to other hubs. Contention, therefore, occurs only at the hub. (100VG AnyLAN cables receive and transmit data at the same time, using four twisted pairs in a scheme called *quartet signaling.*) The advantage is that a mechanism is provided for prioritizing data types.

Demarc—A demarc is the location of the subscriber's premises that marks the dividing line between equipment that is the responsibility of the subscriber and the local loop, which is the responsibility of the local phone company.

Demultiplexing—The recovering of the original separate channels from a multiplexed signal. *See* Multiplexing, Mux.

Desktop—The background area of the screen that contains the My Computer, Network Neighborhood, Recycle Bin, and other icons. This area directly relates to a folder on the hard disk that allows folders and files to be created directly on the desktop. The default location for this folder under '98, and is \windows\desktop.

Device Driver—A program that provides an interface between a piece of hardware and an OS.

Dial Up Server '95—Installed from the Plus pack—only one inbound connection at a time is supported. For more, use NT server.

Dial Up Server '98—'98 can support only one connection at a time for inbound connections. Although it can be configured to take incoming calls on multiple modems, it can use only one at a time. For more than this, use NT Server. You should use **User Level Security** to help avoid any security breaches and also require encrypted passwords. Should the Internet be involved, use a **Firewall** to keep dial up and **Internet** access separate and try to encrypt data when sending it over the Internet. **Note:** If you set up user-level security but leave the password blank, users who connect must also leave the password blank to connect successfully.

Dial Up Networking '98—Supporting the protocols of NetBEUI, IPX/SPX, and TCP/IP. The generic term for remote connection to a network with MS Clients. When connected, you are basically a remote node of your network, with full access if allowed by the Administrator, albeit over a slow connection. Supported line protocols are PPP (point-to-point protocol), SLIP (serial line interface protocol), NT RAS, and NetWare Connect. Supported Network interfaces are NetBIOS, Mailslots, Named RPC (remote procedure call), LAN Manager, TCP/IP, and WinSockets. Install either through Add/Remove Programs in Control Panel or the Internet Connection Wizard.

Digital—Characterized by discrete states of typically "on" and "off" or "one" and "zero." Computer data is a good example. Because computer data is inherently digital, most WANs use some form or other of digital signaling. *See* Analog.

Direct Sequence Modulation—*See also* Spread Spectrum Radio Transmission—Breaks original messages into parts called *chips,* which are transmitted on separate frequencies. To confuse eavesdroppers, decoy data also can be transmitted on other frequencies. The intended recipient knows which frequencies are valid and can isolate the chips and reassemble the message. Because different sets of frequencies can be selected, this technique can operate in environments that support other transmission activity. Operating on a frequency of 900 MHz, a bandwidth of 2–6 Mbps can be supported.

Disk Compression—A feature that lets you free-up hard drive space by compressing the data via an algorithm.

Disk Defragmentor—A utility that is run to defragment the hard disk. Fragmentation happens as files are removed from the hard disk, leaving blank areas that are not contiguous. When files are written to these non-contiguous blank areas, they become fragmented and so are slower to be accessed because more movement of the hard drive's heads are required to read the file in later.

Delete Permission—Allows you to delete files. Equates to the NetWare Erase Right.

DHCP—Dynamic Host Configuration Protocol—A service for dynamically and automatically assigning TCP/IP configuration information to computers on a network. NT server is the only OS package to offer this service.

Differential Backup—A full backup is first made, then all the files since the last Full backup only are saved. The Archive bit is not reset, so the Differential file gets larger and larger until it becomes worthwhile to do another Full backup and go back to a small Differential file. It is in the middle in regards to speed of backing up and restoring because only two tapes are ever needed. *See also* Full Backup, Incremental Backup.

Differential Manchester—A technique for encoding bits on the wire used in Token Ring Networks.

Disk Defragmentation/Fragmentation—A process that happens over time with normal use of the hard drive and in which files end up being not written contiguously. If not attended to, access to files slows down as the heads have to seek more and more to load programs and data. *See* Task Monitor.

DLC—Data Link Control Protocol—This protocol provides connectivity with mainframe computers such as IBM hosts and AS/400s. This protocol is also used to access Hewlett Packard network printers. It is not normally used as a connectivity protocol on Microsoft or Novell Networks.

DLL Dynamic Link Library—A library file containing functions that many different programs can use, allowing those programs to be smaller and to load only the shared execution code into memory when needed.

DMA—Direct Memory Access—Normally, should a device or program need to use RAM in a computer, it must go through the processor for access, which slows things down. DMA assigns up to 8 different channels that allow devices to access memory directly and so remove the processor from the equation, which speeds things up. Not all channels are available right away, although on x86 machines, channel 4 is normally reserved for the DMA controller chip and channel 2 is reserved for the floppy disk controller.

DNS—Domain Naming Service/Server/System—provides resolution of Internet domain names to TCP/IP addresses such as http://freespace.virgin.net/mark.hodgkinson.

Domain—A logical grouping of computers similar to that of workgroup or peer-to-peer network, with the exception that a Windows NT server (installed as a PDC) offers central administration and security.

Domain Admins—A very powerful global group in Windows NT. This group is able to create and delete users, take control of files and other resources, and act as part of the OS. Only authorized persons should be in this group.

Domain Controller—*See* PDC, BDC.

Domain Guests—A global group in NT used to provide Guest-level accounts throughout the Domain to give limited access to roaming users throughout the domain.

Domain Name Server—A server on a network whose job it is to resolve Internet domain names into IP addresses.

Domain Users—A built-in NT group that contains all users in the domain. User rights and resources granted to this group are usually severaly limited.

DPMI—DOS Protected Mode Interface—Most often seen in "Doom" type games to allow easier addressing of memory, etc.

Drop Cable—The cable that might run, for instance, behind a wall socket to a patch panel.

DS-0—Each channel of a T1 line can transmit at 64Kbps. This single channel service is called *DS-0* and it is one of the most popular dedicated line service types.

DS-1—A full T1 line. *See* T1.

DS-2—Four T1 lines. *See* T1.

DSR—Data Set Ready.

DTE—Data Terminal Equipment—In X.25 parlance, a computer or terminal. *See* X.25.

DTR—Data Terminal Ready.

Dual Booting—The process of running more than one OS on a machine at a time and being able to select the OS of choice at boot-up.

DVM—Digital Volt Meter—A hand-held electronic measuring tool that enables you to check the voltage of network cables. You can use a DVM to find a break or short in a network cable. *See* TDR, Oscilloscope.

DWORD value—*See also* Registry—A binary value in the registry, but represented as hexadecimal.

DX—A specification of x86 processors. An example might be 386DX. The DX denotes that the processor contains an enabled maths coprocessor used to help speed the processing used to help speed the processing of extensive maths calculations and graphics.

Dynamic Memory Allocation—Normally seen as a setting in PIF files for MS-DOS apps under Windows. When checked by default, it allows the program to control the amount of memory set aside for displaying graphics. Make sure it is enabled for programs that often switch between text and graphics, but remove it if running Windows on a machine with little RAM, which allows Windows to allocate the memory as it sees fit.

Dynamic Route Selection—Under this selection method, routing cost information is used to select the best route for a given packet. As network conditions change and are reflected in routing tables, the router selects different paths to continue using the lowest cost path. To discover this path, two techniques are used—

	Distance Vector Routing	Link State Routing
How often routing information is sent?	Repeated broadcasts	Only when a route changes
Which routes are transmitted?	All routes, even those learned secondhand	Only routes for directly connected networks or subnets
How long until all routers know about route changes?	At least several minutes and, possibly many minutes	Seconds to tens of seconds
Ease of configuration?	Simple and easy	More complex
Bandwidth overhead?	Substantial in large networks	Small, even in large networks (unless unstable)

Dynamic Routers—These routers have the capability to determine routes dynamically (and to find the optimum path among redundant routes) based on information obtained from other routers via a routing protocol or routing algorithm. *See also* Static Routers, Routers.

E

a b c d e f g h i j k l m n o p q r s t u v w x y z

EBCDIC—Extended Binary-Coded Decimal Interchange Code.

ECP—Extended Capabilities Port—Provides for high-speed printing. An ECP Port, ECP Printer, and IEEE1284 printer cable are all required to make it work.

EIA—Electronic Industries Association.

EISA (Extended Industry Standard Architecture)—An enhanced version of the ISA bus standard, using a 32-bit bus with bus mastering that can transfer data at up to 66Mb/second while still maintaining backward compatibility with ISA cards. EISA is a relatively rare bus standard, found mostly in older server-class systems. EISA bus devices can be configured automatically using software utilities. These slots accept either EISA or ISA cards; however, EISA cards cannot be placed into an EISA slot. *See also* ISA, MicroChannel, PCI

EMF Spooling—Expanded Metafile—Used when you want faster printing and faster return to applications. It is specific to the OS and not the printer as RAW Data output is. Because it is not printer specific, it can be sent across the network and used by other machines using the same OS to whatever kind of printer they use. Most applications written for OSs that support EMF produce the EMF automatically, and this allows faster return to the application because the Windows OS can handle the EMF as a background process. Postscipt printers cannot use EMF spooling.

EMI—Electromagnetic Interference—Electrical background noise that distorts a signal carried by a transmission medium. EMI makes it harder for a station listening to a medium to detect valid data signals on it. Some network media are more susceptible than others. *See also* Crosstalk.

EMS—Expanded Memory—Invented by Lotus and Intel working together. By using the top 384Kb of reserved memory of the old MS-DOS memory model, code could be paged between the conventional and expanded memory area. To do this, an expanded memory manager was needed to do this paging. The most common was EMM386.SYS, which later became EMM386.EXE. *See also* XMS—Extended Memory. Note: Some MS-DOS applications under Windows have difficulty when they cannot see the limit should the Auto settings be used in the PIF file. Try setting to 8,192Kb to cure this.

Encapsulating Bridge—In encapsulating mode, a bridge packages frames of one format in the format of another. For example, Token Ring frames may be "encapsulated" in Ethernet frames and passed out onto the Ethernet Network. Presumably, there would be another Ethernet–Token Bridge would "de-encapsulate" the packets and put them on a second Token Ring network, where they would be read by a destination station. To stations on the intermediate Ethernet, these "encapsulated" frames would be unrecognizable because there is no low-level address translation being performed by the bridge. In the example below, packets from LAN A could be read by nodes on LAN C because they share a common addressing scheme. Nodes on LAN B could not read the packets. Encapsulation is faster than translation. It allows the LAN to pass data more quickly when the packets have to pass through multiple LANs.

ESD—Emergency Startup Disk—Under '98, this is a bootup disk used in troubleshooting and installation.

ESD—Electrostatic discharge.

Ethernet—Can refer to original Ethernet or now the Ethernet II. It is a very popular LAN architecture based on CSMA/CD access methodology. Ethernet networks, depending on the particular variety, will operate at 10, 100, or even 1000 Mbps using baseband transmission. Each of the IEEE 802.3 specifications specifies the type of cabling it supports. Ethernet is generally used on light-to medium-traffic networks and performs best when a network's data traffic transmits in short bursts. It is the most commonly used network standard.

Ethernet II—*See* Ethernet.

Everyone—A local special group in Windows NT that contains all users in local or trusted domains as well as non-recognized users. It is impossible to add or delete users from this special group.

Excessive Paging—There comes a point at which not all applications can be paged out to the hard drive. When this happens, then there is only one solution, and that is to add more physical memory to the system. When excessive paging occurs, the system runs very slowly and the hard drive does nothing but "thrash." This can be damaging to hard drives and should be rectified as soon as possible, especially on demand-heavy servers. *See also* VMM Virtual Memory Manager.

Exclusive Mode—Usually a setting in a .PIF file and locks the mouse within the MS-DOS application Window.

Expanded Memory—Memory in addition to conventional memory that some older MS-DOS programs use. This memory is installed on expanded memory boards and comes with an expanded memory manager. This memory is not part of memory normally installed on PCs.

Explicit Permissions—When a user is specifically given full control to a share, this is deemed an explicit permission. Note: On '98, you can give an explicit permission to a subfolder and you are not limited to the initial permissions given to that initial share as NT is (i.e., when you share a folder with Read permissions on NT, you cannot give users full access to another folder beneath the shared one. Users will be limited to read access throughout the hierarchy of that share). *See also* **Implicit permissions.**

Extended Local Networks—A wireless connection serves as a backbone between two LANs. For instance, a company with office networks in two nearby but separate buildings could connect those networks via a wireless bridge.

Extended Memory—Memory higher than 1Mb on computers with x86 processors. Extended memory is required by Windows and Windows-based programs.

F

a b c d e f g h i j k l m n o p q r s t u v w x y z

F4—Press during a Windows 98 boot, after the message saying "Starting Windows 98." F4 will give you the option to boot into the previous version of MS-DOS so long as BootMulti =1 is set in the MSDOS.SYS file.

F5—Press F5 when the system displays the message "Starting Windows 98" to start the system in Safe Mode. Press **Shift + F5** to boot the system to the command prompt only. **Alt + F5** brings you to the Safe Mode Command Prompt.

F8—Displays the Start up menu under Windows 98. Available when you see the "Starting Windows 98" message. The time you have to press this key is set in MSDOS.SYS and the BootDelay option. Press **Shift + F8** to start Windows with a step by step confirmation.

Fast Infrared—This protocol is used to provide connectivity for infrared wireless networks and works on the same principle as your remote control for your television at home. Connectivity with this protocol is capable of up to 4Mbps.

Fast ROM Emulation—Checked by default in the **PIF** file, and when checked allows the program to use **ROM** drivers in **protected** mode, which can increase video performance. Should your program have difficulty writing to the screen, then disable this check box in your PIF file.

FAT and FAT16—File Allocation Table—Initially a 12-bit file system that works by basically mapping out to the OS where the files physically reside on the disk. As of MSDOS 3.0, it became a 16-bit file system, or FAT16. Partitions can be up to 2GB in size. Supports more file systems than any other file format, including OS2 and

UNIX. Most efficient on drives less than 256Mb when compared against FAT 32 and 400Mb when compared against NT. It has a 32Kb cluster size on a 2GB disk. Disk compression such as Drvspace is supported. Only 512 files are supported in the root directory of a FAT partition. It works as **real mode.** *See* **VFAT.**

FAT32—Introduced with Windows 95 OSR2.1. The latest upgrade over VFAT to a fully 32-bit File Allocation Table format. Allows disk partitions to be formatted up to 2.1 Terabytes in size. Only supports drives larger than 512Mb. Currently only supported by Win95 OSR2.1 and Win98, although NT variants are expected to adopt the file system. Features a very small 4Kb cluster size on a 2GB disk. Disk compression is not currently supported by the current Operating Systems that support it. It is a **protected mode** component. The root directory entry limit is now 65,535 entries, as opposed to 512.

Fax Gateways—Allow users to send and receive faxes from their workstations.

FDDI—Fiber Distributed Data Interface—A fiber-optic network technology used for high-speed WAN interconnectivity, using a token-based access method. FDDI is commonly used for high-speed fault tolerant interconnection between LANs in a wide-area network (WAN).

Fiber-Optic Cable—The ideal cable for data transmission. It supports extremely high bandwidths, presents to problems with EMI, and supports durable cables and cable runs as long as several kilometers. The difficulty lies in the area of cost and installation. The center conductor is made of highly refined glass or plastic used to transmit light signals. This fiber is coated with a cladding that reflects signals back into the fiber to reduce signal loss, and a plastic sheath then protects the cladding and fiber. It can be spliced by means of electric fusion, laser tape, or mechanical connectors.

Fiber-Optic Network Cable—Consists of two strands enclosed in plastic sheaths. One strand sends and the other receives. There are loose and tight cable types, depending on whether there is a space between the fiber sheath and the outer plastic casement filled with gel or other material or when strength wires are between the conductor and outer plastic casement. In both cases, the plastic casement supplies the strength, whereas the gel layer or strength wires protect the delicate fiber from mechanical damage. *See also* Fiber-Optic Cable.

FIFO—First in, first out.

File Server—Computers whose main task is to provide file sharing to computers on the network. These machines can handle multiple and simultaneous requests for file resources and may employ specialized hardware to increase reliability, speed, and capacity (e.g., RAID). File backup and maintenance are easier in a server-based network because only the file servers need to be backed up.

File and Print Sharing '98—Allows you to advertise your files and printers on the network and your machine to participate in Browser elections. However, on large networks it may be advisable to disable this service because not every machine must be a Backup Browser, and having lots of machines participating in elections simply means lots of network traffic. Also, because machines participate in Browser updates every 15 minutes, if a machine is improperly closed down, it takes three updates for a machine to be removed from the Browse list, or 45 minutes. Note: File and Print Sharing for Microsoft Networks and File and Print Sharing for NetWare Networks will not coexist on the same machine; you can only have one or the other. Microsoft's version does not work with Client 32. To find your resources on the network, when entering into a Novell environment, you must have SAP advertising enabled for NETX and VLM based clients.

Firewall—A collection of hardware or software that collectively provides a protected channel between two networks with different security settings. Normally used to allow network connectivity to the Internet (an outside network) without giving complete access to the internal network. Used to limit attacks from a malicious person(s). Firewalls support extensive auditing. You can set up auditing to generate alarms based on filtered traffic, security violation detection, and/or IP address usage monitoring.

Firmware—Firmware is another way of referring to software stored on a ROM chip. This information is nonvolatile, meaning that it remains there when the power is switched off, but can be upgraded, as opposed to true Read Only Memory. Quite often, floppy controller tape backup units require firmware revision to work properly, especially with the Windows 98 MSBackup program.

Folder—A container of objects. A folder usually contains other folders, files, or shortcuts. In '95, '98, and NT, the folder is equivalent to a directory in other OSs.

FQDN—Fully Qualified Domain Name—The full name of a system, consisting of its local hostname and its domain name. For example, support is a hostname and support.Microsoft.com is an FQDN.

Fractional and Multiple T1 or T3—Subdivided channels of a T1 or T3 line or combined channels of a T1 or a T3 line, respectively. Each channel of a T1's 24-channel bandwidth can transmit at 64Kbps. This-single channel service is called DS-0, and is one the most popular service types.

Fragmentation—The process of spreading the contents of a file across the free space on a hard disk rather than placing them in contiguous blocks. Fragmentation generally occurs when a file system inefficiently plans space for file saves and the growth of existing files. Fragmentation affects different file systems differently.

Fragmentation can also take place within the virtual memory address space of the OS, but does not seriously affect OSs with efficient memory managers.

Frame Relay—Designed to support B-ISDN. The specifications for frame relay address some of the limitations of X.25. Frame Relay is a packet switching network service like X.25, only designed around newer, faster fiber-optic networks. Unlike X.25, frame relay assumes a reliable network. This enables it to eliminate much of the X.25 overhead required to provide reliable service on less reliable networks. Frame relay networks rely on higher-level protocol layers to provide error control. Frame Relay is typically implemented as a public data network and so is regarded as a WAN protocol. It provides PVC (Permanent Virtual Circuits), which supply permanent pathways for WAN connections. Frame Relay services typically are implemented at line speeds from 56 KBps up to 1.544 Mbps (T1). Customers typically purchase access to a specific amount of bandwidth on a frame-relay service. This bandwidth is called the CIR (Committed Information Rate), a data rate for which the customer has guaranteed delivery for its data. Customers are usually also permitted to access data rates faster than the CIR by incurring the risk that their additional data might be discarded by the network should it become congested. Many people feel this is an acceptable risk because most applications can tolerate a small level of packet loss. Frame Relay circuits then should be used when they are less expensive than equivalent point-to-point circuits, and the data transported can tolerate the additional delay that packet switching introduces. It has less overhead than other packet switching technologies, such as ATM or X.25.

Frame Type—Applies to IPX/SPX. If a protocol is the language of the network, the frame type is the dialect spoken. Available types are Ethernet 802.2, Ethernet 802.3, Ethernet II, Token Ring, and Token Ring SNAP. If using NetWare 3.1 or earlier, the frame type is 802.3, later is 802.2. Frame Type numbers are not like software version numbers, but relate to different IEEE standards. 802.3 uses a carrier-sense multiple access with collision detection format, whereas 802.2 uses an LLC (Logical Link Control) format. If you set the Frame Type to Auto-detect, then the next time the computer is booted (once only), the frame type will be auto-detected. If no frame type is detected, the default frame type is 802.2. If multiple frame types are selected, then the most prevalent one will be used.

Frequency Hopping—Frequency Hopping switches (hops) between several available frequencies, staying on each frequency for a specified interval of time. The transmitter and receiver must remain synchronized during a process called a *hopping sequence* for the technique to work. Range is up to 2 miles outdoors and 400 feet indoors. Frequency hopping typically transmits at up to 250 Kbps, although some versions can reach as high as 2 Mbps. *See* Spread Spectrum Radio Transmission.

FTP—File Transfer Protocol—Part of the TCP/IP suite of protocols and allows files to be transferred between dissimilar systems. *See* **RSync.**

Full Backup—All files on your system are saved. Quickest to restore but the longest to run initially. *See also* Differential Backup, Incremental Backup.

G

a b c d e f **g** h i j k l m n o p q r s t u v w x y z

Gateway—Functions at the Application layer of the OSI model The term *gateway* was originally used in the Internet protocol suite to refer to a router. Today, the term most commonly refers to a device functioning at the top levels of the OSI model enabling communication between dissimilar protocol systems. A gateway generally is dedicated to a specific conversion, and the exact functioning of the gateway depends on the protocol translation it must perform. They work by removing the layered protocol information of incoming packets and replacing it with the packet information necessary for the dissimilar environment. They can be software, hardware, or both. (NT4 RAS can act as an IP or an IPX router, but RAS's NetBIOS gateway is an even more powerful feature. Not only does NetBIOS gateway forward remote packets to the LAN, but it also acts as Gateway enabling NetBEUI clients access to the LAN, even if the LAN uses only TCP/IP or IPX/SPX.) Gateway Types: There are several different types of gateways, including Asynchronous, Communication with Async host computers and bulletin boards; X.25, allows connection to remote hosts through a PDN using a wide range of terminal emulation; 3270, communication with IBM mainframe or compatible hosts; 5250, communication with IBM System/3X series minicomputers; and TCP/IP, link non-TCP/IP systems with TCP/IP systems. Gateways can operate in both local and remote environments. 3270 gateways may establish a link from a workstation on a LAN to a cluster controller via coaxial cable, or through the placement of a cluster controller or FEP (Front End Processor) in the ring of a Token Ring network. Gateways can provide significant network savings because of the reduction in the number of required cables and computer parts, as opposed to stand-alone microcomputer gateways.

GDI—Graphics Device Interface—Takes care of anything placed on screen. Mainly concerns itself with the mathematical calculations which most graphics are. Under '98, it is composed of two .DLL's for backward compatibility reasons, one 16 bit and the other 32 bit. Windows Management under '98 is still predominantly 16 bit, however, for those compatibility reasons.

Geosynchronous—If a satellite is positioned so that it appears to be motionless to an observer on the Earth, then it is said to be in Geosynchronous orbit.

Global Groups—Domain level groups that can contain only users. These groups may be exported to other domains via trust relationships. *See* Local Groups.

GPF—General Protection Fault—Occurs under '95 or '98 far less than 3.11, but is seen when a program attempts to violate system integrity. This is classed as an application failure. If under '98 and an MS-DOS GPF occurs, the system will carry on quite happily because the app is running in its own Virtual Machine. If a Win16 app generates a GPF, close this application and the other Win16 applications will start again. If a GPF occurs with a Win32 application, then only that application will have stopped and terminating that application will allow everything else to continue. *See* **Local Restart.**

H

a b c d e f g h i j k l m n o p q r s t u v w x y z

Header—The part of a packet that signifies its start. It contains a bundle of important parameters, such as the source and destination address and time/synchronization information.

Heterogeneous (Translating) Bridges—A bridge must read an actual MAC layer frame. Therefore, some bridges may be limited to linking similar MAC layer protocols. In special cases in which physical addressing is similar and the logical link services are identical, hybrid bridges can be developed to allow linkages between dissimilar MAC layer protocols. Because a number of the 802 series of protocols share the common 802.2 LLC layer, it is possible for bridges to interconnect different types of networks, such as Ethernet and Token Ring. One example of this is the IBM Model 8209 Ethernet to Token Ring Bridge. Bridges of this type are also known as *translating bridges*.

High End Spoolers—Provide high-speed image and management (OPI) that enables the clients to work with low-resolution representations of actual images.

HKEY_CLASSES_ROOT—OLE and file associations. Under NT, if you upgraded on top of 3.11 here, you would find the previous system details.

HKEY_CURRENT_CONFIG—Storage of multiple configurations, such as when a laptop is docked or undocked.

HKEY_CURRENT_USER—Information about the currently logged-on user. This key is created at logon.

HKEY_DYN_DATA—The dynamic status of devices used for **PnP** configuration. Created at bootup and not a part of either of the '98 registry files.

HKEY_LOCAL_MACHINE—The registry key that contains all machine specific information, such as drivers and hardware.

HKEY_USERS—Information in the registry about all the users of this computer.

Hooks—A point in the Windows Message–handling system at which an application can install a subroutine to monitor the message traffic in the system and process certain types of messages before they reach the target Window procedure.

Host Drive—The Compressed Disk in DrvSpace, etc. When the CVF is created, it is given the host drive's original letter (e.g. C:) while the uncompressed portion of the drive is given a new letter (e.g. H:). The CVF is duly viewed as the C: drive. *See* **CVF.**

Hosts File—HOSTS.SAM—A manual method of resolving DNS–domain names to IP addresses.

Hot Docking—Connecting a laptop to a docking station where the laptop can be left switched on regardless of it being docked or undocked.

HTTP—Hypertext Transport Protocol—The protocol used to transfer web pages to browsers over the World Wide Web.

Hybrid—A LAN with both wireless- and cable-based components.

I

a b c d e f g h i j k l m n o p q r s t u v w x y z

IBM Cabling—IBM uses its own separate names, standards, and specifications for network cabling and cabling components. The standards roughly parallel standard forms used elsewhere in the industry.

Cable Type	Description	Comment
Type 1	STP	Two twisted pairs of 22 AWG wire in braided shield.
Type 2	Voice and data	Two twisted pairs of 22 AWG wire for data, and two twisted pairs of 26 AWG for voice with braided shield.
Type 3	Voice	Four solid UTP pairs; 22 or 24 AWG.
Type 4	Not defined	
Type 5	Fibre-optic	Two 62.5/125 micron multimode fibres.
Type 6	Data Patch Cable	Two twisted pairs of 26 AWG wire, dual foil and braided shield.
Type 7	Not defined	
Type 8	Carpet Grade	Two twisted pairs of 26 AWG wire, with shield for use under carpets.
Type 9	Plenum Grade	Two twisted pairs, shielded.

Idle Sensitivity—Normally a setting in a .PIF file. How long you want Windows to wait before it suspends an application for inactivity. This can crash some MS-DOS applications. If so, set the the lowest level.

IEAK—Internet Explorer Administration Kit—Can be used to automatically configure Internet Explorer. Downloadable from http://IEAK.Microsoft.com or from '98's Resource Kit.

IEEE 802—As networks became popular, the IEEE began the 802 project in February 1980 to define certain LAN standards. 802 comes from the year and the month the project was started, and it focused on the physical aspects of networking relating to cabling and data transmission on the cable. These fall into the bottom two layers of the OSI model.

IEEE 802.4—Implemented infrequently; defines a bus network that also employs token passing.

IEEE 802.5—Essentially Token Ring.

IEEE1284—Printer cable standard for **bidirectional printing.**

IEEE1394 Serial Bus—FireWire—Created as a way to share real-time information among devices such as digital camcorders, digital VCRs, DVD, and even some high-quality printers. This standard allows transmission speeds of 100, 200, and even 400 Mbps using both Asynchronous and Isochronous transfer modes. It is similar in many ways to **USB.** 63 devices may be connected. The machine need not be restarted or rebooted when a device is added. FireWire can be used with machines that have two processors and requires the use of between 8 and 40 volts. The maximum distance between devices is 4.5M, and there can be no more than 16 hops between devices.

IFS—Installable File System—It routes calls made by applications to the appropriate destination, for example, it determines whether a call to retrieve a file should be sent to the local hard drive or across the network.

Independent Redundant Backup Network—If you have chosen a centralized backup server, huge amounts of data are transferred during the backup. This enormous amount of data can seriously degrade system performance. For that reason, you may want to connect your servers to a central backup server, using a second network card in each server.

Individual Tape Units—This backup system involves a smaller tape unit installed on each server. Individual units may cost less than a larger array required to backup an entire network. However, this solution can be more difficult to manage. For a network spread across several buildings, this may present an attractive alternative to a single backup server if management of the units can be delegated to on-site personnel. *See also* Single Backup Server, Independent Redundant Backup Network.

Infrared Transmission—Similar to a television with a remote control that transmits pulses of infrared light carrying coded instructions to a receiver. This technology has been adapted to network communications. Infrared transmissions are typically limited to 100 feet; within this range, however, it is relatively fast, supporting transmission speeds of up to 100 Mbps. It is immune to RF interference, but reception can be degraded by bright light. Because transmissions are tightly focused, they are fairly immune to electronic eavesdropping. All systems operate in the lowest range of visible light frequencies of between 100GHz and 1000THz. *See also* Broadband Optical Telepoint, Line of Sight Infrared, Reflective Infrared, Scatter Infrared.

Implicit Permissions—When a folder is shared on a client machine and the user has access to the folders below it, the access to subfolders is deemed an implicit permission. *See also* Explicit permissions.

Incremental Backup—A Full backup is first made, and then a following Incremental backup is made the following night. However, the Archive bit is reset, so only altered/new files since last night's Incremental backup are captured the following night. Hence, it is quick to run, and remains so, but could take the longest time to restore because a dozen or more tapes (in theory) could be needed to do a full restore. *See also* Full Backup, Differential Backup.

The Internet—The Internet (capital *I*) is the worldwide network of networks that to many has been the latest craze. An internet (small *i*) is usually several departmental LANs in a company or organization that are connected together to form a WAN. Both are similar here because both are comprised of smaller networks interconnected by Routers. *See also* Intranet.

InterNIC—Inter Network Information Center—The organization that manages IP addresses in use. Contactable at http://www.Internic.net. More commonly though your Internet Service Provider will supply you with the IP addresses you can use. In any event, the IP address must be unique. They also keep track of domain names on the Internet.

Intranet—An internal network that uses the same applications as the Internet. For instance, people on the company's network use Browser's to surf the Intranet in the same way they use browsers to surf the World Wide Web. However, the information is secured from the outside—the Internet by a firewall or other security system and only internal access is allowed. *See* Internet.

Internet Packet Exchange Protocol (IPX)—A network layer protocol that provides connectionless (datagram) service. Responsible for internetwork routing and maintaining network logical addresses. Relies on hardware physical addresses found at lower layers to provide network device addressing, which makes it much easier to manage than TCP/IP.

IO Port—Devices that must perform input and output functions (such as a network or modem card) are allocated space in memory called an *IO port*. These address spaces are referred to using hexadecimal addresses such as 0-FFFF.

IP—Internet Protocol—This portion of TCP/IP takes care of addressing (e.g., 207.68.154.25). This format is usually referred to as *dotted decimal format*. Each IP address is split between a host and network component, the network component being assigned by the InterNIC.

IP Spoofer—Programs that let you simulate a message coming from a physical point on the network by mimicking its IP address

IPC$—Created on '98 when Remote Administration is enabled. It provides an Interprocess Communications Channel for computers to communicate on.

IPX/SPX—Internet Packet Exchange/Sequenced Packet Exchange—The protocol that used to be of choice for Novell Networks until NetWare 5 was introduced with true IP. IPX is a routable protocol, and so is preferable to a protocol like **NetBEUI.** Its configuration is easier than TCP/IP and most configurations such as the **frame type** is automatic. (The default frame type is 802.2 with MS Products.) It is, however, not as capable as TCP/IP on very large networks. Microsoft has their own implementation of IPX/SPX on their OSs called *via means of distinction Microsoft IPX/SPX*. Microsoft's version will work with MS and Novell 32-bit clients but not with NETX or VLM clients. You can only have IPX/SPX bound to one card under a MS operating system.

IrDA—Infrared Data Association—Low-end point-to-point data transmission running at 115 Kbps.

ISA Industry Standard Architecture—The standard data bus architecture; originally developed for the IBM AT.

ISDN—Integrated Services Digital Network—This type of connection can offer speeds of up to 128 Kbps. A special ISDN adapter or modem is required. An ISDN line consists of three channels, one used for switching, on hook/off hook, while the other two are being used for transferring data. The data channels are referred to as *B channels,* each being capable of 64 Kbps. *See* **Multilink** for extra information. It is a term used for a group of ITU (CCITT) standards designed to provide full-featured, next-generation services via a digital telephone network, the original idea being to enable existing telephone lines to carry digital information. As such, it is more like a traditional telephone service than some of other WAN services like X.25 and Frame Relay. It is intended as a dial-up service and not a permanent 24-hour connection; as such, it often includes a per-use fee.

ISDN Types	Description
Basic Rate ISDN (BRI)	The most common type, originally intended to replace standard analogue telephone lines. It uses three channels; two B-channels carry the digital data at 64 Kbps while the third channel (called the *D channel*) provides link and signalling information at 16 Kbps. Commonly described as 2B+D. A single PC transmitting through ISDN can sometimes use both B channel simultaneously, providing a maximum data rate of 128 kbps (or higher with compression).
Primary Rate ISDN (PRI)	Larger scale, supports 23 B channels at 64Kbps and one 64Kbps D channel. Sometimes used in place of a T1 line.
Broadband ISDN (B-ISDN)	A refinement of ISDN that is defined to support higher bandwidth applications. Physical layer support for B-ISDN is provided by ATM (Asynchronous Transfer Mode) and SONET (Synchronous Optical Network). Some typical B-ISDN data rates are 51 Mbps, 155 Mbps, and 622 Mbps. They should be used for places that do not need full-time connectivity or high speeds.
Hybrid	One A channel at 4 KHz analog and one C channel at 8 or 16 Kbps.

If you have many dial-in connections using BRIs, it is most cost-effective to have them dial into different channels on a PRI. These channels are referred to as *bit pipes*. A complete listing would be that Channel A: Runs at 4 KHz using analog channels. Channel B: Runs at 64 Kbps using digital signals. Channel C: Runs at 8 or 16 Kbps using digital signals. Generally used for out-of-band signalling. Channel D: Runs at 16 or 64 Kbps using digital signals for out-of-band signals. Has the following subchannels: -s: Subchannel for call setup and other signalling. -t: Subchannel for telemetry like reading meters. Channel E: Runs at 64 Kbps, using digital signalling for internal ISDN control. Channel H: Runs at 384, 1536, and 1920 Kbps using digital signals.

Isochronous Transfer Mode—Often seen with USB and gives it its fixed data rate of up to 12Mbps. Generally used for middle speed devices such as monitors, which need a guaranteed transfer rate.

ISP—Internet Service Provider.

a b c d e f g h i j k l m n o p q r s t u v w x y z

a b c d e f g h i j k l m n o p q r s t u v w x y z

Kernel—Handles most virtual memory functions, I/O Services, and task scheduling. On recent OSs such as '98 and NT, it is predominantly 32 bit. Under '98, it is made up still of two .DLL's, however, including both a 16- and 32-bit version for backward compatibility reasons.

L

a b c d e f g h i j k l m n o p q r s t u v w x y z

LAN Manager API—Messages are sent to an API, which in turn tells the computer to accomplish a task similar to RPC.

LAPB—Link Access Procedures Balanced—*See* X.25.

Laser—Used with fiber-optic cables for data transmission. The purity of laser light makes lasers ideally suited to data transmissions because they can work at long distances and high bandwidths. They are expensive light sources and are used only when their special characteristics are required. *See* LED.

Laser Tape—Used to splice fiber-optic cabling.

Laser Transmission—High-powered laser transmitters can transmit data for several thousand yards when line-of-sight communication is possible. Lasers can be used in many of the same situations as microwave links without requiring the FCC license. For indoor LANs, laser light technology is rarely used, but is similar to infrared technology. *See* Infrared Transmission.

Leased Dedicated Service—The essential opposite of a dial-up service. Here, the customer is granted exclusive access to some of the bandwidth. Most popular formats are T1, T3, Fractional and multiple T1 and T3, DDS (Digital Data Service), and Switches 56.

LIM—Lotus Intel Memory—*See* Expanded Memory.

Linear Bus—*See* Bus Topology.

Learning Bridge—*See* Bridge.

LED—Light Emitting Diode—LEDs are inexpensive and produce a relatively poorer quality of light than lasers. LEDs are suitable for less stringent applications such as 100 Mbps or slow LAN connections that extend less than 2 Km. *See* Lasers.

Line of Sight Infrared—Transmissions must occur over a clear line-of-sight path between transmitter and receiver. *See* Infrared Transmission.

List Files Permission—On MS Networks, users can see the contents of folders and subfolders. Equates to the NetWare File Scan Right.

LM Announce—Used for older LAN Manager networks. When enabled under File and Print Sharing, the computer announces itself periodically using broadcasts. This setting effectively doubles the amount of announcement traffic your machine makes and allows LAN Manager clients access to your system.

LMHosts file—LMHOSTS.SAM—A manual method of resolving NetBIOS names to IP addresses. Preferable to use WINS if possible. Because the LMHosts file (which contains nothing more than a list of computer names to IP addresses) also has to be copied onto each computer in the network, this just adds another administrative headache for you!

Load Balancing—If using System Policies and only one machine is being used to download the policies, then you can implement load balancing to stop this one machine from being "overrun" by all the requests. Use System Policy Editor in registry mode and select the Local Computer icon and Remote Update check box. Select Load Balance. The machine will now use BDCs and other servers to download its policy.

Local Bridge—When a bridge has a LAN link directly attached on each side, it is referred to as a *local bridge*. Local bridges are characterized by comparable input and output channel capacities. *See* Remote Bridge.

Local Groups—Groups that can contain users and global groups. Local groups are given certain rights and permissions to network resources. *See* Global Groups.

Local Printer—A printer directly connected to your machine, most likely via the LPT1 or parallel port.

Local Restart—By pressing CTRL+ALT+DEL on '95 and '98 systems (once and only once, not 30 times in succession because you're angry) will perform a local restart. You can then terminate the "Not Responding" application and continue with work.

Log On and Restore Network Connections—*See* **Quick Logon.**

Logical Node Name—Identify specific hosts with alphanumeric identifiers, which are easier for users to recall than the numeric IP address. An example of a logical node name is MYHOST.MYCOMPANY.COM.

Long File Names—Have a limit of 255 characters and 260 characters for the entire path name under DOS. It also allows the use of special characters that were not normally supported under MS-DOS, although this support only extends to 32-bit programs. They are not case sensitive, but do preserve case. LFNs incorporate the use of 8.3 file names that, by using a truncation or tilde character, allow 16-bit programs to continue to operate. The algorithm under '98 is to take the first six characters of the file name and places a tilde after them. It then adds the number 1 to 9 at the end to make an eight character name. The three character extension is the same as the LFN's extension—note the last extension where multiple extensions are used. All characters for this alias are uppercase. Should a unique name not be generated, then the first five characters and a two-digit number are used, and this process continues until a unique filename is found. Under NT, the fifth iteration causes a random set of characters to be used. NT Workstations and Servers support long file names of course, but older clients such as MS LAN Manager and Windows for Workgroups do not. They also cannot use shared folders with share names longer than eight characters. LFN support is provided on Netware Servers so long as the OS2 name space component has been loaded onto the NetWare server. Without it, you must use conventional 8.3 file names. Note: To stop MS-DOS listing the filenames, they are given the following set of attributes: System, Hidden, Read-only, and Volume Label.

Long-Range Wireless Bridge—A wireless bridge with a range of up to 25 miles. *See* Wireless Bridge.

Lost Clusters—Sometimes referred to as *Lost File Fragments*. May contain useful data and are deemed another FAT error. Most of the time, however, they only occupy free space.

M

a b c d e f g h i j k l m n o p q r s t u v w x y z

MAC Addresses—MAC (Media Access Control) Addresses are the address that all "FRAMES" are addressed to at the Data Link Layer of the OSI stack. Every station that can Receive/Transmit Frames on a LAN has a MAC address. This address is permanently burnt in (BIA) to a chip of the card and is unique to that card. It is normally 12 Octets long, with the first 6 Octets identifying the manufacturer of the card and the next the serial number of the card. (There is table on the Internet of manufacturers IDs.) On Token Ring, it is possible for the station to be configured in software to use a different MAC address to the BIA. This does not overwrite the original BIA, it is just used instead of. Frames to and from particular MAC addresses are called "Unicast," and there is a special MAC address of all Fs that is called a *Broadcast address*. All stations will accept frames addressed to this address, enabling one frame to be sent to all other stations on the LAN.

Mailslots—This is a one-way connection. The computer sending the message receives no acknowledgment that the receiving machine has received the message.

Mandatory Login—Users are required and must logon to an NT machine before use.

Mandatory Profile—Change USER.DAT under NT or '98 to USER.MAN to make a user have a profile of your choosing. They are able to alter settings while they use the machine, but when they log on again, your profile once again takes effect. If there is a USER.MAN and a USER.DAT on the system, then the USER.MAN takes precedence.

Manual Update—Use this to put your policy files wherever you wish. Use System Policy editor in registry mode to specify where the machine must look for the file, even the local machine itself.

MAU—A central cabling component for IBM Token Ring Networks.

Member Server—An NT server that does not participate in domain security. Normally, to give more resources to applications it may be running such as file or print services on the network.

Memory Region—Some devices reserve memory addresses for storage; others have their own onboard memory. Many legacy devices map memory to the old location below the 1Mb cutoff; newer devices now use protected memory above 1Mb. Even though these devices use that memory in the higher region, they may still map to the lower range in the 386Kb area.

Memory Swapping—*See* Paging.

Message—A message could consist of some function such as an interrupt (the click of a mouse button) or information to be transferred. These messages are used to send information from an application to the rest of the system, regardless of from one program to another, to the processor, or to the screen. *See also* **Message Queue, Message Passing Model.**

Message Passing Model—What '98 uses to control **processes** and **threads**. *See* **Message.**

Message Queue—Messages are passed through one of several message passing queues, which are similar to print queues. When you print something, you must wait until a previous print job has printed; likewise, messages in a message queue must wait for the previous messages to be processed. The number of message queues are directly proportionate to the number and type of applications being run. MS-DOS apps do not use message queues because they are not designed for message passing. Win16 apps share the same address space and so use the same queue, whereas Win32 apps, which can have multiple threads, give each thread it's own message queue. *See also* Message Passing Model.

MicroChannel—A standard proposed by IBM to replace the ISA standard. However, it was not widely accepted and has fallen into misuse.

Microsoft Client for NetWare Networks—The out of the box client with '98 and NT that allows connectivity to NetWare 3.1 and above. On '98, you can also add the Network Service called *Service for NetWare Directory Services*. Note: Should you not need to connect to an NT domain use Client 32 from Novell. If you do, then use Microsoft Client for NetWare with Service for NetWare Directory Services.

Microsoft Domain Services—The previous name for NT Directory Services.

Microsoft Download Libraries—An electronic bulletin board service maintained by Microsoft.

Microwave Transmission—Microwave technology has applications in all three of the wireless networking scenarios—LAN, Extended LAN, and mobile networking. Microwave communication can take two forms, terrestrial (ground links) and satellite links. The frequency of such transmission is higher than that of commercial FM radio stations. *See also* Terrestrial Microwave, Satellite Microwave.

Mini-Driver—Normally found as part of the print subsystem in Windows. A small chunk of code that is printer specific.

MLP—Meridian Lossless Packing—The compression system used in audio DVD disks.

Mobile Computing—A mobile machine connects to the network via cellular or satellite technology. It is a growing technology that provides unlimited range for traveling computers. Typically, it is used with portable PCs or PDA (Personal Digital Assistant) devices. *See also* Packet Radio Networking, Cellular Networking, Satellite Station Networking.

Modem—Modulator/Demodulator.

Modulation—Frequently, information that exists as digital must be converted to analog, and information that exists as analog must be converted to digital. This conversion involves the use of an encoding system that enables the original information.

MPEG—Moving Picture Experts Group—Compression for audio and video now at level 4, which boasts variable bandwidth and compression algorithms that are platform independent, and so, are able to deal with the rigors of mobile channels.

Multiple Master Domain Model—User accounts reside in an account domain and resources reside in resource domains. However, you are essentially taking the single master domain model and multiplying it by establishing a two-way trust between each and every other master (Accounts) domain. Each resource domain continues to trust their corresponding master domain via a one-way trust; however, a one-way trust is applied to each individual master domain. It is recommended that all trust relationships are implemented and the resource access is not denied. By placing your resources in your own domains, you are allowing for centralized or decentralized administration of networking resources. As with a Single Master Domain model, users still must log on to a master domain in order to access resources, but by using Multiple Master domains you can centralize or decentralize network administration. Because two-way trusts exist between all Master Domains, users can still use one account to gain access to all authorized resources. When a user logs on from a domain other than the one to which they are assigned, the domain and the domain that they log into passes the login request to the appropriate domain.

MS-DOS Mode—Provided by '95 and '98 to allow further compatibility with some MS-DOS applications. Use this if your program absolutely will not run (most likely, this will be a game of some sort). Note: Before resorting to this, examine the PIF file for the MS-DOS application and setting the option to stop the MS-DOS application from detecting Windows. Some MS-DOS apps are programmed to work differently if under Windows, and this might be the cause of your problem.

MS-DOS Virtual Machine—Where each MS-DOS programs run under Windows. Each MS-DOS application runs in its own MS-DOS Virtual machine. *See also* **PIF.**

MSAU—A central cabling component for IBM Token Ring networks.

MS Family Log On—Allows a user to choose their name from a list of user names stored on the computer. To have this option available, you must have user profiles enabled on your computer.

Multilink—Under '98 and NT, Multilink allows you to use separate dial-up devices as one to give increased bandwidth (for instance, two 56 Kbps modems on separate phone lines giving 112 Kbps). In particular reference to **ISDN** users, where an ISDN line consists of three channels—one used for switching, on hook/off hook, the other two being used for transferring data. The data channels are referred to as *B channels,* each being capable of 64 Kbps. With Multilink, both can be joined together to enjoy the full benefit. Under '95, only one channel could be used at a time. However, the network you dial into must also support multilink for the above benefits to be gained. Set up Multilink by going to **Dial-Up Networking,** right click the connection you wish to use it with, and select Properties. Select Multilink and use the Enable Use dialog button and click Add. Specify the phone number the second adapter is to dial. **Note:** Different speed modems can be used, but the connection speed for all modems will default to the speed of the slowest modem.

Multiplexing—Enables broadband media to support multiple data channels. *See* DeMultiplexing, Mux.

Multiplexor— *See* Mux.

Multitask—The ability to do more than one task at a time. See **Cooperative** Multitasking, **Preemptive** Multitasking.

Multithreading—Win32 applications can use multithreading to enhance an application by assigning multiple threads to a process. This enhances throughput, increases process responsiveness, and increases background processing.

Mux—Multiplexor—Has both multiplexing and demultiplexing capabilities.

N

a b c d e f g h i j k l m n o p q r s t u v w x y z

N Connectors—Connectors that screw on instead of requiring crimping or a twist lock. As with Thinnet, both ends of the cable must be terminated, and one end must be grounded.

Named Pipe—A connection process similar to NetBIOS. Two machines establish a two-way connection. The session ensures that the data is received properly.

Narrow Band Radio Transmission—In narrow band radio communication (also called *single frequency radio*), the range is better defined than in infrared transmission, effectively enabling mobile computing over a limited area. Neither the receiver nor the transmitter is required to be in direct line of sight and the signal can bounce off walls, buildings, and even the atmosphere. Heavy walls such as steel or concrete enclosures can block the signal, however.

NBP—Name Binding Protocol—AppleTalk Transport Layer protocol.

NCP—Netware Core Protocol—NetWare networks use NCP to communicate. They are akin to MS Sever Message Blocks. Provides remote function calls that support network services, such as file sharing, printing, name management, file locking, and synchronization. NetWare client software uses NCP to access NetWare services. NCP covers aspects of the Session, Presentation, and Application Layers of the OSI Model.

NCP—Network Core Protocol—When the ARPAnet was first created, the closest thing they had to a common protocol was NCP. This was redefined repeatedly until it became two separate components known as TCP and IP.

NDS—NetWare Directory Services—Replaced the bindery with the advent of NetWare 4, is not server centric and allows better administration of resources. It is a scalable database of all resources on a network, allowing a single point of administration from any server and is a loosely associated distributed database that can be partitioned over more than one NetWare Server.

NDIS—Network Datalink Interface Specifications—Developed by 3Com and Microsoft corporation jointly in 1989. Allows various protocols to bind with a single network adapter card. Microsoft's equivalent to ODI. NDIS 2 is real mode and 16 bit. Version 3.0 is protected mode, and thus 32 bit. Version 3.1 introduced PnP. Version 4 added support for PPTP and Version 5 adds support for a wider range of network media such as FDDI, token ring and ATM. Non-NT implementations include a "protocol manager" (PROTMAN) that binds the different protocols with the MAC layer to the correct protocol stack. Under NT, this is accomplished via the registry and a small wrapper around the NDIS code.

NetBEUI—An acronym inside an acronym, it stands for NetBIOS enhanced user interface. It was developed by IBM and introduced in 1985 and is one of the fastest protocols you can use—or at least, it is now. It is self-configuring, which is good, but cannot be routed. It is a transport protocol. Because NetBEUI was developed for an earlier generation of DOS-based PCs, it is small, easy to implement, and relatively fast. Because it was built for small, isolated LANs, NetBEUI is nonroutable, making it somewhat anachronistic in today's diverse and interconnected networking environment. In general you should not design a new LAN to use NetBEUI. Fortunately, NDIS and ODI standards allow NetBEUI to coexist with other routable protocols. For instance, you could configure a network to use NetBEUI for communications on the LAN segment and then use TCP/IP for communications that require routing. Because NetBEUI cannot be routed, you have to use a switch, bridge, or brouter if you must introduce some network isolation on network segments that use it.

NetBIOS—Network Basic Input Out System—A two-way connection to send data. The session ensures that the data is received properly. NetBIOS names are always 15 characters long. Lotus Notes uses NetBIOS to establish logical names on the network, establishing sessions between computers using logical names and the like. It is an interface that provides applications with access to network resources—IT IS NOT A PROTOCOL.

NetBIOS Over IPX—You should enable this option when using IPX/SPX on your MS Clients, otherwise they will not necessarily see each other.

Netlogon—A share created on NT Servers for logon validation purposes. Scripts and Policy files are normally stored here. The local path to the share is normally %systemroot%\system32\repl\import\scripts.

NetStat—Part of the TCP/IP protocol suite. This utility displays the network status.

Network Layer—*See* OSI.

Network Redirectors—Allow access to files on the network.

Network Security (MS)—Can take 3 forms. **Access permissions applied to a shared resource, User-level security applied to a shared resource,** and **File-level security applied to a shared resource.** Access permissions refer to the simple password-protection schemes applied to network shares. A resource is shared with two passwords, one for full access and one for read-only access. The user supplies the relevant password for the relevant access. User-level security, when applied to a resource, is more robust. Here, the network administrator shares a certain resource and adds user groups from the user database to have access, the most common form of access. File-level security is the highest level of security and requires the hard drive to be formatted under NTFS (New Technology File System). In addition, to share level security, the network administrator can grant or deny individual users and groups the ability to read specific files or directories individually.

NFS—Network File System.

Nibble Mode—An asynchronous identification channel between a printer and a computer. It allows the printer to identify itself to the OS using the information stored in the printers ROM.

Nonblocking Spoolers—Can accept jobs from many clients in parallel.

NRN—NetWare Connect—A proprietary NetWare protocol. It gives clients the capability to dial into a NetWare server. In addition, IPX/SPX can be used over an NTN connection.

NTDS—NT Directory Services—A replicated database system used for the creation and management of the NT domain model. *See also* Replicated Database.

NT RAS—*See* Windows NT RAS.

NWGINA—NetWare Graphical Identification and Authentication—Handles the simultaneous login to the NDS tree and local account of the NT workstation. *See also* MSGINA, which this file replaces.

O

---

a b c d e f g h i j k l m n o p q r s t u v w x y z

ODI—Open Datalink Interface—Developed jointly by Novell and Apple and released in 1989. Allows multiple protocols to bind to one card. Essentially, Novell's version of **NDIS.** Before ODI, there was dedicated IPX, which allowed only one protocol (IPX) to use a NIC. It consists of three main components—Protocol Stacks, Multiple Link Interface Drivers (MLIDs), and Link Support Layer (LSL).

OLE—Object Linking and Embedding—Win32 applications in Windows do not use shared memory to share information with each other, as do Win16 applications. Therefore, they use this more sophisticated system to share information between them.

Oscilloscope—This device measures fluctuations in signal voltage and can help find faulty damaged cabling.

OSI—Open Standards Interconnect Model—Remember the layers with this memory jogger—All People Seem To Need Data Processing, or Application, Presentation, Session, Transport, Network, Datalink, and Physical. It gives a standard (although not strictly applied by anyone!) for creating network devices, drivers, etc. In detail, the Application layer is not actually an application. But when an Application communicates on the network, it goes to this layer first. This prevents the application from having to care about whether the resource is on the computer locally or on the network. Presentation Layer—This layer takes the information to the OS and network in a form that they understand. Session Layer—Ensures the communication's process does not break down. It keeps track of the network session and ensures the synchronization of the communication process. Transport Layer—Its sole purpose is to keep track of the network session and to check that the information that has arrived is the same as the information that was sent. Network Layer—This layer is in charge of addressing the information that is sent to the correct destination. Data-Link Layer—Consists of network protocols and network card drivers. The protocols ensure the information is in the correct form and error-free; the network card driver communicates with the card hardware to deliver and receive information from the protocols. Physical Layer—The network card resides at this layer. This is where the information from the computer is generated into electricity to be communicated to the network wire or other media.

OSPF—Open Shortest Path First—A protocol used in dynamically configuring router tables.

Overhead—The number of actual data bits that can be transported per second will always be less than the bandwidth of the media, because we must also use that bandwidth to send control signals, called *overhead*.

P

a b c d e f g h i j k l m n o p q r s t u v w x y z

Packet Radio Networking—The mobile device sends and receives network style packets via satellite. Packets contain a source and destination address, and only the destination device can receive and read the packet. *See also* Mobile Computing.

Packet Services—Many organizations must communicate between several points. Leasing a line between each pair of points can prove too costly. Many telecommunications services are now available to route packets between different sites, such as X.25, Frame Relay, ISDN, and ATM. These services are available on a leased basis from service providers. An organization that must communicate between many sites pays to connect each site to the service, and the service assumes the responsibility of routing the packets. The expense of operating the network is duly shared among all network subscribers. Because the exact switching process is concealed from the subscriber, these networks are normally depicted as a communications cloud. This sort of WAN architecture can be vastly more cost-effective than leasing enough lines to provide equivalent connectivity. *See also* Virtual Circuits.

PAD—Packet Assembler/Disassembler—The equivalent of a modem on an X.25 network.

Paging—A term normally related to memory. Paging is used when a memory manager calls for information and stores it in an area of memory for later retrieval. VMMs, for example, tell each application that it has 4GB of addressable memory on the system, which of course most systems do not have, so where does the deficit come from? It comes from the hard drive. Consider that although you run more than one application at a time, not all must be active at the same time. As such, idle processes can be saved out to the hard drive, freeing up valuable system resources for active programs on the system. This data from the idle process is paged out to the hard drive for later retrieval when needed. This is called **memory swapping** or **demand paging.** *See also* **Excessive Paging.**

Paging Table—How the **VMM** (**Virtual Memory Manager**) keeps track of all the applications it has allocated virtual and physical memory for.

Parity—The use of a checking bit to ensure the integrity of received data. When used with modems, the setting is usually none with newer models.

Pass-Through Authentication—This is the process in which, for example, a user accessing a resource on a '98 client machine using User-Level Security has their access checked by the Server on the network. If the server agrees, access is OK. Then the user is given access and pass-through authentication has occurred. This also happens in trust domain relationships, where accounts are held in the trusted domain. Before a user can log on to the local resource domain, their details must be validated by the (possibly remote) trusted account domain.

Password Protected Share—Now disused term from Windows for Workgroups. The same as Share-Level Security, essentially. *See also* Share-Level Security.

Patch Cable—Term for the cable that runs from a PC to a wall jack with connectors on both ends.

PCI (Peripheral Component Interconnect)—A 64-bit bus in a PC. This is the most popular bus on the PC today. Most Pentium and higher systems include both ISA and PCI slots. They are shorter and more offset than ISA slots.

PDC—Primary Domain Controller—In NT Directory Services, there can be only one PDC per a domain and it provides the main point of administration and security of the network. It is the machine that holds the main copy of the SAM database. Think of it as the policeman of the network. Users are validated at logon and whenever they access a resource. Additional security services such as **policies** and **profiles** can also be used with **NT** Servers. *See also* **BDC.**

Physical Layer—*See* **OSI.**

PIC—Programmable Interrupt Controller—When PCs first came out, they had just a single PIC. This gave just 8 interrupts. The next generation PCs had a second PIC, which cascaded from the first to give 16 interrupts, (normally the cascade is on interrupt 2).

PIF—Program Information File—You can modify the behavior of an **MS-DOS virtual machine** by editing the properties of the PIF of an MS-DOS executable program.

PING—Packet Internet Groper—Part of the **TCP/IP** protocol suite. This command line utility is used to test connectivity. For a continuous PING use it with the -t switch.

Peer-to-Peer Network—Basically a WorkGroup. A logical grouping of computers that offers little security. Not recommended to go for more than 10 users in this environment, otherwise things become very difficult to administer.

Permissions '98—When **User-Level Security** is installed on '98 machines, the **permissions** that can be issued are **Read Files, Write to files, Create Files and Folders, Delete Files, Change File Attributes List Files, and Change Access Control.**

PIF—Program Information File—Windows applications have information stored in the program's .EXE file that tells Windows what resources and memory it needs, along with other attributes about the program. MS-DOS programs running in a Windows environment, however, do not carry this information. To create a PIF, right click the .EXE file in Explorer and look at the program's properties. A file with the same name as the .EXE is automatically created, with the extension of .PIF. The PIF file must reside in the same directory as the .EXE file to work, and more than one PIF can be created for a program, which means different settings can be set up. You must run the program by clicking or setting up a shortcut to the PIF, however, not the .EXE file.

Point and Print—Simply, printers shared on the network, the only point of interest being the ease with which they are set up. It is not a printer or print driver feature, but is simply the way the printer is installed on a machine. The two main features of Point and Print is that you can browse in Network Neighborhood for a printer and drag and drop it into your printers folder, or you can right click it. In either case, you will be asked for nominal information. Secondarily, you can use Drag and Drop Printing. Simply drag a document on top of the printers icon and the machine will be automatically configured to print the document. Handy for one-time temporary access to the printer. Any time a printer is installed under '98, it is installed as a Point-and-Print device. This works through a hidden share PRINTER$, which is the Windows/System folder shared with read-only access. When another machine connects, the drivers are loaded from here. **Note:** If you stop sharing the printer or change the share name, the printer will cease to be Point-and-Print enabled. If the machine is connected to an NT Server or NetWare server, some additional configuration is needed to support Point and Print. The Novell Server must be running Bindery emulation; NDS will not work. The Bindery must be programmed with the Printer Driver file names, Print drivers files location, and the Printer Name and Configuration.

Policy—*See* System Policy.

Policy Templates—Have a .ADM extension. They are not what makes System Policy Editor work, they are simply a default policy that you can begin with. See the relevant file guru section on this site for details of further ADM option files provided with your OS.

Port Scanner—Plugs into the network and listens for information packets between two specific terminals, or the e-mail port of a specific server. This allows hackers to silently intercept memos, find out exactly what someone is doing on the Net during office time, and even send fake messages from an individual that the recipient cannot tell from the real ones. Usually allied with IP spoofers. *See* IP Spoofers.

POST—Power-On Self-Test.

Power Users—A local group in Windows NT Workstation that is given greater rights than domain users are given.

PPI—Print Provider Interface—A modular interface under '98, designed to let you install several print providers at the same time (e.g., MS, Novell). The PPI also allows third-party network vendors to create a print provider for it's own OS. Local printing is handled by **SPOOLSS.DLL** (*see* 98 file guru). **WinNet16** offers backward compatibility with Windows 3.11 and its network drivers and **Network,** which is a 32-bit print provider that converts data to the appropriate network language specific to that vendor, such as Win32 calls for Microsoft.

PPP—Point-to-Point Protocol—The most popular line protocol, because it supports multiple protocols and supports DHCP having been created from TCP/IP. It functions at the Physical and Data-Link layers of the ODI Model. The design goal of PPP as referenced by RFC 1711 was threefold: to provide a method of encapsulating datagrams over serial links, to provide extensible Link Control Protocol (LCP), and a family of Network Control Protocols (NCPs) for establishing and configuring different network layer protocols. Improvements over SLIP include the capability to dynamically negotiate IP addresses, the addition of checksum error checking for each frame, the capability to support multiple protocols over a single serial connection, the addition of NCPs to negotiate choice of network layer protocols, and the addition of LCP to establish link options. It is a bit-oriented protocol that identifies the beginning and the end of a packet with bit patterns referred to as *flags*. This was derived from the HDLC (High Data Link Control) protocol. Including the beginning and end flag, the PPP packet can be as large as 1508 bytes, with a maximum of 1500 bytes for data.

PPTP—Point-to-Point Tunneling Protocol—Supported fully by NT. Under '98, it is an enhancement to Dial-Up Networking that allows it to take advantage of those NT services. It adds the ability to treat the Internet as a Point-to-Point dial-up networking connection. All data can be encrypted and compressed. In addition multiple protocols can be used—for instance, IPX/SPX on your office network sent over a TCP/IP network (the Internet). Over direct connections to the Internet, VPN and PPTP allows your IP addresses to be uncoordinated with the Internet's addresses. You can use whatever IP address scheme you like. All network protocols supported by Dial-Up Networking are supported over PPTP. Networks running combinations of NetBEUI IPXSPX and TCP/IP can be joined. RAS security protocols and policies are used to prevent unauthorized connections. *See also* **VPN.**

Preamble-Section of an Ethernet II Frame—A field that defines the beginning of a frame.

Preemptive Multitasking—Here, the OS is in charge of which program can do what processing and when. Under Windows NT, a preemptive multitasking OS, different priorities can be assigned to different applications, fine-tuning system performance further. 32-bit applications are normally required for preemptive multitasking from the very beginning. Windows 95/98 preemptively multitask 32-bit applications. In a **multitasking** environment, threads are assigned a priority and are then executed in the order of priority they hold. With preemptive multitasking, each thread is run for a preset amount of time, or until another thread with a higher priority comes along. This stops one thread controlling the processor wholly. *See also* Cooperative Multitasking.

Preferred Master Browser—Always a PDC; the machine will win a Browser Election.

Preferred Server—The bindery-based server by which you would like to be validated.

Presentation Layer—*See* OSI.

Primary Scheduler—*See* Scheduling '98.

Print Router—(SPOOLSS.DLL)—In '98. Its job is to route the print job to the appropriate print provider. If destined for a local printer, then it sends the job to the local print provider. If destined for the network, the router sends it to the appropriate network print provider (MS or Novell). Note: If the SPOOLSS.DLL file becomes corrupted or fails, you will not receive an error message that the print job did not reach the printer. The process appears to work properly, but the print job is never printed.

Print Server—Used to maintain a single printer or group of printers across the network. A print server manages access to shared printers making them accessible to users at other network machines. They usually incorporate a Spooler mechanism that can accept jobs from clients faster than the printer can print them. This helps to free up the client machine quickly because the client is fooled into thinking the document has printed. A nonblocking spooler can accept jobs from many clients in parallel.

Printer Driver '98—A printer driver consists of a combination of the universal driver and minidriver.

Printing-3.11—Be aware that when a user prints from an application, he has to wait for the printing to finish before control of the machine is regained. Also, data would be sent to the printer regardless of whether the printer is there. If you try to print from MS-DOS and Windows 3.11, you would normally end up with a locked up spooler and/or lost print data. 95/98/NT fixed all this.

Printing '98—The default location for queued print jobs is \WINDOWS\SPOOL\PRINTERS. The default location for printer drivers is \WINDOWS\SYSTEM. '98 supports bidirectional communications so the printer can report back device-specific information stored in the Printers ROM. '98 printing capabilities include more printer drives (more than 200), better color management, better Web-based printing, and an updated HP JetAdmin utility. As a basic overview of the print process, when you send a document to be printed to a printer you have never used before, the driver is uploaded from the server (which can be another '98 machine with the printer attached). If it exists, then this process is skipped; but if a newer version of driver exists, then it is updated on your machine ('95 did not look for updates). The application generates information about the document to be printed and is sent through the drivers to the GDI (Graphics Device Interface). The GDI produces an EMF file. '98 spools the information to the EMF Spooler if configured, and if so, frees up the application. Otherwise, the application must wait until the print is done. If EMF spooled, then the EMF Spooler passes the document back through the driver, where it is converted to RAW output. It then gets passed to the local Print Spooler. The local Print Spooler communicates with the spooler on the server, where it passes the RAW formatted print job (if printing to a local printer then this bit is skipped). The Print spooler on the server sends the data to the print monitor, where it gets directed to the correct port for printing. The port may be an LPT port or a COM port, or it may pass the print job onto a network provider. *See also* Printer Driver '98

Priority Inheritance Boosting—A function carried out by the **process scheduler.** If a low-priority thread is using a resource that a higher-priority thread needs, it can block the higher-priority thread from being processed and possibly cause the system to lock up. To avoid this problem, the secondary scheduler raises the priority of the lower-priority thread to match that of the higher-priority thread, so that the Primary scheduler gives each thread an equal slice of time. As soon as the thread releases the resource, its priority is lowered back again to what it was in the beginning.

Process—A task an application wants to accomplish. An application may have a number of processes. A process consists of at least one **thread.**

Process Scheduler—Basically, gives the thread with the highest priority use of the processor. If each thread has the same priority, then it gives each thread an equal time slice (quantum) of processor usage. It also reevaluates the priority of threads to increase system performance, if possible. For example, threads waiting for user input, such as a foreground window, get a priority increase, to make the system more responsive; threads that have been waiting voluntarily get an increase; compute-bound threads, such as spreadsheet calculations, get their priorities lowered, and all threads get a priority increase to keep lower-priority threads from holding on to shared resources that a higher-priority thread might need. *See also* **Priority Inheritance Boosting.**

Profile—Gives each user of a machine individualized settings. Profiles affect Applications, including your Address book, Quicklaunch Toolbar, Outlook Mail and News, and Welcome; any cookies downloaded while using your browser package; Active Desktop settings (if enabled); your favorites for your browser are enabled; the history of sites visited from your browser; the contents of the My Documents folder; the Start Menu; documents last accessed; options; and the temporary Internet files folder again is enabled. Profiles can either be locally based or on your server. If local, then they are always available. If on your server, you will be able to maintain your Desktop settings from virtually any machine on the network. *See* **Roving Profiles.**

Propagation Delay—The time required for a signal to reach the farthest point of the network and then return. *See* Satellite Microwave.

Program Resources '98—When an MS-DOS application is terminated, all its resources are returned to the system. However, if a Win16 app, then '98 may not be aware of all the resources being used by any and all Win16 apps running in the same memory space. In short, all Win16 apps must be closed before all resources are returned to the system. Win32 app is terminated, and all resources are returned to the system.

Protected Mode Drivers—Generally loaded from within Windows and get their name from the fact that such components are loaded into memory and are protected from interfering with each other. They are generally 32 bit and preferred to their 16-bit real mode counterparts. Protected Mode Windows drivers usually have a .VXD or .386 extension.

Protocol—In computer terms, a standardized method for two computers to talk. A little like you and I speaking the same language, as opposed to you speaking French and my speaking German. You must have at least one protocol installed for two computers to communicate. Examples of protocols are **TCP/IP, NetBEUI, IPX/SPX, Microsoft DLC and DLC, ATM,** and many others.

Protocol Stack—Consists of the Transport and Network Layers of the OSI model. It is the protocol stack that acts as the interface to the Application Layers. It packages data from the Presentation and Application Layers, then provides network functionality by routing packets to their destination. Multiple ODI protocol stacks can exist on a single system. To distinguish media frames, two IDs are used—the stack ID and the protocol ID.

Proxy Server—One of the key components of a firewall. A proxy server acts on behalf of several client computers that request content from the Internet or an Intranet. The Proxy Server is transparent to both the user and the server unless the user tries to gain access to a disallowed server. It controls the passing of information back and forth from the inside and outside networks, and so can control which information goes both in and out. They also provide increased performance because they can cache frequently requested information.

PSTN—Packet Switched Telephone Network—Sometimes referred to as *POTS* (Plain Old Telephone System). Basically, the normal telephone system.

Pull Installation—Where the user sits at the client and effectively requests the installation. *See* **Push Installation.**

Push Installation—Where the client has no choice but to accept the installation pushed on it from the server. Normally a function of advanced management tools like Novell's ZEN (Zero Effort Networking) works and Microsoft's SMS (Server Management Server).

PVC—Permanent Virtual Circuit—A permanent route through the network that is always available to the customer. With a PVC, charges may or may not be billed on a per-use basis.

PVC—Polyvinyl Chloride—Used as insulation in coax cables.

PWS—Peer Web Server—The package supplied with '98 and NT Workstation lets you quickly share information in a corporate environment by hosting your own web pages. If going out to the Internet, you should really consider IIS (Internet Information Server), which is the full version of this product and is more suited to the open Internet. PWS includes a Web Publishing Wizard and a Home Page Wizard. IIS 4 supports FTP, an unlimited number of clients, and full web security.

a b c d e f g h i j k l m n o p q r s t u v w x y z

Quick Logon—An option in '98 with Microsoft Client for Networks. You are logged onto the network with this option, but network drives are not reconnected until you use them. The alternative is **Logon and Restore Network Connections,** which is self-explanatory.

a b c d e f g h i j k l m n o p q r s t u v w x y z

RAID - Redundant Array of Inexpensive Disks—There are six levels of RAID, from 0 to 5. Level 0 is Disk Striping; level 1 is Disk Mirroring; level 2 is Disk Striping with ECC; level 3 is ECC stored as parity; level 4 is Disk Striping with Large Blocks; and level 5 is Disk Striping with Parity. Only three are in common use today—levels 0, 1, and 5—providing varying levels of fault tolerance. **Raid Level 0** is used on systems that require the fastest access without regard to redundancy. The disks are arranged to provide parallel access. If a file contains four sectors worth of data, then sectors 1 and 3 are stored on one disk while 2 and 4 are stored on the other. Level 1 (Disk Mirroring) is used when data redundancy and lack of downtime is critical. Here, when a write is issued to one disk, the other disk receives the exact same command. Should one disk become defective, the other can take over without shutting down the computer. Level 5 provides faster disk access and a better utilization of disks than level 1 and indeed becomes better as the number of disks increases. A parity sector is maintained across all disks and can be used to rebuild a defective disk when replaced.

RAM—Random Access Memory.

RAW Data Output—When printing, this is the natural language of the printer, such as its escape codes or post-script, and this data is usually printer specific. *See also* EMF Data Output.

RD—Receive Data.

Read Ahead Caching—Allows the computer to read ahead a specified amount of space on the hard drive and load this information in RAM for faster access.

Read Permission—On Microsoft Networks, users with this permission are able to read and execute files. Equates to the NetWare Read Right.

Real Mode Drivers—Generally 16-bit drivers associated with MS-DOS and usually loaded prior to Windows starting. Usually specified in CONFIG.SYS and are not as robust and stable as Protected Mode Drivers. Real Mode MS-DOS drivers usually have a .SYS extension. Real Mode Windows drivers usually have a .DRV extension and are listed in the SYSTEM.INI file.

Reflective Infrared—Wireless PCs transmit toward a common central unit, which then directs communication to each of the nodes. *See also* Infrared Transmission.

Registry '98—The registry is a hierarchical database of all the computers settings, including hardware, software, and users. It replaces the need for AUTOEXEC.BAT, CONFIG.SYS, and .INI files found in 3.11 and MS-DOS. Different OSs, have different forms of registry, however. The registry under Windows '98 consists of two files, SYSTEM.DAT and USER.DAT, located in the Windows folder. When a profile is enabled, however, each user gets their own USER.DAT. *See* HKEY_CLASSES_ROOT, HKEY_USERS, HKEY_CURRENT_USER, HKEY_LOCAL_MACHINE, HKEY_CURRENT_CONFIG, and HKEY_DYN_DATA.

Remote Administration '98—To administer another '98 computer remotely, you must enable the Remote Registry Service, which also makes the System Monitor possible. To add this, right click Network Neighborhood, select Properties, and add the service that you find in the \tools\reskit\netadmin\remotreg directory of the CD. The file to click on is regsrv.inf. This adds a Delta properties screen when browsing the Network and on clicking that machine's properties. There are three options: Net Watcher, System Monitor, and Administer. Note: It must be installed on both machines for full functionality.

Remote Bridge—When a bridge must link a local network across a wide area segment, it is referred to as *remote bridge*. The output channel from the remote bridge is usually of dramatically lower bandwidth capacity. The difference in relative bandwidth capacity on the input and output channels makes the remote bridges significantly more complex to design and manage. They must be able to buffer inbound traffic and manage time-out errors. It is also necessary to design your network to keep traffic requirements over the remote link to a minimum. *See also* Local Bridge.

Repeater—Allows the distance between two devices to be increased by regenerating the signal. It works at the physical layer of the OSI model. Their characteristics are that they regenerate a signal that comes in on one port and transmit it out through the other repeater ports. They do not filter or interpret; they merely repeat a signal, including errors in one direction, even if it is a broadcast storm caused by a malfunctioning adapter. They do not require any addressing information from the data frame, and are inexpensive and simple devices. Although they cannot connect dissimilar networks (i.e., Ethernet and Token Ring), they can connect segments with similar frame types but dissimilar cabling.

Replication (Database)—Keeping more than one copy of information might make sense if we want to have a local copy of distant information. When multiple copies of the database exist, we run the risk of basing decisions on outdated information. Replicating the database and synchronizing it across the network cuts down on the risk of retrieving faulty data. Currently, there are two methods for replicating the database. The first method uses a master database to store all additions and changes. The database server is responsible for updating both the master and copying the master to the portions of the database that are stored in the individual departments. The second method distributes the responsibility for additions and changes to the local database engines. This second method also makes the local database servers responsible for updating any copies of the database across the network. The first method is easier to implement, but the second method may be needed for networks that include unreliable links and portable computers.

Replicated Database—Example: NTDS. The PDC replicates or copies the master directory database to the BDCs.

Replication (MS)—Replication is a feature somewhat specific to LAN–Manager based networking products, specifically Windows NT Server. It is also supported on Windows NTAS v3.1 (NT Advanced Server). Through replication, data is automatically copied from a source system (exporter) to a destination system (importer). This can be a one-to-one relationship, several exporters to one importer, or one exporter sending to several importers. Only Windows NT Servers (including NTAS) and LAN Manager v2.x servers may act as exporters. An export directory is specified as the data source. Any number of direct subdirectories and up to 32 levels of subdirectories below the export directory are supported. Windows NT Workstation, Windows NT Servers, and LAN Manager servers can act as importers. An importer does not have to be located in the same domain as the exporter with which it communicates. Replication occurs when a file is modified then closed. This gives you a near immediate backup of volatile data files, or propagates files as needed between network servers.

Resource Arbitration—A process in which two conflicts during the PnP process are (it is hoped) resolved. This arbitration process assigns resources to boot devices first (IDE drives if they boot first, SCSI Controller if SCSI boot disk); then assigns resources to legacy devices, which are generally inflexible and unprogrammable by the PnP OS; and finally, PnP devices are configured.

RI—Ring Indicator.

Rights—Under Novell, these are the equivalent to Microsoft's Permissions. Under Microsoft, Rights refer to specific abilities a user or group may possess. Unlike MS Permissions, Rights focus on the user or group instead of the resource. Typical rights include the ability to Create Accounts, Log on to a particular computer, or log on as a service. Basically, MS Rights modify the abilities of a particular user or group, enabling the person or group to carry out administrative functions.

Ring 0—Within this ring, the processor provides the most protection against other programs as well as giving the most privileges to system resources. For instance, components such as the GDI, Kernel, and user interface run within ring 0 on a '95/'98 system. If anything crashes within Ring 0, then the whole system will probably crash. *See also* Ring Architecture.

Ring 3—Most components and software run within this ring. The processor does not provide any protections for the software running here. Any protection would have to be supplied by the OS.

Ring Architecture—Processors that rely on the Intel x86 processor, including AMD and Cyrix, use Intel Ring Architecture. Each software component runs in one of four rings, numbered from 0 to 3. The lower the number, the more critical the software component to the system. Switching between rings costs the processor time and a lot of memory; therefore, OSs like '95 and '98 just use rings 0 and 3. *See also* **Ring 0, Ring 3.**

Ring Topology—Wired in a circle. Each device incorporates a receiver and transmitter and serves as a repeater that passes the signal on to the next device in the ring in one direction only. Because the signal is regenerated at each device, signal degradation is low and longer distances are generally supported. They are ideally suited to token passing access methods. However, ring physical topologies are quite rare. The ring topology is almost always implemented as a logical topology. For example, Token Ring (the most popular) arranges the nodes in a physical star while passing the data in a logical ring.

RIP—Routing Interface Protocol—Used in Dynamic Router Configuration. *See also* OSPF.

RJ-45 Connector—Similar to the RJ-11 often used in telephone installations. These are 8 pin connectors that are simple to install on cables and are easy to connect and disconnect. Used primarily for category 5 or category 3 twisted-pair cable.

RJ-56 Connector—Used to connect coaxial cable on 10Base 2 networks.

ROM—Read Only Memory.

Route—Part of the TCP/IP protocol suite. Used to configure routing.

Router—Whenever computers in different TCP/IP subnet want to communicate, they must send their messages to a router, which then forwards the messages to their correct destination subnet. It may be a hardware device or even an NT Server, which is the only OS in the MS family that can act as a TCP/IP router. They work at the Network layer of the OSI model. Because each network has its own internetwork address, each network can be considered logicaly separate. Internetwork connectivity devices, such as routers, use network address information to assist in the efficient delivery of messages. Using network address information to deliver messages is called *routing*. The common feature that unites Internetwork connectivity devices (routers and brouters) is that these devices can perform routing. They organize the large network in terms of logical network segments. Each network segment is assigned an address so that every packet has both a destination network address and a destination device address. They are more intelligent than bridges. Not only do routers build tables of network locations, but they also use algorithms to determine the most efficient path for sending a packet to any given network. Even if a network segment is not directly attached to the router, the router knows the best way to send a packet to a device on the network. After route costs are established, routers can select routes either statically or dynamically. *See also* Static Routers, Dynamic Routers, Static Route Selection, Dynamic Route Selection, Brouters.

Roving Profiles—Profiles held on a server so that they can be downloaded by users regardless of where they log on. However, due to differences in registries, '98 profiles will not work on NT machines and vice versa. A local profile always exists in case the server is not available, of course. Also, be aware that the most recent profile at the time of the user logging off is the one written up to the server, so if they are logged into a number of machines, the profile, if altered, may not be what the user expects if they are not thinking about what they are doing. You must also make sure your local machines are synchronized with the server. Use **NET TIME \\servername /SET /Y.** Furthermore, the OS root directory should be the same on local machines (i.e., one machine should not have '98 installed in c:\Windows while another has it installed in c:\Win98, or they just will not work). On NT Servers, remote profiles are stored in the user's home directory. On Novell Servers they are stored in the user's Mail directory, unless your '98 client uses MS Services for Netware Directory Services and your Novell Server runs NDS, in which case, it resides once again in the user's home directory. If not all servers log onto the same type of NetWare server, this may cause a problem. Note: Screen saver settings under '98 are not a part of a roving profile because the password is saved locally on the system.

Roving Users—By using profiles, users can keep their desktop preferences the same no matter where they log on to the network from.

RPC—Remote Procedure Call—Allows one computer to send a message telling another computer to accomplish some task.

RPM—Red Hat Package Manager—Makes software almost as easy to install under UNIX as it is on a Windows PC.

RS232—The original RS232 connector was designed to specify the point at which the PC and modem interact. The EIA developed the RS232, and this specification helped relieve the problems of signal connection incompatibilities that arose because no prior standard existed. Each of the pins that comprise the RS232 standard communicates the status information from the modem to the serial port of the computer. It is also referred to as *RS232-C,* and in 1987 was renamed *EIA232-D.* It was slightly redesigned at this time to be compatible with several internal specifications: ISO 2110, V.24, and V.28. Cable lengths should be limited to 50M, and the DB25 is its standard connector, although the DB9 is a popular alternative.

RS449—The EIA sponsored this standard to overcome the speed and distance limitations of RS232. It relies on the RS422 (balanced transmission, two wires, no common ground) and RS423 (unbalanced, common ground shared by all circuits) standards to specify the electrical characteristics of the pins. RS449 allows for more signal lines, which enables several lines to be used for return signals. It uses two connectors, which makes it expensive compared to RS232. The main channel connector is a 37-pin connector; the secondary is 9 pin. The EIA also released the EIA530, which is electrically compatible to RS449 but uses a DB25 connector.

Rsync—A file transfer program for UNIX systems. Rsync uses the "rsync algorithm," which provides a very fast method for bringing remote files into sync. It does this by sending just the differences in the files across the link without requiring that both sets of files be present at one of the ends of the link beforehand.

RTS—Request To Send.

a b c d e f g h i j k l m n o p q r s t u v w x y z

Safe Mode—Under '98 and '95 you boot into safe mode so that any troublesome drivers and start-up files are bypassed to aid in your troubleshooting efforts.

SAM Database—Holds all account and security information on a PDC. Limited to 40Mb in size. It is a replicated database.

SAP—Server Advertising Protocol—Used by NetWare servers to make other computers aware of resources. For example, to see another computer sharing resources, you could use Network Neighborhood. However, machines running a NETX or VLM clients can see computers only with SAP advertising enabled. Check if your machine has this enabled by using the **nlist server /b** command at the DOS prompt. A SAP packet is broadcast by a Novell Server every 60 seconds and contains the name of the server and the resources it is sharing. As the number of servers increases, so do the number of SAP broadcasts. By default, SAP advertising with clients is switched off, but without it, you do not show up on the Browse Lists of other NetWare clients. *See also* Workgroup Advertising.

SATAN—Systems Administrator Tool for Analyzing Networks—A computer program designed to probe all the vulnerable parts of a network to show weaknesses that might allow a hacker entrance. Armed with this knowledge, you can plug the holes in your network. Hackers, of course, use this program to show the weaknesses that they can then use easily.

Satellite Microwave—Relay transmissions through communication satellites that operate in geosynchonous orbits 22,300 miles above the earth. Satellites orbiting at this distance remain located above a fixed point on earth. Earth stations use parabolic antennas to communicate with satellites. These satellites can then retransmit signals in broad or narrow beams, depending on the locations intended to receive the signals. Because no cables are required, satellite microwave communication is possible with most remote sites and with mobile devices (e.g., ships, motor vehicles). The distances involved in satellite communication result in a phenomenon where because all signals must travel up 22,300 miles and then down 22,300 miles, the time required to transmit a signal independent of distance; therefore, it takes as long to transmit a signal to a receiver in the same county as it does a third of the way around the world. This is called *propagation delay*, and ranges from 0.5 to 5 seconds. *See also* Microwave Transmission.

Satellite Station Networking—Satellite mobile networking stations use satellite microwave technology. *See also* Microwave Technology, Mobile Computing.

Scandisk—Disk checking utility in Windows 95 and 98. On a FAT32 partition, it is limited by the partition not being more than 127.53GB. It checks long file names, directory tree structure, Doublespace and Drivespace volumes, lost clusters, invalid file names and dates, etc.

Scatter Infrared—Transmissions reflect off doors, walls, and ceilings until (theoretically) they finally reach the receiver. Due to the imprecise trajectory, data transfer rates are low and the maximum reliable distance is about 100 ft.

Scheduling '98—Windows '98 actually has two schedulers that determine which thread has use of the processor at a given time: the Primary and Secondary schedulers. All threads are given a base priority, of which there are 32. User applications and noncritical functions are given a priority of 0–15, whereas critical system functions range from 16 to 32. Most user apps run at 7, and base priorities can be moved by as much as two levels up or down to increase system performance. *See also* **Process Scheduler.**

Secondary Scheduler—*See* **Scheduling '98.**

Sector Allocation Granularity—Programs such as DriveSpace 3 uses the sector as their allocation unit.

Security Provider—A characteristic of an advanced Network server. For '95 and '98, using User Level security, a security provider must be used to provide pass through authentication. NetWare Servers, NT Server, and NT Workstation can all be used as security authorities.

Service for NetWare Directory Services—*See* Microsoft Client for NetWare Networks.

Session Layer—*See* OSI.

SET Variables—Things such as the PATH statement, the TEMP folder, and the like are configured using the SET statement.

Share Level Security—Should only be used in peer-to-peer networks in which there is no advanced server such as Netware or NT. This is because each share/resource has its own security and password. This quickly becomes a nightmare. Note: When you share a folder on a '95/'98 computer, users accessing this folder also gain access to any subfolders it contains. There are three choices with this type of security for users over the network: **read-only, Full Control** (or just Full in '98), and **Depends on Password.** *Depends on password* means that depending on the password the user gives determines whether they gain full control or read-only access to the resource. *See also* User Level Security.

Shift + F5—Starts a Win'95/'98 system at the command prompt.

Shift + F8—Held down at bootup on a Win'95/'98 machine causes the system to start in step-by-step confirmation mode.

Shutdown—The process of closing the machine down properly so that all relevant data can be written to disk and services can be terminated gracefully, as well as to avoid any disk errors a sudden shutdown can cause, especially in relation to **FAT**-based file systems. For example, if a '98 machine is not closed down properly, its name can appear in browse lists for up to 45 minutes before being cleared!

SID—Security Identifier—Each time you create a user or group account in an NT domain, the account is assigned a security identifier, which is a unique number. It is also used to access rights to resources. Each resource has an ACL (Access Control List) that contains a list of SIDs that are permitted access to the resource.

SIDF—System Independent Data Format.

Signal Propagation—*See* Propagation Delay.

Signaling—Amounts to communicating information. This information can take one of two forms. *See also* Analog or Digital.

Single Backup Server—This backup strategy uses a centralized backup server with a very large tape drive or an array of multiple tape drives. Each server is connected to the backup server over the network. If there are large amounts of data to backup, you may need a large array of multigigabyte tape drives to store all the data. This can become expensive. If your servers are spread out over multiple floors, you can administer and monitor all the backup activity from a central location. *See also* Individual Tape Units, Independent, Redundant Backup Network.

Single Domain Model—This model places all accounts, user groups, and computer and resources into a single domain. It is suitable for small organizations only. As such, the following advantages are gained. Easy installation—the only requirement is to install the first server as PDC. Simple configuration—because all resources and accounts reside in one domain, there are no trust relationships to configure. All of your networks servers and workstations join the same domain. Centralized administration of user accounts—because there is only one Directory Database (SAM), local and global groups need be defined only once, whereas in a multidomain network, you must define multiple groups, depending on which domain the resources were located. Centralized administration of network resources—all of the network resources reside in the same domain, so resource administration is centralized. Ideal for small organizations. Relatively small organizations will not require additional servers as would be needed by a multidomain network. Note: MS recommends at least one BDC be installed for fault tolerance. Number of user accounts—the SAM is limited to 40Mb in size. This restriction limits the number of users and group and computer accounts that can be handled at any one time. No Decentralized Administration of User Accounts—larger organizations will want to distribute the management of user accounts to assist in cutting down the workload involved. The single domain model permits only centralized administration. No decentralized administration of network resources—larger organizations typically distribute resource management among departmental or geographic boundaries, much as with user accounts. Political issues within the company may also require decentralized resource management.

Single Frequency Radio—*See* Narrow Band Radio Transmission.

Single Master Domain Model—Here you have two or more domains, of which one containing the user accounts for the entire domain. The resource domains communicate via way of single one-way trust relationships. Here you have centralized administration with users benefiting by having a single user account and password because everyone logs on through the master domain. However, there is a single SAM, and so a limit to the size of the database. Local groups must also be defined in both master and resource domains. Resource administration can either be centralized or decentralized; however, if you chose to decentralize resource administration, the administrator of the resource domain has no control over global group membership. Therefore, there must be some coordination of trust between account and resource administrators in order to keep the network secure.

SLIP—Serial Line Interface Protocol—Used to connect to UNIX servers primarily and created from TCP/IP and originated in 1984. Now losing favor and popularity against PPP. This protocol operates only at the OSI physical layer and does not provide any automated method of registering IP addresses. If addresses are assigned by DHCP, with SLIP, it must be manually assigned or done via a logon script. It also has the following disadvantages: operator intervention is required to register IP addresses; SLIP can support only one protocol over a serial link; it does not perform any error checking for bad frames. NT RAS supports the client end of SLIP, but the server component is not provided.

SMB—Server Message Blocks—Windows-based networks use SMBs to communicate. Akin to Novell's NetWare Core Protocol (NCP).

SMDS—Switched Megabit Data Service—Developed by Bellcore in 1991 out of technical requirements by RBOCs. SMDS is based on the IEEE 802.6 MAN specification. SMDS provides a connectionless, point-to-multipoint service at the network layer. At the physical layer, SMDS uses high-speed links such as SONET, FDDI, and DQDB. Like ATM, it uses a 53-byte data cell at the data-link layer. The typical data rates available for SMDS are between 1.544 Mbps (T1 speed) and 45 Mbps (T3 speed).

SMTP—Simple Mail Transfer Protocol.

SNMP—Simple Network Management Protocol—Part of the TCP/IP suite. Allows computers and devices to be managed by a standard set of network tools.

SONET—Synchronous Optical Network—First introduced by Bell Communications Research or Bellcore in 1984 and standardized by ANSI. By the end of 1988, the CCITT/ITU published a similar set of standards called Synchronous Digital Hierarchy (SDH). Different regions use different variations of SDH, such as SDH Europe, SDH-Japan, and SDH-SONET (for North America). It is a physical layer specification. SONET starts at a data rate of 51.84 Mbps (called *Optical Carrier level 1*, or *OC-1*) and goes up to OC-8 or 2.488 Gbps. The ANSI SONET and CCITT SDH specifications are similar enough to be compatible. The only major difference lies in the data rates. With proper coordination among the standard bodies, SONET/SDH will soon have worldwide compatibility. *See also* ISDN, ATM.

Source Routing—Used in Token Ring Networks, the option specifies the cache file size. Source Routing is found in IBM's Token Ring Networks. In this environment, a workstation determines the routes to other workstations with which it wishes to communicate by transmitting an all-routes broadcast frame, which propagates throughout the network. The second station's reply to the broadcast includes the route that the original frame took, and from that point on, that route is specified by the initial station (the source) for the duration of communications between the stations. As a result, bridges in a Token Ring environment rely on the source to supply the routing information in order to be able to forward frames to other networks. Source Routing Transparent bridges have been proposed by IBM that would forward other types of frames as well as those with source routing information.

Spanning Tree Algorithm—Bridges that support the Spanning Tree Algorithm are able to communicate with each other and negotiate which bridge(s) will remain in blocking mode (not forwarding packets) to prevent the formation of loops in the network. The bridges in the blocking mode continue to monitor the network. When they notice that another bridge has failed, they come back on-line and maintain the network connections.

Splitter—Allows the sharing of a device by two computers, and vice versa.

Spread Spectrum Radio Transmission—A technique originally developed by the military to solve several communication problems. Spread spectrum improves the reliability, reduces the sensitivity to interference and jamming, and is less vulnerable to eavesdropping than single frequency radio. As its name suggests, spread spectrum transmission uses multiple frequencies to transmit messages. Two techniques are used. *See also* Frequency Hopping, Direct Sequence Modulation.

SPX—Sequenced Packet Exchange—A transport layer protocol that provides connection-orientated services with reliable delivery. Reliable delivery is ensured by transmitting packets in the event of an error. SPX is used in situations in which reliable transmission of data is needed.

SSH—Secure Shell—A program to log into another computer over a network, execute commands on a remote machine, and move files from one machine to another. It provides strong authentication and secure communications over insecure channels. It is intended as a replacement for rlogin, rsh, and rcp. In addition, ssh provides secure X connections and secure forwarding of arbitrary TCP connections.

Star Bus Topology—Here, hubs in a Star network are interconnected with linear bus trunks. This enables networks to grow beyond the number of ports on a single hub.

Star Topology—Where all devices are connected directly to a central hub. This type of network is usually easier to troubleshoot because each device can be individually removed from the hub. The first star network was ArcNet, invented by Datapoint in 1977. This Token Passing network used RG62 cable which is roughly the same size as RG58 but with higher impedance. The most common Star topology used today is 10Base-T. 10BaseT is an Ethernet network running over Cat 3 unshielded twisted pair UTP. The data rate is 10 Mbits per second. Logically, star topology sends data like the bus topology. A newcomer to the star topology is 100Base T, which has a data rate of 100 MBits per second. This requires Cat 5 UTP cable. *See also* Star Bus Topology.

Static Route Selection—This selection method uses only routes that have been programmed by the network administrator. Static Routers can use this method and no other. *See also* Dynamic Route Selection, Routers.

Static Routers—Do not determine paths dynamically. Instead, you must configure the routing table, specifying potential routes for packets. If the network connections change, a router using statically configured routes must be manually reconfigured. These preprogrammed routers cannot adjust to changing network conditions. *See also* Dynamic Routers, Routers.

Stop Bits—Used to identify the end of a packet. In dial-up networking, the default is 1.

STP—Shielded Twisted Pair—Consists of one or more twisted pairs of wire enclosed on a foil wrap and woven copper shielding. Often used in early Token Ring Networks. It was used because the shield reduces the tendency of the cable to radiate EMI, and thus reduces the cable's sensitivity to outside interference. Various types of STP exist; some shield each pair individually, and others shield several pairs. Each cable type is appropriate to a given situation. It has a theoretical capacity of 500 Mbps, although no common technologies exceed 155 Mbps with 100M cable runs. The most common speeds are 4 and 16 Mbps, which are those used by Token Ring Networks. Common connectors are RJ45, DIN-type connectors, and Token Ring Networks use the IBM Hermaphroditic Data Connector.

Subnet—TCP/IP terminology. When a network becomes too large, it can be broken down into a different subnet and "routed" together. If you had three computers on the same piece of wire, each programmed with a different subnet address, then none could see the others. If the machines were on the same subnet, then all would be well.

Subnet Mask—Determines which part of an IP address is the network portion and which part is the host portion. Default subnet masks for class A addresses are 255.0.0.0, class B addresses are 255.255.0.0, and class C addresses are 255.255.0.0.

SWC—Switched Virtual Circuit—Created for a specific communication session and then disappears after the session has ended. The next time the computers communicate, a different virtual circuit might be used. *See also* Virtual Circuit.

Switched 56—A full duplex, wide area digital data line offering 56Kbits/s service on a dial-up basis.

Switches—Operate at the Data-Link level of the OSI model. It performs the same function as a learning bridge, but typically had many more ports and higher throughput rates. Again, they do not block broadcast packets. Their primary use is to increase network performance by reducing levels of network contention. A switch filters out traffic that is not destined for a station on a given port, so the station sees less traffic and works better. It has become common practice to replace repeaters (often called *hubs*) with switches to increase network performance without requiring changes to the computers on the network.

.SYS—Indicative of a real mode MS-DOS driver.

System Monitor—The utility in '98 that allows you to track most of the system components and any program running at that time. Install it through Add/Remove programs in Control Panel.

System Policies '98—You can use System Policies on a '98 machine to lock down the machine on a per-user, per-group, or per-computer basis. (You must copy GROUPPOL.DLL onto each machine for this to work properly, and no User Policy must exist for an individual or it will override the Group policy option. You install Grouppol.DLL through the Add/Remove programs icon in Control Panel.) You can affect such things as Access to the Control Panel, Remove the Run option from the Start menu, show no drives in My Computer, and deny access to the My Settings tab in the Display properties. They are basically registry changes that happen automatically when the user logs on. Although the machine may know your logon or computer name, it cannot work out your group membership. To use this feature, you must have user-level security enabled. In addition, the CONFIG.POL file must be stored on the server in the NETLOGON share of the NT domain controller that validates the user. If using NetWare, it must be stored in the SYS:PUBLIC directory. You will find the System Policy Editor under the \TOOLS\RESKIT\NETADMIN\POLEDIT\ directory, and it is installed through the Add/Remove programs icon of the Control Panel. Note: Local Machine edits SYSTEM.DAT and Local User edits USER.DAT. *See also* Policy Templates, Registry '98.

System Virtual Machine—Term for where all system processes, Windows 16-bit applications and Windows 32-bit applications run in a single **virtual machine.** All the base components run here, including the **user interface, kernel,** and **GDI.** Also in '98, all Win16 and Win32 applications run here. *See also* **MS-DOS Virtual Machine.**

T

a b c d e f g h i j k l m n o p q r s t u v w x y z

T1—Lines used by many businesses to achieve transfer rates of 1.544 Mbps. It is a very popular digital line and provides point-to-point connection using a total of 24 channels across two wire pairs, one pair for sending and the other for receiving, giving a transmission rate of 1.544Mbps in each direction. DS-1 service is a full T1 line. DS-2 is four T1 lines, but is uncommon—it would typically be ordered as fractional T3.

T3—T3 is similar to T1, but T3 has greater capacity, being able to transmit at up to 45Mbps. *See also* Fractional and Multiple T1 and T3.

TAPI—Telephony Application Programming Interface—Designed to give programmers a standard way of defining how their applications will work with Windows.

Task Monitor—The program in Windows '98 that works with disk defragmentor by tracing the DLL files associated with programs and that at time of defragmentation puts them closer together to give programs a faster start-up time. It creates log files, typically stored using the application's name as the filename with an extension that denotes the drive where the application resides.

TCP—Transport/Transmission Control Protocol—It is TCP's job to ensure that the data is received reliably. IP handles only the sending of the packets and their addressing. For instance, it does not tell the sending computer to resend the packet if it was damaged in transit. It is basically the error-checking part of the TCP/IP protocol suite, although it does do various other things.

TCP/IP—Transport/Transmission Control Protocol/Internet Protocol—(Microsoft has their own version as well, deemed *Microsoft TCP/IP,* which is a full implementation of this standard.)—TCP/IP is the protocol synonymous with the Internet. Unlike NetBEUI and IPX/SPX, TCP/IP is a suite of protocols and was developed as part of the ARP Anet (Advanced Research Projects Agency Network), which needed some way for computers to communicate. It later became part of the Berkeley Standard Distribution (BSD) of UNIX. As with the Internet, no one owns this protocol. Reasons for using TCP/IP are that it is the Internet protocol; it can connect dissimilar systems, such as UNIX, IBM Mainframes, VMS, etc.; and many utilities common to other networks come as standard, such as **PING, FTP, Telnet,** and **NETSTAT.**

TD—Transmit Data.

TDR—Time Domain Reflectometer—TDRs send sound waves along a cable to look for imperfections that might be caused by a break or short in the line. From a technical standpoint, a TDR is a highly sophisticated line-test device. In most cases, however, modern TDRs have been designed as small, easy-to-use devices that many technicians consider an invaluable troubleshooting tool. A TDR works by sending out a signal and looking at the characteristics of the return signal. With most modern devices, you do not need to concern yourself with the way the device works, but with the information it provides. Typically, a TDR can tell you if your cable has a short or open, the distance to a short, the distance to an open, overall cable length, termination value, and line impedance. Although useful in troubleshooting any type of cable, a TDR is often critical for locating and resolving coaxial cable faults. *See also* DVM, Oscilloscope.

Technet—Available by monthly subscription, this disk features many technical articles and a searchable index of error reports and problems.

Telnet—Part of the TCP/IP protocol suite. This utility allows you to connect to mainframes by emulating a terminal. A problem with Telnet you should be aware of is that it sends your password over the network in plain text format. *See* **SSH.**

Terrestrial Microwave—Employs Earth-based transmitters and receivers. Uses low-gigahertz range frequencies, which limits communications to line of sight. These typically use parabolic antennas that produce a narrow, highly directional signal. A similar antenna at the receiving site is sensitive to signals only within a narrow focus. The highly focused antennas require careful adjustment. Costs are highly variable, depending on requirements. Long distance microwave systems can be quite expensive, but might be less costly than alternatives with their recurring monthly expenses. When line-of-sight transmission is possible, a microwave link is a one-time expense that can offer greater bandwidth than a leased circuit. Attenuation characteristics are determined by transmitter power, frequency, and antenna size. Properly designed systems are not affected by attenuation under normal operating conditions. Rain and fog, however, can cause attenuation of higher frequencies. Because microwave signals are vulnerable to electronic eavesdropping, signals transmitted through microwaves are frequently encrypted.

Thick Client—The thick client, such as Windows '95/'98 and NT Workstation, is in wider use today. Basically, they execute programs in their own RAM and have local disks on which to store files.

Thicknet—Thicker than Thinnet and approximately 0.5 inches (13 mm) in diameter. Due to it being thicker, it cannot bend as well as other cables types like Thinnet, and so is harder to work with. A thicker center core and better shielding means Thicknet can carry signals a longer distance; Ethernet for approximately 500M (1,650 feet).

Thin Client—This type of client features very little hardware and typically consists of software such as browsers to access network servers to which they are attached. This type of client is receiving much press as the NetPC. *See* Thick Client.

Thinnet—A light and flexible cabling medium that is inexpensive and easy to install. Thinnet is similar enough to members of the RG58 family of cables that have a 50-ohm impedance, and they are often substituted for each other. Thinnet is approximately .25 inches (6 mm) thick. It can reliably transmit Ethernet for 185M (610 feet). *See also* Bus Topology.

Thrashing—A phenomena seen with hard drives when there is too little memory on the system and too much memory paging occurring. It can damage hard drives if left to continue for long periods of time.

Thread—The smallest unit of code that can be executed by a process/application.

Thunking—The process of translating 32-bit calls into 16-bit calls and vice versa. A process performed by the **Kernel.**

Token Conversion—A subprocess of disk compression utilities such as DriveSpace 3. Here, the compression utility interrupts attempts to write to the compressed drives and searches for duplicate data strings.

Token Passing—A system developed by Datapoint initially in 1977. ArcNet computers use this system by not transmitting data until they own a token. The token is passed sequentially to each computer on the network based on the Network Card's ID. ArcNet networks require the System Administrator to set the NIC ID between 1 and 255. Duplicate IDs can be a major difficulty in ArcNet networks. If a computer has no data to transmit, that computer simply passes the token along. ArcNet supports both active and passive hubs. Token-passing networks use station priorities and other methods to prevent any one station from monopolizing the network. Because each computer has a chance to transmit each time the token travels around the network, each station is guaranteed a chance to transmit at some minimum time interval.

Token Ring—The most common token-passing standard, embodied in IEEE802.5. It uses a token-passing architecture, and the topology is always physically a star while using a logical ring to pass the token from station to station. Each node must be attached to a concentrator called a *Multistation Access Unit* (MSAU). The MSAU is used to bypass token-ring stations that are not active. Although 4 Mbps token-ring networks interface cards can only run at that data rate, 16-Mbps cards can be configured to run at 4 Mbps or 16 Mbps. All cards on a given ring must run at the same rate. Each node acts as a repeater that receives tokens and data frames from its Nearest Active Upstream Neighbor (NAUN). After the node processes a frame, the frame transmits it downstream to the next attached node. Each frame makes one trip around the entire ring and then returns to the originating node, which removes it from the ring and releases the token. Workstations that detect problems send a beacon to identify the fault domain of the potential failure.

TraceRt—Part of the TCP/IP protocol suite. This command line utility is used to check the route (number of hops) to another computer.

Trailer—The trailer marks the end of a packet and typically contains error-checking (Cyclical Redundancy Check, or CRC) information.

Tranceiver—Converts a computer's digital signals into electrical or optical signals that can then travel on network cable.

Transport Layer—*See* **OSI.**

Trust, or Trust Relationship—A communications and administrative link between two domains that permits the sharing of resources and account information. Without the trust relationship, individual domains have no method of communicating or sharing resources. An NT Server box is limited to 128 incoming trust relationships, but can establish an unlimited number of trusting relationships because the workload is not on that NT box. *See also* Trusted, Trusting, Single Domain Model, Single Master Domain Model, Multiple Master Model, Complete Trust Model.

Trusted—The domain that contains the accounts. Also then known as the *Account domain.*

Trusting—The domain that contains the resources. The resource domain must trust the account domain.

TSR—Terminate and Stay Resident—A program that visibly finishes running, but carries on with a process in the background. This style of program will not run on Windows '98 machines. Should you need to use such a program, specify it in the AUTOEXEC.BAT file, as opposed to a Novell Login Script.

Tunnel Server—A server running **PPTP.** *See also* **Tunneling, Dial-Up Networking.**

Tunneling—The networking term describing the encapsulation of one protocol within another protocol. It is typically used when joining two networks using an intermediate network running an incompatible protocol or that might be under the administrative control of a third-party. Using the tunneling concept, you can set up private connections across the Internet using **PPTP.**

Twisted Pair—Has become the dominant cable type for all new network designs that use copper cable. It is inexpensive to install and offers the lowest cost per foot of any cable type. It consists of two strands of copper wire twisted together, and is used to carry a single signal. This twisting reduces the sensitivity of the cable to EMI and also reduces the tendency of the cable to radiate RF noise that would also interfere with nearby cables and components. This is because the radiated signals from the twisted wires tend to cancel each other out and controls the tendency of the wires in the pair to cause EMI in each other.

U

a b c d e f g h i j k l m n o p q r s t u v w x y z

UART—Universal Asynchronous Receiver/Transmitter.

UDF—Universal Disk Format—As used in DVD disks.

UDF—Uniqueness Database File—Used in multiple NT Workstation installation to automatically specify unique settings for a batch of machines.

UDP—User Datagram Protocol—Part of the TCP/IP protocol suite. Gives a computer the capability of sending information without actually establishing a connection between the two machines.

UNC—Universal Naming Convention—Actually a Microsoft naming convention. Note: If you use spaces, then MS-DOS-based clients may well have difficulty connecting. In fact, I have seen Windows '98 struggle with this on NT networks. An example format of UNC naming is \\Server\share (the two \\ is not a typo).

Unimodem Driver—The driver provided by MS that when combined with minidrivers gives Windows full functionality with modems.

Universal Serial Bus (USB)—USB technology was designed to help alleviate the problems of having more peripherals than ports and interrupts. The USB standard developed by COMPAQ, Digital Equipment, IBM, Intel, Microsoft, NEC, and Northern Telecom allows you to connect up to 127 devices to a PC without internal cards or specialized slots. It gives Plug and Play configuration of your peripherals as soon as they are attached to your PC. Resource settings do not cause a problem because of the USB's design and by it allocating each device a USB address. You can attach or detach any device without powering down or rebooting the system, and all peripherals use a single type of connector, so no need to worry about running out of a certain type of port. Devices can be externally powered or gain their power directly from the bus, so long as it does not draw more than 100 milliamps. You can use hubs to expand the setup of your USB devices, but two bus-powered hubs should not be connected together. Externally powered hubs are available to support higher-power devices. Bus-powered hubs can support a maximum of four downstream ports. Each device should be located no more than 5M away from the hub. Data is transferred in either an Asynchronous or Isochronous modes. It does not support devices with firmware version 000, which use an A1 stepping chip.

Universal Driver—Part of the team concept used in '98 printing. A universal driver must be accompanied by a minidriver. The universal driver is a generic driver created by MS and is designed to work with many different types of printers. Because there are three main types of printers, there are three main types of universal drivers: **MS Universal Drivers,** used for all but Post Script printers and some HP Inkjet Printers; **MS Postscript Drivers,** used for all PS Printers; and **Monolothic** drivers, primarily used for HP Inkjet and compatible printers. The monolothic driver must be provided by the manufacturer.

Upper Memory Blocks—Referring to the old MS-DOS memory model where you had 640K of conventional RAM with an additional 384K that was split into UMBs. Used for loading some types of driver because some reserved areas were not always used.

UPS—Uninterruptable Power Supply—A special device that continues to supply power after a power failure. They are most commonly used with network servers to prevent a disorderly shutdown that could damage data on the server. The UPS can talk to the server via software. Should the power fail, the UPS and server can be configured to shut down the server gracefully, send a broadcast message to all users about the power failure and notify them that the server will be shut down, send an administrative alert to the system administrator about the power loss, and even page the administrator.

User Interface—Manages keyboard, mouse, and other I/O devices. Most of the user portion of code in '98 is 16 bit for compatibility reasons.

User Level Security—A better way of handling security on '95/'98 networks. A security provider gives you access to all resources so long as you log on properly, as opposed to Share Level Security. Note: '95 or '98 cannot be a security provider—you need an advanced server such as NetWare or NT. You also are no longer limited to just **Read Only** or **Full-control access**. **Custom** access can be granted to resources. Instead of needing a password for each shared resource, you have only one password. Note: When setting up user-level security, you should specify where the list of users and groups will be obtained from. You should specify the name of the domain, preferably as opposed to a particular server because better load balancing is used by looking at PDCs and BDCs as opposed to just one server. It is a requirement to have user-level security when using File and Print Sharing for NetWare Networks.

UTP—Unshielded Twisted Pair—Does not incorporate the braided shield of STP into its structure. It is similar to STP in many ways, although it differs in its characteristics of attenuation and EMI. Several twisted pairs are often included in the same cable and color-coded. Telephone systems typically use UTP. Network engineers can even sometimes use existing telephone cabling (if of high enough quality) for network cabling. There are five main standards, as follows:

Categories 1 and 2	Voice-grade cabling only suitable for voice and low data rates (below 4Mbps). The growing need for data-ready cabling systems has caused Cat's 1 and 2 to be supplanted by Cat 3 for new installations.
Category 3	The lowest data-grade cable, this is generally suited for data rates up to 10 Mbps, although some encoding schemes allow up to 100 Mbps. It features four twisted pairs with three twists per foot and is now the standard cable used for most telephone installations. It is also the minimum grade that supports 10Base-T.
Category 4	This data-grade cable consists of four twisted pairs and is suitable for data rates up to 16 Mbps, but is not widely used.
Category 5	This data-grade cable also consists of four twisted pairs, and is suitable for data rates up to 100 Mbps. Most new cabling systems for 100 Mbps or faster are designed around Cat 5 cable.

a b c d e f g h i j k l m n o p q r s t u v w x y z

V Series—The CCITT V series recommendations are the most widely used international specifications for defining the transmission of data between computers via the telephone network. Almost every modem and multiplexor in use today conforms to V Series protocols. Among the areas defined in the V series recommendations are physical layer interfaces such as V. 24, which is similar to RS-232; voice-band and wide-band modems; error control for modems; diagnostic procedures; and Internetworking (including ISDN).

Vampire Tap—Used to connect a transceiver to Thicknet cable. It has pins that penetrate the insulation and so remove the need to cut it.

Vcache—Replaced the Smart Drive Caching Utility in Windows '95 and '98. It encompasses **Read Ahead** and **Write Behind Caching.**

VESA Local Bus—An enhancement to the ISA bus. An additional slot continues after the ISA slot for VESA local bus devices. VESA local bus cards can be installed in ISA slots, but will only function in ISA mode.

VFAT—Virtualized Fat Allocation Table. Introduced with Windows '95, which no longer needed **FAT16** because it was finally an OS in its own right. It is a **protected mode** component and 32 bit.

Virtual 8088 Mode—When a 16-bit driver is used in a 32-bit OS like '98, it must switch from protected mode to virtual 8088 mode to simulate an old 8088 machine. This may have to be done several times during an I/O operation, vastly slowing the system.

Virtual Circuits—Packet switching networks use virtual circuits to route data from the source to the destination. A virtual circuit is a specific path through the network—a chain of communication links leading from the source to the destination (as opposed to a scheme in which each packet finds its own path). Virtual circuits enable the network to provide better error checking and flow control. *See also* SVC (Switched Virtual Circuit), PVC (Permanent Virtual Circuit).

Virtual Communication—Where each layer of the OSI model acts as if it is communicating with its associated layer on the other machine.

Virtual Machine—Simulates an entire computer's resources to an application. *See also* **System Virtual Machine, MS DOS Virtual Machine.**

VMM—Virtual Memory Manager—As used in the latest OSs for their memory model. Gives each application a 4GB address space, of which 2GB is assigned for the program and 2GB is assigned for the system. The VMM takes the information from virtual memory addresses for the applications and maps them to actual spaces in physical memory. It tricks the application into thinking its information is stored somewhere it is not, essentially! Two MSDOS apps, for instance, are run at the same time. They both want to use 0 to 640Kb address space and the OS gives them both exactly that! Only when even the applications calls on the VMM, the VMM puts the information where it wants as well, hiding this from the applications. In extended memory, blocks were also 64Kb in size; here, they are only 4Kb in size, which is much more efficient, and pages of memory need no longer be contiguous.

VPN—Virtual Private Network—In an attempt to cut the cost of long distance remote connections, people are now using the Internet to connect to their private networks. VPN technologies primarily deal with the security implications of doing this. **PPTP** is often used to achieve this. Under '98, VPN is supported out of the box—note Dial-Up Adapter #2 (VPN Support) and MS Virtual Private Networking Adapter. (For '95 to support this, you must have the dial-up networking upgrade from the MS Website at http://www.Microsoft.com/windows/get ISDN/dload.htm.) To use PPTP and VPNs, two types of connections must be configured: a connection to the **Internet** via your **ISP** and a **tunnel** connection to the PPTP server on the target network. Depending on how this will be used, you might not need to configure both types of connections. If using a serial **modem** or **ISDN,** you will need both connections. If connecting to a tunnel server on your LAN, you will only need one connection. You must first connect to your ISP. If you already have Internet connection, then you do not need to do this again. The next step is to create a Dial-Up Networking connection for the PPTP server using the Microsoft VPN adapter. You will need to specify the host name or **IP address** of the PPTP server.

.VXD—Indicative of a Protected Mode Windows Driver.

a b c d e f g h i j k l m n o p q r s t u v w x y z

.WLG—created by Dr. Watson utility. Viewable by MS System Information View.

Warm Docking—A portable computer can be docked or undocked with a docking station while in sleep or suspended mode at best.

WDM—Windows 32 Driver Model—A new standard introduced with Windows '98 that gives a standard driver design for all future MS OSs. It uses a standard set of I/O services and no longer includes **hooks,** as device drivers have in the past.

WHOAMI—A Novell command that tells you under which server you are logged in and what your user name is. Also, right-click Network Neighborhood and choose the WHOAMI option. When using the Microsoft for Novell Client with Services for NetWare Directory Services installed, you will also get NDS information.

Windows '95—The first ever PnP compliant OS. Comes in a variety of guises and versions, the latest being OSR2.1. Minimum system requirements are 4Mb of RAM 8Mb recommended and 386DX processor.

Windows '98—Microsoft's revision to their hugely successful Windows '95 OS, including many new revisions and features. Minimum machine requirements are a 486DX processor and 16Mb of RAM. MS recommends 24 Mb RAM. Mutliple displays are supported. Support for Novell versions 3.11 or later is provided.

Windows '98 Resource Kit—An MS Press publication that contains valuable information designed for use by IS professionals.

Windows NT RAS—The protocol designed for earlier versions of NT and is included for backward compatibility in '98. It uses AsyBEUI (Asynchronous NetBEUI), which is much slower and less flexible than PPP or even SLIP.

WinIPCfg—A utility that allows you to view your TCP/IP configuration even when the addresses are assigned automatically via a DHCP server.

WINS—Windows Internet Naming Service—Provides resolution of NetBIOS computer names to TCP/IP addresses.

WINS Proxy Agents—Not all systems running NetBIOS over TCP/IP can be configured to use WINS servers. They can, however, be given access to WINS server databases through the use of proxy agents. Windows NT, Windows for Workgroups, and Windows '95 stations can be configured as WINS proxy agents, provided they are not WINS servers. If a system is configured as a WINS Server, it cannot also act as a WINS proxy agent. WINS proxy agents provide indirect access to the WINS database. When a system broadcasts a name query, a WINS proxy agent intercepts the broadcast. The broadcast is answered, if possible, from the local cache or a WINS server query. There is a possibility of a strain being placed on the WINS proxy agent, depending on the number of name query broadcasts occurring. Only a limited number of WINS proxy agents should be defined on each domain, preferably systems that are not already supporting resource intensive tasks.

WinSockets—A two-way connection specific to PCs but based on the Berkley UNIX Sockets standard.

Wireless Bridging—Wireless technology can connect LANs in two different buildings into an extended LAN. This capability is also available through other technologies such as T1 lines, etc., but depending on the conditions, a wireless solution is sometimes more cost-effective. A wireless connection between two buildings also provides a solution to the problem of ground potential. A wireless bridge acts as a network bridge, merging two local LANs over a wireless connection. They typically use spread spectrum radio technology to transmit data for up to 3 miles. Antennae at each end of the bridge should be placed in an appropriate location, such as a rooftop. *See also* Long-Range Wireless Bridge.

Workgroup Advertising—With Novell **SAP,** this accomplishes the same tasks as the Browse Master options, such as whether and how to participate in Browser elections. If you do not enable some form of Workgroup advertising, then MS-based computers (ones not running Novell's real mode stack) will not see your machine in the Browse list.

Write Behind Caching—Speeds up the time it takes for an application to return control to the user by saving the file in RAM until the system is not as busy. This can be dangerous in areas where there are frequent power cuts or on a portable machine.

Write Permission—Users with this permission of Microsoft Networks can edit files within a folder. Equates to the NetWare Write right.

a b c d e f g h i j k l m n o p q r s t u v w x y z

X Series—The CCITT's X series applies to public data networks. Every day, communications networks send billions of bytes of data around the world. The X series is designed to ensure compatibility, which is essential for the tremendous data exchange. Among the areas defined in the X series recommendations are services and facilities; Interfaces (including X.25 data packets, X.29 PAD, X.31 internetworking, and X.32 dial and answer); transmission and signalling; the OSI models X.200 standard; ISDN, telephone, packet, and satellite networks through X.300; message handling system X.400; and directory applications X.500.

X.25—This type of connection is part of a worldwide network made up of X.25 nodes, responsible for forwarding packets to their destination. At that destination, a PAD (Packet Assembler/Disassembler) is used instead of a modem. It is a packet switching network standard referred to as *recommendation x.25* and implemented most commonly in old or international WANs. It is one level of a three-level stack that spans the network, datalink, and physical layers of the OSI model. The middle layer, Link Access Procedures Balanced (LAPB), is a bit-orientated, full-duplex synchronous data-link layer LLC protocol. Physical layer connectivity is provided by a variety of standards including X.21, X.21 bis, and V.32. X.25 packet switching networks offer the choice of Permanent or Switched Virtual Circuits, and X.25 is required to provide reliable service and end-to-end flow control. Because each device on a network can operate more than one virtual circuit, X.25 must provide error and flow control for each virtual circuit. Generally implemented with line speed of up to 64 Kbps and as such is suitable for mainframe terminal activity, which was what X.25 was designed for. However, this makes X.25 a poor choice for LAN-based services that require 1 Mbps or better. In X.25, a computer or terminal called a DTE (Data Terminal Equipment) that could also act as a gateway providing access to the local network. DCE (Data Communications Equipment) provides access to the PSTN (Public Switched Telephone Network). A PSE is a packet switching exchange, also called a switch or switching node. X.25 also defines the communications between the DCE and DTE. A device called a PAD (Packet Assembler/Disassembler) translates asynchronous input from the DTE into packets suitable for the PDN. It should be used for international data circuits in which other, higher-speed technologies are not available or cost-effective.

XMS—Extended Memory—The next move from Expanded Memory—Anything above the 1Mb memory limit on early PCs. Again, software was needed to implement this memory, the most common of which was MS's HIMEM.SYS. Memory was stored in 64Kb blocks, which was very inefficient because a small piece of code could end up sitting in a 64Kb address space. Also, memory had to be contiguous, so you might have two 64Kb blocks free, but unless they were next to each other, you would never feel the benefit. **Note:** Some MS-DOS applications under Windows have difficulty when they cannot see the limit should the Auto settings be used in the PIF file. Try setting to 8,192Kb to cure this. *See also* **VMM—Virtual Memory Manager, EMS—Expanded Memory.**

a b c d e f g h i j k l m n o p q r s t u v w x y z

Z

a b c d e f g h i j k l m n o p q r s t u v w x y z

Zip Drive—100Mb removeable disk format by Iomega. Available in Internal/External IDE and SCSI variants.

Metrification

Contents

3.B0.0   Introduction to the 1975 Metric Conversion Act

As the federal government moves to convert the inch–pound units to the metric system, in accordance with the 1975 Metric Conversion Act, various parts of the construction industry will begin the conversion to this more universal method of measurement.

Metric units are often referred to as *SI units*, an abbreviation taken from the French Le Système International d'Unités. Another abbreviation that will be seen with more frequency is ISO—the International Standards Organization—charged with supervising the establishment of a universal standards system. For everyday transactions, it may be sufficient to gain only the basics of the metric system.

Name of metric unit	Symbol	Approximate size (length/pound)
meter	m	39½ inches
kilometer	km	0.6 mile
centimeter	cm	width of a paper clip
millimeter	mm	thickness of a dime
hectare	ha	2½ acres
square meter	m2	1.2 square yards
gram	g	weight of a paper clip
kilogram	kg	2.2 pounds
metric ton	t	long ton (2240 pounds)
liter	L	one quart and two ounces
milliliter	mL	⅕ teaspoon
kilopascal	kPa	atmospheric pressure is about 100 kPa

The Celsius temperature scale is used. Instead of referring to its measurement as *degree centigrade*, the term *degree Celsius* is the correct designation. Using this term, familiar points are:

- Water freezes at 0 degrees
- Water boils at 100 degrees
- Normal body temperature is 37 degrees (98.6 F)
- Comfortable room temperature 20 to 35 (68 to 95 F)

3.B1.0   What Will Change and What Will Stay the Same?

Metric Module and Grid

What Will Change

- The basic building module, from 4 inches to 100 mm.
- The planning grid, from 2′ × 2′ to 600 × 600 mm.

What Will Stay the Same

- A module and grid based on rounded, easy-to-use dimensions. The 100 mm module is the global standard.

Drawings

What Will Change

- Units, from feet and inches to millimeters for all building dimensions and to meters for site plans and civil engineering drawings. Unit designations are unnecessary: if there's no decimal point, it's millimeters; if there's a decimal point carried to one, two, or three places, it's meters. In accordance with ASTM E621, centimeters are not used in construction because (1) they are not consistent with the preferred use of multiples of 1000, (2) the order of magnitude between a millimeter and centimeter is only 10 and the use of both units would lead to confusion and require the use of unit designations, and (3) the millimeter is small enough to almost entirely eliminate decimal fractions from construction documents.

- Drawing scales, from inch-fractions-to-feet to true ratios. Preferred metric scales are:

  1:1 (full size)

  1:5 (close to 3″ = 1′-0″)

  1:10 (between 1″ = 1′-0″ and 1½″ = 1′-0″)

  1:20 (between ½″ = 1′-0″ and ¾″ = 1′-0″)

  1:50 (close to ¼″ = 1′-0″)

  1:100 (close to ⅛″ = 1′-0″)

  1:200 (close to 1/16″ = 1′-0″)

  1:500 (close to 1″ = 40′-0″)

  1:1000 (close to 1″ = 80′-0″)

As a means of comparison, inch-fraction scales may be converted to true ratios by multiplying a scale's divisor by 12; for example, for ¼″ = 1′-0″, multiply the 4 by 12 for a true ratio of 1:48.

- Drawing sizes, to ISO "A" series:

  A0 (1189 × 841 mm, 46.8 × 33.1 inches)

  A1 (841 × 594 mm, 33.1 × 23.4 inches)

  A2 (594 × 420 mm, 23.4 × 16.5 inches)

  A3 (420 × 297 mm, 16.5 × 11.7 inches)

  A4 (297 × 210 mm, 11.7 × 8.3 inches)

Of course, metric drawings can be made on any size paper.

What Will Stay the Same

- Drawing contents

Never use dual units (both inch-pound and metric) on drawings. It increases dimensioning time, doubles the chance for errors, makes drawings more confusing, and only postpones the learning process. An exception is for construction documents meant to be viewed by the general public.

Specifications

What Will Change

- Units of measure, from feet and inches to millimeters for linear dimensions, from square feet to square meters for area, from cubic yards to cubic meters for volume (except use liters for fluid volumes), and from other inch-pound measures to metric measures as appropriate.

What Will Stay the Same

- Everything else in the specifications

Do not use dual units in specifications except when the use of an inch–pound measure serves to clarify an otherwise unfamiliar metric measure; then place the inch–pound unit in parentheses after the metric. For example, "7.5 kW (10 horsepower) motor." All unit conversions should be checked by a professional to ensure that rounding does not exceed allowable tolerances.

For more information, see the July–August 1994 issue of *Metric in Construction*.

Floor Loads

What Will Change

- Floor load designations, from "psf" to kilograms per square meter (kg/m^2) for everyday use and kilonewtons per square meter (kN/m^2) for structural calculations.

What Will Stay the Same

• Floor load requirements

Kilograms per square meter often are used to designate floor loads because many live and dead loads (furniture, filing cabinets, construction materials, etc.) are measured in kilograms. However, kilonewtons per square meter or their equivalent, kilopascals, are the proper measure and should be used in structural calculations.

Construction Products

What Will Change

• Modular products: brick, block, drywall, plywood, suspended ceiling systems, and raised floor systems. They will undergo "hard" conversion; that is, their dimensions will change to fit the 100 mm module.

• Products that are custom-fabricated or formed for each job (for example, cabinets, stairs, handrails, ductwork, commercial doors and windows, structural steel systems, and concrete work). Such products usually can be made in any size, inch–pound or metric, with equal ease; therefore, for metric jobs, they simply will be fabricated or formed in metric.

What Will Stay the Same

• All other products, since they are cut-to-fit at the jobsite (for example, framing lumber, woodwork, siding, wiring, piping, and roofing) or are not dimensionally sensitive (for example, fasteners, hardware, electrical components, plumbing fixtures, and HVAC equipment). Such products will just be "soft" converted—that is, relabeled in metric units. A $2^3/_4'' \times 4^1/_2''$ wall switch face plate will be relabeled 70×115 mm and a 30 gallon tank, 114 L. Manufacturers eventually may convert the physical dimensions of many of these products to new rational "hard" metric sizes but only when it becomes convenient for them to do so.

"2-By-4" Studs and Other "2-By" Framing (Both Wood and Metal)

What Will Change

• Spacing, from 16″ to 400 mm, and 24″ to 600 mm.

What Will Stay the Same

• Everything else.

"2-bys" are produced in "soft" fractional inch dimensions so there is no need to convert them to new rounded "hard" metric dimensions. 2-by-4s may keep their traditional name or perhaps they'll eventually be renamed 50 by 100 (mm), or, more exactly, 38×89.

Drywall, Plywood, and Other Sheet Goods

What Will Change

• Widths, from 4′-0″ to 1200 mm.
• Heights, from 8′-0″ to 2400 mm, 10′-0″ to 3000 mm.

What Will Stay the Same

• Thicknesses, so fire, acoustic, and thermal ratings won't have to be recalculated.

Metric drywall and plywood are readily available but may require longer lead times for ordering and may cost more in small amounts until their use becomes more common.

Batt Insulation

What Will Change

• Nominal width labels, from 16″ to 16″/400 mm and 24″ to 24″/600 mm.

What Will Stay the Same

- Everything else.

Batts will not change in width; they'll just have a tighter "friction fit" when installed between metric-spaced framing members.

Doors

What Will Change

- Height, from 6'-8" to 2050 mm or 2100 mm and from 7'-0" to 2100 mm.
- Width, from 2'-6" to 750 mm, from 2'-8" to 800 mm, from 2'-10" to 850 mm, from 3'-0" to 900 mm or 950 mm, and from 3'-4" to 1000 mm.

What Will Stay the Same

- Door thicknesses.
- Door materials and hardware.

For commercial work, doors and door frames can be ordered in any size since they normally are custom fabricated.

Ceiling Systems

What Will Change

- Grids and lay-in ceiling tile, air diffusers and recessed lighting fixtures, from 2' × 2' to 600 × 600 mm and from 2' × 4' to 600 × 1200 mm.

What Will Stay the Same

- Grid profiles, tile thicknesses, air diffuser capacities, fluorescent tubes, and means of suspension.

On federal building projects, metric recessed lighting fixtures may be specified if their total installed costs are estimated to be no more than for inch–pound fixtures.

Raised Floor Systems

What Will Change

- Grids and lay-in floor tile, from 2' × 2' to 600 × 600 mm.

What Will Stay the Same

- Grid profiles, tile thicknesses, and means of support.

HVAC Controls

What Will Change

- Temperature units, from Fahrenheit to Celsius.

What Will Stay the Same

- All other parts of the controls.

Controls are now digital so temperature conversions can be made with no difficulty.

Brick

What Will Change

- Standard brick, to 90 × 57 × 190 mm.

- Mortar joints, from $^3/_8''$ and $^1/_2''$ to 10 mm.
- Brick module, from $2' \times 2'$ to 600×600 mm.

What Will Stay the Same

- Brick and mortar composition.

Of the 100 or so brick sizes currently made, 5 to 10 are within a millimeter of a metric brick so the brick industry will have no trouble supplying metric brick.

For more information, see the March–April 1995 issue of *Metric in Construction*.

Concrete Block

What Will Change

- Block sizes, to $190 \times 190 \times 390$ mm.
- Mortar joints, from $^1/_2''$ to 10 mm.
- Block module, from $2' \times 2'$ to 600×600 mm.

What Will Stay the Same

- Block and mortar composition.

On federal building projects, metric block may be specified if its total installed cost is estimated to be no more than for inch–pound block. The Construction Metrication Council recommends that, wherever possible, block walls be designed and specified in a manner that permits the use of either inch–pound or metric block, allowing the final decision to be made by the contractor.

Sheet Metal

What Will Change

- Designation, from "gage" to millimeters.

What Will Stay the Same

- Thickness, which will be soft-converted to tenths of a millimeter.

In specifications, use millimeters only or millimeters with the gage in parentheses.

Concrete

What Will Change

- Strength designations, from "psi" to megapascals, rounded to the nearest 5 megapascals per ACI 318M as follows:

  2500 psi to 20 MPa

  3000 psi to 25 MPa

  3500 psi to 25 MPa

  4000 psi to 30 MPa

  4500 psi to 35 MPa

  5000 psi to 35 MPa

Depending on exact usage, however, the above metric conversions may be more exact than those indicated.

What Will Stay the Same

- Everything else.

For more information, see the November–December 1994 issue of *Metric in Construction*.

Rebar

What Will Change

- Rebar will not change in size but will be renamed per ASTM A615M-96a and ASTM A706M-96a as follows:

No. 3 to No. 10	No. 9 to No. 29
No. 4 to No. 13	No. 10 to No. 32
No. 5 to No. 16	No. 11 to No. 36
No. 6 to No. 19	No. 14 to No. 43
No. 7 to No. 22	No. 18 to No. 57
No. 8 to No. 25	

What Will Stay the Same

- Everything else.

For more information, see the July–August 1996 issue of *Metric in Construction*.

Glass

What Will Change

- Cut sheet dimensions, from feet and inches to millimeters.

What Will Stay the Same

- Sheet thickness; sheet glass can be rolled to any dimension and often is rolled in millimeters now.

See ASTM C1036.

Pipe and Fittings

What Will Change

- Nominal pipe and fitting designations, from inches to millimeters

What Will Stay the Same

- Pipe and fitting cross sections and threads.

Pipes and fittings are produced in "soft" decimal inch dimensions but are identified in nominal inch sizes a matter of convenience. A 2-inch pipe has neither an inside nor an outside diameter of 2 inches, a 1-inch fitting has no exact 1-inch dimension, and a $1/2$-inch sprinkler head contains no $1/2$-inch dimension anywhere; consequently, there is no need to "hard" convert pipes and fittings to rounded metric dimensions. Instead, they will not change size but simply be relabeled in metric as follows:

$1/8''$ = 6 mm	$11/2''$ = 40 mm
$3/16''$ = 7 mm	$2''$ = 50 mm
$1/4''$ = 8 mm	$21/2''$ = 65 mm
$3/8''$ = 10 mm	$3''$ = 75 mm

$\frac{1}{2}'' = 15$ mm $3\frac{1}{2}'' = 90$ mm

$\frac{5}{8}'' = 18$ mm $4'' = 100$ mm

$\frac{3}{4}'' = 20$ mm $4\frac{1}{2}'' = 115$ mm

$1'' = 25$ mm $1'' = 25$ mm for all larger sizes

$1\frac{1}{4}'' = 32$ mm

For more information, see the September–October 1993 issue of *Metric in Construction*.

Electrical Conduit

What Will Change

- Nominal conduit designations, from inches to millimeters.

What Will Stay the Same

- Conduit cross sections.

Electrical conduit is similar to piping: it is produced in "soft" decimal inch dimensions but is identified in nominal inch sizes. Neither metallic nor nometallic conduit will change size; they will be relabeled in metric units as follows:

$\frac{1}{2}'' = 16$ (mm) $2\frac{1}{2}'' = 63$ (mm)

$\frac{3}{4}'' = 21$ (mm) $3'' = 78$ (mm)

$1'' = 27$ (mm) $3\frac{1}{2}'' = 91$ (mm)

$1\frac{1}{4}'' = 35$ (mm) $4'' = 103$ (mm)

$1\frac{1}{2}'' = 41$ (mm) $5'' = 129$ (mm)

$2'' = 53$ (mm) $6'' = 155$ (mm)

These new metric names were assigned by the National Electrical Manufacturers Association.

Electrical Wire

What Will Change

- Nothing at this time.

What Will Stay the Same

- Existing American Wire Gage (AWG) sizes.

Structural Steel

What Will Change

- Section designations, from inches to millimeters and from pounds per foot to kilograms per meter, in accordance with ASTM A6M.
- Bolts—to metric diameters and threads per ASTM A325M and A490M.

What Will Stay the Same

- Cross sections.

Like pipe and conduit, steel sections are produced in "soft" decimal inch dimensions (with actual depths varying by weight) but are named in rounded inch dimensions so there is no need to "hard" convert them to metric units. Rather, their names will be changed to metric designations, and rounded to the nearest 10 mm. Thus, a 10-inch section is relabeled as a 250-mm section and a 24-inch section is relabeled as a 610-mm section.

3.B2.0　How Metric Units Will Apply in the Construction Industry

	Quantity	Unit	Symbol
Masonry	length	meter, millimeter	m, mm
	area	square meter	m²
	mortar volume	cubic meter	m³
Steel	length	meter, millimeter	m, mm
	mass	megagram (metric ton) kilogram	Mg (t) kg
	mass per unit length	kilogram per meter	kg/m
Carpentry	length	meter, millimeter	m, mm
Plastering	length	meter, millimeter	m, mm
	area	square meter	m²
	water capacity	liter (cubic decimeter)	L (dm³)
Glazing	length	meter, millimeter	m, mm
	area	square meter	m²
Painting	length	meter, millimeter	m, mm
	area	square meter	m²
	capacity	liter (cubic decimeter) milliliter (cubic centimeter)	L (dm³) mL (cm³)
Roofing	length	meter, millimeter	m, mm
	area	square meter	m²
	slope	percent ratio of lengths	% mm/mm, m/m
Plumbing	length	meter, millimeter	m, mm
	mass	kilogram, gram	kg, g
	capacity	liter (cubic decimeter)	L (dm³)
	pressure	kilopascal	kPa
Drainage	length	meter, millimeter	m, mm
	area	hectare (10 000 m2) square meter	ha m²
	volume	cubic meter	m³
	slope	percent ratio of lengths	% mm/mm, m/m
HVAC	length	meter, millimeter	m, mm
	volume (capacity)	cubic meter liter (cubic decimeter)	m³ L (dm³)
	air velocity	meter/second	m/s
	volume flow	cubic meter/second liter/second (cubic decimeter per second)	m³/s L/s (dm³/s)
	temperature	degree Celsius	°C
	force	newton, kilonewton	N, kN
	pressure	pascal, kilopascal	Pa, kPa
	energy	kilojoule, megajoule	kJ, MJ
	rate of heat flow	watt, kilowatt	W, kW
Electrical	length	millimeter, meter, kilometer	mm, m, km
	frequency	hertz	Hz
	power	watt, kilowatt	W, kW
	energy	megajoule kilowatt hour	MJ kWh

Quantity	Unit	Symbol
electric current	ampere	A
electric potential	volt, kilovolt	V, kV
resistance	milliohm, ohm	mΩ, Ω

3.B3.0 Metrification of Pipe Sizes

Pipe diameter sizes can be confusing because their designated size does not correspond to their actual size. For instance, a 2-inch steel pipe has an inside diameter of approximately $2\frac{1}{8}$ inches and an outside diameter of about $2\frac{5}{8}$ inches.

The *2 inch* designation is very similar to the 2″×4″ designation for wood studs, neither dimensions are "actual", but they are a convenient way to describe these items.

Pipe sizes are identified as *NPS (nominal pipe size)* and their conversion to metric would conform to ISO (International Standards Organization) criteria and are referred to as *DN (diameter nominal)*. These designations would apply to all plumbing, mechanical, drainage, and miscellaneous pipe commonly used in civil works projects.

NPS size	DN size	NPS size	DN size
$\frac{1}{8}″$	6 mm	8″	200 mm
$\frac{3}{16}″$	7 mm	10″	250 mm
$\frac{1}{4}″$	8 mm	12″	300 mm
$\frac{3}{8}″$	10 mm	14″	350 mm
$\frac{1}{2}″$	15 mm	16″	400 mm
$\frac{5}{8}″$	18 mm	18″	450 mm
$\frac{3}{4}″$	20 mm	20″	500 mm
1″	25 mm	24″	600 mm
$1\frac{1}{4}″$	32 mm	28″	700 mm
$1\frac{1}{2}″$	40 mm	30″	750 mm
2″	50 mm	32″	800 mm
$2\frac{1}{2}″$	65 mm	36″	900 mm
3″	80 mm	40″	1000 mm
$3\frac{1}{2}″$	90 mm	44″	1100 mm
4″	100 mm	48″	1200 mm
$4\frac{1}{2}″$	115 mm	52″	1300 mm
5″	125 mm	56″	1400 mm
6″	150 mm	60″	1500 mm

For all pipe over 60 inches nominal, use 1 inch equals 25 mm.

3.B4.0 Metrification of Standard Lumber Sizes

Metric units: ASTM Standard E 380 was used as the authoritative standard in developing the metric dimensions in this standard. Metric dimensions are calculated at 25.4 millimeters (mm) times the actual dimension in inches. The nearest mm is significant for dimensions greater than $\frac{1}{8}$ inch, and the nearest 0.1 mm is significant for dimensions equal to or less than $\frac{1}{8}$ inch.

The rounding rule for dimensions greater than $\frac{1}{8}$ inch: If the digit in the tenths of mm position (the digit after the decimal point) is less than 5, drop all fractional mm digits; if it is greater than 5 or if it is 5 followed by at least one nonzero digit, round one mm higher; if 5 followed by only zeroes, retain the digit in the unit position (the digit before the decimal point) if it is even or increase it one mm if it is odd.

The rounding rule for dimensions equal to or less than $\frac{1}{8}$ inch: If the digit in the hundredths of mm position (the second digit after the decimal point) is less than 5, drop all digits to the right of the tenth position; if greater than or it is 5 followed by at least one nonzero digit, round one-tenth mm higher; if 5 followed by only zeros, retain the digit in the tenths position if it is even or increase it one-tenth mm if it is odd.

In case of a dispute on size measurements, the conventional (inch) method of measurement shall take precedence.

3.B5.0 Metric Rebar Conversions

A615 M-96a & A706M-96a Metric Bar Sizes	Nominal Diameter	A615-96a & A706-96a Inch-Pound Bar Sizes
#10	9.5 mm/0.375"	#3
#13	12.7 mm/0.500"	#4
#16	15.9 mm/0.625"	#5
#19	19.1 mm/0.750"	#6
#22	22.2 mm/0.875"	#7
#25	25.4 mm/1.000"	#8
#29	28.7 mm/1.128"	#9
#32	32.3 mm/1.270"	#10
#36	35.8 mm/1.410"	#11
#43	43.0 mm/1.693"	#14
#57	57.3 mm/2.257"	#18

3.B6.0 Metric Conversion of ASTM Diameter and Wall Thickness Designations

METRIC CONVERSION OF ASTM DIAMETER DESIGNATIONS

in	mm	in	mm	in	mm	in	mm
6	150	30	750	57	1425	96	2400
8	200	33	825	60	1500	102	2550
10	250	36	900	63	1575	108	2700
12	300	39	975	66	1650	114	2850
15	375	42	1050	69	1725	120	3000
18	450	45	1125	72	1800	132	3300
21	525	48	1200	78	1950	144	3600
24	600	51	1275	84	2100	156	3900
27	675	54	1350	90	2250	168	4200

METRIC CONVERSION OF ASTM WALL THICKNESS DESIGNATIONS

in	mm	in	mm	in	mm	in	mm
1	25	3-1/8	79	5	125	8	200
1-1/2	38	3-1/4	82	5-1/4	131	8-1/2	213
2	50	3-1/2	88	5-1/2	138	9	225
2-1/4	56	3-3/4	94	5-3/4	144	9-1/2	238
2-3/8	59	3-7/8	98	6	150	10	250
2-1/2	63	4	100	6-1/4	156	10-1/2	263
2-5/8	66	4-1/8	103	6-1/2	163	11	275
2-3/4	69	4-1/4	106	6-3/4	169	11-1/2	288
2-7/8	72	4-1/2	113	7	175	12	300
3	75	4-3/4	119	7-1/2	188	12-1/2	313

3.B7.0   Metric Conversion Scales (Temperature and Measurements)

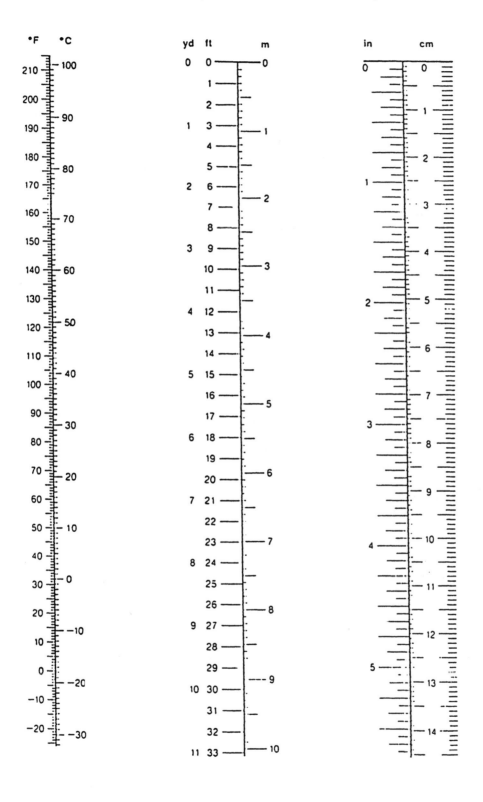

3.B8.0 Approximate Metric Conversions

Symbol	When You Know	Multiply by	To Find	Symbol
LENGTH				
mm	millimeters	0.04	inches	in
cm	centimeters	0.4	inches	in
m	meters	3.3	feet	ft
m	meters	1.1	yards	yd
km	kilometers	0.6	miles	mi
AREA				
cm^2	square centimeters	0.16	square inches	in^2
m^2	square meters	1.2	square yards	yd^2
km^2	square kilometers	0.4	square miles	mi^2
ha	hectares $(10,000\ m^2)$	2.5	acres	
MASS (weight)				
g	grams	0.035	ounces	oz
kg	kilograms	2.2	pounds	lb
t	metric ton $(1,000\ kg)$	1.1	short tons	
VOLUME				
mL	milliliters	0.03	fluid ounces	fl oz
mL	milliliters	0.06	cubic inches	in^3
L	liters	2.1	pints	pt
L	liters	1.06	quarts	qt
L	liters	0.26	gallons	gal
m^3	cubic meters	35	cubic feet	ft^3
m^3	cubic meters	1.3	cubic yards	yd^3
TEMPERATURE (exact)				
°C	degrees Celsius	multiply by 9/5, add 32	degrees Fahrenheit	°F

°C   -40   -20   0   20   37   60   80   100

°F   -40   0   32   80   98.6   160   212

water freezes   body temperature   water boils

(U.S. Department of Commerce Technology Administration, Office of Metric Programs, Washington, DC 20230)

3.B8.0   Approximate Metric Conversion (Continued)

Symbol	When You Know	Multiply by	To Find	Symbol
LENGTH				
in	inches	2.5	centimeters	cm
ft	feet	30	centimeters	cm
yd	yards	0.9	meters	m
mi	miles	1.6	kilometers	km
AREA				
in^2	square inches	6.5	square centimeters	cm^2
ft^2	square feet	0.09	square meters	m^2
yd^2	square yards	0.8	square meters	m^2
mi^2	square miles	2.6	square kilometers	km^2
	acres	0.4	hectares	ha
MASS (weight)				
oz	ounces	28	grams	g
lb	pounds	0.45	kilograms	kg
	short tons (2000 lb)	0.9	metric ton	t
VOLUME				
tsp	teaspoons	5	milliliters	mL
Tbsp	tablespoons	15	milliliters	mL
in^3	cubic inches	16	milliliters	mL
fl oz	fluid ounces	30	milliliters	mL
c	cups	0.24	liters	L
pt	pints	0.47	liters	L
qt	quarts	0.95	liters	L
gal	gallons	3.8	liters	L
ft^3	cubic feet	0.03	cubic meters	m^3
yd^3	cubic yards	0.76	cubic meters	m^3
TEMPERATURE (exact)				
°F	degrees Fahrenheit	subtract 32, multiply by 5/9	degrees Celsius	°C

(United States. Department of Commerce Technology Administration, National Institute of Standards and Technology, Metric Program, Gaithersburg, MD 20899)

3.B9.0   Quick Imperial (Metric Equivalents)

Distance

Imperial	Metric			Metric	Imperial	
1 inch	= 2.540	centimetres		1 centimetre	= 0.3937	inch
1 foot	= 0.3048	metre		1 decimetre	= 0.3281	foot
1 yard	= 0.9144	metre		1 metre	= 3.281	feet
1 rod	= 5.029	metres			= 1.094	yard
1 mile	= 1.609	kilometres		1 decametre	= 10.94	yards
				1 kilometre	= 0.6214	mile

Weight

1 ounce (troy)	= 31.103 grams		1 gram	= 0.032 ounce (troy)	
1 ounce (avoir)	= 28.350 grams		1 gram	= 0.035 ounce (avoir)	
1 pound (troy)	= 373.242 grams		1 kilogram	= 2.679 pounds (troy)	
1 pound (avoir)	= 453.592 grams		1 kilogram	= 2.205 pounds (avoir)	
1 ton (short)	= 0.907 tonne*		1 tonne	= 1.102 ton (short)	

*1 tonne = 1000 kilograms

Capacity

Imperial		U.S.	
1 pint	= 0.568 litre	1 pint (U.S.)	= 0.473 litre
1 gallon	= 4.546 litres	1 quart (U.S.)	= 0.946 litre
1 bushel	= 36.369 litres	1 gallon (U.S.)	= 3.785 litres
1 litre	= 0.880 pint	1 barrel (U.S.)	= 158.98 litres
1 litre	= 0.220 gallon		
1 hectolitre	= 2.838 bushels		

Area

1 square inch	= 6.452 square centimetres
1 square foot	= 0.093 square metre
1 square yard	= 0.836 square metre
1 acre	= 0.405 hectare*
1 square mile	= 259.0 hectares
1 square mile	= 2.590 square kilometres
1 square centimetre	= 0.155 square inch
1 square metre	= 10.76 square feet
1 square metre	= 1.196 square yard
1 hectare	= 2.471 acres
1 square kilometre	= 0.386 square mile

*1 hectare = 1 square hectometre

Volume

1 cubic inch	= 16.387 cubic centimetres
1 cubic foot	= 0.0283 cubic decimetres
1 cubic yard	= 0.765 cubic metre
1 cubic centimetre	= 0.061 cubic inch
1 cubic decimetre	= 35.314 cubic foot
1 cubic metre	= 1.308 cubic yard

3.B10.0   Metric Conversion Factors

The following list provides the conversion relationship between U.S. customary units and SI (International System) units. The proper conversion procedure is to multiply the specified value on the left (primarily U.S. customary values) by the conversion factor exactly as given below and then round to the appropriate number of significant digits desired. For example, to convert 11.4 ft to meters: 11.4 × 0.3048 = 3.47472, which rounds to 3.47 meters. Do not round either value before performing the multiplication, as accuracy would be reduced. A complete guide to the SI system and its use can be found in ASTM E 380, Metric Practice.

To convert from	to	multiply by
Length		
inch (in.)	micron (μ)	25,400 E*
inch (in.)	centimeter (cm)	2.54 E
inch (in.)	meter (m)	0.0254 E
foot (ft)	meter (m)	0.3048 E
yard (yd)	meter (m)	0.9144
Area		
square foot (sq ft)	square meter (sq m)	0.09290304 E
square inch (sq in.)	square centimeter (sq cm)	6.452 E
square inch (sq in.)	square meter (sq m)	0.00064516 E
square yard (sq yd)	square meter (sq m)	0.8361274
Volume		
cubic inch (cu in.)	cubic centimeter (cu cm)	16.387064
cubic inch (cu in.)	cubic meter (cu m)	0.00001639
cubic foot (cu ft)	cubic meter (cu m)	0.02831685
cubic yard (cu yd)	cubic meter (cu m)	0.7645549
gallon (gal) Can. liquid	liter	4.546
gallon (gal) Can. liquid	cubic meter (cu m)	0.004546
gallon (gal) U.S. liquid**	liter	3.7854118
gallon (gal) U.S. liquid	cubic meter (cu m)	0.00378541
fluid ounce (fl oz)	milliliters (ml)	29.57353
fluid ounce (fl oz)	cubic meter (cu m)	0.00002957
Force		
kip (1000 lb)	kilogram (kg)	453.6
kip (1000 lb)	newton (N)	4,448.222
pound (lb) avoirdupois	kilogram (kg)	0.4535924
pound (lb)	newton (N)	4.448222
Pressure or stress		
kip per square inch (ksi)	megapascal (MPa)	6.894757
kip per square inch (ksi)	kilogram per square centimeter (kg/sq cm)	70.31
pound per square foot (psf)	kilogram per square meter (kg/sq m)	4.8824
pound per square foot (psf)	pascal (Pa)†	47.88
pound per square inch (psi)	kilogram per square centimeter (kg/sq cm)	0.07031
pound per square inch (psi)	pascal (Pa)†	6,894.757
pound per square inch (psi)	megapascal (MPa)	0.00689476
Mass (weight)		
pound (lb) avoirdupois	kilogram (kg)	0.4535924
ton, 2000 lb	kilogram (kg)	907.1848
grain	kilogram (kg)	0.0000648

To convert from	to	multiply by
Mass (weight) per length		
kip per linear foot (klf)	kilogram per meter (kg/m)	0.001488
pound per linear foot (plf)	kilogram per meter (kg/m)	1.488
Mass per volume (density)		
pound per cubic foot (pcf)	kilogram per cubic meter (kg/cu m)	16.01846
pound per cubic yard (lb/cu yd)	kilogram per cubic meter (kg/cu m)	0.5933
Temperature		
degree Fahrenheit (°F)	degree Celsius (°C)	$t_C = (t_F - 32)/1.8$
degree Fahrenheit (°F)	degree Kelvin (°K)	$t_K = (t_F + 459.7)/1.8$
degree Kelvin (°K)	degree Celsius (C°)	$t_C = t_K - 273.15$
Energy and heat		
British thermal unit (Btu)	joule (J)	1055.056
calorie (cal)	joule (J)	4.1868 E
Btu/°F · hr · ft²	W/m² · °K	5.678263
kilowatt-hour (kwh)	joule (J)	3,600,000. E
British thermal unit per pound (Btu/lb)	calories per gram (cal/g)	0.55556
British thermal unit per hour (Btu/hr)	watt (W)	0.2930711
Power		
horsepower (hp) (550 ft-lb/sec)	watt (W)	745.6999 E
Velocity		
mile per hour (mph)	kilometer per hour (km/hr)	1.60934
mile per hour (mph)	meter per second (m/s)	0.44704
Permeability		
darcy	centimeter per second (cm/sec)	0.000968
feet per day (ft/day)	centimeter per second (cm/sec)	0.000352

*E indicates that the factor given is exact.
**One U.S. gallon equals 0.8327 Canadian gallon.
†A pascal equals 1.000 newton per square meter.

Note:
One U.S. gallon of water weighs 8.34 pounds (U.S.) at 60°F.
One cubic foot of water weighs 62.4 pounds (U.S.).
One milliliter of water has a mass of 1 gram and has a volume of one cubic centimeter.
One U.S. bag of cement weighs 94 lb.

The prefixes and symbols listed below are commonly used to form names and symbols of the decimal multiples and submultiples of the SI units.

Multiplication Factor	Prefix	Symbol
$1,000,000,000 = 10^9$	giga	G
$1,000,000 = 10^6$	mega	M
$1,000 = 10^3$	kilo	k
$1 = 1$	—	—
$0.01 = 10^{-2}$	centi	c
$0.001 = 10^{-3}$	milli	m
$0.000001 = 10^{-6}$	micro	μ
$0.000000001 = 10^{-9}$	nano	n

Useful Tables, Charts, Formulas

Contents

3.C.0   Nails: Penny Designations ("d") and Lengths (U.S. and Metric)

Nail - Penny Size	Length in Inches	Length in Millimeters
2d	1	25.40
3d	1 1/4	31.75
4d	1 1/2	38.10
5d	1 3/4	44.45
6d	2	50.80
7d	2 1/4	57.15
8d	2 1/2	63.50
9d	2 3/4	69.85
10d	3	76.20
12d	3 1/4	82.55
16d	3 1/2	88.90
20d	3 3/4	95.25
30d	4 1/2	114.30
40d	5	127.00
50d	5 1/2	139.70
60d	6	152.40

3.C.1   Stainless Steel Sheets (Thicknesses and Weights)

Gauge	Thickness Inches	Mm.	Weight lbs/ft2	kg/m2
8	0.17188	4.3658	7.2187	44.242
10	0.14063	3.5720	5.9062	28.834
11	0.1250	3.1750	5.1500	25.6312
12	0.10938	2.7783	4.5937	22.427
14	0.07813	1.9845	3.2812	16.019
16	0.06250	1.5875	2.6250	12.815
18	0.05000	1.2700	2.1000	10.252
20	0.03750	0.9525	1.5750	7.689
22	0.03125	0.7938	1.3125	6.409
24	0.02500	0.6350	1.0500	5.126
26	0.01875	0.4763	0.7875	3.845
28	0.01563	0.3970	0.6562	3.1816
Plates				
3/16"	0.1875	4.76	7.752	37.85
1/4"	0.25	6.35	10.336	50.46
5/16"	0.3125	7.94	12.920	63.08
3/8"	0.375	9.53	15.503	75.79
1/2"	0.50	12.70	20.671	100.92
5/8"	0.625	15.88	25.839	126.15
3/4"	0.75	19.05	31.007	151.38
1"	1.00	25.4	41.342	201.83

3.C.2   Comparable Thicknesses and Weights of Stainless Steel, Aluminum, and Copper

STAINLESS STEEL			ALUMINUM				COPPER		
Thickness (Inch)	Gauge (U.S. Standard)	Lb. sq. ft.	Thickness (Inch)	Gauge (B&S)	Lb. sq. ft.		Thickness (Inch)	Oz. sq. ft.	Lb. sq. ft.
.010	32	.420	.010	30	.141		.0108	8	.500
.0125	30	.525	.0126	28	.177		.0121	9	.563
							.0135	10	.625
.0156	28	.656	.0156		.220		.0148	11	.688
			.0179	25	.253		.0175	13	.813
.0187	26	.788							
.0219	25	.919	.020	24	.282		.021	16	1.000
.025	24	1.050	.0253	22	.352				
							.027	20	1.250
.031	22	1.313	.0313	—	.441		.032	24	1.500
.0375	20	1.575	.032	20	.451		.0337	28	1.750
			.0403	18	.563		.0431	32	2.000
			.0453	17	.100				
.050	18	2.100	.0506	16	.126				

Note that U.S. Standard Gauge (stainless sheet) is not directly comparable with the B&S Gauge (aluminum). A 20-gauge stainless averages .0375˝ thick; while a 20-gauge aluminum averages .032˝ thick; and 20-ounce copper is .027˝ thick. The higher strength of stainless steel permits use of thinner gauges than required for aluminum or copper, which makes stainless more competitive with aluminum on a weight-to-coverage basis and provides stainless with a substantial weight saving compared to copper. For example, 100 sq. ft. of .032˝ aluminum will weigh about 45 pounds, .021˝ (16-ounce) copper will weigh about 100 pounds, and .015˝ stainless will weigh about 66 pounds.

3.C.3 Wire and Sheet-Metal Gauges and Weights

Name of Gage	*United States Standard Gage		The United States Steel Wire Gage	American or Brown & Sharpe Wire Gage	New Birmingham Standard Sheet & Hoop Gage	British Imperial or English Legal Standard Wire Gage	Birmingham or Stubs Iron Wire Gage	Name of Gage
Principal Use	Uncoated Steel Sheets and Light Plates		Steel Wire except Music Wire	Non-Ferrous Sheets and Wire	Iron and Steel Sheets and Hoops	Wire	Strips, Bands, Hoops and Wire	Principal Use
Gage No.	Weight Oz. per Sq. Ft.	Approx. Thickness Inches	Thickness, Inches					Gage No.
7/0's			.4900		.6666	.500		7/0's
6/0's			.4615	.5800	.625	.464		6/0's
5/0's			.4305	.5165	.5883	.432	.550	5/0's
4/0's			.3938	.4600	.5416	.400	.454	4/0's
3/0's			.3625	.3648	.500	.372	.425	3/0's
2/0's			.3310	.3249	.4452	.348	.380	2/0's
1/0			.3065	.2893	.3964	.324	.340	1/0
1			.2830	.2576	.3532	.300	.300	1
2			.2625	.2294	.3147	.276	.284	2
3	160	.2391	.2437	.2043	.2804	.252	.259	3
4	150	.2242	.2253	.1819	.250	.232	.238	4
5	140	.2092	.2070	.1620	.2225	.212	.220	5
6	130	.1943	.1920	.1443	.1981	.192	.203	6
7	120	.1793	.1770	.1285	.1764	.176	.180	7
8	110	.1644	.1620	.1144	.1570	.160	.165	8
9	100	.1495	.1483	.1019	.1398	.144	.148	9
10	90	.1345	.1350	.0907	.1250	.128	.134	10
11	80	.1196	.1205	.0808	.1113	.116	.120	11
12	70	.1046	.1055	.0720	.0991	.104	.109	12
13	60	.0897	.0915	.0641	.0882	.092	.095	13
14	50	.0747	.0800	.0571	.0785	.080	.083	14
15	45	.0673	.0720	.0508	.0699	.072	.072	15
16	40	.0598	.0625	.0453	.0625	.064	.065	16
17	36	.0538	.0540	.0403	.0556	.056	.058	17
18	32	.0478	.0475	.0359	.0495	.048	.049	18
19	28	.0418	.0410	.0320	.0440	.040	.042	19
20	24	.0359	.0348	.0285	.0392	.036	.035	20
21	22	.0329	.0317	.0253	.0349	.032	.032	21
22	20	.0299	.0286	.0226	.0313	.028	.028	22
23	18	.0269	.0258	.0201	.0278	.024	.025	23
24	16	.0239	.0230	.0179	.0248	.022	.022	24
25	14	.0209	.0204	.0159	.0220	.020	.020	25
26	12	.0179	.0181	.0142	.0196	.018	.018	26
27	11	.0164	.0173	.0126	.0175	.0164	.016	27
28	10	.0149	.0162	.0113	.0156	.0148	.014	28
29	9	.0135	.0150	.0100	.0139	.0136	.013	29
30	8	.0120	.0140	.0089	.0123	.0124	.012	30
31	7	.0105	.0132	.0080	.0110	.0116	.010	31
32	6.5	.0097	.0128	.0071	.0098	.0108	.009	32
33	6	.0090	.0118	.0063	.0087	.0100	.008	33
34	5.5	.0082	.0104	.0056	.0077	.0092	.007	34
35	5	.0075	.0095	.0050	.0069	.0084	.005	35
36	4.5	.0067	.0090	.0045	.0061	.0076	.004	36
37	4.25	.0064	.0085	.0040	.0054	.0068		37
38	4	.0060	.0080	.0035	.0048	.0060		38
39			.0075	.0031	.0043	.0052		39
40			.0070		.0039	.0048		40

* U.S. Standard Gage is officially a weight gage, in oz. per sq. ft. as tabulated. The Approx. Thickness shown is the "Manufacturers' Standard" of the American Iron and Steel Institute, based on steel as weighing 501.81 lb. per cu. ft. (489.6 true weight plus 2.5 per cent for average over-run in area and thickness).

3.C.4 Weights and Specific Gravities of Common Materials

Substance	Weight Lb. per Cu. Ft.	Specific Gravity	Substance	Weight Lb. per Cu. Ft.	Specific Gravity
METALS, ALLOYS, ORES			**TIMBER, U. S. SEASONED**		
Aluminum, cast, hammered	165	2.55-2.75	Moisture Content by Weight:		
Brass, cast, rolled	534	8.4-8.7	Seasoned timber 15 to 20%		
Bronze, 7.9 to 14% Sn.	509	7.4-8.9	Green timber up to 50%		
Bronze, aluminum	481	7.7	Ash, white, red	40	0.62-0.65
Copper, cast, rolled	556	8.8-9.0	Cedar, white, red	22	0.32-0.38
Copper ore, pyrites	262	4.1-4.3	Chestnut	41	0.66
Gold, cast, hammered	1205	19.25-19.3	Cypress	30	0.48
Iron, cast, pig	450	7.2	Fir, Douglas spruce	32	0.51
Iron, wrought	485	7.6-7.9	Fir, eastern	25	0.40
Iron, spiegel-eisen	468	7.5	Elm, white	45	0.72
Iron, ferro-silicon	437	6.7-7.3	Hemlock	29	0.42-0.52
Iron ore, hematite	325	5.2	Hickory	49	0.74-0.84
Iron ore, hematite in bank.	160-180	....	Locust	46	0.73
Iron ore, hematite loose	130-160	....	Maple, hard	43	0.68
Iron ore, limonite	237	3.6-4.0	Maple, white	33	0.53
Iron ore, magnetite	315	4.9-5.2	Oak, chestnut	54	0.86
Iron slag	172	2.5-3.0	Oak, live	59	0.95
Lead	710	11.37	Oak, red, black	41	0.65
Lead ore, galena	465	7.3-7.6	Oak, white	46	0.74
Magnesium, alloys	112	1.74-1.83	Pine, Oregon	32	0.51
Manganese	475	7.2-8.0	Pine, red	30	0.48
Manganese ore, pyrolusite.	259	3.7-4.6	Pine, white	26	0.41
Mercury	849	13.6	Pine, yellow, long-leaf	44	0.70
Monel Metal	556	8.8-9.0	Pine, yellow, short-leaf	38	0.61
Nickel	565	8.9-9.2	Poplar	30	0.48
Platinum, cast, hammered	1330	21.1-21.5	Redwood, California	26	0.42
Silver, cast, hammered	656	10.4-10.6	Spruce, white, black	27	0.40-0.46
Steel, rolled	490	7.85	Walnut, black	38	0.61
Tin, cast, hammered	459	7.2-7.5	Walnut, white	26	0.41
Tin ore, cassiterite	418	6.4-7.0			
Zinc, cast, rolled	440	6.9-7.2			
Zinc ore, blende	253	3.9-4.2	**VARIOUS LIQUIDS**		
			Alcohol, 100%	49	0.79
			Acids, muriatic 40%	75	1.20
VARIOUS SOLIDS			Acids, nitric 91%	94	1.50
			Acids, sulphuric 87%	112	1.80
Cereals, oats .... bulk	32	....	Lye, soda 66%	106	1.70
Cereals, barley .... bulk	39	....	Oils, vegetable	58	0.91-0.94
Cereals, corn, rye .... bulk	48	....	Oils, mineral, lubricants	57	0.90-0.93
Cereals, wheat .... bulk	48	....	Water, 4°C. max. density	62.428	1.0
Hay and Straw .... bales	20	....	Water, 100°C.	59.830	0.9584
Cotton, Flax, Hemp	93	1.47-1.50	Water, ice	56	0.88-0.92
Fats	58	0.90-0.97	Water, snow, fresh fallen	8	.125
Flour, loose	28	0.40-0.50	Water, sea water	64	1.02-1.03
Flour, pressed	47	0.70-0.80			
Glass, common	156	2.40-2.60			
Glass, plate or crown	161	2.45-2.72	**GASES**		
Glass, crystal	184	2.90-3.00			
Leather	59	0.86-1.02	Air, 0°C. 760 mm.	.08071	1.0
Paper	58	0.70-1.15	Ammonia	.0478	0.5920
Potatoes, piled	42	....	Carbon dioxide	.1234	1.5291
Rubber, caoutchouc	59	0.92-0.96	Carbon monoxide	.0781	0.9673
Rubber goods	94	1.0-2.0	Gas, illuminating	.028-.036	0.35-0.45
Salt, granulated, piled	48	....	Gas, natural	.038-.039	0.47-0.48
Saltpeter	67	....	Hydrogen	.00559	0.0693
Starch	96	1.53	Nitrogen	.0784	0.9714
Sulphur	125	1.93-2.07	Oxygen	.0892	1.1056
Wool	82	1.32			

The specific gravities of solids and liquids refer to water at 4°C., those of gases to air at 0°C. and 760 mm. pressure. The weights per cubic foot are derived from average specific gravities, except where stated that weights are for bulk, heaped or loose material, etc.

3.C.4   Weights and Specific Gravities of Common Materials (Continued)

Substance	Weight Lb. per Cu. Ft.	Specific Gravity	Substance	Weight Lb. per Cu. Ft.	Specific Gravity
ASHLAR MASONRY			**MINERALS**		
Granite, syenite, gneiss	165	2.3-3.0	Asbestos	153	2.1-2.8
Limestone, marble	160	2.3-2.8	Barytes	281	4.50
Sandstone, bluestone	140	2.1-2.4	Basalt	184	2.7-3.2
			Bauxite	159	2.55
MORTAR RUBBLE MASONRY			Borax	109	1.7-1.8
			Chalk	137	1.8-2.6
Granite, syenite, gneiss	155	2.2-2.8	Clay, marl	137	1.8-2.6
Limestone, marble	150	2.2-2.6	Dolomite	181	2.9
Sandstone, bluestone	130	2.0-2.2	Feldspar, orthoclase	159	2.5-2.6
			Gneiss, serpentine	159	2.4-2.7
DRY RUBBLE MASONRY			Granite, syenite	175	2.5-3.1
Granite, syenite, gneiss	130	1.9-2.3	Greenstone, trap	187	2.8-3.2
Limestone, marble	125	1.9-2.1	Gypsum, alabaster	159	2.3-2.8
Sandstone, bluestone	110	1.8-1.9	Hornblende	187	3.0
			Limestone, marble	165	2.5-2.8
BRICK MASONRY			Magnesite	187	3.0
Pressed brick	140	2.2-2.3	Phosphate rock, apatite	200	3.2
Common brick	120	1.8-2.0	Porphyry	172	2.6-2.9
Soft brick	100	1.5-1.7	Pumice, natural	40	0.37-0.90
			Quartz, flint	165	2.5-2.8
CONCRETE MASONRY			Sandstone, bluestone	147	2.2-2.5
Cement, stone, sand	144	2.2-2.4	Shale, slate	175	2.7-2.9
Cement, slag, etc.	130	1.9-2.3	Soapstone, talc	169	2.6-2.8
Cement, cinder, etc.	100	1.5-1.7			
VARIOUS BUILDING MATERIALS			**STONE, QUARRIED, PILED**		
Ashes, cinders	40-45	....	Basalt, granite, gneiss	96	———
Cement, portland, loose	90	....	Limestone, marble, quartz	95	———
Cement, portland, set	183	2.7-3.2	Sandstone	82	———
Lime, gypsum, loose	53-64	....	Shale	92	———
Mortar, set	103	1.4-1.9	Greenstone, hornblende	107	———
Slags, bank slag	67-72	....			
Slags, bank screenings	98-117	....			
Slags, machine slag	96	....			
Slags, slag sand	49-55	....	**BITUMINOUS SUBSTANCES**		
			Asphaltum	81	1.1-1.5
EARTH, ETC., EXCAVATED			Coal, anthracite	97	1.4-1.7
Clay, dry	63	....	Coal, bituminous	84	1.2-1.5
Clay, damp, plastic	110	....	Coal, lignite	78	1.1-1.4
Clay and gravel, dry	100	....	Coal, peat, turf, dry	47	0.65-0.85
Earth, dry, loose	76	....	Coal, charcoal, pine	23	0.28-0.44
Earth, dry, packed	95	....	Coal, charcoal, oak	33	0.47-0.57
Earth, moist, loose	78	....	Coal, coke	75	1.0-1.4
Earth, moist, packed	96	....	Graphite	131	1.9-2.3
Earth, mud, flowing	108	....	Paraffine	56	0.87-0.91
Earth, mud, packed	115	....	Petroleum	54	0.87
Riprap, limestone	80-85	....	Petroleum, refined	50	0.79-0.82
Riprap, sandstone	90	....	Petroleum, benzine	46	0.73-0.75
Riprap, shale	105	....	Petroleum, gasoline	42	0.66-0.69
Sand, gravel, dry, loose	90-105	....	Pitch	69	1.07-1.15
Sand, gravel, dry, packed	100-120	....	Tar, bituminous	75	1.20
Sand, gravel, wet	118-120	....			
EXCAVATIONS IN WATER			**COAL AND COKE, PILED**		
Sand or gravel	60	——.	Coal, anthracite	47-58	———
Sand or gravel and clay	65	——.	Coal, bituminous, lignite	40-54	———
Clay	80	——.	Coal, peat, turf	20-26	———
River mud	90	——.	Coal, charcoal	10-14	———
Soil	70	——.	Coal, coke	23-32	———
Stone riprap	65	——.			

The specific gravities of solids and liquids refer to water at 4°C., those of gases to air at 0°C. and 760 mm. pressure.   The weights per cubic foot are derived from average specific gravities, except where stated that weights are for bulk, heaped or loose material, etc.

3.C.5   Useful Formulas

Circumference of a circle = π × *diameter* or 3.1416 × *diameter*
Diameter of a circle = *circumference* × 0.31831
Area of a square = *length* × *width*
Area of a rectangle = *length* × *width*
Area of a parallelogram = *base* × *perpendicular height*
Area of a triangle = $\frac{1}{2}$ *base* × *perpendicular weight*
Area of a circle = π *radius squared* or *diameter squared* × 0.7854
Area of an ellipse = *length* × *width* × 0.7854
Volume of a cube or rectangular prism = *length* × *width* × *height*
Volume of a triangular prism = *area of triangle* × *length*
Volume of a sphere = *diameter cubed* × 0.5236 (*diameter* × *diameter* × *diameter* × 0.5236)
Volume of a cone = π × *radius squared* × $\frac{1}{3}$ *height*
Volume of a cylinder = π × *radius squared* × *height*
Length of one side of a square × 1.128 = *diameter of an equal circle*
Doubling the diameter of a pipe or cylinder increases its capacity 4 times
Pressure (in lb/sq in.) *of a column of water* = *height of the column* (in feet) × 0.434
Capacity of a pipe or tank (in U.S. gallons) = *diameter squared* (in inches) × *length* (in inches) × 0.0034
1 gal water = $8\frac{1}{3}$ lb = 231 cu in.
1 cu ft water = $62\frac{1}{2}$ lb = $7\frac{1}{2}$ gal.

3.C.6   Decimal Equivalents of Inches in Feet and Yards

Inches	Feet	Yards
1	.0833	.0278
2	.1667	.0556
3	.2500	.0833
4	.333	.1111
5	.4166	.1389
6	.5000	.1667
7	.5833	.1944
8	.6667	.2222
9	.7500	.2500
10	.8333	.2778
11	.9166	.3056
12	1.000	.3333

3.C.7   Conversion of Fractions to Decimals

Fractions	Decimal	Fractions	Decimal
1/64	.015625	33/64	.515625
1/32	.03125	17/32	.53125
3/64	.046875	35/64	.546875
1/16	.0625	9/16	.5625
5/64	.078125	37/64	.578125
3/32	.09375	19/32	.59375
7/64	.109375	38/64	.609375
1/8	.125	5/8	.625
9/64	.140625	41/64	.640625
5/32	.15625	21/32	.65625
11/64	.1719	43/64	.67187
3/16	.1875	11/16	.6875
13/64	.2031	45/64	.70312
7/32	.2188	23/32	.71875
15/64	.234375	47/64	.734375
1/4	.25	3/4	.75
17/64	.265625	49/64	.765625
9/32	.28125	25/32	.78125
19/64	.296875	51/64	.796875
5/16	.3125	13/10	.8125
21/64	.328125	53/64	.828125
11/32	.34375	27/32	.84375
23/64	.359375	55/64	.859375
3/8	.375	7/8	.875
25/64	.398625	57/64	.890625
13/32	.40625	29/32	.90625
27/64	.421875	60/64	.921875
7/16	.4375	15/16	.9375
20/64	.453125	61/64	.953125
15/32	.46875	31/32	.96875
31/64	.484375	63/64	.984375
1/2	.50	1"	1.000000

3.C.7.1   Decimals of a Foot for Each $^1/_{32}''$

Inch	0	1	2	3	4	5
0	0	.0833	.1667	.2500	.3333	.4167
$^1/_{32}$	.0026	.0859	.1693	.2526	.3359	.4193
$^1/_{16}$	.0052	.0885	.1719	.2552	.3385	.4219
$^3/_{32}$	.0078	.0911	.1745	.2578	.3411	.4245
$^1/_8$	.0104	.0938	.1771	.2604	.3438	.4271
$^5/_{32}$	.0130	.0964	.1797	.2630	.3464	.4297
$^3/_{16}$	.0156	.0990	.1823	.2656	.3490	.4323
$^7/_{32}$	.0182	.1016	.1849	.2682	.3516	.4349
$^1/_4$	.0208	.1042	.1875	.2708	.3542	.4375
$^9/_{32}$	.0234	.1068	.1901	.2734	.3568	.4401
$^5/_{16}$	.0260	.1094	.1927	.2760	.3594	.4427
$^{11}/_{32}$	.0286	.1120	.1953	.2786	.3620	.4453
$^3/_8$	.0313	.1146	.1979	.2812	.3646	.4479
$^{13}/_{32}$	.0339	.1172	.2005	.2839	.3672	.4505
$^7/_{16}$	.0365	.1198	.2031	.2865	.3698	.4531
$^{15}/_{32}$	.0391	.1224	.2057	.2891	.3724	.4557
$^1/_2$	.0417	.1250	.2083	.2917	.3750	.4583
$^{17}/_{32}$	.0443	.1276	.2109	.2943	.3776	.4609
$^9/_{16}$	.0469	.1302	.2135	.2969	.3802	.4635
$^{19}/_{32}$	.0495	.1328	.2161	.2995	.3828	.4661
$^5/_8$	.0521	.1354	.2188	.3021	.3854	.4688
$^{21}/_{32}$	.0547	.1380	.2214	.3047	.3880	.4714
$^{11}/_{16}$	.0573	.1406	.2240	.3073	.3906	.4740
$^{23}/_{32}$	.0599	.1432	.2266	.3099	.3932	.4766
$^3/_4$	.0625	.1458	.2292	.3125	.3958	.4792
$^{25}/_{32}$	.0651	.1484	.2318	.3151	.3984	.4818
$^{13}/_{16}$	.0677	.1510	.2344	.3177	.4010	.4844
$^{27}/_{32}$	.0703	.1536	.2370	.3203	.4036	.4870
$^7/_8$	.0729	.1563	.2396	.3229	.4063	.4896
$^{29}/_{32}$	.0755	.1589	.2422	.3255	.4089	.4922
$^{15}/_{16}$	.0781	.1615	.2448	.3281	.4115	.4948
$^{31}/_{32}$	.0807	.1641	.2474	.3307	.4141	.4974

3.C.7.2   Decimals of an Inch for each $\frac{1}{64}''$, with Millimeter Equivalents

Fraction	$\frac{1}{64}$ths	Decimal	Millimeters (Approx.)	Fraction	$\frac{1}{64}$ths	Decimal	Millimeters (Approx.)
...	1	.015625	0.397	...	33	.515625	13.097
$\frac{1}{32}$	2	.03125	0.794	$\frac{17}{32}$	34	.53125	13.494
...	3	.046875	1.191	...	35	.546875	13.891
$\frac{1}{16}$	4	.0625	1.588	$\frac{9}{16}$	36	.5625	14.288
...	5	.078125	1.984	...	37	.578125	14.684
$\frac{3}{32}$	6	.09375	2.381	$\frac{19}{32}$	38	.59375	15.081
...	7	.109375	2.778	...	39	.609375	15.478
$\frac{1}{8}$	8	.125	3.175	$\frac{5}{8}$	40	.625	15.875
...	9	.140625	3.572	...	41	.640625	16.272
$\frac{5}{32}$	10	.15625	3.969	$\frac{21}{32}$	42	.65625	16.669
...	11	.171875	4.366	...	43	.671875	17.066
$\frac{3}{16}$	12	.1875	4.763	$\frac{11}{16}$	44	.6875	17.463
...	13	.203125	5.159	...	45	.703125	17.859
$\frac{7}{32}$	14	.21875	5.556	$\frac{23}{32}$	46	.71875	18.256
...	15	.234375	5.953	...	47	.734375	18.653
$\frac{1}{4}$	16	.250	6.350	$\frac{3}{4}$	48	.750	19.050
...	17	.265625	6.747	...	49	.765625	19.447
$\frac{9}{32}$	18	.28125	7.144	$\frac{25}{32}$	50	.78125	19.844
...	19	.296875	7.541	...	51	.796875	20.241
$\frac{5}{16}$	20	.3125	7.938	$\frac{13}{16}$	52	.8125	20.638
...	21	.328125	8.334	...	53	.828125	21.034
$\frac{11}{32}$	22	.34375	8.731	$\frac{27}{32}$	54	.84375	21.431
...	23	.359375	9.128	...	55	.859375	21.828
$\frac{3}{8}$	24	.375	9.525	$\frac{7}{8}$	56	.875	22.225
...	25	.390625	9.922	...	57	.890625	22.622
$\frac{13}{32}$	26	.40625	10.319	$\frac{29}{32}$	58	.90625	23.019
...	27	.421875	10.716	...	59	.921875	23.416
$\frac{7}{16}$	28	.4375	11.113	$\frac{15}{16}$	60	.9375	23.813
...	29	.453125	11.509	...	61	.953125	24.209
$\frac{15}{32}$	30	.46875	11.906	$\frac{31}{32}$	62	.96875	24.606
...	31	.484375	12.303	...	63	.984375	25.003
$\frac{1}{2}$	32	.500	12.700	1	64	1.000	25.400

3.C.8   Solutions of the Right Triangle

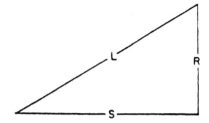

To find side	When you know side	Multiply side	For 45 Ells-By	For 22 1/2 Ells-By	For 67 1/2 Ells-By	For 72 Ells-By	For 60 Ells-By	For 80 Ells-By
L	S	S	1.4142	2.6131	1.08	1.05	1.1547	2.00
S	L	L	.707	.3826	.92	.95	.866	.50
R	S	S	1.000	2.4142	.414	.324	.5773	.1732
S	R	R	1.000	.4142	2.41	3.07	1.732	.5773
L	R	R	1.4142	1.0824	2.61	3.24	2.00	1.1547
R	L	L	.7071	.9239	.38	.31	.50	.866

(*By permission of Cast Iron Soil Pipe Institute*)

3.C.9 Area and Other Formulas

Parallelogram	*Area = base × distance between the two parallel sides*
Pyramid	*Area = $\frac{1}{2}$ perimeter of base × slant height + area of base*
	Volume = area of base × $\frac{1}{3}$ of the altitude
Rectangle	*Area = length × width*
Rectangular prisms	*Volume = width × height × length*
Sphere	*Area of surface = diameter × diameter × 3.1416*
	Side of inscribed cube = radius × 1.547
	Volume = diameter × diameter × diameter × 0.5236
Square	*Area = length × width*
Triangle	*Area = one half of height times base*
Trapezoid	*Area = one half of the sum of the parallel sides × height*
Cone	*Area of surface = one half of circumference of base × slant height + area of base*
	Volume = diameter × diameter × 0.7854 × one third of the altitude
Cube	*Volume = width × height × length*
Ellipse	*Area = short diameter × long diameter × 0.7854*
Cylinder	*Area of surface = diameter × 3.1416 × length + area of the two bases*
	Area of base = diameter × diameter × 0.7854
	Area of base = volume + length
	Length = volume + area of base
	Volume = length × area of base
	Capacity in gallons = volume in inches + 231
	Capacity of gallons = diameter × diameter × length × 0.0034
	Capacity in gallons = volume in feet × 7.48
Circle	*Circumference = diameter × 3.1416*
	Circumference = radius × 6.2832
	Diameter = radius × 2
	Diameter = square root of = (area + 0.7854)
	Diameter = square root of area × 1.1283

3.C.10   Volume of Vertical Cylindrical Tanks (in Gallons Per Foot of Depth)

Diameter in Feet	Inches	U. S. Gallons	Diameter in Feet	Inches	U. S. Gallons	Diameter in Feet	Inches	U. S. Gallons
1	0	5.875	3	6	71.97	6	0	211.5
1	1	6.895	3	7	75.44	6	3	220.5
1	2	7.997	3	8	78.99	6	6	248.2
1	3	9.180	3	9	82.62	6	9	267.7
1	4	10.44	3	10	86.33	7	0	287.9
1	5	11.79	3	11	90.13	7	3	308.8
1	6	13.22	4	0	94.00	7	6	330.5
1	7	14.73	4	1	97.96	7	9	352.9
1	8	16.32	4	2	102.0	8	0	376.0
1	9	17.99	4	3	106.1	8	3	399.9
1	10	19.75	4	4	110.3	8	6	424.5
1	11	21.58	4	5	114.6	8	9	449.8
2	0	23.50	4	6	119.0	9	0	475.9
2	1	25.50	4	7	123.4	9	3	502.7
2	2	27.58	4	8	127.9	9	6	530.2
2	3	29.74	4	9	132.6	9	9	558.5
2	4	31.99	4	10	137.3	10	0	587.5
2	5	34.31	4	11	142.0	10	3	617.3
2	6	36.72	5	0	146.9	10	6	647.7
2	7	39.21	5	1	151.8	10	9	679.0
2	8	41.78	5	2	156.8	11	0	710.9
2	9	44.43	5	3	161.9	11	3	743.6
2	10	47.16	5	4	167.1	11	6	777.0
2	11	49.98	5	5	172.4	11	9	811.1
3	0	52.88	5	6	177.7	12	0	846.0
3	1	55.86	5	7	183.2	12	3	881.6
3	2	58.92	5	8	188.7	12	6	918.0
3	3	62.06	5	9	194.2	12	9	955.1
3	4	65.28	5	10	199.9			
3	5	68.58	5	11	205.7			

(By permission of Cast Iron Soil Pipe Institute)

3.C.11   Volume of Rectangular Tank Capacities (in U.S. Gallons Per Foot of Depth)

Width Feet	LENGTH OF TANK — IN FEET						
	2	2 1/2	3	3 1/2	4	4 1/2	5
2	29.92	37.40	44.88	52.36	59.84	67.32	74.81
2 1/2	—	46.75	56.10	65.45	74.81	84.16	93.51
3	—	—	67.32	78.55	89.77	101.0	112.2
3 1/2	—	—	—	91.64	104.7	117.8	130.9
4	—	—	—	—	119.7	134.6	149.6
4 1/2	—	—	—	—	—	151.5	168.3
5	—	—	—	—	—	—	187.0
	5 1/2	6	6 1/2	7	7 1/2	8	8 1/2
2	82.29	89.77	97.25	104.7	112.2	119.7	127.2
2 1/2	102.9	112.2	121.6	130.9	140.3	149.6	159.0
3	123.4	134.6	145.9	157.1	168.3	179.5	190.8
3 1/2	144.0	157.1	170.2	183.3	196.4	209.5	222.5
4	164.6	179.5	194.5	209.5	224.4	239.4	254.3
4 1/2	185.1	202.0	218.8	235.6	252.5	269.3	286.1
5	205.7	224.4	243.1	261.8	280.5	299.2	317.9
5 1/2	226.3	246.9	267.4	288.0	308.6	329.1	349.7
6	—	269.3	291.7	314.2	336.6	359.1	381.5
6 1/2	—	—	316.1	340.4	364.7	389.0	413.3
7	—	—	—	366.5	392.7	418.9	445.1
7 1/2	—	—	—	—	420.8	448.8	476.9
8	—	—	—	—	—	478.8	508.7
8 1/2	—	—	—	—	—	—	540.5
	9	9 1/2	10	10 1/2	11	11 1/2	12
2	134.6	142.1	149.6	157.1	164.6	172.1	179.5
2 1/2	168.3	177.7	187.0	196.4	205.7	215.1	224.4
3	202.0	213.2	224.4	235.6	246.9	258.1	269.3
3 1/2	235.6	248.7	261.8	274.9	288.0	301.1	314.2
4	269.3	284.3	299.2	314.2	329.1	344.1	359.1
4 1/2	303.0	319.8	336.6	353.5	370.3	387.1	403.9
5	336.6	355.3	374.0	392.7	411.4	430.1	448.8
5 1/2	370.3	390.9	411.4	432.0	452.6	473.1	493.7
6	403.9	426.4	448.8	471.3	493.7	516.2	538.6
6 1/2	437.6	461.9	486.2	510.5	534.9	559.2	583.5
7	471.3	497.5	523.6	549.8	576.0	602.2	628.4
7 1/2	504.9	533.0	561.0	589.1	617.1	645.2	673.2
8	538.6	568.5	598.4	628.4	658.3	688.2	718.1
8 1/2	572.3	604.1	635.8	667.6	699.4	731.2	763.0
9	605.9	639.6	673.2	706.9	740.6	774.2	807.9
9 1/2	—	675.1	710.6	746.2	781.7	817.2	852.8
10	—	—	748.1	785.5	822.9	860.3	897.7
10 1/2	—	—	—	824.7	864.0	903.3	942.5
11	—	—	—	—	905.1	946.3	987.4
11 1/2	—	—	—	—	—	989.3	1032.0
12	—	—	—	—	—	—	1077.0

(By permission of Cast Iron Soil Pipe Institute)

3.C.12   Capacity of Horizontal Cylindrical Tanks

% Depth Filled	% of Capacity	% Depth Filled	% of Capacity	% Depth Filled	% of Capacity	% Depth Filled	% of Capacity
1	.20	26	20.73	51	51.27	76	81.50
2	.50	27	21.86	52	52.55	77	82.60
3	.90	28	23.00	53	53.81	78	83.68
4	1.34	29	24.07	54	55.08	79	84.74
5	1.87	30	25.31	55	56.34	80	85.77
6	2.45	31	26.48	56	57.60	81	86.77
7	3.07	32	27.66	57	58.86	82	87.76
8	3.74	33	28.84	58	60.11	83	88.73
9	4.45	34	30.03	59	61.36	84	89.68
10	5.20	35	31.19	60	62.61	85	90.60
11	5.98	36	32.44	61	63.86	86	91.50
12	6.80	37	33.66	62	65.10	87	92.36
13	7.64	38	34.90	63	66.34	88	93.20
14	8.50	39	36.14	64	67.56	89	94.02
15	9.40	40	37.36	65	68.81	90	94.80
16	10.32	41	38.64	66	69.97	91	95.50
17	11.27	42	39.89	67	71.16	92	96.26
18	12.24	43	41.14	68	72.34	93	96.93
19	13.23	44	42.40	69	73.52	94	97.55
20	14.23	45	43.66	70	74.69	95	98.13
21	15.26	46	44.92	71	75.93	96	98.66
22	16.32	47	46.19	72	77.00	97	99.10
23	17.40	48	47.45	73	78.14	98	99.50
24	18.50	49	48.73	74	79.27	99	99.80
25	19.61	50	50.00	75	80.39	100	100.00

(*By permission of Cast Iron Soil Pipe Institute*)

3.C.13   Round-Tapered Tank Capacities

$$Volume = \frac{h^3}{3} \frac{[(Area_{Top} + Area_{Base}) + \sqrt{(Area_{Top} + Area_{Base}]}}{231}$$

If inches are used.

$$Volume = \frac{h}{3} [(Area_{Base} + Area_{Top}) + \sqrt{(Area_{Base} + Area_{Top}]} \times 7.48$$

If feet are used.

3.C.14   Circumferences and Areas of Circles

	Of One Inch				Of Inches or Feet				
Fract.	Decimal	Circ.	Area	Dia.	Circ.	Area	Dia.	Circ.	Area
1/64	.015625	.04909	.00019	1	3.1416	.7854	64	201.06	3216.99
1/32	.03125	.09818	.00077	2	6.2832	3.1416	65	204.20	3318.31
3/64	.046875	.14726	.00173	3	9.4248	7.0686	66	207.34	3421.19
1/16	.0625	.19635	.00307	4	12.5664	12.5664	67	210.49	3525.65
5/64	.078125	.24545	.00479	5	15.7080	19.635	68	213.63	3631.68
3/32	.09375	.29452	.00690	6	18.850	28.274	69	216.77	3739.28
7/64	.109375	.34363	.00939	7	21.991	38.485	70	219.91	3848.45
1/8	.125	.39270	.01227	8	25.133	50.266	71	223.05	3959.19
9/64	.140625	.44181	.01553	9	28.274	63.617	72	226.19	4071.50
5/32	.15625	.49087	.01917	10	31.416	78.540	73	229.34	4185.50
11/64	.171875	.53999	.02320	11	34.558	95.033	74	232.48	4300.84
3/16	.1875	.58905	.02761	12	37.699	113.1	75	235.62	4417.86
13/64	.203125	.63817	.03241	13	40.841	132.73	76	238.76	4536.46
7/32	.21875	.68722	.03757	4	43.982	153.94	77	241.90	4656.63
15/64	.234375	.73635	.04314	15	47.124	176.71	78	245.04	4778.36
1/4	.25	.78540	.04909	16	50.265	201.06	79	248.19	4901.67
17/64	.265625	.83453	.05542	17	53.407	226.98	80	251.33	5026.55
9/32	.28125	.88357	.06213	18	56.549	254.47	81	254.47	5153.0
10/64	.296875	.93271	.06922	19	59.690	283.53	82	257.61	5281.02
5/16	.3125	.98175	.07670	20	63.832	314.16	83	260.75	5410.61
21/64	.328125	1.0309	.08456	21	65.973	346.36	84	263.89	5541.77
11/32	.34375	1.0799	.09281	22	69.115	380.13	85	267.04	5674.50
23/64	.35975	1.1291	.10144	23	72.257	415.48	86	270.18	5808.80
3/8	.375	1.1781	.11045	24	75.398	452.39	87	273.32	5944.68
25/64	.390625	1.2273	.11984	25	78.540	490.87	88	276.46	6082.12
13/32	.40625	1.2763	.12962	26	81.681	530.93	89	279.60	6221.14
27/64	.421875	1.3254	.13979	27	84.823	572.56	90	282.74	6361.71
7/16	.4375	1.3744	.15033	28	87.965	615.75	91	258.88	6503.88
29/64	.453125	1.4236	.16126	29	91.106	660.52	92	289.03	6647.61
15/32	.46875	1.4726	.17257	30	94.248	706.86	93	292.17	6792.91
31/64	.484375	1.5218	.18427	31	97.389	754.77	94	295.31	6939.78
1/2	.5	1.5708	.19635	32	100.53	804.25	95	298.45	7088.22

(By permission of Cast Iron Soil Pipe Institute)

Index

R

S

About the Author

Sidney M. Levy is a construction consultant in Baltimore, Maryland, with more than 40 years experience in the industry. He is the author of 11 books, including several devoted to international construction. Mr. Levy is the author of *Project Management in Construction,* published in both English and Spanish editions.

Sidney M. Levy